TODAY'S TECHNICIAN

SHOP MANUAL

FOR BASIC AUTOMOTIVE SERVICE & SYSTEMS

TODAY'S TECHNICIAN™

SHOP MANUAL FOR BASIC AUTOMOTIVE SERVICE & SYSTEMS

5TH EDITION

Chris Hadfield
Central Lakes College
Brainerd, Minnesota

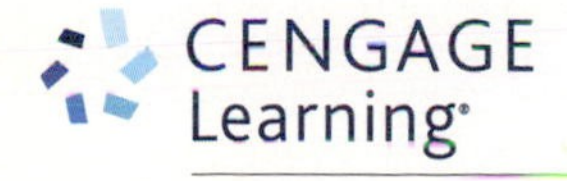

Australia • Brazil • Mexico • Singapore • United Kingdom • United States

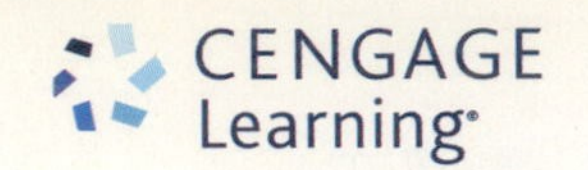

Today's Technician™: Shop Manual for Basic Automotive Service & Systems, 5th Edition
Chris Hadfield

SVP, GM Skills & Global Product Management: Dawn Gerrain

Product Team Manager: Erin Brennan

Senior Director, Development: Marah Bellegarde

Senior Product Development Manager: Larry Main

Content Developer: Mary Clyne

Product Assistant: Maria Garguilo

Vice President, Marketing Services: Jennifer Ann Baker

Marketing Director: Michele McTighe

Senior Marketing Manager: Jenn Barbic

Senior Production Director: Wendy Troeger

Production Director: Andrew Crouth

Senior Content Project Manager: Cheri Plasse

Senior Art Director: Bethany Casey

Cover Image: © cla78/Shutterstock

Library of Congress Control Number: 2014938772

Book-only ISBN: 978-1-285-44231-0

Package ISBN: 978-1-285-44229-7

Cengage Learning
20 Channel Center Street
Boston, MA 02210
USA

Cengage Learning is a leading provider of customized learning solutions with office locations around the globe, including Singapore, the United Kingdom, Australia, Mexico, Brazil, and Japan. Locate your local office at:
www.cengage.com/global

Cengage Learning products are represented in Canada by Nelson Education, Ltd.

To learn more about Cengage Learning, visit **www.cengage.com**

Purchase any of our products at your local college store or at our preferred online store **www.cengagebrain.com**

Printed in the United States of America
Print Number: 01 Print Year: 2014

CONTENTS

CONTENTS

PHOTO SEQUENCES

JOB SHEETS

JOB SHEETS

PREFACE

Thanks to the support the *Today's Technician*™ Series has received from those who teach automotive technology, Cengage Learning, the leading publisher in automotive-related textbooks, is able to live up to its promise to provide new editions regularly. We have listened and responded to our critics and our fans and present this new updated and revised fifth edition. By revising our series regularly, we can and will respond to changes in the industry, changes in technology, changes in the certification process, and to the ever-changing needs of those who teach automotive technology.

The *Today's Technician*™ Series features textbooks that cover all mechanical and electrical systems of automobiles and light trucks (while the Heavy-Duty Trucks portion of the series does the same for heavy-duty vehicles). The individual titles correspond to the main areas of ASE (National Institute for Automotive Service Excellence) certification. Additional titles include remedial skills and theories common to all of the certification areas and advanced or specific subject areas that reflect the latest technological trends. Each text is divided into two volumes: a Classroom Manual and Shop Manual.

This new edition, like the previous one, is designed to give students a chance to develop the same skills and gain the same knowledge that today's successful technician has. This edition also reflects the changes in the guidelines established by the National Automotive Technicians Education Foundation (NATEF) in 2013.

The purpose of NATEF is to evaluate technician training programs against standards developed by the automotive industry and recommend qualifying programs for certification (accreditation) by ASE. Programs can earn ASE certification upon the recommendation of NATEF. NATEF's national standards reflect the skills that students must master. ASE certification through NATEF evaluation ensures that certified training programs meet or exceed industry-recognized, uniform standards of excellence.

The technician of today and of the future must know the underlying theory of all automotive systems and be able to service and maintain these systems. Dividing the material into two volumes provides the reader with the information needed to begin a successful career as an automotive technician without interrupting the learning process by mixing cognitive and performance learning objectives into one volume.

The design of Cengage's *Today's Technician*™ Series is based on features that are known to promote improved student learning. The design is further enhanced by a careful study of survey results, in which the respondents were asked to value particular features. Some of these features can be found in other textbooks, while others are unique to this series.

Each Classroom Manual contains the principles of operation for each system and subsystem. The Classroom Manual also contains discussions on design variations of key components used by the different vehicle manufacturers. This volume is organized to build upon basic facts and theories. The primary objective of this volume is to allow the reader to gain an understanding of how each system and subsystem operates. This understanding is necessary to diagnose the complex automobiles of today and tomorrow. Although the basics contained in the Classroom Manual provide the knowledge needed for diagnostics, diagnostic procedures appear only in the Shop Manual. An understanding of the basics is also a requirement for competence in the skill areas covered in the Shop Manual.

PREFACE

A spiral-bound Shop Manual covers the "how-tos." This volume includes step-by-step instructions for diagnostic and repair procedures. Photo sequences are used to illustrate some of the common service procedures. Other common procedures are listed and are accompanied by line drawings and photos that allow the reader to visualize and conceptualize the finest details of the procedure. This volume also contains the reasons for performing the procedures, as well as when that particular service is appropriate.

The two volumes are designed to be used together and are arranged in corresponding chapters. Not only are the chapters in the volumes linked together, the contents of the chapters are also linked. This linking of content is evidenced by margin callouts that refer the reader to the chapter and page where the same topic is addressed in the other volume. This feature is valuable to instructors. Users of other two-volume textbooks without this feature must search the index or table of contents to locate supporting information in the other volume. This is not only cumbersome, but also creates additional work for an instructor when planning the presentation of material and when making reading assignments. It is also valuable to the students. With the page references, they also know exactly where to look for supportive information.

Both volumes contain clear and thoughtfully selected illustrations, many of which are original drawings or photos specially prepared for inclusion in this series. This means that the art is a vital part of each textbook and not merely inserted to increase the numbers of illustrations.

The page layout is designed to include material that would otherwise break up the flow of information presented to the reader. The main body of the text includes all of the "need-to-know" information and illustrations. In the wide side margins of each page are many of the special series features. Items that are truly "nice-to-know" information include simple examples of concepts just introduced in the text, explanations or definitions of terms that may not be in the glossary, examples of common trade jargon used to describe a part or operation, and exceptions to the norm explained in the text. This type of material is placed in the margin, out of the normal flow of information. Many textbooks attempt to insert this type of information in the main body of the text. This tends to interrupt the thought process and cannot be pedagogically justified. Since this complementary information is off to the side of the main text, the reader can choose when to refer to it.

Highlights of This Edition—Classroom Manual

Fully updated to align with the new NATEF accreditation model introduced in 2013, the fifth edition is primarily coordinated with NATEF's Maintenance and Light Repair program option.

This edition's full-color design with new photographs and updated line art helps readers better visualize processes and procedures. The entire text was updated with information about new tools, equipment, and vehicle systems in use today and expected in the near future. A new chapter has been added to expand coverage of electrical systems and components. Coverage of hybrid vehicles and emerging technologies, along with revised and updated coverage of electronic engine controls, prepares students for today's workplace.

Highlights of This Edition—Shop Manual

Along with the Classroom Manual, the Shop Manual was revised and updated in both text and figures to emphasize the changing technology in tools, equipment, and procedures. The chapters parallel and complement the Classroom Manual's layout and closely match a typical automotive school's sequence of instruction. The fifth edition includes even more photo sequences in full color and all-new Job Sheets that are focused on skill building.

New content cross-references current industry practices to help the student understand and practice real-world workplace competencies, and employer-focused soft skills are integrated into each topic.

The revised edition features in-depth basic and preventative maintenance information with a focus on skill building, and expanded coverage of electrical theory and diagnosis.

SHOP MANUAL

To stress the importance of safe work habits, the Shop Manual dedicates one full chapter to safety. Other important features of this manual include:

PERFORMANCE-BASED OBJECTIVES

These objectives define the contents of the chapter and what the student should have learned upon completion of the chapter.

AUTHOR'S NOTE

This feature includes simple explanations, stories, or examples of complex topics. These are included to help students understand difficult concepts.

PHOTO SEQUENCES

Many procedures are illustrated in detailed Photo Sequences. These photographs show the students what to expect when they perform particular procedures. They also familiarize students with a system or type of equipment that the school may not have.

BASIC TOOLS LISTS

Each chapter begins with a list of the basic tools needed to perform the tasks included in the chapter.

SPECIAL TOOLS LISTS

Whenever a special tool is required to complete a task, it is listed in the margin next to the procedure.

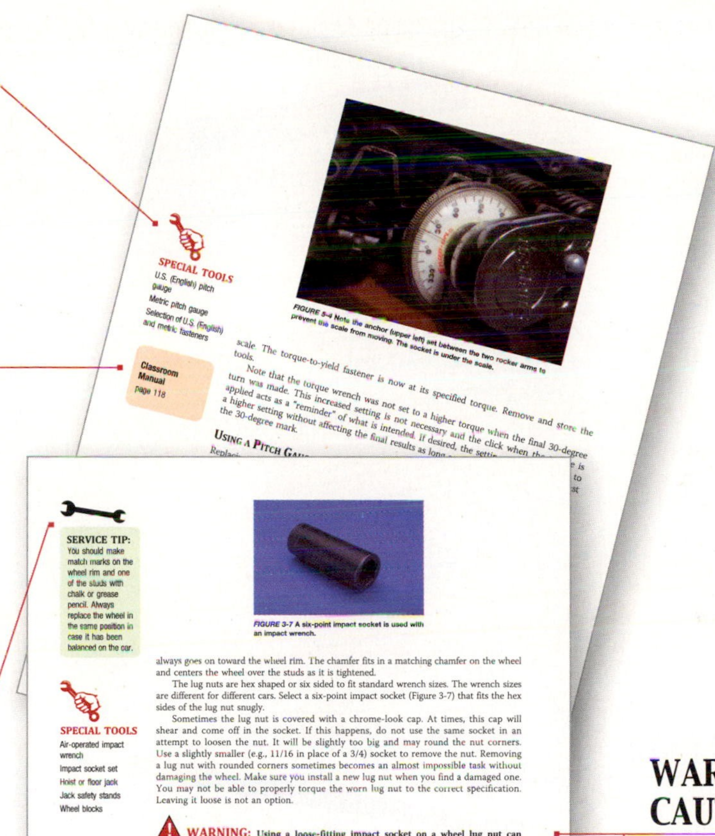

REFERENCES TO THE CLASSROOM MANUAL

Reference to the appropriate page in the Classroom Manual is given whenever necessary. Although the chapters of the two manuals are synchronized, material covered in other chapters of the Classroom Manual may be fundamental to the topic discussed in the Shop Manual.

SERVICE TIPS

Whenever a shortcut or special procedure is appropriate, it is described in the text. Generally, these tips are things commonly done by experienced technicians.

WARNINGS AND CAUTIONS

Throughout the text, cautions are given to alert the reader to potentially hazardous materials or unsafe conditions. Warnings are also given to advise the student of things that can go wrong if instructions are not followed or if a nonacceptable part or tool is used.

MARGIN NOTES

New terms are pulled out and defined. Common trade jargon also appears in the margins and gives some of the common terms used for components. This allows the student to understand and speak the language of the trade, especially when conversing with an experienced technician.

SHOP MANUAL

CASE STUDIES

Case Studies concentrate on the ability to properly diagnose the systems. Each chapter ends with a Case Study in which a vehicle has a problem, and the logic used by a technician to solve the problem is explained.

ASE-STYLE REVIEW QUESTIONS

Each chapter contains ASE-Style Review Questions that reflect the performance objectives listed at the beginning of the chapter. These questions can be used to review the chapter as well as to prepare for the ASE certification exam.

ASE-STYLE CHALLENGE QUESTIONS

Each technical chapter ends with five ASE challenge questions. These are not more review questions; rather, they test the students' ability to apply general knowledge to the contents of the chapter.

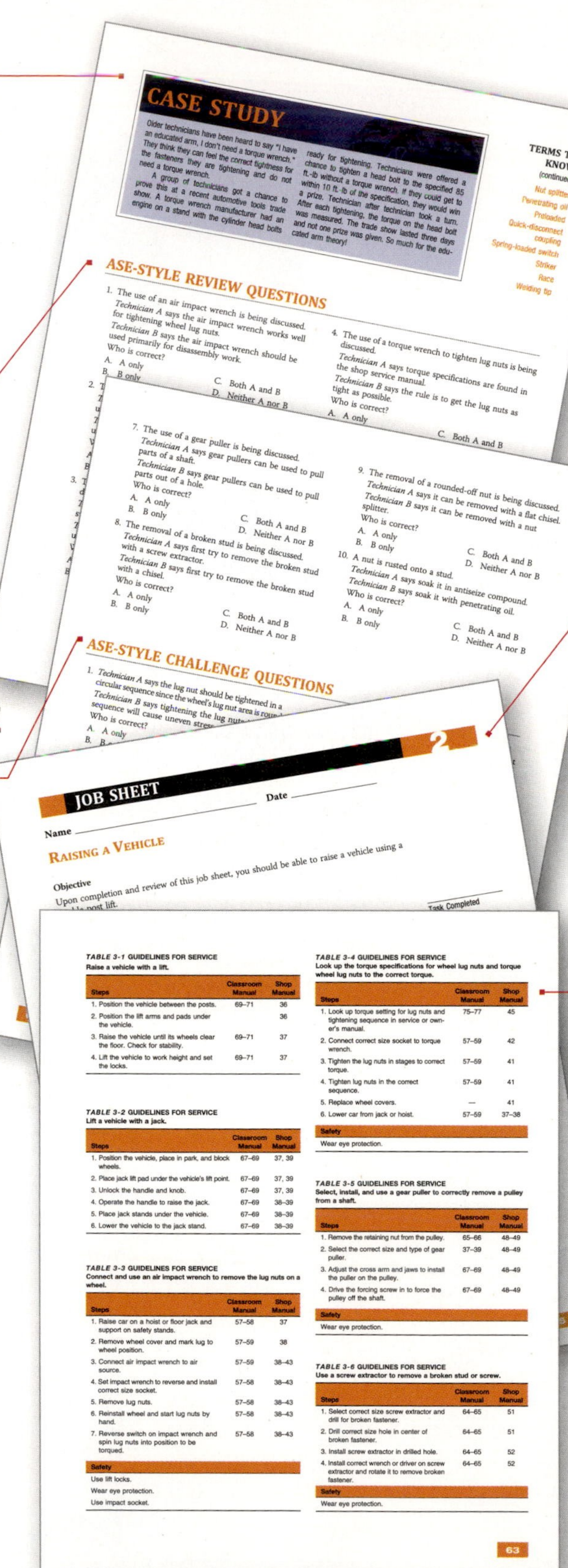

TABLE 3-1 GUIDELINES FOR SERVICE
Raise a vehicle with a lift.

Steps	Classroom Manual	Shop Manual
1. Position the vehicle between the posts.	69–71	36
2. Position the lift arms and pads under the vehicle.		36
3. Raise the vehicle until its wheels clear the floor. Check for stability.	69–71	37
4. Lift the vehicle to work height and set the locks.	69–71	37

TABLE 3-2 GUIDELINES FOR SERVICE
Lift a vehicle with a jack.

Steps	Classroom Manual	Shop Manual
1. Position the vehicle, place in park, and block wheels.	67–69	37, 39
2. Place jack lift pad under the vehicle's lift point.	67–69	37, 39
3. Unlock the handle and knob.	67–69	37, 39
4. Operate the handle to raise the jack.	67–69	38–39
5. Place jack stands under the vehicle.	67–69	38–39
6. Lower the vehicle to the jack stand.	67–69	38–39

TABLE 3-3 GUIDELINES FOR SERVICE
Connect and use an air impact wrench to remove the lug nuts on a wheel.

Steps	Classroom Manual	Shop Manual
1. Raise car on a hoist or floor jack and support on safety stands.	57–58	37
2. Remove wheel cover and mark lug to wheel position.	57–59	38
3. Connect air impact wrench to air source.	57–59	38–43
4. Set impact wrench to reverse and install correct size socket.	57–58	38–43
5. Remove lug nuts.	57–58	38–43
6. Reinstall wheel and start lug nuts by hand.	57–58	38–43
7. Reverse switch on impact wrench and spin lug nuts into position to be torqued.	57–58	38–43
Safety		
Use lift locks.		
Wear eye protection.		
Use impact socket.		

TABLE 3-4 GUIDELINES FOR SERVICE
Look up the torque specifications for wheel lug nuts and torque wheel lug nuts to the correct torque.

Steps	Classroom Manual	Shop Manual
1. Look up torque setting for lug nuts and tightening sequence in service or owner's manual.	75–77	45
2. Connect correct size socket to torque wrench.	57–59	42
3. Tighten the lug nuts in stages to correct torque.	57–59	41
4. Tighten lug nuts in the correct sequence.	57–59	41
5. Replace wheel covers.	—	41
6. Lower car from jack or hoist.	57–59	37–38
Safety		
Wear eye protection.		

TABLE 3-5 GUIDELINES FOR SERVICE
Select, install, and use a gear puller to correctly remove a pulley from a shaft.

Steps	Classroom Manual	Shop Manual
1. Remove the retaining nut from the pulley.	65–66	48–49
2. Select the correct size and type of gear puller.	37–39	48–49
3. Adjust the cross arm and jaws to install the puller on the pulley.	67–69	48–49
4. Drive the forcing screw in to force the pulley off the shaft.	67–69	48–49
Safety		
Wear eye protection.		

TABLE 3-6 GUIDELINES FOR SERVICE
Use a screw extractor to remove a broken stud or screw.

Steps	Classroom Manual	Shop Manual
1. Select correct size screw extractor and drill for broken fastener.	64–65	51
2. Drill correct size hole in center of broken fastener.	64–65	51
3. Install screw extractor in drilled hole.	64–65	52
4. Install correct wrench or driver on screw extractor and rotate it to remove broken fastener.	64–65	52
Safety		
Wear eye protection.		

63

TERMS TO KNOW LIST

Terms in this list can be found in the Glossary at the end of the manual.

JOB SHEETS

Located at the end of each chapter, the Job Sheets provide a format for students to perform procedures covered in the chapter. A reference to the ASE Task addressed by the procedure is referenced on the Job Sheet.

DIAGNOSTIC CHARTS

Some chapters include detailed diagnostic charts that list common problems and most probable causes. They also list a page reference in the Classroom Manual for better understanding of the system's operation and a page reference in the Shop Manual for details on the procedure necessary for correcting the problem.

Features of the Classroom Manual include the following:

COGNITIVE OBJECTIVES

These objectives define the contents of the chapter and define what the student should have learned upon completion of the chapter. Each topic is divided into small units to promote easier understanding and learning.

MARGIN NOTES

New terms are pulled out and defined. Common trade jargon also appears in the margin and gives some of the common terms used for components. This allows the student to understand and speak the language of the trade, especially when conversing with an experienced technician.

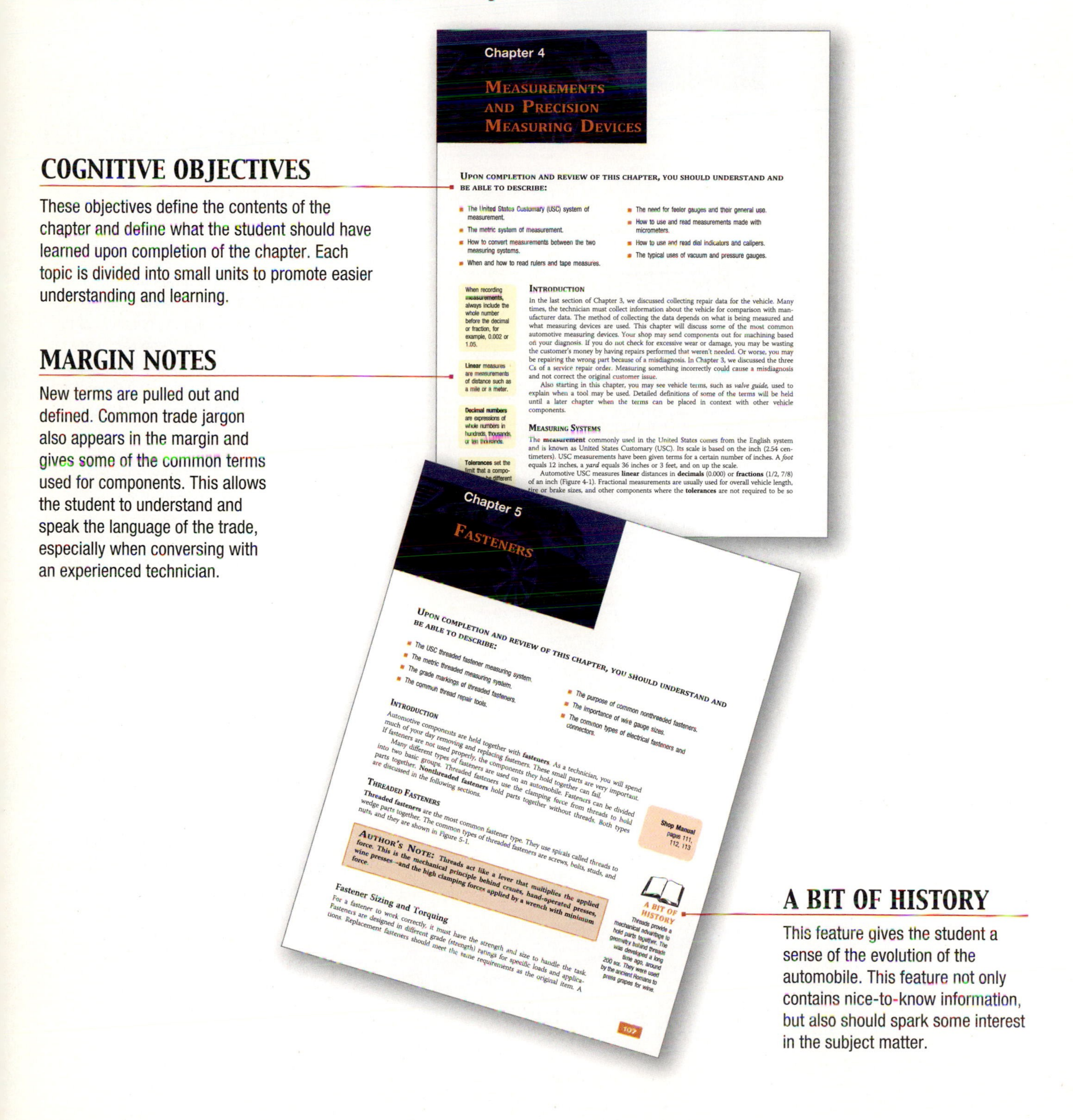

Chapter 4

MEASUREMENTS AND PRECISION MEASURING DEVICES

UPON COMPLETION AND REVIEW OF THIS CHAPTER, YOU SHOULD UNDERSTAND AND BE ABLE TO DESCRIBE:

- The United States Customary (USC) system of measurement.
- The metric system of measurement.
- How to convert measurements between the two measuring systems.
- When and how to read rulers and tape measures.
- The need for feeler gauges and their general use.
- How to use and read measurements made with micrometers.
- How to use and read dial indicators and calipers.
- The typical uses of vacuum and pressure gauges.

When recording **measurements**, always include the whole number before the decimal or fraction, for example, 0.002 or 1.05.

Linear measures are measurements of distance such as a mile or a meter.

Decimal numbers are expressions of whole numbers in hundreds, thousands, or ten thousands.

Tolerances set the limit that a compo- ... be different

INTRODUCTION

In the last section of Chapter 3, we discussed collecting repair data for the vehicle. Many times, the technician must collect information about the vehicle for comparison with manufacturer data. The method of collecting the data depends on what is being measured and what measuring devices are used. This chapter will discuss some of the most common automotive measuring devices. Your shop may send components out for machining based on your diagnosis. If you do not check for excessive wear or damage, you may be wasting the customer's money by having repairs performed that weren't needed. Or worse, you may be repairing the wrong part because of a misdiagnosis. In Chapter 3, we discussed the three Cs of a service repair order. Measuring something incorrectly could cause a misdiagnosis and not correct the original customer issue.

Also starting in this chapter, you may see vehicle terms, such as *valve guide*, used to explain when a tool may be used. Detailed definitions of some of the terms will be held until a later chapter when the terms can be placed in context with other vehicle components.

MEASURING SYSTEMS

The **measurement** commonly used in the United States comes from the English system and is known as United States Customary (USC). Its scale is based on the inch (2.54 centimeters). USC measurements have been given terms for a certain number of inches. A *foot* equals 12 inches, a *yard* equals 36 inches or 3 feet, and on up the scale.

Automotive USC measures **linear** distances in **decimals** (0.000) or **fractions** (1/2, 7/8) of an inch (Figure 4-1). Fractional measurements are usually used for overall vehicle length, tire or brake sizes, and other components where the **tolerances** are not required to be so

Chapter 5

FASTENERS

UPON COMPLETION AND REVIEW OF THIS CHAPTER, YOU SHOULD UNDERSTAND AND BE ABLE TO DESCRIBE:

- The USC threaded fastener measuring system.
- The metric threaded measuring system.
- The grade markings of threaded fasteners.
- The common thread repair tools.
- The purpose of common nonthreaded fasteners.
- The importance of wire gauge sizes.
- The common types of electrical fasteners and connectors.

INTRODUCTION

Automotive components are held together with **fasteners**. As a technician, you will spend much of your day removing and replacing fasteners. These small parts are very important. If fasteners are not used properly, the components they hold together can fail.

Many different types of fasteners are used on an automobile. Fasteners can be divided into two basic groups. Threaded fasteners use the clamping force from threads to hold parts together. **Nonthreaded fasteners** hold parts together without threads. Both types are discussed in the following sections.

Shop Manual pages 111, 112, 113

THREADED FASTENERS

Threaded fasteners are the most common fastener type. They use spirals called threads to wedge parts together. The common types of threaded fasteners are screws, bolts, studs, and nuts, and they are shown in Figure 5-1.

AUTHOR'S NOTE: Threads act like a lever that multiplies the applied force. This is the mechanical principle behind cranes, hand-operated presses, wine presses—and the high clamping forces applied by a wrench with minimum force.

A BIT OF HISTORY

Threads provide a mechanical advantage to hold parts together. The geometry behind threads was developed a long time ago, around 200 BCE. They were used by the ancient Romans to press grapes for wine.

Fastener Sizing and Torquing

For a fastener to work correctly, it must have the strength and size to handle the task. Fasteners are designed in different grade (strength) ratings for specific loads and applications. Replacement fasteners should meet the same requirements as the original item. A

107

A BIT OF HISTORY

This feature gives the student a sense of the evolution of the automobile. This feature not only contains nice-to-know information, but also should spark some interest in the subject matter.

CLASSROOM MANUAL

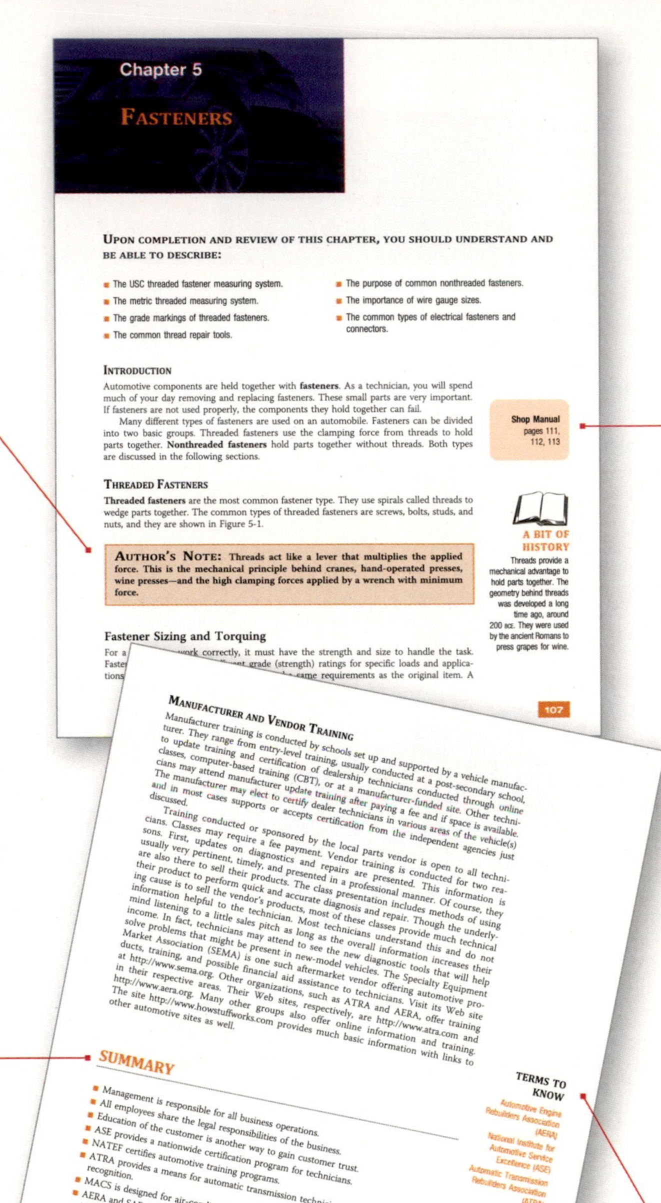

Chapter 5

FASTENERS

UPON COMPLETION AND REVIEW OF THIS CHAPTER, YOU SHOULD UNDERSTAND AND BE ABLE TO DESCRIBE:

- The USC threaded fastener measuring system.
- The metric threaded measuring system.
- The grade markings of threaded fasteners.
- The common thread repair tools.
- The purpose of common nonthreaded fasteners.
- The importance of wire gauge sizes.
- The common types of electrical fasteners and connectors.

INTRODUCTION

Automotive components are held together with **fasteners**. As a technician, you will spend much of your day removing and replacing fasteners. These small parts are very important. If fasteners are not used properly, the components they hold together can fail.

Many different types of fasteners are used on an automobile. Fasteners can be divided into two basic groups. Threaded fasteners use the clamping force from threads to hold parts together. **Nonthreaded fasteners** hold parts together without threads. Both types are discussed in the following sections.

Shop Manual pages 111, 112, 113

THREADED FASTENERS

Threaded fasteners are the most common fastener type. They use spirals called threads to wedge parts together. The common types of threaded fasteners are screws, bolts, studs, and nuts, and they are shown in Figure 5-1.

A BIT OF HISTORY

Threads provide a mechanical advantage to hold parts together. The geometry behind threads was developed a long time ago, around 200 BCE. They were used by the ancient Romans to press grapes for wine.

AUTHOR'S NOTE: Threads act like a lever that multiplies the applied force. This is the mechanical principle behind cranes, hand-operated presses, wine presses—and the high clamping forces applied by a wrench with minimum force.

Fastener Sizing and Torquing

For a ... work correctly, it must have the strength and size to handle the task. Faste... grade (strength) ratings for specific loads and applications ... same requirements as the original item. A

107

MANUFACTURER AND VENDOR TRAINING

Manufacturer training is conducted by schools set up and supported by a vehicle manufacturer. They range from entry-level training, usually conducted at a post-secondary school, to update training and certification of dealership technicians conducted through online classes, computer-based training (CBT), or at a manufacturer-funded site. Other technicians may attend manufacturer update training after paying a fee and if space is available. The manufacturer may elect to certify dealer technicians in various areas of the vehicle(s) and in most cases supports or accepts certification from the independent agencies just discussed.

Training conducted or sponsored by the local parts vendor is open to all technicians. Classes may require a fee payment. Vendor training is conducted for two reasons. First, updates on diagnostics and repairs are presented. This information is usually very pertinent, timely, and presented in a professional manner. Of course, they are also there to sell their products. The class presentation includes methods of using their product to perform quick and accurate diagnosis and repair. Though the underlying cause is to sell the vendor's products, most of these classes provide much technical information helpful to the technician. Most technicians understand this and do not mind listening to a little sales pitch as long as the overall information increases their income. In fact, technicians may attend to see the new diagnostic tools that will help solve problems that might be present in new-model vehicles. The Specialty Equipment Market Association (SEMA) is one such aftermarket vendor offering automotive products, training, and possible financial aid assistance to technicians. Visit its Web site at http://www.sema.org. Other organizations, such as ATRA and AERA, offer training in their respective areas. Their Web sites, respectively, are http://www.atra.com and http://www.aera.org. Many other groups also offer online information and training. The site http://www.howstuffworks.com provides much basic information with links to other automotive sites as well.

SUMMARY

- Management is responsible for all business operations.
- All employees share the legal responsibilities of the business.
- Education of the customer is another way to gain customer trust.
- ASE provides a nationwide certification program for technicians.
- NATEF certifies automotive training programs.
- ATRA provides a means for automatic transmission technicians to gain specific skill recognition.
- MACS is designed for air-conditioning technicians.
- AERA and SAE provide technical and educational support for various establishments and technicians.
- Manufacturer and vendor training is essential for technicians wishing to keep up to date on various aspects of vehicle repair.

TERMS TO KNOW

Automotive Engine Rebuilders Association (AERA)
National Institute for Automotive Service Excellence (ASE)
Automatic Transmission Rebuilders Association (ATRA)
Business ethics
Express service
Franchise
General manager
Investment
Mobile Air Conditioning Society (MACS)
NATEF key
Vendors

33

AUTHOR'S NOTE

This feature includes simple explanations, stories, or examples of complex topics. These are included to help students understand difficult concepts.

CROSS-REFERENCES TO THE SHOP MANUAL

Reference to the appropriate page in the Shop Manual is given whenever necessary. Although the chapters of the two manuals are synchronized, material covered in other chapters of the Shop Manual may be fundamental to the topic discussed in the Classroom Manual.

SUMMARIES

Each chapter concludes with summary statements that contain the important topics of the chapter. These are designed to help the reader review the contents.

TERMS TO KNOW LIST

A list of new terms appears next to the Summary. Definitions for these terms can be found in the Glossary at the end of the manual.

REVIEW QUESTIONS

Short-answer essays, fill-in-the-blanks, and multiple-choice-type questions follow each chapter. These questions are designed to accurately assess the student's competence in the stated objectives at the beginning of the chapter.

CAUTIONS AND WARNINGS

Throughout the text, warnings are given to alert the reader to potentially hazardous materials or unsafe conditions. Cautions are given to advise the student of things that can go wrong if instructions are not followed or if a nonacceptable part or tool is used.

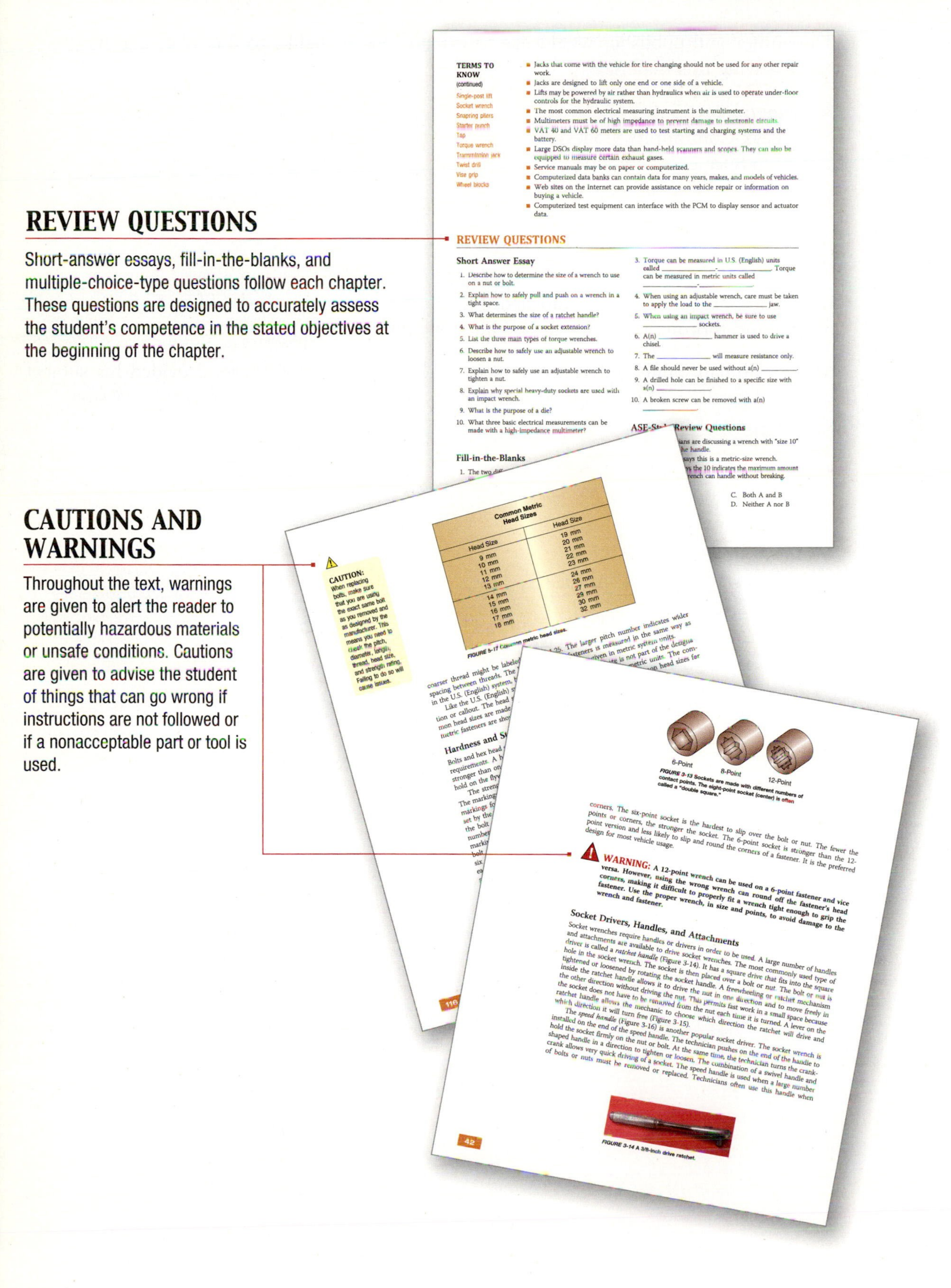

TERMS TO KNOW (continued)

Single-post lift
Socket wrench
Snapring pliers
Starter punch
Tap
Torque wrench
Transmission jack
Twist drill
Vise grip
Wheel blocks

- Jacks that come with the vehicle for tire changing should not be used for any other repair work.
- Jacks are designed to lift only one end or one side of a vehicle.
- Lifts may be powered by air rather than hydraulics when air is used to operate under-floor controls for the hydraulic system.
- The most common electrical measuring instrument is the multimeter.
- Multimeters must be of high impedance to prevent damage to electronic circuits.
- VAT 40 and VAT 60 meters are used to test starting and charging systems and the battery.
- Large DSOs display more data than hand-held scanners and scopes. They can also be equipped to measure certain exhaust gases.
- Service manuals may be on paper or computerized.
- Computerized data banks can contain data for many years, makes, and models of vehicles.
- Web sites on the Internet can provide assistance on vehicle repair or information on buying a vehicle.
- Computerized test equipment can interface with the PCM to display sensor and actuator data.

REVIEW QUESTIONS

Short Answer Essay

1. Describe how to determine the size of a wrench to use on a nut or bolt.
2. Explain how to safely pull and push on a wrench in a tight space.
3. What determines the size of a ratchet handle?
4. What is the purpose of a socket extension?
5. List the three main types of torque wrenches.
6. Describe how to safely use an adjustable wrench to loosen a nut.
7. Explain how to safely use an adjustable wrench to tighten a nut.
8. Explain why special heavy-duty sockets are used with an impact wrench.
9. What is the purpose of a die?
10. What three basic electrical measurements can be made with a high-impedance multimeter?

Fill-in-the-Blanks

1. The two di...

3. Torque can be measured in U.S. (English) units called ______-______. Torque can be measured in metric units called ______-______.
4. When using an adjustable wrench, care must be taken to apply the load to the ______ jaw.
5. When using an impact wrench, be sure to use ______ sockets.
6. A(n) ______ hammer is used to drive a chisel.
7. The ______ will measure resistance only.
8. A file should never be used without a(n) ______.
9. A drilled hole can be finished to a specific size with a(n) ______.
10. A broken screw can be removed with a(n) ______.

ASE-St... Review Questions

...ians are discussing a wrench with "size 10" ...he handle.
...says this is a metric-size wrench.
...s the 10 indicates the maximum amount ...rench can handle without breaking.

C. Both A and B
D. Neither A nor B

CAUTION: When replacing bolts, make sure that you are using the exact same bolt as you removed and as designed by the manufacturer. This means you need to check the pitch, diameter, length, thread, head size, and strength rating. Failing to do so will cause issues.

Common Metric Head Sizes	
Head Size	Head Size
9 mm	19 mm
10 mm	20 mm
11 mm	21 mm
12 mm	22 mm
13 mm	23 mm
14 mm	24 mm
15 mm	26 mm
16 mm	27 mm
17 mm	29 mm
18 mm	30 mm
	32 mm

FIGURE 5-17 Common metric head sizes.

Hardness and St...

116

6-Point
8-Point
12-Point

FIGURE 3-13 Sockets are made with different numbers of contact points. The eight-point socket (center) is often called a "double square."

corners. The six-point socket is the hardest to slip over the bolt or nut. The fewer the points or corners, the stronger the socket. The 6-point socket is stronger than the 12-point version and less likely to slip and round the corners of a fastener. It is the preferred design for most vehicle usage.

WARNING: A 12-point wrench can be used on a 6-point fastener and vice versa. However, using the wrong wrench can round off the fastener's head corners, making it difficult to properly fit a wrench tight enough to grip the fastener. Use the proper wrench, in size and points, to avoid damage to the wrench and fastener.

Socket Drivers, Handles, and Attachments

Socket wrenches require handles or drivers in order to be used. A large number of handles and attachments are available to drive socket wrenches. The most commonly used type of driver is called a *ratchet handle* (Figure 3-14). It has a square drive that fits into the square hole in the socket wrench. The socket is then placed over a bolt or nut. The bolt or nut is tightened or loosened by rotating the socket handle. A freewheeling or ratchet mechanism inside the ratchet handle allows it to drive the nut in one direction and to move freely in the other direction without driving the nut. This permits fast work in a small space because the socket does not have to be removed from the nut each time it is turned. A lever on the ratchet handle allows the mechanic to choose which direction the ratchet will drive and which direction it will turn free (Figure 3-15).

The *speed handle* (Figure 3-16) is another popular socket driver. The socket wrench is installed on the end of the speed handle. The technician pushes on the end of the handle to hold the socket firmly on the nut or bolt. At the same time, the technician turns the crank-shaped handle in a direction to tighten or loosen. The combination of a swivel handle and crank allows very quick driving of a socket. The speed handle is used when a large number of bolts or nuts must be removed or replaced. Technicians often use this handle when

FIGURE 3-14 A 3/8-inch drive ratchet.

42

REVIEWERS

The author and publisher would like to extend special thanks to the following instructors for reviewing this material:

Joe Magnuson
Pikes Peak Community College
Colorado Springs, Colorado

Don Sykora
Morton College
Plainfield, Illinois

Eric Pang
Leeward Community College
Kapolei, Hawaii

In addition, the publishers acknowledge this text's previous author, Clifton E. Owen, for his patient and thorough review of copyedited manuscript and page proofs.

Michael Ronan, State University of New York, Alfred State College, provided his subject-matter expertise and contributed many of the new, four-color photos in this revision.

TODAY'S TECHNICIAN™

Shop Manual

for Basic Automotive Service & Systems

Chapter 1

Safety Issues and First Aid

Upon completion and review of this chapter, you should understand and be able to describe:

- Personal safety and work practices.
- How personal protection equipment applies to you.
- Common safety hazards in an automotive repair business.
- The technical and legal requirements and expectations of the business and the technician.
- The legal requirements and practices for control of hazardous material and waste.
- The purposes for the Environmental Protection Agency (EPA) and Occupational Safety and Health Administration (OSHA).
- The information contained on Material Safety Data Sheets (MSDS) and container labels.
- How to inspect fire extinguishers and understand operating procedures.
- Basic first aid steps.
- Hybrid vehicle safety.
- The safety layout of a shop.
- How to work safely on and around a vehicle.
- How to disarm an air bag system.

Introduction

Accidents cost employers and employees millions of dollars and work hours each year. An accident affects the individual employee, the family, the employer, and possibly all coworkers. Paying attention to the task, practicing safe habits, and applying common sense will lessen the chance of being injured. Knowing the environment—the workplace—also helps prevent loss of time from mechanical, chemical, and burn accidents.

Many repair facilities and car dealerships have raised insurance rates because of employee accidents. This added cost has to be either passed onto the consumer or taken out of the profit margin. As a prospective employee (or maybe even an owner), it is important to understand safety because a serious accident can make a significant difference at the workplace.

Governmental agencies at all levels provide guidance and legal requirements to assist in accident prevention. However, because all technicians are human, there may be a time when some first aid is needed. Understanding some basic steps may reduce the amount of injury and possibly save a life.

Personal and Coworker Safety

The way you and your coworkers dress and act in the shop reflects your professionalism. Wearing professional clothing and acting in a safe manner helps keep you healthy and makes your customers feel happier about hiring you to repair their vehicles.

Clothing, Hair, and Jewelry

A **logo** is artwork used by a business to promote its name and product more easily. The "Golden Arches" is a visual logo for McDonald's restaurants. The "blue seal" is a visual logo for technicians who are certified.

Professional clothing ranges from lab coats to jump suits to clean "civilian" work clothes (Figure 1-1). To present a better image, most shops supply or require you to wear uniforms with the business **logo**. Uniforms are fitted to make sure no loose sleeves or shirttails can become tangled. If uniforms are not required, follow the same procedures with civilian work clothes. Extra looseness is not beneficial in the automotive shop environment.

Most uniforms have polyester as the main material. This material is more resistant to battery acid and cleans easier. Cotton clothing does not go well with battery acid.

Long hair in a shop can be dangerous. If the hair is long enough to get tangled in rotating engine or drive train parts, it must be pinned or fastened securely out of the way. Never assume it is short enough. Under some circumstances, hair that is slightly more than shoulder length can cause problems.

Jewelry is one item that technicians do not like to remove, particularly wedding rings and religious images. Do not wear rings when working in an automotive shop. Rings are made of metal, and they tend to conduct electricity and react in strange ways when exposed to chemicals. Typical of personal religious symbols is a long necklace with the jewelry hanging inside the shirt or blouse. Long necklaces should be removed before beginning work. Ask the supervisor if there are any storage areas that may be used for storage of personal valuable jewelry.

Eye and Face Protection

A person can go to almost any shop and see technicians working without eye or face protection. Most of them feel the protection is uncomfortable or cuts visibility. Most of them also have not seen serious eye or face injuries firsthand. The wearing of eye protection

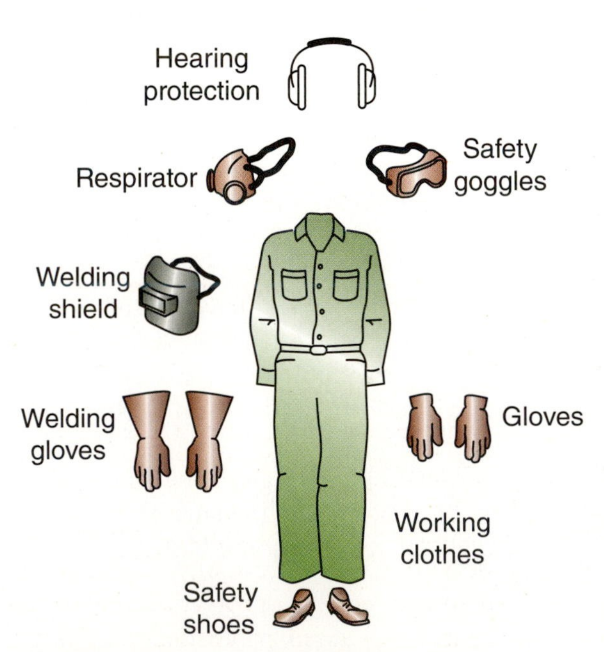

FIGURE 1-1 **Work clothes should fit neatly without binding or looseness.**

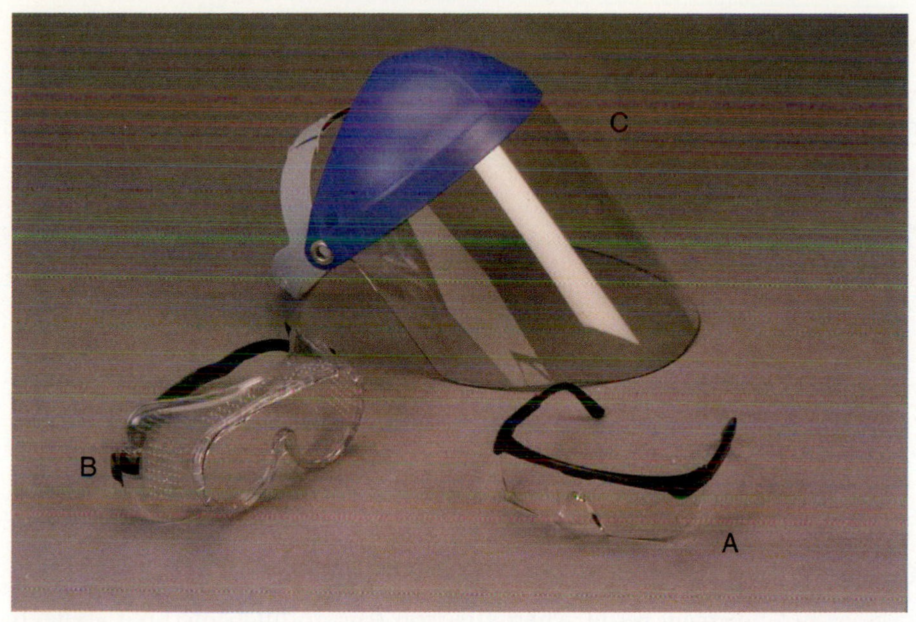

FIGURE 1-2 **The only way to protect the eyes during work is to wear proper eye or face protection.**

during work is the safest method for protecting sight (Figure 1-2). Proper safety glasses are rated by the American National Standards Institute and have a stamping on the frames or lens. This stamping recognizes that the safety lens will not break during a "concentrated high impact." This rating is Z.87. Most eye injuries in the automotive shop come from a piece of rust or metal from a routine job such as exhaust work. Face injuries often happen during cleaning or grinding parts.

Though it may feel uncomfortable when first wearing eye protection, the technician will become accustomed to it with daily wear and eventually feel naked without it. Remember your first ring or your first watch? For a few days, it worried you to wear it. Now, forgetting that ring or watch throws the whole day off. The same applies to eye and face protection. The vision problem is corrected in two ways. First, buy a set of safety glasses that are worth the money, and second, wear them every day. Technicians' tools are the best investment in their career. Eye and face protection devices are part of the tool set.

CAUTION:
Ordinary prescription glasses are not safety glasses even though they may offer some impact protection. Order safety prescription glasses to include the frame. You are not protecting yourself if the glasses do not meet occupational safety glasses requirements.

Hearing and Hand Protection

With some of the loud music being played on vehicle audio equipment, it is hard to imagine that an automotive shop can cause ear damage. A typical repair business does not have a great deal of *continuous noise* as in a body shop, but there are instances when hearing needs to be protected. Earmuffs or plugs are excellent for temporary protection on the job (Figure 1-3).

Hands are subject to cuts and abrasions by sharp tools and burning by chemicals and heat from the exhaust system. Be careful how the hands are worked into those crevices on the vehicle and use good hand tools. When cleaning parts, wear chemical protection gloves to prevent burns. One thing about hearing and hand protection is that most technicians use these types of protection devices without prompting. The pain and discomfort of exposure is immediate, as opposed to eye and face protection where the danger is not readily apparent. Lack of eye, face, hearing, or hand protection may put the technician's career on hold for some time.

FIGURE 1-3 **Ear protection should be worn during any metal working, such as grinding or drilling.**

Lifting and Carrying

Back pain does not normally result when a person tries to lift something above his or her capabilities. It usually happens when the body is in a position that will not accept the weight being moved. Before lifting any object, always position the body so the additional weight is placed over the legs. If lifting from the floor, bend at the knees and attempt to keep the back straight (Figure 1-4). This will allow the legs to perform the lifting motion. The reverse applies if an object is to be placed on the floor or a lower surface. Secure the object and lower it to the floor by bending the knees. This may be awkward and could cause a loss of control with heavier objects. Before moving heavy objects, ask another person to help distribute the weight and maintain control. If the object slips or moves off balance, the best method is to step back and allow the object to drop.

CAUTION:
Trying to stop a falling object can result in injury to you without regaining control of the object. Step back and allow the object to drop. If possible, you may try to control it after it hits the floor, but do not let yourself be placed in jeopardy.

Also, if a change of direction is needed, such as moving an object from one table to another, do not twist the body. Turn the entire body with the object. Keeping feet and legs in place and twisting at the waist places the weight of the object at an angle to the muscles. Back and shoulder muscles may not be able to fully support the weight and may be overexerted.

Carrying heavier objects is dangerous to the technician, the object, and personnel working nearby. If the object cannot be carried comfortably, ask a second person to help. Do not wait until the object becomes unstable.

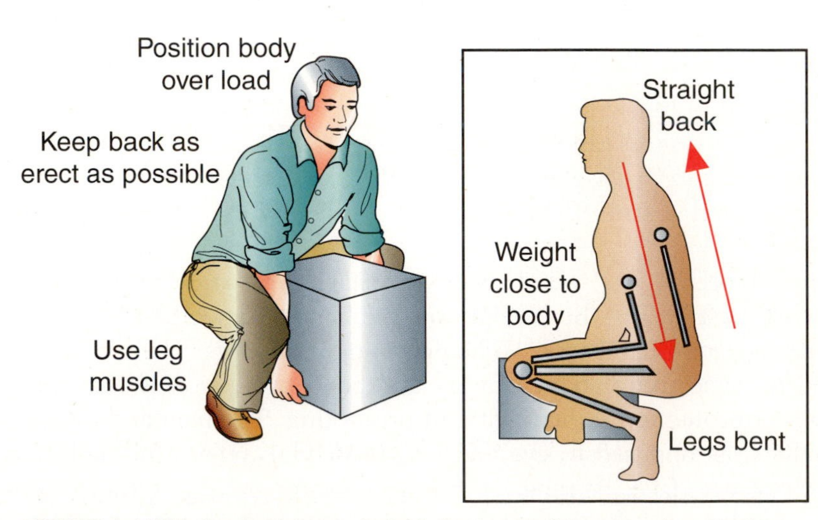

FIGURE 1-4 **With the back kept straight, the load will be placed on the legs.**

AUTHOR'S NOTE: One of the worst back injuries I have endured was the result of picking up a 2-pound motorcycle battery. Down on one knee, I twisted to the left, bent slightly, grabbed the battery, and straightened up to slide it in place on the bike. About halfway around through the straightening and twisting, I felt an extremely sharp pain in the vicinity of my right shoulder blade. The battery and I hit the ground about the same time. I was out of work for 2 weeks, had trouble breathing, and was not fully recovered for about 5 weeks. Remember: It's not always the weight so much as the position of the body that can cause injury.

Coworker Safety

Another safe practice is to watch out for coworkers. Each person needs to have an idea of what is happening around him or her. The person in the next bay may be busy moving something, not knowing someone is approaching from the blind side. In addition, coworkers are like everyone else. Do them a favor and remind them of eye protection and other safety violations they may have missed.

HAZARDOUS MATERIAL AND WASTE CONTROL

Right-to-know laws place a great deal of responsibility on employers to ensure that employees know about all hazardous materials they may be using. It is the employee's responsibility to utilize that training.

Environmental Protection Agency (EPA)

The Environmental Protection Agency, or EPA, is a federal agency charged with instituting and enforcing regulations that assist in protecting the environment. It was formed in the early 1970s to reduce air pollution caused by vehicle and manufacturing emissions. Inherent within that charter was the control and disposal of waste products from almost all businesses, including local automotive repair shops and individuals. The main issue with the EPA is the storage and disposal of hazardous waste from major manufacturers, plants, the local garbage dump, and everything in between. Though its formation met with much resistance, the results some 40 years later are cleaner air and less ground and water pollution. The EPA has had a very positive effect on businesses and the environment and will continue to do so in the foreseeable future. The EPA's Web site is http://www.epa.gov.

Environment Canada

Environment Canada is the Canadian version of the U.S. EPA. It has requirements that relate to Canada's more northern environment and citizens. Within Environmental Canada, there are subagencies that perform the same governmental functions as their sister agencies with the U.S. EPA. As far as the automotive industry is concerned, however, the legal and environmental control requirements are almost exactly the same. Section 7 of the Canadian Environmental Protection Act specifically covers the Canadian automotive industry. The Web site best suited for obtaining information on this agency is http://www.ceaa-acee.gc.ca.

Occupational Safety and Health Administration (OSHA)

Occupational Safety and Health Administration (OSHA) was formed to help protect employees and ultimately employers. It has the legal authority to inspect businesses and ensure that working areas are safe for the employees. Some safety concerns of utmost importance are the control of chemicals within the workplace, the equipment and facilities to store or use those chemicals, the equipment and tools used within the facility, and the general working environment. It should be noted that since the formation of OSHA, accidents resulting from unsafe working environments have been reduced, with an increase in production associated with lowered lost man-hours and fewer accidents. Visit OSHA's Web site at http://www.osha.gov/SLTC/index.html for more detailed information.

Canadian Center for Occupational Health and Safety (CCOHS)

This agency is similar to the U.S. OSHA, with a similar mandate, responsibility, and authority. It performs inspections, determines administrative fines, may file criminal charges, and directs training programs in much the same manner as the U.S. OSHA. Its Web site is http://www.ccohs.ca/.

It should be noted that each of the four agencies listed operate across borders because many pollutants tend to cross borders. The automotive manufacturing, vehicle repair, and vehicle operations are shared by the United States and Canada, and many associated problems are the result of actions in one country affecting the environment of its neighbor. Each of the listed Web sites contains information pertaining to many environmental and safety issues.

Right-to-Know

Each of these agencies enforces what are known as "right-to-know" laws or regulations. Basically, right-to-know laws require the employer to notify employees of dangerous materials that are housed or used on-site. They also require the initial training of new employees, annual (or more often) refresher training of all employers, and employer-designated personnel with specific authority to train, maintain records, and in some instances act as first responders to fires or accidents. Of direct interest to all employees are the three main informational documents pertaining to on-site chemicals, described next.

Material Safety Data Sheets

The *Material Safety Data Sheet (MSDS)* is most important for an employee. In Figure 1-5, the chemical listed is a common brake cleaner (Sections 1 and 2). Search the sheet for the safety hazards that may cause harm to humans and the environment (Sections 3 and 4). Using that information, it should be easy to decide what personal protection measures must be taken to protect people. Go to Section 5 on Figure 1-5 and compare those measures with the items listed there. Analyze the first-aid treatment that may be needed in case of accident. Survey the shop and locate the items that may be needed to prevent injury and the items needed as first aid for this chemical. The minimum would be an eyewash station and a place to wash any flesh that may be heavily exposed to this chemical.

Container Label

Every chemical must have a label on its container listing the same items found on an MSDS. A container from a manufacturer will have that label, but many times chemicals are stored in a large container and transferred to a smaller (transfer) container for ease of use. Sometimes this transfer container doesn't get labeled. It is required by law, and common sense, that hazardous materials not be stored in an unmarked or improperly marked container. Surprises can lead to serious injuries or damage. The label on the transfer container must list exactly the same information shown on the manufacturer's label.

```
HEXANE
=======================================================
MSDS Safety Information
=======================================================
Ingredients
=======================================================
Name: HEXANE (N_HEXANE)
% Wt: >97
OSHA PEL: 500 PPM
ACGIH TLV: 50 PPM
EPA Rpt Qty: 1 LB
DOT Rpt Qty: 1 LB
=======================================================
Health Hazards Data
=======================================================
LD50 LC50 Mixture: LD50:(ORAL,RAT) 28.7 KG/MG
Route Of Entry Inds _ Inhalation: YES
Skin: YES
Ingestion: YES
Carcinogenicity Inds _ NTP: NO
IARC: NO
OSHA: NO
Effects of Exposure: ACUTE:INHALATION AND INGESTION ARE HARMFUL AND MAY BE FATAL.
INHALATION AND INGESTION MAY CAUSE HEADACHE, NAUSEA, VOMITING, DIZZINESS, IRRITATION
OF RESPIRATORY TRACT, GASTROINTESTINAL IRRITATION AND UNCONSCIOUSNESS. CONTACT
W/SKIN AND EYES MAY CAUSE IRRITATION. PROLONGED SKIN MAY RESULT IN DERMATITIS (EFTS
OF OVEREXP)
Signs And Symptoms Of Overexposure: HLTH HAZ:CHRONIC:MAY INCLUDE CENTRAL
NERVOUS SYSTEM DEPRESSION.
Medical Cond Aggravated By Exposure: NONE IDENTIFIED.
First Aid: CALL A PHYSICIAN. INGEST:DO NOT INDUCE VOMITING. INHAL:REMOVE TO FRESH AIR. IF
NOT BREATHING, GIVE ARTIFICIAL RESPIRATION. IF BREATHING IS DIFFICULT, GIVE OXYGEN.
EYES:IMMED FLUSH W/PLENTY OF WATER FOR AT LEAST 15 MINS. SKIN:IMMED FLUSH W/P LENTY
OF WATER FOR AT LEAST 15 MINS WHILE REMOVING CONTAMD CLTHG & SHOES. WASH CLOTHING
BEFORE REUSE.
=======================================================
Handling and Disposal
=======================================================
Spill Release Procedures: WEAR NIOSH/MSHA SCBA & FULL PROT CLTHG. SHUT OFF
IGNIT SOURCES:NO FLAMES, SMKNG/FLAMES IN AREA. STOP LEAK IF YOU CAN DO SO W/OUT
HARM. USE WATER SPRAY TO REDUCE VAPS. TAKE UP W/SAND OR OTHER NON_COMBUST MATL &
PLACE INTO CNTNR FOR LATER (SU PDAT)
Neutralizing Agent: NONE SPECIFIED BY MANUFACTURER.
Waste Disposal Methods: DISPOSE IN ACCORDANCE WITH ALL APPLICABLE FEDERAL, STATE AND
LOCAL ENVIRONMENTAL REGULATIONS. EPA HAZARDOUS WASTE NUMBER:D001 (IGNITABLE
WASTE).
Handling And Storage Precautions: BOND AND GROUND CONTAINERS WHEN TRANSFERRING LIQUID.
KEEP CONTAINER TIGHTLY CLOSED.
Other Precautions: USE GENERAL OR LOCAL EXHAUST VENTILATION TO MEET
TLVREQUIREMENTS. STORAGE COLOR CODE RED (FLAMMABLE).
=======================================================
Fire and Explosion Hazard Information
=======================================================
Flash Point Method: CC
Flash Point Text: _9F,_23C
Lower Limits: 1.2%
Upper Limits: 77.7%
Extinguishing Media: USE ALCOHOL FOAM, DRY CHEMICAL OR CARBON DIOXIDE. (WATER MAY BE
INEFFECTIVE.)
Fire Fighting Procedures: USE NIOSH/MSHA APPROVED SCBA & FULL PROTECTIVE
  EQUIPMENT (FP N).
Unusual Fire/Explosion Hazard: VAP MAY FORM ALONG SURFS TO DIST IGNIT SOURCES & FLASH
BACK. CONT W/STRONG OXIDIZERS MAY CAUSE FIRE. TOX GASES PRDCED MAY INCL:CARBON
MONOXIDE, CARBON DIOXIDE.
=======================================================
```

FIGURE 1-5 **Each section of the MSDS lists specific information about the properties of the material.**

Hazardous Material Inventory Roster

This roster is not readily available to every employee every day, but the employee should know its location. The roster is a list of every hazardous material, mostly chemicals with their individual hazards, that is stored or used on the business property. It is usually stored away from the "dangerous" areas of the property so it will be available to first responders in case of accident or fire. It shows the location of each material, and several copies may be located in different places, especially in large, single-site businesses. The roster is available to every employee and may be required reading for selected employees.

Storage of Hazardous Materials and Waste

Hazardous *waste* is a different matter from hazardous *materials.* Waste is something to be discarded; it is not reusable and must be stored in marked and completely sealed containers. In most cases, oil and antifreeze are the most common hazardous waste in an

FIGURE 1-6 **Hazardous material storage areas should be marked as flammable and should be enclosed or locked.**

automotive shop. The EPA does not consider most used materials as hazardous waste until they are picked up for recycling or disposal.

Since hazardous materials are used almost daily, they need to be stored close to the work area, usually inside the shop. The storage device should be a closed metal cabinet of some type that is marked with fire warnings (Figure 1-6). Some materials, such as fuel, should be *stored outside the building* in approved specialty metal containers. Remember, the storage and safe use of hazardous materials is one of the safety requirements of the shop and employees. Anyone using a hazardous material must use it safely and store it properly. Any material remaining in the container after use must be returned to the storage area or disposed of properly.

Used material such as oil must be stored under certain conditions (Figure 1-7). Each material must have its own container. The container must be labeled with the material's name and be marked as USED.

FIGURE 1-7 **Flammable fluids should be stored in sealed marked containers only.**

Another type of hazardous waste is the can used to store propellant-driven material such as brake cleaner. Once empty with all of its pressure released, the can must be split open before its disposal. The EPA considers a container of this type to be hazardous waste if it is not split open.

Fires and Fire Extinguishers

The danger of fire in an automotive shop is present at most times. Fire can be caused by oily rags improperly stored, hazardous materials exposed to heat, fuel leaks, or poor storage of hazardous material or waste. There are four general classifications of fires and a type of fire extinguisher to match the burning materials (Figure 1-8). Each class of fire is matched with a type of fire extinguisher containing the best material for controlling or extinguishing that fire. In most cases, the type of fire in an automotive shop will not be paper or metallic such as magnesium.

AUTHOR'S NOTE: Be aware that some of the wheels offered for sale contain magnesium. They do not usually present a great fire hazard to the shop or the employees, but there is always that possibility.

A fire in a shop has a good chance of being a Class B type, which involves flammables such as fuel, oil, and solvents. However, the technician must understand the types of fires being combated and know whether the fire extinguishers in the shop are right kind and operable.

	Class of Fire	Typical Fuel Involved	Type of Extinguisher
Class A Fires (green)	For Ordinary Combustibles Put out a class A fire by lowering its temperature or by coating the burning combustibles.	Wood Paper Cloth Rubber Plastics Rubbish Upholstery	Multipurpose dry chemical
Class B Fires (red)	For Flammable Liquids Put out a class B fire by smothering it. Use an extinguisher that gives a blanketing, flame-interrupting effect; cover whole flaming liquid surface.	Gasoline Oil Grease Paint Lighter fluid	Carbon dioxide Halogenated agent Standard dry chemical Purple K dry chemical Multipurpose dry chemical
Class C Fires (blue)	For Electrical Equipment Put out a class C fire by shutting off power as quickly as possible and by always using a nonconducting extinguishing agent to prevent electric shock.	Motors Appliances Wiring Fuse boxes Switchboards	Carbon dioxide Halogenated agent Standard dry chemical Purple K dry chemical Multipurpose dry chemical
Class D Fires (yellow)	For Combustible Metals Put out a class D fire of metal chips, turnings, or shavings by smothering or coating with a specially designed extinguishing agent.	Aluminum Magnesium Potassium Sodium Titanium Zirconium	Dry power extinguishers and agents only

FIGURE 1-8 **Class B and C type fires present the greatest fire concern in an automotive shop.**

FIGURE 1-9 **The fire extinguisher in the center is a multipurpose fire extinguisher for use against Class A, B, and C fires.**

The extinguisher's **tag** is usually a small card tied directly to the extinguisher. The inspector may enter his or her initials, date, and status of the extinguisher.

Upwind means the wind is blowing on the person's back. If the wind is in the firefighter's face (downwind), the flame and the fire-fighting material can be blown toward the fighter.

The fire extinguisher normally found in an automotive shop is either a Class B or a multipurpose extinguisher that can be used on different types of fires. Look for a "B" enclosed in a square on the fire extinguisher (Figure 1-9). It may have more than one symbol indicating that it is a multipurpose extinguisher. Obviously, inspection of a fire extinguisher cannot be done while extinguishing the fire. Fire extinguishers must be inspected periodically, preferably by the same person. There are usually several fire extinguisher points throughout the shop. Look at the **tag** and gauge of each to make sure the extinguisher is charged and ready for work. The safety clip or wire must be in place, and there must be no damage to the fire extinguisher. The inspector initials and dates the tag if the extinguisher is ready for use. Study the different types of extinguishers available; learn their locations and the proper use of each style during the inspection. Most fire codes call for an inspection of fire extinguishers on a periodic basis.

The use of a Class B fire extinguisher is discussed because this is the most common type of fire in an automotive shop. Class B fires must have their oxygen supply cut off before they go out. Water or other spreading material is not applicable on Class B fires because it spreads the fuel. When a fire is discovered, sound the alarm and locate a Class B fire extinguisher. Position yourself about 8 feet from the base of the fire, **upwind** if possible. Remove the safety clip, aim the nozzle at the base of the fire, and pull the trigger. Continue using the extinguisher until the fire is out or the extinguisher is empty. If necessary, spray the fire area to cool any hot spots that may reignite the fire. If you or any of your coworkers feel the fire is too large for the shop's fire extinguishers, leave the area and notify the local fire department.

Although Class B fires are the most common in an auto shop, it is important to know the other types of fires and the classes. A Class A fire is a combustible products fire. This includes wood, paper, or a cloth seat. These fires can be extinguished by water. A Class C fire is an electrical fire. On some occasions you may experience a Class C fire. The smell of a Class C fire is very distinct, and you will never forget it. A Class D fire is very rare in an auto shop. Class D fires are caused when pyrophoric metals explode. An example of this metal is magnesium. Most shops today have multipurpose fire extinguishers.

The easiest way to remember how to use a fire extinguisher is to remember the acronym **PASS**. This stands for Pull, Aim, Squeeze, and Sweep. Pull the trigger, aim the nozzle, squeeze the lever, and sweep the hose side to side. Photo Sequence 1 explains the use of a dry chemical multipurpose fire extinguisher.

USING A DRY CHEMICAL FIRE EXTINGUISHER

P1-1 **Multipurpose dry chemical fire extinguisher.**

P1-2 **Hold the fire extinguisher in an upright position.**

P1-3 **Pull the safety pin from the handle.**

P1-4 **Stand 8 to 10 feet upwind of the fire. Do not go any closer to the fire.**

P1-5 **Free the hose from its retainer and aim it at the *base* of the fire.**

P1-6 **Squeeze the lever while sweeping the hose from side to side. Keep the hose aimed at the *base* of the fire.**

FIRST AID

This section is not intended to make the reader a certified medical first responder. *First aid* is defined as actions to protect the victim and reduce the injury until trained medical personnel are on the scene. Remember that we are not discussing extreme cases, but common and usually non-life-threatening injuries. There are four basic first aid steps that can be applied by any person in the shop and in whatever order is appropriate to the situation:

1. Stop the bleeding and protect the wound.
2. Treat burn.
3. Treat for shock.
4. Protect the victim.

For the most part, injuries during an automotive repair will be cuts or burns. However, shock may kill the victim faster than the injury. Some people will go into deep shock at the sight of blood—whether it's their own or others'. In event of an accident, call or have some one contact medical personnel. In the United States, this is usually accomplished by dialing "911" and giving necessary information for quick response. Don't forget that most schools and businesses require a prefix number, usually "9" in the United States, to get an outside telephone line. Most schools also have a medical person or department that is in charge of administrating first aid. Schools also have a first-aid procedure in their school policies or handbook. Make it a point to check the first-aid procedures where you are.

Follow the appropriate step(s) listed earlier by using: sterile bandages and pressure for bleeding, burn treatment or cooling the area with clean water for burns, talking quietly to the individual for shock, and, when necessary, evacuate the person from the area in cases such as a nearby fire or chemical spill. One of the best things a first aid provider can do is remain calm and keep the injured calm.

First Aid for Eye Injury

One of the most common injuries that occurs in an automotive repair shop is injuries to the eyes. These are all too often caused by the failure, intentional or otherwise, to wear protective face and eye protection. Common protection devices to help prevent this were discussed earlier in this chapter. However, many technicians, smart in every other way, fail to use them for various reasons. Eye injuries are painful, distracting, and can be permanent. In almost every case of eye injury, certain steps should be followed immediately.

First, do not rub the eye(s). This will only cause more damage and pain. If necessary, guide the person (or be guided) to the eye wash station. It is very important to know the location of the eye wash area, in case you may need it and not be able to see well to get there. There are two general types. One is an attachment to a water source by which low-pressure, high-volume water can be directed over the eyes (Figure 1-10). After flushing the eyes, the water naturally flows back off the face. Some state safety codes call for this water to be warmed for a certain amount of time that it is on. The second common type is the use of sterile water in a throwaway plastic bottle (Figure 1-11). The head is tilted backward, and the water is poured from the bottle over the affected eye(s). Tilting the head toward the side of the injured eye helps the water flush the chemical or debris from the eye over the cheek. Both methods work well but may require some assistance from another person. Regardless of the method used, do not rub the eye(s), and seek medical attention as soon as possible. Metal debris can cut the eye and cause infection, and chemicals can cause immediate serious burns. It is best if

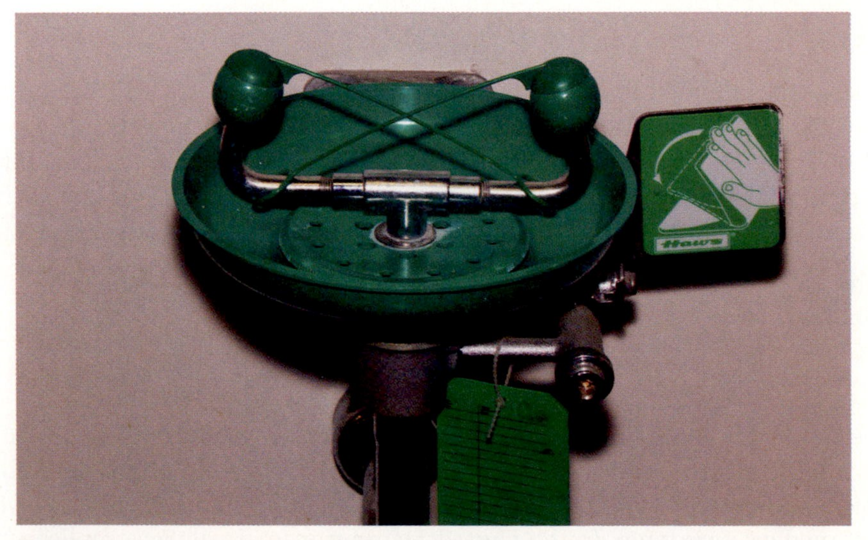

FIGURE 1-10 **This type of eye wash can be taken to the injured worker. Once opened, any remaining liquid should be discarded.**

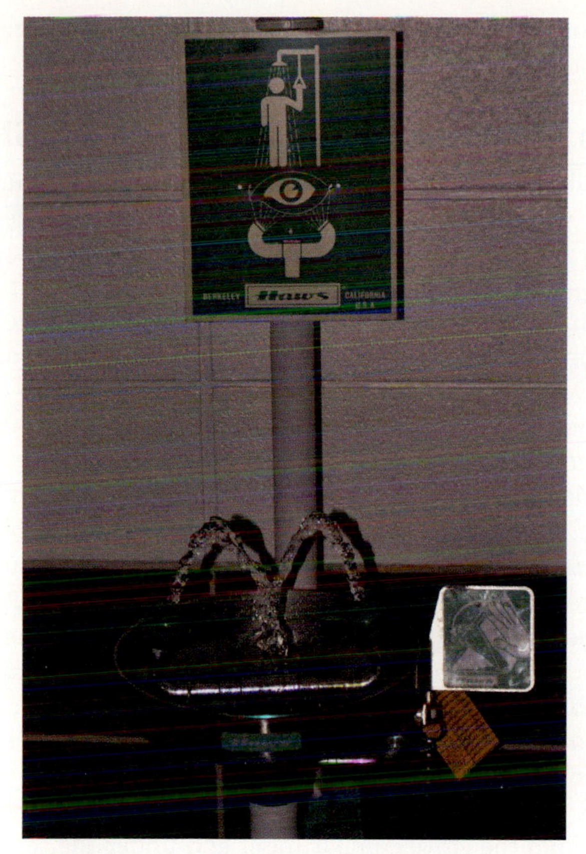

FIGURE 1-11 **A typical eye wash station.**

water is constantly flushed over the eye until medical assistance arrives or a visit is made to the nearest medical facility. Chemical intrusion into the eye should be flushed out for at least 15 minutes.

Supervisors and teachers should not, and usually don't, accept the "I'm OK, I don't need a doctor" idea. Supervisors must maintain an accident report, and more important, the technician's health and well-being is more of a concern than the medical cost or lost time. In this instance, the injured should flush the eyes, refrain from the natural tendency to rub them, and accept medical assistance as soon as possible.

CAUTION:
For some repair tasks on a hybrid vehicle, it may be necessary to disconnect the drive batteries from the rest of the vehicle. This should be done while wearing insulated gloves similar to those worn by power company linemen. Very serious injury or death could result if proper person protection equipment (PPE) is not worn. See Chapter 9 for further details.

Hybrid Vehicle Safety and High-Voltage Safety

With the increasing sales of hybrid vehicles, a specific safety issue has emerged for the automotive technician. Greater details on this issue are discussed in Chapter 8 of this manual and the Classroom Manual, but they basically concern the much higher available electrical current from the drive batteries or power pack. This higher available current in the vehicle's drive batteries require much larger conductors and cables. The positive cable of the drive battery will be orange in color and equal to or larger than typical 12-volt battery cables. In some cases the voltage can be above 300 volts. The negative cable will be the same size and black in color. Paying close attention and using common sense can prevent most, if not all, of the electrical safety concerns with hybrid vehicle. If the cable is big and orange, don't touch it without proper equipment and training.

SERVICE TIP:
Waiting time for the capacitor to discharge varies. Consult the service manual for specifics.

Some nonhybrid electric vehicles have high-voltage circuits. Some vehicles have high-intensity headlights. These headlights are brighter and provide a better focus for the driver because they illuminate the road brighter, especially during bad weather. These headlights operate on voltage higher than the battery. They range from 14 to 24 volts. When servicing this voltage level, it is important to follow the manufacturer's suggestions and safety precautions. Just like other parts of the vehicle, there are always safety warnings and labels under the hood.

CAUTION:
Since 1998, most pickup trucks have been equipped with a switch to turn off the passenger-side air bag. If it is switched off, the owner has a reason. Ensure that the switch is set to the same function (on/off) before returning the vehicle to the owner. If appropriate, notify the customer that the switch is left in the off position. Death and serious injury could result if this air bag functions and the passenger does not fit the normal passenger profile.

Working Around Supplemental Restraint Systems (SRS)

Supplemental restraint systems (SRS) are commonly known as air bags. Though many lives have been saved with SRS, some lives have been lost or injuries incurred by technicians and emergency personnel working near or around the air bag. Like all computers, the one for the air bags is "dumb." Its sole purpose in life is to wait for a signal and then ignite the air bag chemicals. It does not know or care if that signal comes from an air bag arming switch or from a technician probing the wires under the dash or an emergency medical technician (EMT) trying to extract an individual from the vehicle. It just senses a signal and activates the air bag, sometimes with drastic results.

Before beginning any work inside the vehicle's passenger compartment, check to see how many air bags are installed. Naturally, there is the driver-side air bag and probably a passenger-side air bag, but there may be side or curtain bags at the outside edge of the seat and possibly in the headrests. (Some vehicles are now equipped with air bags to protect rear-seat passengers.) There are two definite steps that must be taken with any air bag system. First, disconnect the negative battery cable and place it so that accidental contact cannot be made with the battery. Second, wait a specific time for the air bag's **capacitor** to discharge.

A **capacitor** is a small electrical storage device.

If work is being done on the electrical system inside the passenger compartment, all air bag connections should be disconnected before work begins. With the capacitor discharged, look near the lower part of the steering column and disconnect all bright yellow connections. Air bags are the only systems now using bright yellow connections in this area. Check in and around the glove compartment and under the seats for additional bright yellow connections. Disconnect all of them. Consult the service manual for the location of all air bag electrical connections of this type. Once this is done, the battery can be reconnected to make electrical checks. Before reconnecting any of the air bag connectors, disconnect the battery again. Reconnect all air bag connectors and then reconnect the battery. This prevents any surges from activating the air bags.

SUMMARY

TERMS TO KNOW

capacitor
Logo
PASS
Tag
Upwind

- Eye and face protection provides excellent protection against injuries.
- Gloves should always be worn when working with chemicals.
- Lifting should be done with the legs while keeping the back straight to avoid back injury.
- Most businesses, particularly automotive repair, keep hazardous material available on site.
- Right-to-know regulations require the business owners and managers to inform employees of the presence of hazardous materials and waste on the site.
- The EPA is a federal agency charged with regulating possible pollutants that may harm the environment.
- OSHA is a federal agency primarily concerned with safety at the work site.
- Each container must have a container label.
- Canada has environmental and safety agencies that are similar to the ones in the United States.
- Each chemical sold in the United States requires a Material Safety Data Sheet to be kept at the user's location.
- A Hazardous Material Inventory Roster is kept in a secure location and is available to all emergency first responders.

- There are four classes of fires and a matching type of fire extinguisher.
- Class B fires are most common in automotive repair shops.
- A combination fire extinguisher is the best type for an automotive shop.
- First aid is the immediate assistance given to a victim before the arrival of medical personnel.
- Hybrid vehicles usually operate on a combination of electric power and internal combustion engines.
- Hybrid vehicles have high-energy electrical systems requiring additional safety measures.

ASE-STYLE REVIEW QUESTIONS

1. Firefighting is being discussed.
   Technician A says to use a fire extinguisher that has a circle and a "B" on it to fight a cleaner solvent fire.
   Technician B says no spreading agent is to be used for fighting a gasoline fire.
   Who is correct?
   A. A only C. Both A and B
   B. B only D. Neither A nor B

2. Hazardous material and waste are being discussed.
   Technician A says both should be stored the same way.
   Technician B says waste is not considered to be waste until it has been picked up for disposal or thrown into a dumpster.
   Who is correct?
   A. A only C. Both A and B
   B. B only D. Neither A nor B

3. *Technician A* says that an oil fire is a Class B fire. *Technician B* says that you can use a water hose to extinguish a Class A fire.
   Who is correct?
   A. A only C. Both A and B
   B. B only D. Neither A nor B

4. Fire extinguisher use is being discussed.
   Technician A says to stand 8 feet from the base of a fire and downwind if possible.
   Technician B says to squeeze or pull the trigger and then pull out the safety clip.
   Who is correct?
   A. A only C. Both A and B
   B. B only D. Neither A nor B

5. The MSDS and container label DOES NOT contain which of the following information?
   A. Reactivity data
   B. Transportation method
   C. Special fire control
   D. Storage procedures

6. The route of entry into a human body may be:
   A. Inhalation C. Ingestion
   B. Skin D. All of the above

7. *Technician A* says the container label of a cleaning material should be read before using the cleaner. *Technician B* says an empty cleaning container should be split before it is thrown away.
   Who is correct?
   A. A only C. Both A and B
   B. B only D. Neither A nor B

8. Class B fires are being discussed.
   Technician A says a fire extinguisher labeled with B is to be used on this type of fire.
   Technician B says a spreading agent fire extinguisher can be used to separate the burning materials.
   Who is correct?
   A. A only C. Both A and B
   B. B only D. Neither A nor B

9. *Technician A* says gloves should be worn when using caustic cleaning materials.
   Technician B says face and eye protection should be worn when cleaning parts.
   Who is correct?
   A. A only C. Both A and B
   B. B only D. Neither A nor B

10. Each of the following hybrid vehicle electrical characteristics is true EXCEPT:
    A. The positive conductor will be orange in color.
    B. The positive cable will be equal to or smaller than the typical 12-volt battery cable.
    C. The drive battery cables carry more current.
    D. Insulated gloves should be worn when disconnecting the drive batteries from the vehicle electrical system.

Chapter 2

Shop Operations and the Technician

Upon completion and review of this chapter, you should understand and be able to describe:

- The basic duties of key shop personnel.
- The technician's technical responsibilities.
- How technicians are paid for the work they perform.
- Some of the shop's and technician's legal concerns.
- The differences between warranties and recalls.
- The process for completion of a service repair order.

Introduction

A shop employs personnel in several key positions. As stated in the Classroom Manual, the service manager is responsible for the service department. To make the operation run smoothly, he or she hires service writers, accountants, cashiers, parts managers, personnel, and technicians. We will discuss shop operations that would be typical of dealerships and large independents.

Managers and Administrative Offices

There are several levels of management in the larger shops, especially at dealerships. In smaller shops, the owner may be the only manager and may also cover billing and writing service. The **owner** or **manager** may help out by turning wrenches when necessary. In this section, the operation of a typical local dealership will be used to outline the common positions and responsibilities of each. A typical manager usually has training and experience in business operations, whereas persons in other positions have the knowledge for their particular job such as the accountant's knowledge of debits and credits. Each person is assigned a specific job, including technicians, but there are times, particularly in a smaller shop, when the assignments may have to overlap to satisfy a sudden short-term increase in customers.

The **owner** or **manager** is responsible for the daily operation of a segment of a business and reports to the general manager.

Without a concerted effort by all members of an organization, a customer may buy a vehicle or parts from one business, then go elsewhere to have it installed or maintained. This reduces profit and essentially loses a customer.

The managers of each department must work together to ensure the customer receives first-class service from any employee they may contact. Some dealerships introduce their new-car customers to the service department within a short time of purchase. The customer's perception of receiving individual, personalized treatment boosts the prestige of the company and its employees. At many dealerships service department, you are greeted as soon as you get out of the car. Some dealerships even have the customers drive their vehicle into the service department so that they are not outside in the weather.

Independents, on the other hand, are usually smaller and locally owned operations. Most of the employees and the owner are from the local area and know their customers

GENERAL MANAGERIAL DUTIES

General Manager	Sales Manager	Service Manager
Reports to owner	Reports to general manager	Reports to general manager
Delegates authority Supervises department managers	Supervises sales staff	Supervises service staff
General budget reports	Sales budget reports	Service budget reports
Hires, fires, and trains managers and general staff personnel	Hires, fires, and trains sales and staff personnel	Hires, fires, and trains parts personnel, staff, and technicians
Supervises interactions between department	Coordinates with general manager and service to gain and keep customers for the business	Coordinates with general manager and sales to gain and keep customers for the business

FIGURE 2-1 **Managers share some common duties so the various departments can work together to build and maintain the business.**

The **service manager** controls the operation of the service department, including parts, technicians, support staff, and the training of all department employees.

The **service writer** meets the customer, writes the repair order, and assigns a technician based on company policy.

firsthand. Many times, independents will work with local dealerships to handle each other's work and capabilities. For example, a local independent may specialize in transmission repair. A dealership that has an older vehicle or a trade-in vehicle with a bad transmission may subcontract some of the repair of the transmission after it has been removed, and vice versa: the independent may send a customer to the dealership when the computer needs reprogramming, which can be an expensive and not profitable repair for an independent. Both operations profit from this arrangement in monetary terms and customer trust.

Service Departments

The service writer is usually the first person to greet the customer and the **accountant** is the last. Both are essential to the shop's operation.

Usually, a **service manager,** who is responsible for the **service writer,** technicians, department support staff, and parts, heads the service department. The service manager works with vehicle sales in a dealership to bring new (and pre-owned) car customers back for service when needed (Figure 2-1). The parts department has a manager who reports to the service manager. In an independent shop, the number of people actually involved in these types of duties and responsibilities is often much less than that of a dealership.

Service Writers and Accountants

Classroom Manual
page 25

The **repair order** is a document that tracks the vehicle from its arrival at the shop for repairs to the time the repairs are paid by the customer.

The service writer greets the customer and initiates the **repair order** (Figure 2-2). The service writer must be able to communicate with a wide variety of customers and must understand enough of the vehicle's operating systems to interpret the customer's complaint. The diagnosis goes much quicker if the complaint is clear and concise on the repair order. The service writer may create an estimate of the repair. In many instances, the customer only conducts business with the service writer and the accountant.

The service writer usually assigns the repair order to the first available technician experienced in that repair. However, an automatic transmission service, such as a filter change, may not be assigned to the "transmission" tech but to the entry-level or maintenance light repair tech. This allows the higher-paid technician to perform high-value jobs while the hourly person performs the low-profit, quick-turnaround jobs. This results in an efficient use of personnel while providing excellent training for the less experienced tech.

Express service is a term used to emphasize the speed of basic and regularly schedule maintenance. As mentioned in the Classroom Manual, express service gives the customer a

FIGURE 2-2 **The service writer initiates the repair order.**

quick turnaround and a detailed inspection of the vehicle. To speed things up more, some dealerships have service attendants. These are people who assist the service writer. The attendant has several duties. One is greeting the customer if the writer is busy. One protocol is to never let the customer stand around waiting to be noticed. A quick greeting and "How may we help you?" immediately upon the customer's arrival indicate that person is recognized and welcomed. A second duty consists of placing disposable floor mats and seat covers in the car and retrieving the odometer mileage and the **vehicle identification number (VIN).** The third duty, but not necessarily the last, is to begin the vehicle inspection. Dealership service attendants check all lighting, tire wear and air pressure, parking brakes, horn, and warning lights and perform a walk-around visual check for damage to the vehicle body. The attendant starts the inspection sheet and marks the result of this first phase. This is done while the service order is being prepared and will save time in the express service bay. In addition, any defects found can now be provided immediately to the customer by the writer. This express service job may become a bigger repair job for items such as tires or lighting problem. Part of this third duty is to remove the ignition key from the customer's key ring and return all other keys to the customers. Explain, if questioned, that the shop can replace a lost ignition key easily, but personal ones will present a problem—the customer being locked out of his or her home. This shows that the business takes care of the customer. With that said, the single most important duty is still greeting the customer and making that customer feel appreciated.

The **vehicle identification number (VIN)** is a number assigned to a particular vehicle. Information on the engine size and type, transmission, year of production, and other vehicle data is encoded into the VIN.

The accountant may also act as the cashier for the service department (Figure 2-3). The customer will pick up his or her vehicle keys and pay the repair cost at the cashier's window. However, the accountant does much more. The information on the customer, the vehicle, the repair accomplished, and so on is usually collected and entered into a computerized database. The information is used to remind customers of scheduled maintenance and may be used to resolve a customer's complaint. In larger shops, there may be several individuals working in a service department's accounting office.

FIGURE 2-3 **Good customer relations is important for all who work in the automotive industry.**

FIGURE 2-4 **The parts department stocks parts and materials that are required on a routine basis.**

Parts

The parts manager usually reports to the service manager. The parts department is responsible for ensuring that sufficient routine maintenance items are present (Figure 2-4). Also, parts personnel must order, deliver, pick up, and enter parts on the repair order. The department operates a pickup window for walk-in customers. A running inventory is kept at all times to prevent shortages of routine items. In addition, dealerships are linked by telephone and satellite communications to other dealerships. In this manner, it may be possible to get an out-of-stock part from a dealer on the other side of town instead of waiting several days to get one from the manufacturer's warehouse. The parts are charged against the repair order or a separate invoice, which is sent to the accountant for payment collection.

Technician Technical Responsibilities and Pay

The technician performs the basic profit-making portion of the shop's operations: vehicle repairs. The technician may also have the most personal money invested in the position. Most shops will not hire a technician who does not have personal tools (Figure 2-5). A basic tool set may cost over a thousand dollars, and it is not unusual for an experienced technician to have over $20,000 worth of tools collected through his or her career. Remember that a quality tool set is the good way to make sure that you can make money.

Classroom Manual
page 29

FIGURE 2-5 **A typical tool set for an experienced technician is far more robust than a student's tool set.**

FIGURE 2-6 **Though this is the type of entry-level task performed by new technicians, it is only the first step to a successful career.**

Entry-level technicians are usually recent graduates or may still be attending school. This is their first official job as automotive technicians. An entry-level technician may have different titles when he or she gets employed. Such titles often are lube technician, maintenance light repair technician, express service technician, or general service technician.

Commission pay is when the technician receives a percentage of the labor and part sales as part or all of his/her pay.

Flat rate is the time it takes to perform a job. It is an average based on research and practice. Some technicians are paid by flat rate.

A **labor guide** is a book or guide that the service writer uses to estimate the amount of time a repair will take. This rate is then used to calculate the amount that the customer is charged.

Good tools don't break and get the job done faster and with high quality. This is the recipe for success and a good pay check.

The technician must follow all regulations concerning automotive environmental hazards and safety. The repair should be performed correctly the first time. The best customer relations possible are quick, fair, and accurate repairs to the vehicle.

Entry-level technicians are hired to perform basic services such as oil and filter changes, tires, brakes, and other routine maintenance (Figure 2-6). As the technicians gain experience, they gain more and more responsibilities. In order to stay abreast of vehicle operations, it is necessary for the technician to attend training at the manufacturer school, vendor-sponsored training, or a technical school.

Technicians get paid through several methods. The most common are hourly, **commission,** and **flat rate.** Hourly is usually reserved for entry-level technicians and is paid by the clock hours that the technician is at the shop, regardless of the amount of repair business. This will give the person a certain pay that is consistent week to week and provides time to become proficient at the job. Technicians drawing commission will usually be guaranteed a minimum hourly pay plus a percentage of the shop's charges for labor and parts. In most cases, though the pay can be excellent, the technician is only considered to have worked specific hours per week.

Flat rate is a means to motivate technicians' speed and proficiency by paying them a per-hour rate that is based on a **labor guide.** A labor guide is a publication, paper or computerized, that shows the amount of time an average technician will take to complete the repair. Another name for the per-hour rate found in the labor guide is flat rate. A trained, motivated technician can beat the time allowed in many instances. The guide may state it takes 2 hours to complete a repair, but the trained and experienced technician only takes 1 hour, thus earning 2 hours of pay. Also, it allows an additional clock hour that day to do another job. It is not unusual for an experienced technician to turn in 60 to 80 labor hours for a 40-clock-hour week. However, if a "comeback" (shop warranty) repair comes in, the technician isn't paid for that repair. The lack of pay for a comeback provides an excellent incentive to fix the vehicle right the first time. Many shops offer the best technicians a guaranteed weekly check even when business is slow. This also applies to those working on commission.

AUTHOR'S NOTE: **Flat rate and commission pay also provides an incentive for managing money. There are several times during the year, such as the weeks between Thanksgiving and Christmas and right after school session ends, when repair business drops drastically. Recall, warranty, and preventive services will still be performed, but many customers will be out buying gifts or going on vacation. They tend to put off any repairs they deem not immediately necessary. Then there are times, such as tax refund season, when business is booming. There are also times when the weather changes drastically and business is busy from the drop in temperature. A sudden drop in temperature can bring in vehicles with dead batteries, flooded engines, flat tires, and more. Auto collision repair facilities are busiest just after a weather storm passes through. A technician that is on commission or working flat rate quickly learns to put aside a little pay to cover those times when work can be seasonally slower than normal.**

SHOP AND TECHNICIAN LEGAL CONCERNS

Regardless of the bad publicity about crooked shops and technicians sometimes seen on television, probably at least 99 percent of the repair shops are honest and work hard to satisfy all customers. But like all of life, sometimes mistakes are made. Mistakes in automotive repair can range from a simple broken wire to mistakes in brake or wheel repairs that could be life threatening. Sometimes a vehicle inspection reveals a very unsafe condition that the customer for whatever reason does not want the shop to fix. This could be because of lack of funds on the customer's part. This places the shop and the technician in a vulnerable position. The problem vehicle could be unsafe when operated. Some courts have held the shop responsible for damage and injury because the shop and its employees are considered to be professional in their jobs and should have known not to release the vehicle for use on public roads. This problem should be handled by the shop's management and the customer; it is not a decision for the technician to make. Fortunately, this would only happen in an extreme case. However, the shop and the technician do have some legal obligations in this area. State and local laws dictate those obligations, and moral guidance should do the rest.

With regard to the situation noted in the above paragraph, the author is not a legal expert, but it is recommended that the shop and technician do at least one thing: Write a detailed comment concerning the safety situation on the repair order and have the customer sign it in acknowledgement of the problem, the shop's position, and the customer's refusal to allow the repair work. This may not have any bearing in a trial situation, but it will at least show the shop's and owner's state of mind at the time of disagreement. There are few cases overall that require court or arbitration actions, but it is the customer's vehicle, and he or she is paying the bill and has a right to expect good work. Many shops will voluntarily lose money for repairs on warranty work, even if the complaint may be unfair, to gain good public relations and the good word-of-mouth advertising they receive from a satisfied customer. Experienced service managers know how to gain customer confidence and prevent or minimize incidents from occurring in the first place. With time and a good mentor, you will learn techniques to deal with, and hopefully avoid, many of the difficult situations.

Environmental Protection

Most laws and regulations on environmental protection are set by the local, state, and federal statutes. Hazardous materials and waste are controlled or regulated by federal environmental laws and supplemented by lower levels of government. The shop must also consider damaged parts, such as vehicle computers, as hazardous waste because of the chemicals contained in the circuit boards. Most computers and many other damaged parts are recycled or rebuilt by domestic and foreign companies. At a minimum, removed parts should not just be thrown in the trash bin. They should be sent to a scrap yard for recycling.

Repairs

Both the shop and its employees at all levels should strive to repair the vehicle right the first time and charge a fair price. If a mistake should happen, it should be repaired at the shop's expense. Charging to fix something you broke is not a good business tactic. This could cause lawsuits, legal costs, and generally poor public relations. Prime examples of absorbing the cost of a mistake are the recalls issued by the vehicle manufacturer. A mistake was made either by the engineering and design team or by the part manufacturer. Though it may seem to take a long time to get a government-mandated recall in place, most manufacturers issue recalls without outside pressure.

At the shop level, local arbitration boards or committees consisting of members from outside the industry review and settle disputes between a shop and an unhappy customer. The board's decision is usually binding on the shop, but the customer has the right to appeal the decision or even go to court for a settlement. Most states have a "lemon" law stating that after a certain number or types of repairs are unsuccessful on a given vehicle within a reasonable time frame, the shop or the manufacturer is required to properly fix the vehicle or replace the vehicle at minimum cost to the owner. In almost all cases, the lemon law applies to the business seller of the vehicle. Violations of this law are not usually considered a criminal act unless there is compelling evidence of intentional fraud. Over the years, this law has been applied to new and used vehicle sales.

Warranties and Recalls

Warranties are offered to the customer either as part of the business or at cost, similar to life insurance. Warranties are issued by the vehicle manufacturer for a specified time and mileage. The one that seems the best is 10 years or 100,000 miles from the date of sale. Based on the driver's lifestyle and commuting distance, the mileage limit may expire before the time limit. In large cities and suburbs where the job may be up to 70 miles from home, 100,000 miles is reached in 2 or 3 years or less. Compared to the old days when a warranty on a new vehicle was 12/12 (12 months or 12,000 miles), modern-day vehicle warranties are generous. Many warranties are bumper-to-bumper for 36 months or 36,000 miles, so that any repair is covered. Bumper-to-bumper means every component between the front and rear bumpers—other than routine maintenance such as oil changes, belts, and brakes—is fixed free. Many new car sales, shops, and used vehicle sales offer some type of extended warranty that covers specific repairs for a given time or mileage period. Most extended warranties require the customer to pay the warranty premium and a fixed fee or deductible for the repair. Most repair shops guarantee or warrant their labor and parts for a fixed time or mileage. This has led to a change in the business practice of many shops. At one time a repair shop would overhaul or rebuild an engine or transmission. However, with the growth of Jasper Engine and Transmission Rebuilders and similar operations, it is more profitable for the shop to replace the engine or transmission with a rebuilt unit. If the component fails within its warranty period, the vendor replaces the component free of cost and usually reimburses the shop for some, if not all, of the labor costs. This allows the shop to offer excellent guarantees without having to absorb the full cost of an expensive replacement. In turn, it frees that bay and technician quickly for other work.

Recalls are issued only by the vehicle manufacturer and are usually safety or emission related and apply to any new vehicle sold in the United States and Canada. However, the manufacturer usually attempts to locate any recalled vehicle worldwide. At times, the manufacturer is forced to issue a recall by the National Safety and Transportation Board (NSTB), a federal agency. Other times, a recall may be issued when a problem is found through recurring defects or by the engineering or quality control sections of the manufacturer. Recalls may be necessary on a certain batch of vehicles because of an error at the parts manufacturer or at the vehicle assembly plant. A single recall usually only concerns a certain brand, year, and model of vehicle. Recalls can also be to update the software program of a vehicle computer. Recall repairs are performed by manufacture certified technicians, generally at a dealership. To accomplish a recall, the manufacturer must notify the recorded owner of the vehicle. If the recall is for vehicles that are several years old, the manufacturer's records may not be current. The person receiving the mailed notice is supposed to either take the vehicle in for repairs or complete the recall form with respect to the present owner of the vehicle. In many cases, the vehicle was traded for another or sold and the present owner is unknown to the original owner. The notified owner is expected to provide as much information as possible on the vehicle and return the card to the manufacturer. Unfortunately, if the vehicle is no longer owned by the original individual, the card often goes in the trash, and thus the recall is not completed. The manufacturer is protected to some extent provided it made a reasonable effort to locate and fix the vehicle.

Recall listings may be found at the dealerships, on paper manuals, or on the Internet site of the NSTB, http://www.nstb.gov/cars/problems.

Service Repair Orders

A completed repair order lists the customer's complaint, the diagnosis, repair parts, labor, and costs of each. Some may include a history of the most recent services performed on the vehicle (Figure 2-7). While the individual form may differ from shop to shop, the same information is normally used.

Upon receipt of the repair order, the technician retrieves the vehicle and performs the diagnosis. Assuming that parts are available, the technician performs the repairs and conducts a test drive as needed. Parts personnel charge the parts to the repair order. The technician includes the **labor time** and other information about the repairs required by the shop. This may include the method or type of testing performed and the diagnosis.

The repair order is routed to the accountant, who computes the bill based on information supplied by the technician, parts, and shop policy. Warranty repairs do not cost the customer, but the same information is entered on the order and the customer receives a copy. The repair order information is entered into the database and the order is filed after the customer pays. The manufacturer reimburses the dealership for warranty recall repairs.

Completing a Repair Order

The following steps are based on the form in Figure 2-7. Most of the information will be included on almost any shop's repair order.

The most important parts of a service repair order are the **3 C's.** The 3 C's are concern, cause, and correction. The concern is the reason the customer brought the vehicle in. The cause is the diagnosis of the concern. The correction is the actual repair that was made as a result of the concern. Although this seems like a simple concept, it can be easily forgotten when you are busy and performing multiple repairs on multiple vehicles each day. Always remember that your job is to determine why the customer brought their vehicle to you, what it will take to bring the vehicle back to correct operating conditions, and to link the 3 C's to each other.

The **3 C's** of automotive repair are concern, cause, and correction. The correction (actual repair the technician does) needs to match the concern (the reason it was brought in for service in the first place).

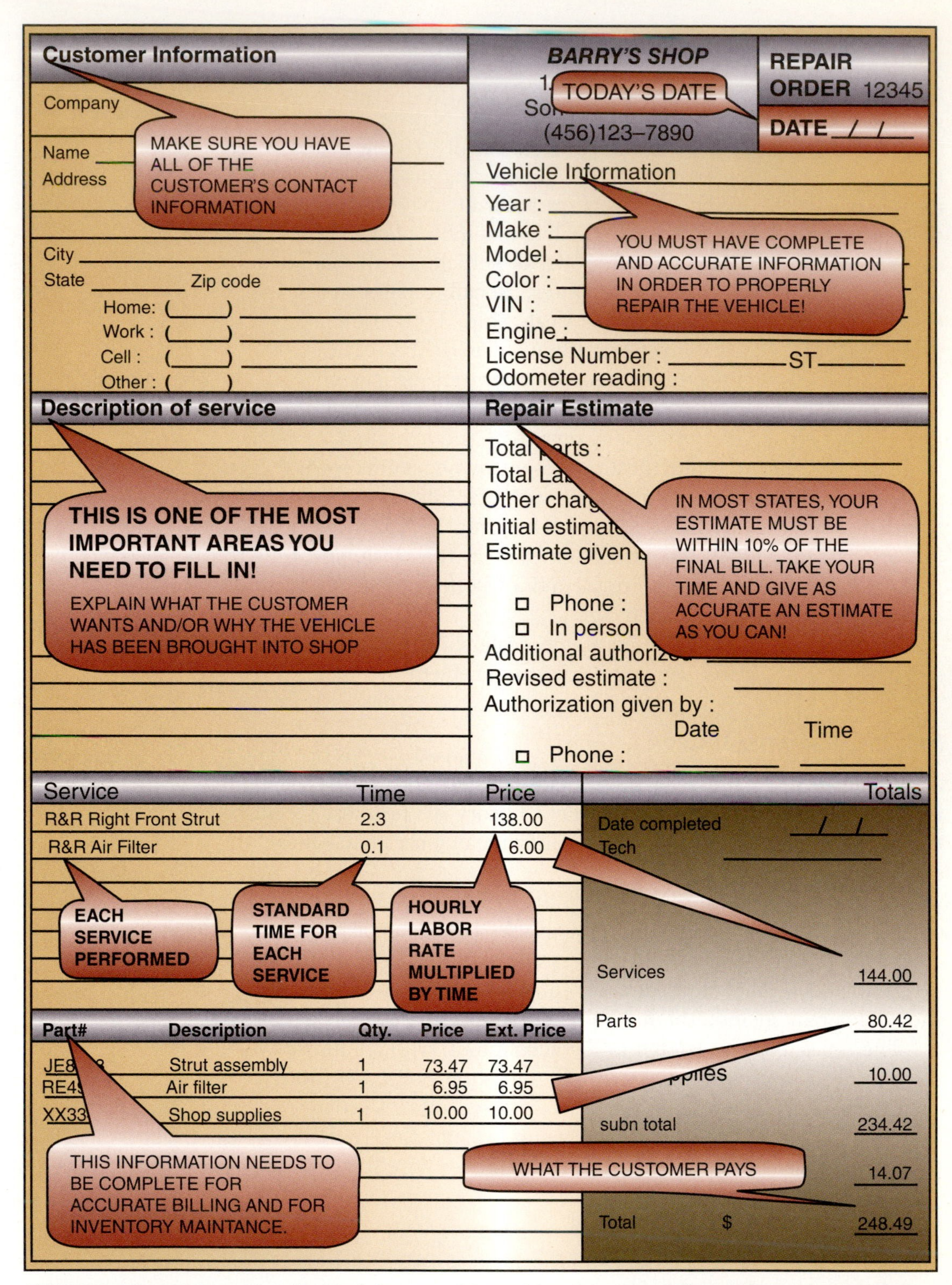

FIGURE 2-7 **A service repair order blank form used for training. It is very similar to those used by most automotive shops.**

Complete the service writer's portion of the repair order by entering the date, VIN, mileage, make and type of vehicle, customer personal information, and complaint (Figure 2-8). Enter the complaint as clearly and concisely as possible. Be sure to secure the customer's signature authorizing the repairs. If requested, provide an estimate of the cost and time of completion to the customer.

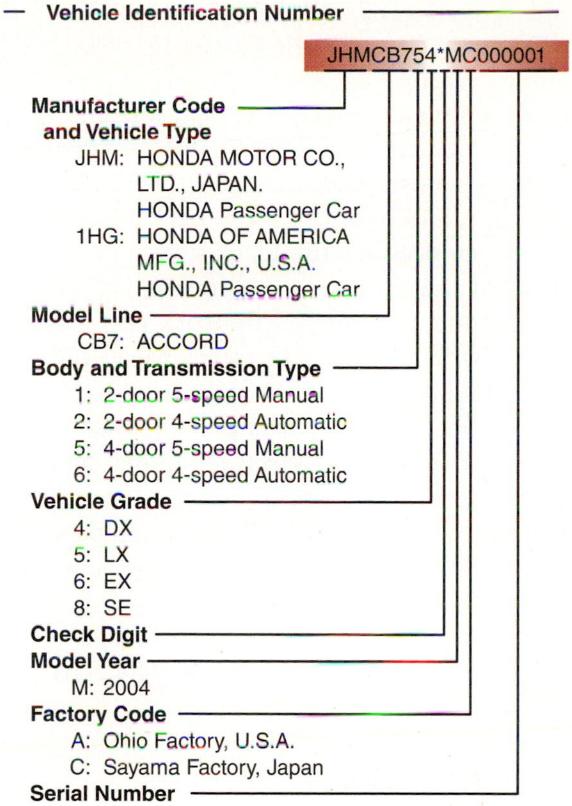

Vehicle Identification Number

JHMCB754*MC000001

Manufacturer Code and Vehicle Type
JHM: HONDA MOTOR CO., LTD., JAPAN. HONDA Passenger Car
1HG: HONDA OF AMERICA MFG., INC., U.S.A. HONDA Passenger Car

Model Line
CB7: ACCORD

Body and Transmission Type
1: 2-door 5-speed Manual
2: 2-door 4-speed Automatic
5: 4-door 5-speed Manual
6: 4-door 4-speed Automatic

Vehicle Grade
4: DX
5: LX
6: EX
8: SE

Check Digit

Model Year
M: 2004

Factory Code
A: Ohio Factory, U.S.A.
C: Sayama Factory, Japan

Serial Number

FIGURE 2-8 **The VIN is a 13- to 17-digit number that identifies the vehicle and certain components of that vehicle. Character interpretation of a typical VIN is shown here.**

At this point it is necessary to record any repair history this vehicle had in the past and locate any **Technical Service Bulletins (TSBs),** warranties, and recalls that pertain to this vehicle or this repair. This information may be entered by the service writer or the technician after confirming the customer complaint. TSBs are bulletins issued by the manufacturer based on problems found on the same model vehicle. Repairs performed under TSBs may be covered by warranty, but usually the customer must pay. The title or code listings for any recalls, TSBs, or warranty should be recorded on the repair order before the actual repair is started. This information may give the technician a head start on the diagnosis. During the actual diagnosis and repairs, each step should be listed according to shop rules. In most cases, if the technician does not list the diagnostic steps taken, that time will not be paid. This is particularly true when dealing with recalls or warranties.

Usually, the shop has some type of computer software or manual that has numerical codes that are used instead of entering the complete procedures. Most dealerships have manufacturer-approved codes for this purpose. Before turning the repair order over to the accountant or service writer, ensure that *all* repair labor and parts are noted on the repair order. The number of labor, guide, or clock hours and the number of parts, such as eight spark plugs, must be listed for the technician to receive pay and the shop to collect the final cost from the customer.

Many shops offer the removed part to the customer as "proof" that a new part was installed. In some cases, a customer will demand the old part. If the part is under warranty, the dealership or parts vendor may hold it in an area of the shop until a representative can look at it and verify that it was bad. If the new part has a core charge, that can be passed to the customer if he or she desires to take the old part. A core is a component that can be rebuilt for future sales. The core charge is the cost of replacing the base part, such as the housing. This is a common practice for items such as engines, starters, alternators, and brake shoes, which can be refurnished much more cheaply and quickly than recasting and manufacturing the entire component.

Enter the technician's diagnosis based on the complaint. Most shops require the technician to enter a starting and ending time on the repair order to track the actual time the vehicle was in the shop. Enter the labor time and, if not already provided, the **hourly rate.** Once the repair is complete, provide a short, concise statement about the repair that will give the customer an idea of the work performed. The information may be handwritten or printed by a computer terminal based on numerical codes entered by the technician. The technician must enter the employee identification in order to receive credit and payment for the repair. Many shops use computer-generated forms that allow the service writer, technician, parts personnel, and accountant to process a complete repair order by entering codes into a terminal without any paper used.

Hourly pay is the amount paid by clock hour to the technician regardless of the amount or type of work done in that hour. Entry level technicians are usually paid by the hour.

Complete the parts department information by entering part numbers, quality, cost per unit, and total cost of parts. The repair order is given to the accountant for final calculations including taxes before the customer pays.

Some shops require a foreman or supervisor to sign orders for major repairs. The supervisor may choose to test drive the vehicle. The repair order should not go to the accountant until the required signature is obtained. The accountant is the last stop for the repair order before filing. With the repair order complete, notify the service writer that the vehicle is ready for the customer to pick up.

Additional Vehicle Data Plates

There are several places on the vehicle with plates or tags containing specific information about that component or assembly. The engine normally has a casting number forged into the engine block. This number can be used to correctly identify the engine size, manufacture date, and location of the manufacturing plant. There should be a plate or tag on the transmission or transaxle identifying that assembly. The same applies to the differential of a rear-wheel-drive vehicle. Some older Fords had a calibration tag in the glove box and on the radiator support bracket showing the calibration for that vehicle's emission system. A label is sometimes placed on the radiator support or under the hood listing which emission devices are installed on the vehicle. This same label lists the vehicle's emission certifications and other information related to the engine's performance.

Within the engine compartment, there is usually a decal listing the type and amount of refrigerant used in the climate control system, one showing the routing of the serpentine belt, one showing the remote jump points for the concealed battery, and any one of several others showing information the manufacturer or technician deem necessary for the maintenance and operation of the vehicle.

Other data plates and tags are located on the driver door pillar listing gross vehicle weight, individual axle weight, tire size, and pressure. Recently, the VIN number has been added to major components, such as the transmission and the unibody, to assist in identifying the vehicle if stolen, used in a crime, or involved in an accident. Most service data will list the location of specific tags, plates, or decals for a specific component or assembly.

The major problem with tags as opposed to plates is the loss during repair procedures. It is not uncommon to find that the metal tag on the differential or transmission has disappeared because it was not reinstalled at a former service. Most plates are fastened in such a way that they cannot be easily lost. Decals, however, can be soiled with grease and other material that covers or erases the ink.

AUTHOR'S NOTE: **My students were having a problem determining how much refrigerant to be installed in an air-conditioning system. The decal was present, but the lettering could not be read because of grime. Being the brilliant person that I am, I simply grabbed a rag and wiped off the grime. It disappeared, as did all of the printing on the decal. What was left was a nice, clean, perfectly blank white decal. Fortunately, a service manual was available. Moral of the story: Don't waste your time trying to wipe a dirty decal. Just get the service manual.**

SUMMARY

- Automotive repair shops have personnel ranging from janitorial services to managers and owners.
- Service managers, service writers, and accountants generally provide the direct customer contact.
- The parts manager is responsible for storing, ordering, issuing, and charging parts.
- The technician most directly influences the profits for the shop.
- Technicians should be trained to the basic level and then attend continuous additional training to keep current with changing automotive technology.
- Technicians can receive certifications in specific systems through several national agencies.
- Technicians may receive a salary based on an hourly, commission, or flat rate.
- Technicians usually are required to have their own tool set.
- All shop employees and the owners share responsibility in environmental protection.
- Warranties cover the parts and services required that may be due to errors in manufacturing, design, or previous repairs.
- Recalls are issued by the vehicle manufacturers to correct some deficiency in design or manufacturing.
- A repair order is the first step in the diagnostic and repair procedure.
- The VIN is a number that identifies a particular vehicle and will contain information pertinent to that vehicle's manufacture.
- Additional data plates and tags are placed in various locations on the vehicle.

TERMS TO KNOW

Accountant
Commission
Entry-level technician
Flat rate
Hourly rate
Labor guide
Labor time
Manager
Recalls
Repair order
Service manager
Service writer
Technical Service Bulletin (TSB)
The 3 C's
Vehicle identification number (VIN)
Warranties

ASE-STYLE REVIEW QUESTIONS

1. The duties of the service writer are being discussed.
   Technician A says the service writer completes the repair order.
   Technician B says the service writer may provide a repair estimate to the customer.
   Who is correct?
   A. A only
   B. B only
   C. Both A and B
   D. Neither A nor B

2. *Technician A* says a service writer usually has only business administration training.
   Technician B says most technicians have a background in business.
   Who is correct?
   A. A only
   B. B only
   C. Both A and B
   D. Neither A nor B

3. The parts department is being discussed.

   Technician A says all parts are sold on the shop's repair orders.

   Technician B says the technician performing the service must charge the parts to the *order*.

   Who is correct?

   A. A only
   B. B only
   C. Both A and B
   D. Neither A nor B

4. *Technician A* says based on a labor guide a technician may receive up to 80 hours of pay per 40-hour week under the commission pay scale.

   Technician B says flat-rate pay does not strictly rely on actual clock-hours worked.

   Who is correct?

   A. A only
   B. B only
   C. Both A and B
   D. Neither A nor B

5. The position of an entry-level technician is being discussed.

   Technician A says this individual is usually a graduate or student of a post-secondary automotive program.

   Technician B says most individuals in this position are expected to diagnose a vehicle using a scan tool.

   Who is correct?

   A. A only
   B. B only
   C. Both A and B
   D. Neither A nor B

6. Service repair orders are being discussed.

   Technician A says recall information is entered only by the technician assigned the order.

   Technician B says it is the responsibility of the parts department to make sure all parts are entered onto the order.

   Who is correct?

   A. A only
   B. B only
   C. Both A and B
   D. Neither A nor B

7. Arbitration boards are being discussed.

   Technician A says the board's decision is only binding on the customer filing the complaint.

   Technician B says the board's decision is binding upon the shop.

   Who is correct?

   A. A only
   B. B only
   C. Both A and B
   D. Neither A nor B

8. Repair orders are being discussed.

   Technician A says the technician is responsible for computing the labor cost.

   Technician B says the accountant usually enters the labor time on the order.

   Who is correct?

   A. A only
   B. B only
   C. Both A and B
   D. Neither A nor B

9. *Technician A* says the service manager is responsible for the company's accounting office.

   Technician B says the parts manager usually reports to the service manager.

   Who is correct?

   A. A only
   B. B only
   C. Both A and B
   D. Neither A nor B

10. *Technician A* says the service writer initiates the repair order.

    Technician B says the technician enters the VIN onto the repair order.

    Who is correct?

    A. A only
    B. B only
    C. Both A and B
    D. Neither A nor B

JOB SHEET 1

Name ______________________________ Date ______________

Filling out a Work Order

NATEF Correlation

This job sheet addresses the following NATEF task:

A.1. Complete work order to include customer information, vehicle identifying information, customer concern, related service history, cause, and correction.

Objective

Upon completion and review of this job sheet, you should be able to prepare a service work order based on customer input, vehicle information, and service history.

Tools and Materials Needed

An assigned vehicle or the vehicle of your choice

Service work order or computer-based shop management package

Parts and labor guide

Work Order Source

Describe the system used to complete the work order. If a paper repair order is being used, describe the source.

Procedures

Task Completed

1. Prepare the shop management software for entering a new work order or obtain a blank paper work order. ☐
2. Enter customer information, including name, address, and phone numbers on the work order. ☐
3. Locate and record the vehicle's VIN. ☐
4. Enter the necessary vehicle information, including year, make, model, engine type and size, transmission type, license number, and odometer reading. ☐
5. Does the VIN verify that the information about the vehicle is correct? ____________ ☐
6. Normally, you would interview the customer to identify his or her concerns. However, to complete this job sheet, assume the only concern is that the valve (cam) cover is leaking oil. This concern should be added to the work order. ☐
7. The history of service to the vehicle can often help diagnose problems as well as indicate possible premature part failure. Gathering this information from the customer can provide some of this information. For this job sheet, assume the vehicle has not had a similar problem and was not recently involved in a collision. Service history is further obtained by searching files based on customer name, VIN, and license number. Check the files for any related service work. ☐
8. Search for TSBs on this vehicle that may relate to the customer's concern. ☐

Task Completed

☐ 9. Based on the customer's concern, service history, TSBs, and your knowledge, what is the likely cause of this concern?

___

___

☐ 10. Enter this information onto the work order.

☐ 11. Prepare to make a repair cost estimate for the customer. Identify all parts that may be needed to be replaced to correct the concern. List them here.

___

___

☐ 12. Describe the task(s) that will be necessary to replace the part.

___

___

☐ 13. Using the parts and labor guide, locate the cost of the parts that will be replaced and enter the cost of each item onto the work order at the appropriate place for creating an estimate.

☐ 14. Now, locate the flat-rate time for work required to correct the concern. List each task and its flat-rate time.

___

___

☐ 15. Multiply the time for each task by the shop's hourly rate and enter the cost of each item onto the work order at the appropriate place for creating an estimate.

☐ 16. Many shops have a standard amount they charge each customer for shop supplies and waste disposal. For this job sheet, use an amount of $10 for shop supplies.

☐ 17. Add the total costs and insert the sum as the subtotal of the estimate.

☐ 18. Taxes must be included in the estimate. What is the sales tax rate in your area and does it apply to both parts and labor, or just one of these?

___

___

☐ 19. Enter the appropriate amount of taxes to the estimate, then add this to the subtotal. The end result is the estimate to give the customer.

☐ 20. By law, how accurate must your estimate be?

___

___

☐ 21. Generally speaking, the work order is complete and is ready for the customer's signature. However, some businesses require additional information; make sure you enter that information on the work order. On the work order there is a legal statement that defines what the customer is agreeing to. Briefly describe the contents of that statement.

___

___

Instructor's Response ___

___

___

___

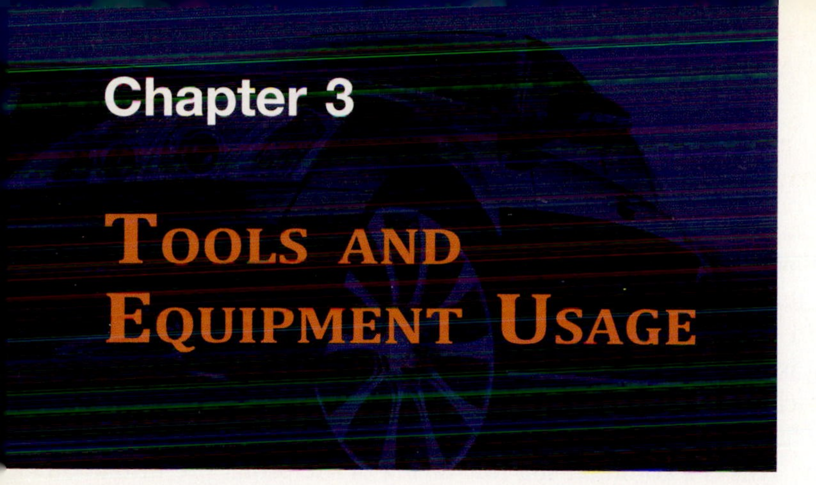

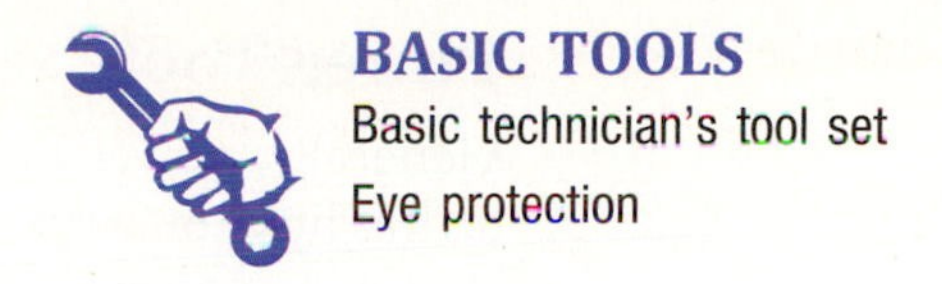

UPON COMPLETION AND REVIEW OF THIS CHAPTER, YOU SHOULD UNDERSTAND AND BE ABLE TO DESCRIBE:

- How to connect and use an air impact wrench to remove the lug nuts on a wheel.
- How to look up the torque specifications for wheel lug nuts.
- How to torque wheel lug nuts to the correct torque using a beam, dial readout, and break-over hinge-type torque wrench.
- How to select, install, and use a gear puller to correctly remove a pulley from a shaft.
- The use of a screw extractor to remove a broken stud or screw.
- The use of a flat chisel to split and remove a stuck nut from a stud or bolt.
- The use of an oxyacetylene torch for heating.
- How to collect vehicle service information.

INTRODUCTION

In the Classroom Manual, we described the types and use of basic tools the technicians use to repair cars. This chapter focuses on some of the most common tool tasks that the technicians perform. We describe how to connect and use an impact wrench, how to adjust and use a torque wrench, how to use a gear puller to remove parts, how to remove a broken stud or screw, and how to remove a damaged nut. Specific safety measures are also discussed in this chapter.

TOOL AND EQUIPMENT INSPECTION AND MAINTENANCE

Before beginning work with any power tool or equipment, it is best to perform a general inspection of the devices and their power connection. Though you may be comfortable picking up one of your own wrenches or screwdrivers and starting work, using a tool or piece of equipment powered by electricity, air, or hydraulics presents a chance to do more damage than a set of skinned knuckles. Plugging the tool into its power source is not part of the pre-use inspection. If there is damage, it may become obvious in a harmful manner. A general inspection at the item in question takes less than a minute and could prevent damage or injury.

Power Tools

Shop Manual
page 57

Generally, power tools in an automotive shop use either air or electricity for power. Look at the hose or cord for any damage or cuts. A ruptured air hose can cause almost as much injury as a shorted 110-volt electrical cord (Figure 3-1)—it just takes longer. Inspect the hose for any swelling along its length with emphasis near the connections. The end connections of the hose or cord should be firmly fastened without damage. On an electric cord, ensure the third prong is present and tight in the connector.

The tool itself should be clean and clear of any burrs or cover cracks, and the connector should be tight and fully functional. On an electric-powered tool, check the area where the cord enters the tool body. This is a common site for fraying of the protective insulation and the wires. The air fittings must be tightened into the cavity.

Though most shops do not use tools driven by hydraulics, technicians working on certain types of equipment may be exposed to tools powered by up to 10,000 pounds per inch (psi) hydraulic pressure. Do not work or disconnect these types of tools until you are sure the pressure is at zero psi and you know how the tool operates.

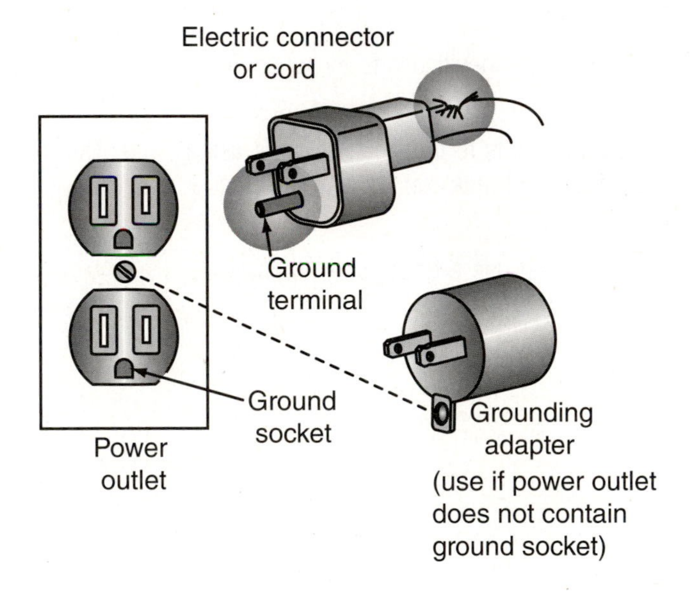

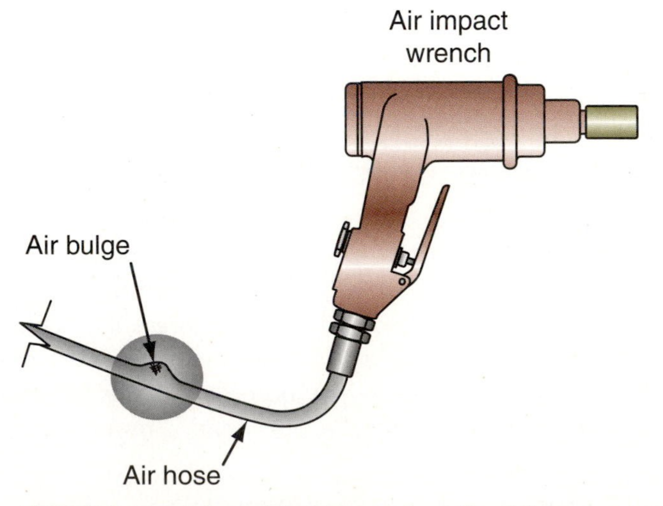

FIGURE 3-1 **Look for bulging hose around the air hose connector and frayed or exposed wiring around electrical connections.**

AUTHOR'S NOTE: **Many electric utility trucks use drills that operate on 10,000 psi. During a typical leak repair on a drill of this type, the technician did not check to ensure the pressure was off. A small leak was found at the hose/connector crimp. There appeared to be no pressure behind the leak; however, when the technician started to disconnect the tool, he twisted the hose "just right" and opened the leak up to about a 1-inch stream of highly pressurized fluid. The fluid entered his skin at the bottom of his wrist just behind the heel of the hand. Before he could move, several ounces of hot fluid were injected under the skin to a point above the elbow. The lower arm was swollen to over double its natural size. He was evacuated to the emergency room where the fluid was sucked out. Luckily, he suffered no lasting damage and was back at work in 2 days. Never trust what appears to be a minor problem when dealing with hydraulic, electrical, and pneumatic systems.**

Equipment

Most automotive equipment is powered by either electricity or hydraulics. In most cases, electricity is used to power the hydraulic pump for hydraulic pressure to drive the cylinder. The most common type is the vehicle lifts (Figure 3-2). The system is a simple one that can injure people or damage vehicles. Inspect the visible hoses and around the area where the lift cylinders are located. Any fluid present on the floor or the lift mechanism covers indicates a leak. In most cases, the lift cylinders are not completely visible, so the leaked fluid may be the only indication of a problem. The electrical cords for a lift are usually encased in a conduit and cannot be inspected, but the conduit itself can be checked.

FIGURE 3-2 This is a lift cylinder for a drive-on scissor lift. The hoses can be damaged by careless work or maintenance, as well as old age. Always inspect for leaks before using lift.

FIGURE 3-3 **This lift arm lock was apparently damaged during careless operation. It must be replaced before further use of the lift.**

Walk around the lift and make an inspection of the lift arms, pads, and pin connections. Of particular interest are the arm locks. When the arms are moved into lift position and moved a few inches up, the mechanical locks drop in place to prevent the arms from swinging or kicking out when the vehicle weight is applied (Figure 3-3). The trouble here may be caused by a technician removing or disabling the locks to make it easier to position the pads. If the lift has not been used for a period of time or has just been repaired, operate the lift until the mechanical locks on the lift cylinder drop in place. Never work under a vehicle if the lift locks are not engaged.

Hoists have to be periodically maintained. Check the owner's manual on the hoist for the specific requirements. Most of the time a hoist with large pistons needs to be greased once a month or more frequently. Sometimes, pulleys, wheels, and cables that move need to be lubricated with a light oil. Hoists should be inspected by the technician every time he or she uses it, but calling in a professional hoist repair company to do an annual inspection of the shop's hoist is usually done.

Lifting and Supporting a Vehicle

Classroom Manual
pages 67, 68, 69, 70

To work safely and quickly, it is sometimes required to raise the vehicle to a good working height. If brakes or tires are being done, a good working height has the wheels at about chest level. Engine and transmission oil changes can be done more easily with the vehicle high enough for the technician to walk under it.

Double-Post Lifts

SPECIAL TOOLS

Vehicle double-post above-the-ground lift

Service manual showing lift points

Double- and single-post lifts work similarly, with the primary difference being the position of the lift cylinder(s). Procedures discussed here can be used to some extent with single-post lifts. Before lifting any vehicle with any type of lift or jack, consult the driver's manual or service manual for the location of the vehicle's lift or jack points (Figure 3-4).

WARNING: Never guess at the vehicle's lifting points. What appears to be solid metal may bend or be crushed when the vehicle's weight is transferred to it. Damage to the vehicle and possible injury could occur.

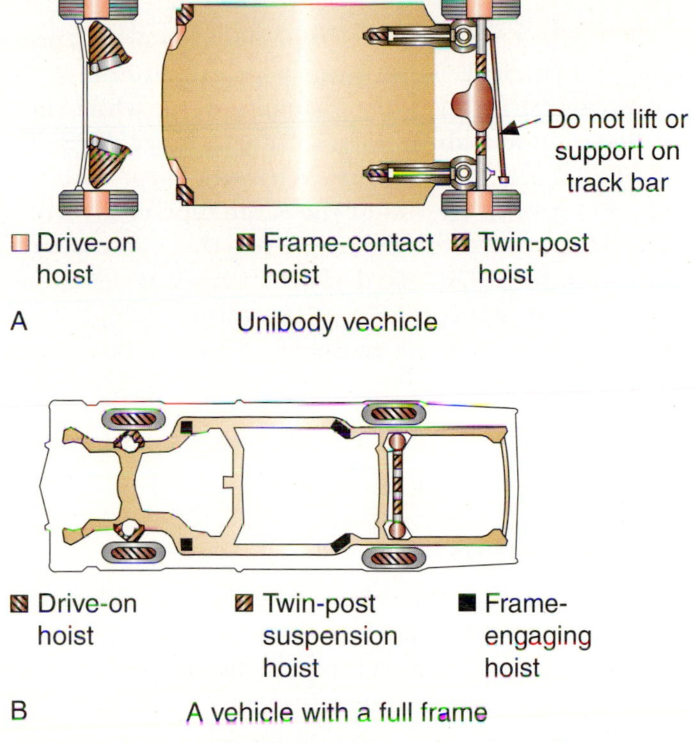

FIGURE 3-4 **The highlighted points are typical lift and jack points.**

Move the vehicle so it is centered between the posts. The front doors should be roughly even with the posts. Most lifts of this type have shorter lift arms at the front. This places most of the vehicle's weight closer to the lifting cylinder and provides better balance. Most double-post lifts have a **spring-loaded switch** button that operates the motor and the pump. Pressing and holding the button down will power the motor. Any time the switch is released for any reason, the spring will automatically push the switch to OFF.

This type of **spring-loaded switch** is sometimes referred to as a "dead-man switch."

After identifying the lift points, rotate the lift arms under the vehicle and position the pads directly under the lift points. On longer cars and trucks, it may be better to raise the lift arms until the pads are just below the vehicle. This makes it easier to properly place the pads.

With the pads aligned, lift the arms until there is contact between the pad and vehicle. Check the contact points to ensure there is good proper alignment between the pads and lift points. Make adjustments as necessary.

As the vehicle is being raised, note that both ends of the vehicle are moving upward evenly. Pickup trucks are hard to keep even because of their length and the position of the lift points. If the vehicle is not even from end to end and side to side, lower it back to the ground and move the vehicle or arms until both ends rise at the same time. Continue to raise the vehicle until all wheels are off the floor. Attempt to rock the vehicle by hand. If positioned properly on the lift, it should not rock at all. In fact, it will probably have a more solid side-to-side movement than when sitting on its wheels.

When the vehicle is at the desired work height, stop the upward movement. Lower the vehicle until the safety locks on the lift engage. The vehicle can now be repaired.

To lower the vehicle, first raise it high enough to clear the locks. There is usually a lock on each post. Move both the locks to the unlock position. Some lifts require the technician to hold an unlocking lever in place while the lowering valve is operated. A small lever attached to another spring-loaded switch usually operates the lowering valve. Lower the vehicle until the pads break contact. Continue lowering the lift arms until they are at the bottom of their travel. Swing the arms from under the vehicle.

SPECIAL TOOLS

Floor jack

Wheel blocks

Service manual showing jack point

Jack stands

CAUTION: Never perform any work under a vehicle supported only by a jack. Serious injury or death could result if the vehicle slips from the jack or the jack moves.

CAUTION: Never perform any work other than a tire change using a tire-changing jack. This jack is designed to raise one corner of the vehicle to change a tire. It is not made for other repairs. Serious injury could result if the jack and vehicle should move.

CAUTION: Never work under a vehicle supported only by a jack of any type. Injuries could occur if the jack slips.

AUTHOR'S NOTE: There are two types of two-post, above-ground lifts: asymmetrical and symmetrical. The primary visual difference is that the four lift arms on a symmetrical lift are of the same length, while the front arms on an asymmetrical lift are noticeably shorter than the rear arms. Each type of lift can be used with almost any vehicle, but each works best with a particular type of vehicle. The shop may have all lifts of the same type or a variety. Honda dealerships have a special-purpose lift designed for the express service bays. The vehicle can be driven onto the lift head. The lift head is designed to make contact along the vehicle's undercarriage and the ramps drop down during lifting so that the wheel assemblies can be removed. This lift saves time and reduces possible damage to the vehicle.

Floor Jacks and Jack Stands

Some work can be done underneath the vehicle using a floor jack and jack stands. It may be more difficult, but for some jobs it may be quicker than using a lift.

Before jacking the vehicle, decide where the jack's lift pad will be placed and how much of the vehicle is being lifted. Shift the transmission to PARK (forward gear on manual transmissions) and set the parking brake. Position the wheel blocks behind and in front of at least one wheel that is being left on the floor. The surface should be level enough for the jack wheels to roll.

Unlock the jack handle and align the jack's lift pad under the jacking point of the vehicle. Rotate the valve knob on the end of the handle clockwise to close the valve. Move the handle up and down to raise the jack pad until it makes contact with the vehicle. Check the contact and if it is correct, continue to jack up the vehicle. The jack should roll further under the vehicle as it is raised. This is to keep the pad directly under the jacking point. If the jack cannot roll, it will either pull the vehicle toward the jack handle or slide the pad from its contact point.

When the vehicle is at the desired height, slide the jack stands under the vehicle. They should be placed at a point that will support the weight of the vehicle. This is usually under the frame, under the lower end of the suspension, or under the axle housing. The stands should be placed opposite to each other and at the same height to keep the vehicle level. With the stands in place, slowly rotate the jack's valve knob counterclockwise. The knob will probably stick a little because of the hydraulic pressure against the valve. As the knob begins to turn, the jack will begin to lower. Proceed slowly until the vehicle is resting securely on the jack stands. Remove the jack, stand the handle straight up, and lock it in place. Do not leave the jack in a walking area, and *never* leave the handle lying down.

To remove the stands and lower the vehicle, carefully place the jack in position to make the same contact as when raising the vehicle. Unlock the handle and close the valve. As before, once the pad makes contact with the vehicle, check it before raising the vehicle. Raise the vehicle just enough to remove the stands. With the stands removed, open the jack valve slowly and lower the vehicle to the ground. Store the jack properly. Ensure that the parking brake is set before removing and storing the wheel blocks. See Photo Sequence 2.

Connecting and Using an Impact Wrench

The air-operated impact wrench is frequently used by all technicians. Photo Sequence 3 shows a typical procedure for connecting and using an impact wrench to remove **lug nuts**. The impact wrench is often misused by technicians who do not understand that its main function is disassembly. The typical air-operated impact wrench is not torque controlled. This means

PHOTO SEQUENCE 2

TYPICAL PROCEDURE FOR JACKING AND SUPPORTING A CAR

P2-1 Drive the car on to a solid, level surface.

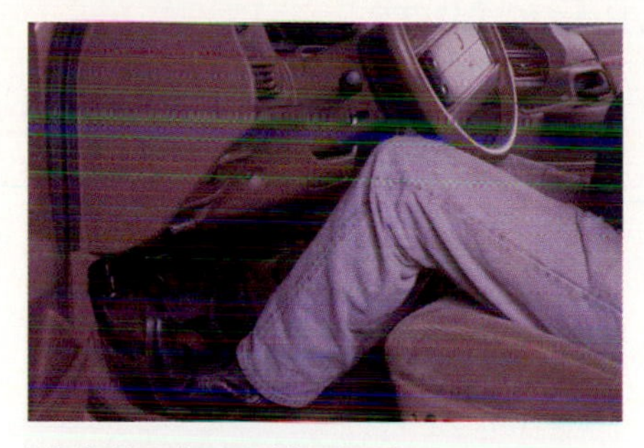

P2-2 Set the parking brake.

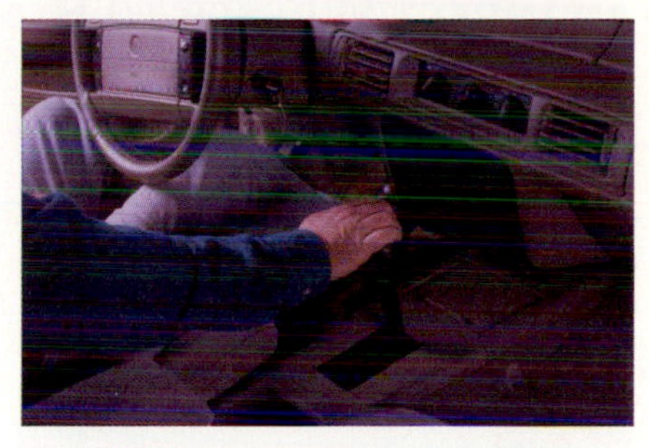

P2-3 Put the transmission in PARK.

P2-4 Block the front and rear of one of the wheels that is not being lifted.

P2-5 Position the jack pad on a major strength point under the car. Check the service manual as needed.

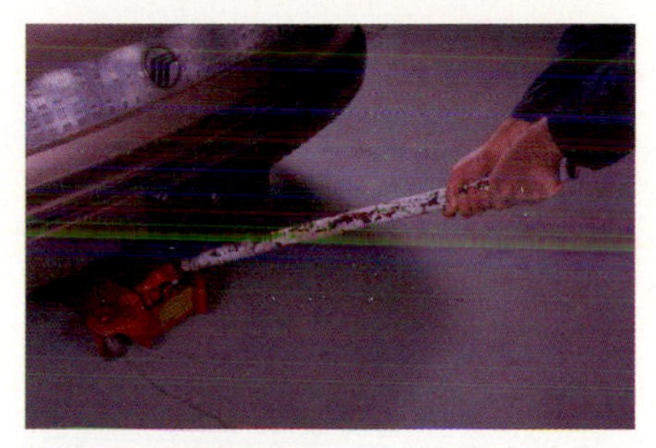

P2-6 Turn the jack handle valve to the lifting position.

P2-7 Raise the car carefully.

P2-8 Adjust and position jack safety stands to properly support the car.

P2-9 Slowly turn the jack handle to the lowering position and lower the car down on the jack safety stands.

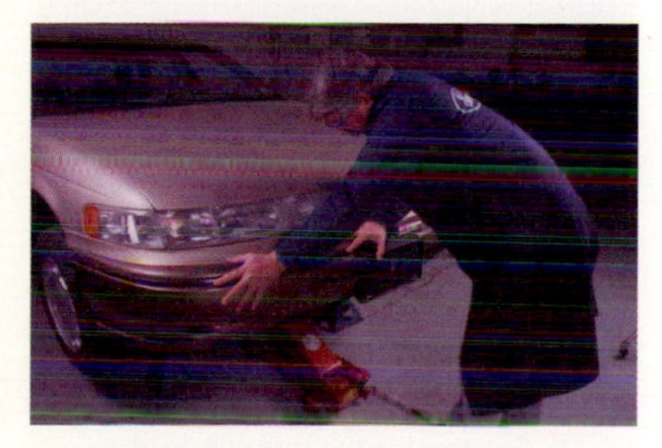

P2-10 Shake the car to test its stability on the jack safety stands.

PHOTO SEQUENCE 3

Typical Procedure for Connecting and Using an Impact Wrench to Remove Lug Nuts

P3-1 Raise the car on a hoist or jack and jack safety stands.

P3-2 Remove the wheel cover from the wheel.

P3-3 Connect the air impact wrench to the air hose quick-disconnect coupling.

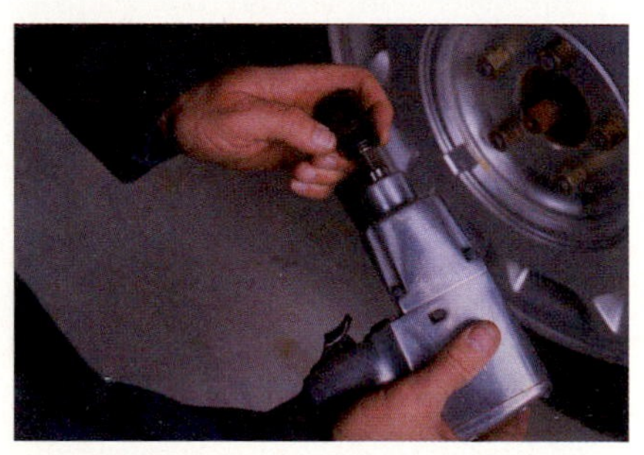

P3-4 Install the correct size impact socket on the wrench drive.

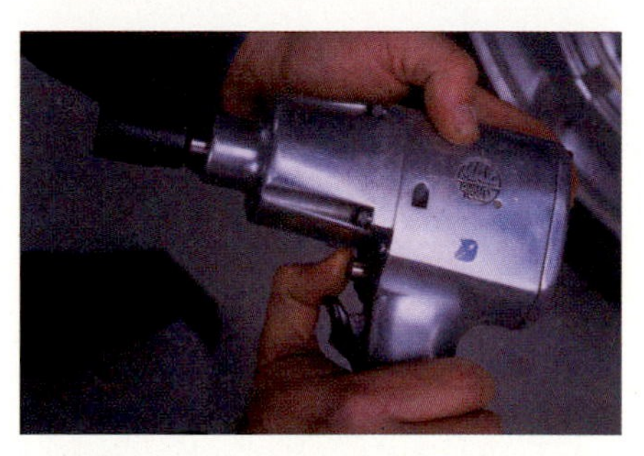

P3-5 Switch the impact wrench to the reverse (counterclockwise) drive direction.

P3-6 Place the wrench and socket over the lug nut and pull the trigger to remove the lug nut.

P3-7 Remove each of the other lug nuts and remove the wheel.

P3-8 Install the wheel back on the car and start each lug nut by hand.

P3-9 Switch the impact wrench to the forward (clockwise) direction.

P3-10 Carefully run the nuts up but do not tighten.

P3-11 Look up the tightening torque and tightening sequence for the lug nuts in a shop service manual.

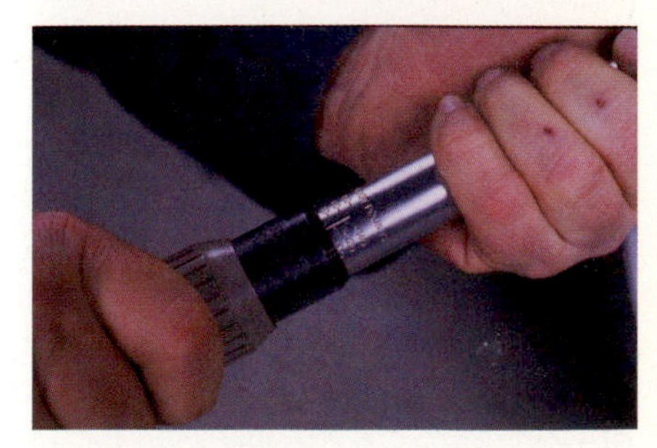

P3-12 Set the torque wrench to the specified torque.

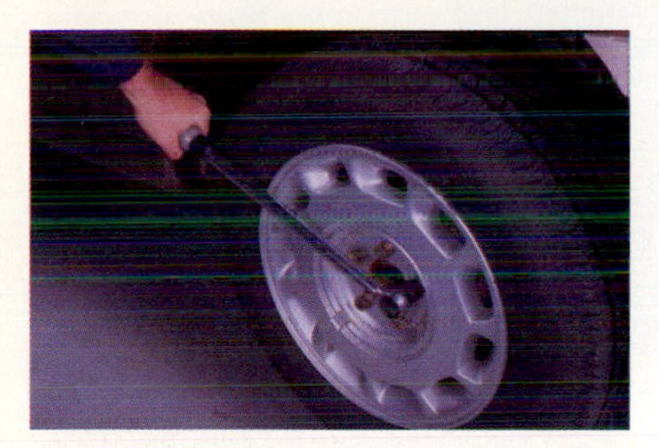

P3-13 Tighten each lug nut in the correct order to the specified torque.

P3-14 Replace the wheel cover and lower the car.

P3-15 A torque stick may be used instead of a torque wrench to torque lug nuts only.

the actual torque applied is not known when this tool is used. Using an air-operated impact wrench to assemble components can result in overtightened or undertightened fasteners.

Lug nuts are the hex-sided fasteners used to hold the wheel rim on the studs of the brake rotor or drum.

A common use of the air-operated impact wrench is to take off wheel lug nuts to remove a wheel from a car. First raise the wheel to be removed safely off the floor. Follow the safe procedure described earlier in this chapter to raise the vehicle on a hoist or to raise and support the car on a floor jack and jack safety stands.

Classroom Manual
pages 57, 58, 59, 60

Many cars use a wheel cover that must be removed to access the lug nuts. Other wheels use the lug nuts to hold on the wheel cover. If the car has wheel covers that must be removed first, use a prying tool to gently pry the wheel cover off the wheel rim as shown (Figure 3-5). Some cars have locks that prevent wheel theft. And some vehicles use plastic screw caps to hold the wheel cover tight on the lug nuts. Be careful not to damage these. Use a ratchet and socket to remove them, not the air impact. The lock mechanism typically fits over one of the lug nut studs. Use the key to remove the lock, then pry off the cover. Be careful not to bend or distort the wheel cover or it will not fit back on properly.

The lug nuts hold the wheel onto studs. Passenger cars typically use either four or five lug nuts to hold on the wheel. Passenger trucks typically use four to eight lug nuts. These studs are attached to the brake rotor, axle flange, or hub. There are several designs of lug nuts (Figure 3-6). Lug nuts that hold on steel wheels are often different than those that hold on aluminum wheels. One side of the lug nuts are chamfered. The chamfered side

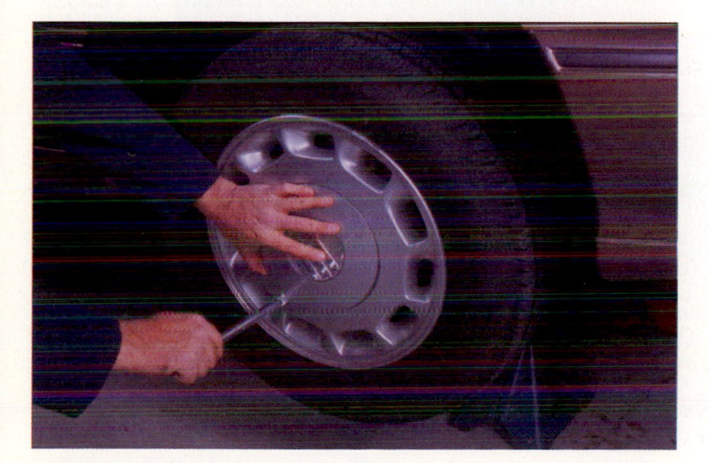

FIGURE 3-5 Removing the wheel cover.

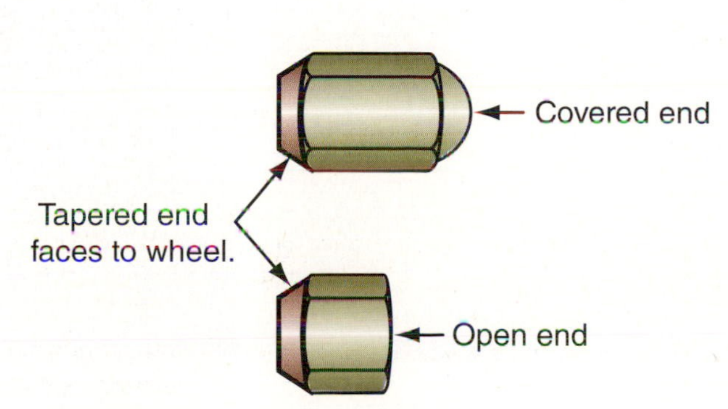

FIGURE 3-6 Parts of a typical lug nut.

SPECIAL TOOLS

Air-operated impact wrench

Impact socket set

Hoist or floor jack

Jack safety stands

Wheel blocks

FIGURE 3-7 **A six-point impact socket is used with an impact wrench.**

always goes on toward the wheel rim. The chamfer fits in a matching chamfer on the wheel and centers the wheel over the studs as it is tightened.

The lug nuts are hex shaped or six sided to fit standard wrench sizes. The wrench sizes are different for different cars. Select a six-point impact socket (Figure 3-7) that fits the hex sides of the lug nut snugly.

Sometimes the lug nut is covered with a chrome-look cap. At times, this cap will shear and come off in the socket. If this happens, do not use the same socket in an attempt to loosen the nut. It will be slightly too big and may round the nut corners. Use a slightly smaller (e.g., 11/16 in place of a 3/4) socket to remove the nut. Removing a lug nut with rounded corners sometimes becomes an almost impossible task without damaging the wheel. Make sure you install a new lug nut when you find a damaged one. You may not be able to properly torque the worn lug nut to the correct specification. Leaving it loose is not an option.

SERVICE TIP: You should make match marks on the wheel rim and one of the studs with chalk or grease pencil. Always replace the wheel in the same position in case it has been balanced on the car.

WARNING: Using a loose-fitting impact socket on a wheel lug nut can cause the corners of the hex to be rounded off and not fit a wrench properly.

At this time, set up and connect the air-operated impact wrench (Figure 3-8). The wrench has a socket drive on the end. Install the impact socket onto the wrench drive.

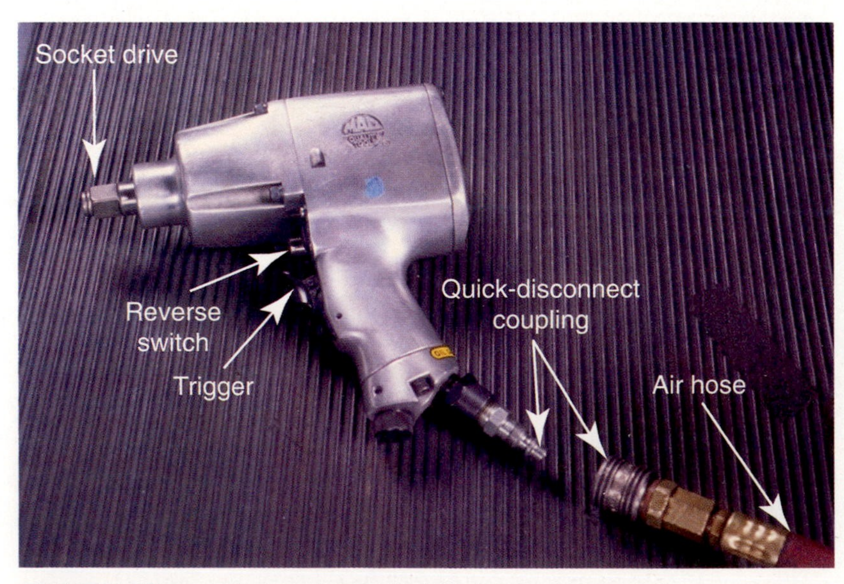

FIGURE 3-8 **Parts and controls on a typical air impact wrench.**

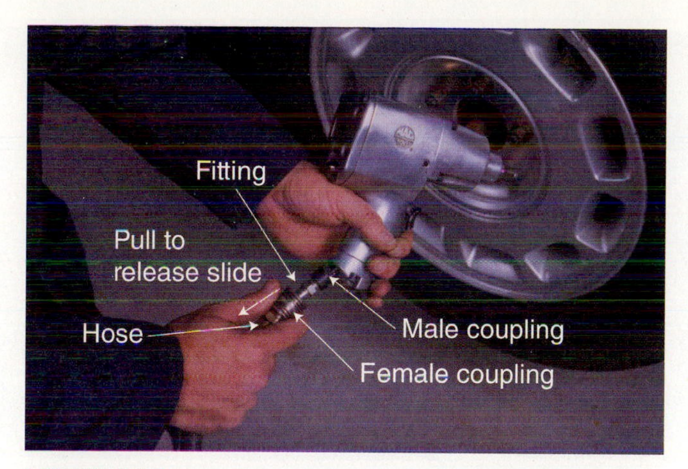

FIGURE 3-9 **Parts and operation of a quick-disconnect air line coupling.**

A **female coupling** houses the locks that hold the male coupling in place after hookup. This coupling also has the release mechanism.

A **male coupling** slides into a mated female unit similar to the manner in which a ratchet's drive stud fits into a socket. The male is fitted with grooves and ridges for locking.

The wrench will need to be connected to the shop air source to get its power. The shop air hose is connected to the impact wrench with a part called a **quick-disconnect coupling**. This coupling allows air lines to be attached to tools quickly without shutting off any air source. The coupling has two basic parts, as shown in Figure 3-9. The part attached to the air hose is called the **female coupling**. The part attached to the wrench is called the **male coupling**. To attach the hose to the gun, pull back on the female coupling slide and insert it over the male coupling. Release the slide and it will lock the two parts together in an airtight fit. The coupling may be disconnected in the same way. Just pull the slide and the two parts of the coupling apart.

With the line connected, the wrench is ready for use. Depress the trigger. The wrench drive will spin the socket at a high speed. Notice the direction the socket spins. In order to remove most lug nuts, set the reverse switch on the wrench so the socket is driven counterclockwise. Try the reversing switch to see how it works, then set the wrench in the correct direction to remove the lug nuts.

CAUTION: Always use impact sockets with an impact wrench. Standard sockets can fracture and explode when used on an impact wrench. Always wear eye protection when using an impact wrench.

WARNING: Some cars made in the 1940s through the 1960s used left-handed threads on the lug nuts on the left side of the car. The designers did this so that if the nuts were loose, they would be automatically tightened by the rotation direction of the wheel. Look for this on old cars and do not mix up the lug nuts.

Insert the impact socket over one of the lug nuts. Pull the trigger and spin off the lug nut (Figure 3-10). Repeat this procedure on each of the other lug nuts. Support the wheel as the last lug nut is removed to prevent the wheel from falling off the studs. Carefully lift the wheel off the studs. Use the proper lifting techniques described in Chapter 1 of the Shop Manual.

To reinstall the wheel, first place it in position on the studs. Always start the lug nuts by hand (Figure 3-11). Make sure they thread onto the stud several turns before using any wrench. Using a wrench to start the lug nuts could damage the threads.

Use the air impact wrench to spin the nuts in contact with the wheel, but do not use a typical impact wrench to tighten the nuts. Reverse the direction of the impact wrench with the reverse switch. If the impact wrench has a pressure switch, set it to the lowest setting. Carefully spin each nut on the stud until it just begins to contact the wheel. Tightening must be done with a torque wrench, as explained in the next section.

CAUTION: Do not let the impact wrench spin off the lug nut. The nut will fly out of the socket and can damage the vehicle's finish or your face, body, or that of a bystander. Slow the wrench's rotation speed by easing up on the trigger once the nut is broken loose to prevent possible damage or injury.

FIGURE 3-10 **Removing lug nuts with an air impact wrench.**

FIGURE 3-11 **Always start lug nuts by hand.**

FIGURE 3-12 **Wheel torque sockets can be used with an impact wrench to torque lug nuts.**

SPECIAL TOOLS

Impact wrench

Wheel torque socket

Vehicle lift or jack

SERVICE TIP: Install each lug nut just until it pulls the rim to the hub or axle. This will prevent the rim from "cocking" and preventing a flat fit against the hub or axle. Once all of the rim seems to be centered and in place, the lug nuts can be spun into place and then torqued.

If wheel torque sockets are available, select the correct one for the torque desired. Place the torque socket in the impact wrench in place of the regular impact socket used to remove the lug nuts (Figure 3-12). Do not hold the bar with your hand. The vibration is part of the calculated torque. Use the impact wrench to torque the lug nuts afterward. If the torque sockets are not available, set the air impact wrench on its lowest setting and then hand torque them with a torque wrench afterward.

Adjusting and Using a Torque Wrench

Installing and tightening the lug nuts is a good way to learn how to use a torque wrench. Lug nuts should *always* be torqued with a torque wrench. Using the correct torque setting ensures that the lug nuts are tightened the right amount and evenly. If the lug nuts are tightened unevenly, the drum or rotor can be warped.

The first step is to look up the torque specifications for the lug nuts on the car being serviced. An example of a wheel lug nut torque specification table is shown in Figure 3-13. This particular car has a recommended wheel nut torque of 140 N · m or 100 ft.-lb. The symbol N · m is for the metric system measurement newton-meter. The symbol ft.-lb is for

FIGURE 3-13 **Many torque sticks come in a set and are color coded to the designed torque speeding up the process of selecting the correct torque stick.**

the conventional (English) system measurement foot-pound. The measurement can be made with either a metric or a conventional reading torque wrench.

Most charts for wheel nuts will have a column labeled "foot-pounds." However, some of the torque measurements may have a small number or mark that indicates this particular torque is different from the others in this column. The number or mark refers to a note at the bottom of the torque chart. This note may state that this torque specification is in "inch-pounds" instead of foot-pounds. Many fasteners have been broken by not paying attention to the details.

SPECIAL TOOLS

Dial readout-type wrench
Break-over hinge-type torque wrench

The most common automotive torque wrench in use today is the break-over or hinge-type (Figure 3-14). They are available in 1/4-inch, 3/8-inch, 1/2-inch, and larger size drives. The smaller sizes usually have inch-pound scales, while the larger use foot-pounds. Most 3/8-inch drives can be purchased scaled in either inch-pounds or foot-pounds. Scale settings commonly range from 20 to 200 inch-pounds or 50 to 250 foot-pounds. A 1/2-inch drive torque wrench is commonly used to torque lug nuts on light vehicles.

Set the wrench up by adjusting the torque wrench setting to the specified torque. This is done by turning the end of the handle until the zero index mark is aligned with the proper torque. If the torque is to be 100 ft.-lb, then align the zero mark with 100 ft.-lb (Figure 3-15). The wrench is now set to click or break at 100 ft.-lb of torque.

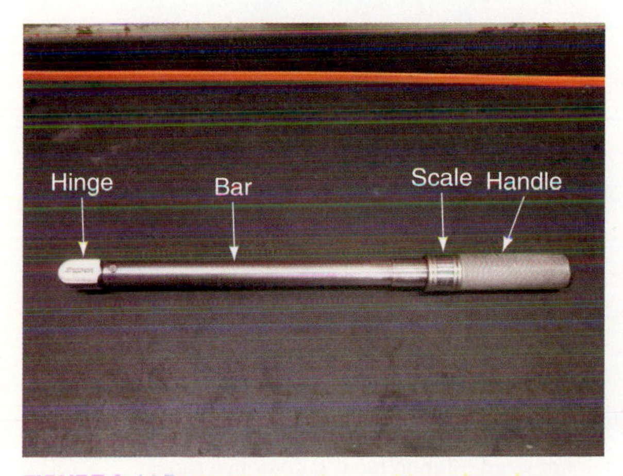

FIGURE 3-14 **Parts of a break-over hinge-type torque wrench.**

FIGURE 3-15 **A break-over hinge handle setting of 100 ft.-lb.**

After setting the wrench to the required torque, attached the proper size socket and tighten the lug nuts in a star pattern. It is best to tighten each nut a few turns until all are almost tightened to specifications. This pulls the wheel in square with the axle or wheel hub. Once the nut reaches the wrench torque setting, the wrench will "click," indicating that the torque is correct. One thing to remember when applying low torque to smaller fasteners is that the click may not be very audible but you should be able to feel the hinge "break over."

Electronic torque wrenches are also available and much more accurate. Switch the wrench on and use the buttons to select the desired torque setting. Make sure the correct measurement—such as foot-pounds—is selected. Torque the fastener until a buzz or vibration is heard or felt. This means the desired torque is achieved. Note the reading on the wrench's display. The torque shown is the actual torque applied. For instance, the wrench may be set for 100 ft.-lb, but after torquing the display indicates that 102 ft.-lb were applied. Normally, this overtorque is not a problem unless the fastener has a very low torque setting. Most torque specifications allow for some tolerances.

Dial readout torque wrenches are not commonly used in automotive repair shops except for certain types of repairs (Figure 3-16). One type of repair involves measuring the turning torque of a gear or shaft. When a technician rebuilds rear-wheel-drive (RWD) differential gears, the bearings must be **preloaded**. Preloading means a certain amount of force is applied against both sides of the bearing to ensure fit within the bearing **race** for quiet, long-life operation.

The term **race** refers to the machined area(s) of the bearing, shaft, or gear where the bearing's rollers or balls contact and roll on.

To do this, the bearing retaining nut is tightened until all back and forth, side-to-side movement is removed. Do not overtighten or you will have to start over and possibly install a new part. A dial torque wrench is attached in some manner to the shaft or gear and is used to turn that shaft or gear (Figure 3-17). The amount of force required to turn the shaft or gear is shown on the dial readout. If too low, the nut is tightened (a turn or two) and another measurement is taken. Continue repeating these steps until the specified turning torque is reached. The dial readout torque wrench is one of the few tools readily available to perform this type of measurement.

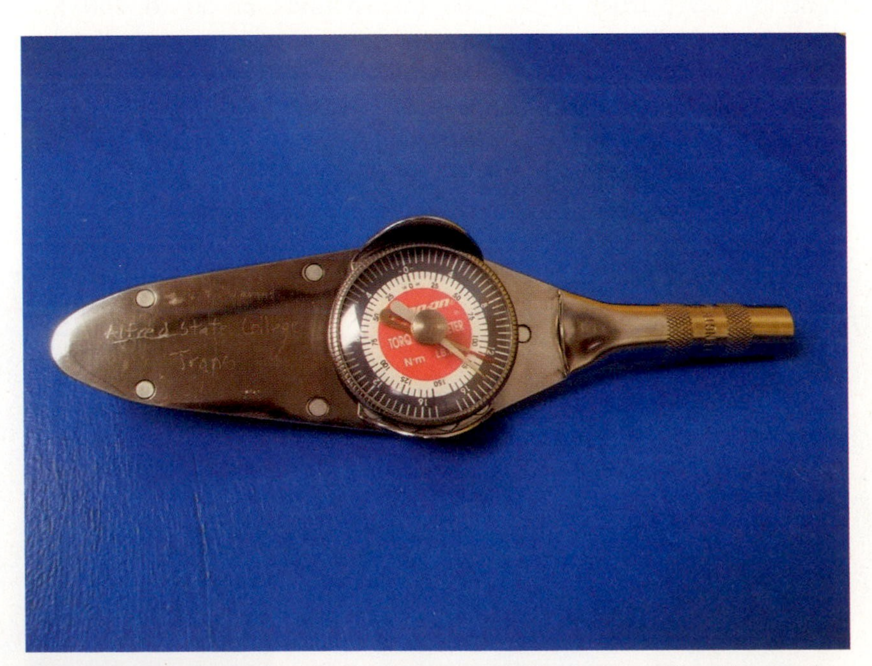

FIGURE 3-16 **A dial torque wrench is not used often in most automotive repair shops and is only used for special jobs.**

FIGURE 3-17 **A dial torque wrench is commonly used to check bearing preload by measuring the amount of force needed to turn the shaft.**

Using a Gear and Bearing Puller

As noted in the Classroom Manual, pullers come in many different sizes and shapes and many can be used for only certain types of pulling operations. The puller in Figure 3-18 is designed to remove the front crankshaft pulley or harmonic balancer on certain models of Chrysler vehicles. It will not work well on other vehicles. Figure 3-19 shows a puller set that is commonly used to remove steering wheels or the harmonic balancer except for certain Chrysler vehicles. However, both types work in a similar manner.

SPECIAL TOOL

Gear and bearing pullers

Removing a Harmonic Balancer

For this exercise, the puller set in Figure 3-19 will be used. This is the most commonly used puller in an independent shop. Refer to Photo Sequence 4 for details. The harmonic balancer was selected for this exercise because it is usually easy to reach, especially if the engine is out of the vehicle for training purposes. Other gears and pulleys can be removed in a manner

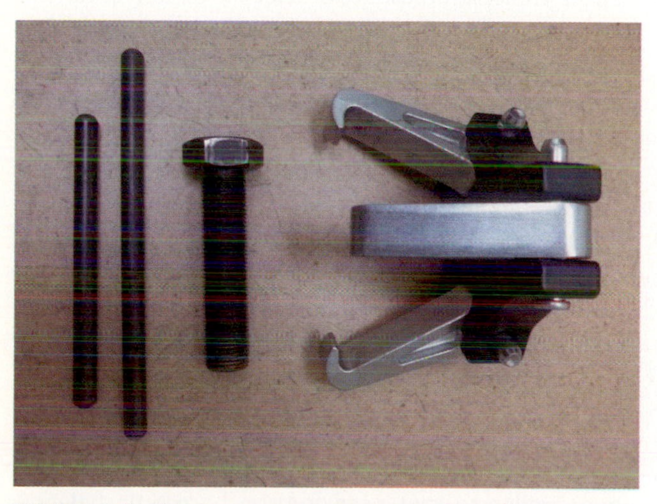

FIGURE 3-18 **This is a special-purpose tool used to remove the harmonic balancer from certain Chrysler engines.**

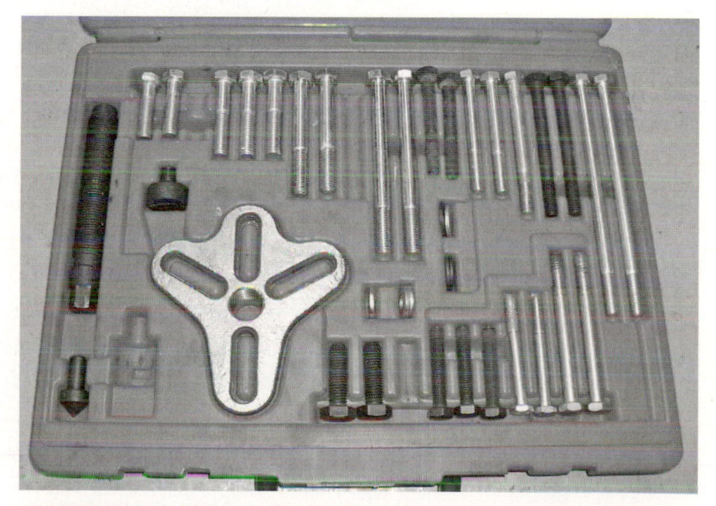

FIGURE 3-19 **A typical puller set used to remove steering wheels and harmonic balancers. It can be used in other applications as well.**

PHOTO SEQUENCE 4

Typical Procedure for Lifting a Vehicle on a Hoist

P4-1 **Refer to the service information to determine the proper lift points for the vehicle you are lifting.**

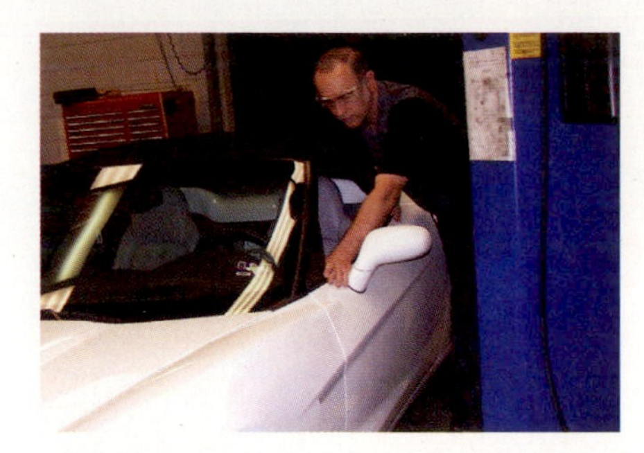

P4-2 **Center the vehicle over the hoist, considering the vehicle's center of gravity and balance point.**

P4-3 **Locate the hoist pads under the lift points. Adjust the pads so that the vehicle will be lifted level.**

P4-4 **Lift the vehicle a couple of inches from the floor. Shake the vehicle while observing it for signs of any movement. If the vehicle is not secure on the hoist or unusual noises are heard while lifting, lower it to the floor and reset the pads.**

P4-5 **Once the vehicle is at the desired height, lock the hoist. Do not get under the vehicle until the hoist locks have been set.**

P4-6 **To lower the vehicle, release the locks, and put the control valve into the "lower" position. Once the vehicle is returned to the floor, push the contact pads to a location that is out of the path of the tires.**

similar to that discussed here. Remove the large fastener securing the harmonic balancer. Inspect the harmonic balancer and select the two or three bolts (depends on the manufacturer) in the set that will screw into the outer treaded holes. Next select a center cone that fits into the large center threaded hole. This cone will help protect the interior threads during the pulling operation. Remove the large pointed threaded bolt from the set and screw it into the center of the puller body. Rest the point of the large puller bolt into the cone and install one of the side bolts selected earlier. Screw this bolt partially into one of the outer holes before installing the other one or two bolts in a similar manner. Screw these bolts in until they exit the back side of the balancer. They may be visible at that point, but they should become much easier to turn. A wrench may be required to screw them all the way. Each of the two (or three) bolts should be of equal length and the puller body should be squared with the face of the balancer. Tighten the large center puller bolt—a wrench will be needed—and the balancer should begin to slide off the crankshaft. Continue tightening until the balancer is clear. Remove and store the puller in its case.

Classroom Manual
pages 65, 66

Removing a Gear

Pulling a gear from a shaft uses a technique similar to that used for the harmonic balancer remover. There are some steps that must be done correctly, including selecting the

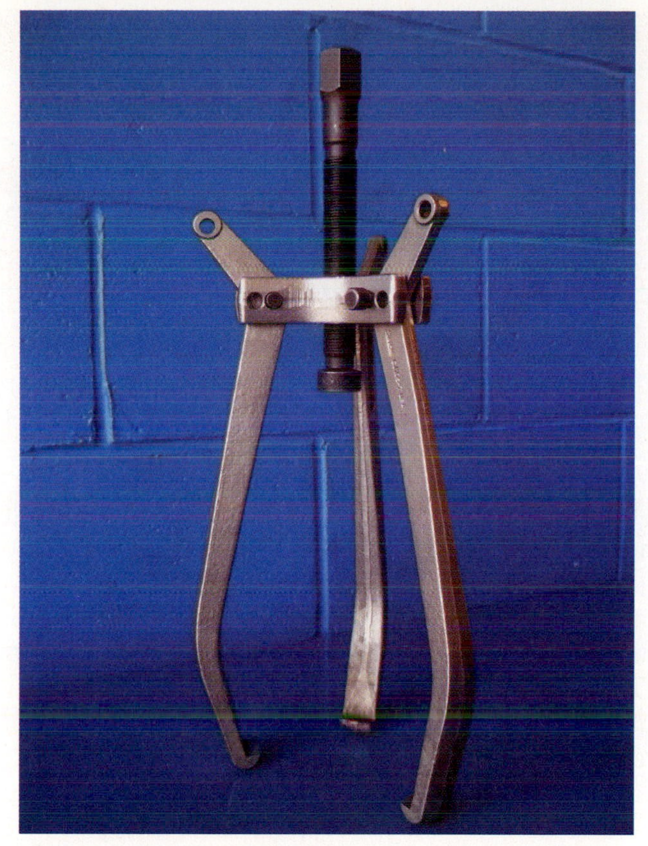

FIGURE 3-20 **A typical three-jaw puller commonly used to remove bearings or gears from a shaft.**

correct puller, or the gear or shaft may be damaged. A three-jawed puller is used to pull many gears (Figure 3-20). Select a puller or reposition the jaws so the pulling ends fit completely around the gear. The jaws should not be pulling on the gear teeth but on the gear body.

Center the long pulling bolt at the center of the end of the shaft and tighten it until the puller holds itself in place. It may be necessary to anchor the shaft and gear to prevent it from turning. Do not clamp the shaft at any bearing journals. Bearing journals are the shiny machined surfaces and can be damaged with a clamping device. Use a wrench to turn the puller center bolt until the gear slips off the shaft.

Removing a Broken Stud or Screw

Regardless of how carefully a technician works, sooner or later the misfortune of breaking a screw or stud occurs. The broken stud or screw will have to be removed before the service work can be completed. Broken screw removal is a very valuable skill. We suggest trying some of the techniques described here on some scrap parts before doing it on a customer's car.

Screws fail for many reasons. Sometimes they are defective when manufactured. Often they are overtightened by a technician not using a torque wrench. Sometimes they are driven too deeply into a threaded hole and bottom out. Regardless of the cause, a broken screw has to be removed and a new fastener used in its place.

There are a couple of removal techniques that should be used before any special tools are tried. Sometimes the screw or stud breaks with a part of the fastener sticking above the surface. This often happens if a screw is bottomed out in the hole. First, try a pair of vise grips. Adjust them to the proper size and grip the broken fastener as shown in Figure 3-21. Then turn the fastener in the correct direction for removal.

SPECIAL TOOLS

Vise grips
Small cape chisel
Set of screw extractors
3/8-inch portable drill

CAUTION:
If the puller jaws do not fit the part, or the cross arm will not allow proper placement of the jaws, select another size puller. Using an incorrect size puller or not adjusting the puller correctly can allow the jaws to slip off during pulling and cause injury. Always wear eye protection when using a puller. Parts can break and cause eye injury.

CAUTION:
Always wear eye protection when drilling and using a screw extractor. Parts could fracture and injure your eyes.

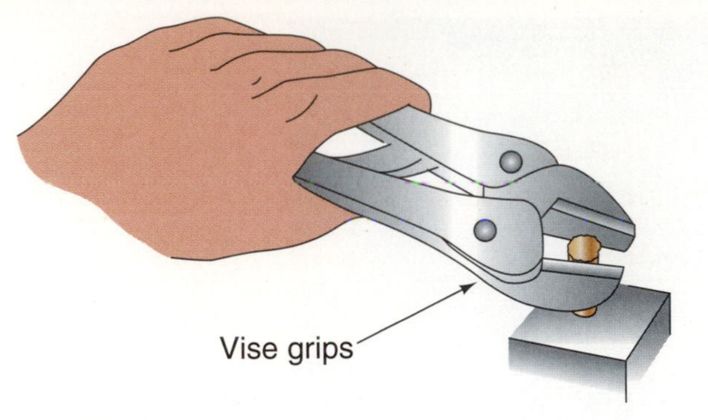

FIGURE 3-21 **Using vise grips to remove a broken fastener.**

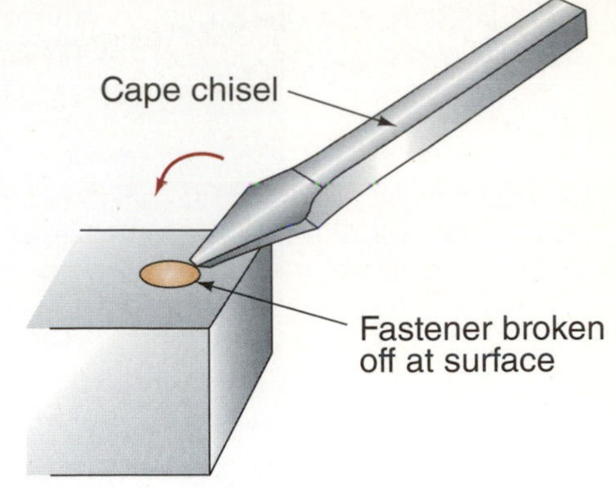

FIGURE 3-22 **Using a chisel to back out a broken fastener.**

SERVICE TIP:
Sometimes difficult-to-remove broken fasteners can be removed by carefully heating or cooling the area. Because the fastener and the threaded hole are made of different materials, they will expand or contract at different rates. This can cause the fastener to loosen up. Heating can be done with an oxyacetylene torch. This technique can only be used where there is no fire danger or around components made of soft metals like aluminum. Heat the area around the fastener. Try to keep as much heat as possible from the fastener itself. This will speed the process of separating the fastener threads from the component.

If the fastener is broken off even with or below the surface, the removal job will be more difficult. Sometimes the fastener breaks even though it is not extremely tight. A fast removal technique uses a small cape chisel. Place the chisel on the broken surface of the fastener. Use a hammer to tap it in a direction for removal as shown in Figure 3-22. If the fastener is loose enough, it may begin to rotate. Tap the fastener around until it sticks above the surface. Then grab and continue turning it with vise grips.

If these techniques fail to work, a screw extractor must be used. There are many different types of screw extractors. They often come in sets and are often supplied with the necessary size of drill, as shown in Figure 3-23.

Two common types of screw extractors are shown in Figure 3-24. One has *reverse threads* and the other has a set of *flutes.* In either case, the screw extractor is made to fit in a hole drilled in the fastener. Insert the screw extractor and turn it with a tap wrench. The reverse threads or flutes grip the fastener and allow it to be backed out of the threads.

When using a screw extractor, always follow the instructions supplied with the set. Also, make sure the correct size drill and screw extractor are selected for the job. The correct size to use is typically shown on the set's instructions.

FIGURE 3-23 **Typical extractors that fit over a nut like a socket and are driven by an impact wrench or ratchet. Extractors may be purchased individually or in sets.**

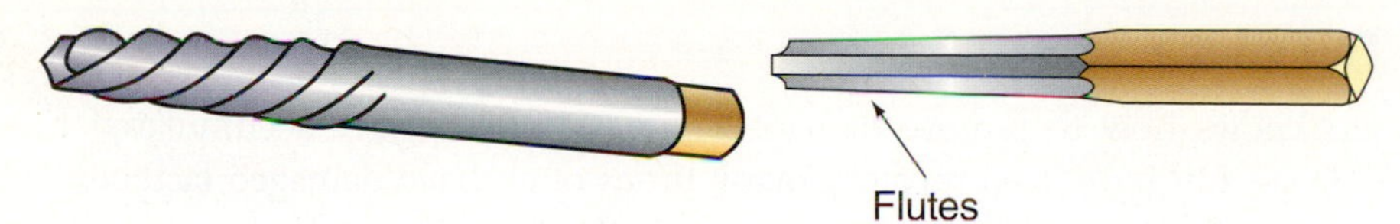

FIGURE 3-24 **The extractor at the top uses a reverse (left hand) thread type screw to grip the fasteners. Both types can be used in most situations. One key point is using heat to loosen a fastener. If heat is used, DO NOT use an extractor because heat will warp it and ruin the extractor.**

The first step is to center punch the exact center of the broken fastener as shown in Figure 3-25. Then install the correct size drill bit in a 3/8-inch size portable drill motor. Drill through the center of the fastener to a depth recommended in the screw extractor instructions, as shown in Figure 3-26. Be careful not to drill too deeply and damage the automotive part, or to drill off to an angle.

With the hole drilled properly, install the extractor. Be sure to choose the correct size extractor for the hole drilled. Install the extractor as shown in Figure 3-27. Some extractors must be tapped into the hole with a hammer.

Use a tap wrench to turn the extractor in a direction to back the fastener out of the hole as shown in Figure 3-28. Some screw extractor sets have drivers that slide over the extractor and fit a standard automotive wrench.

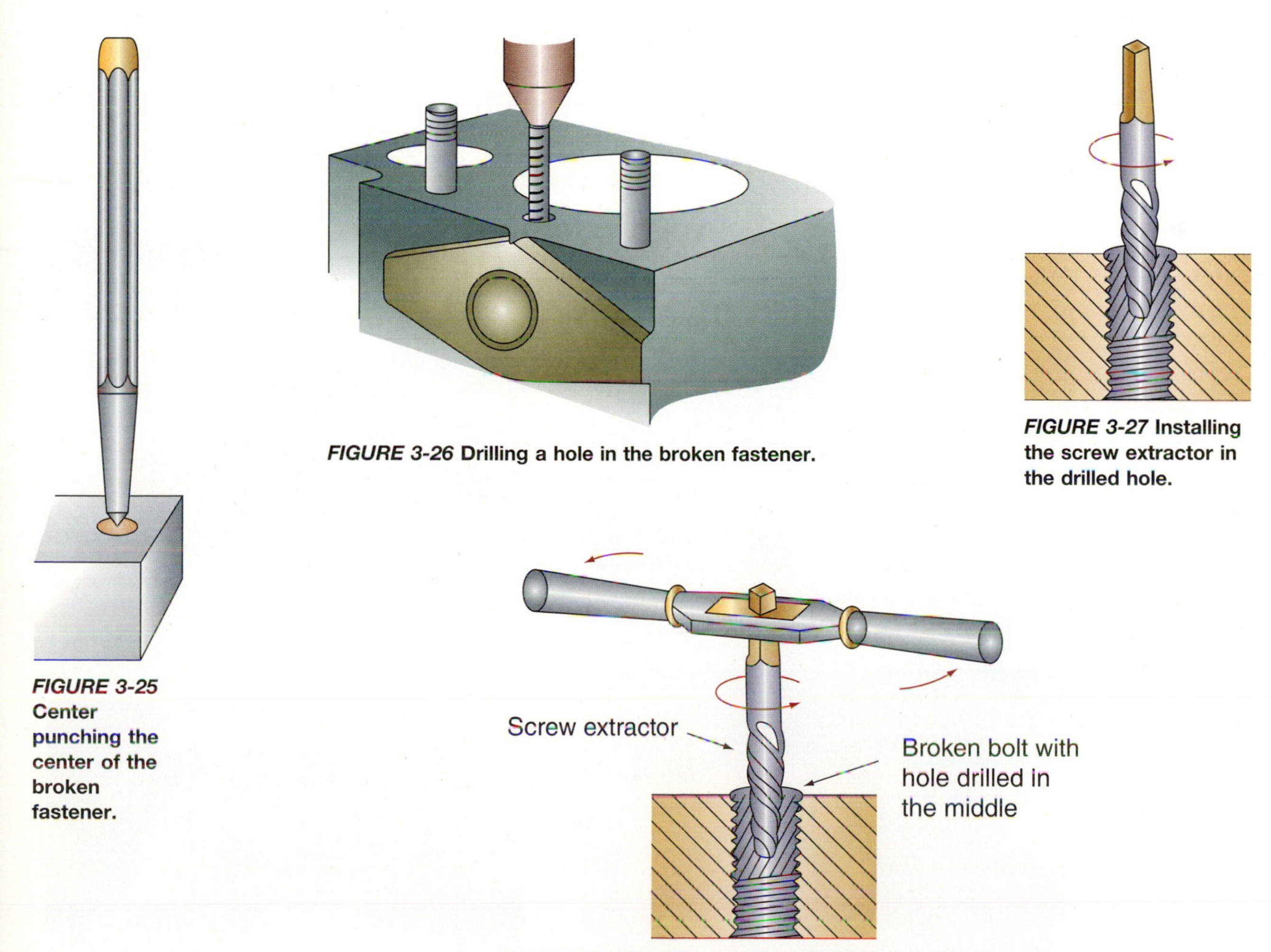

FIGURE 3-25 **Center punching the center of the broken fastener.**

FIGURE 3-26 **Drilling a hole in the broken fastener.**

FIGURE 3-27 **Installing the screw extractor in the drilled hole.**

FIGURE 3-28 **Using a tap wrench and extractor to remove the broken fastener.**

Classroom Manual
pages 59, 60, 61, 62, 63, 64, 65

SPECIAL TOOLS

Cold chisel

Vise grips

Penetrating oil

Nut splitter

Antiseize compound

SERVICE TIP: Stuck fasteners can often be removed by heating them. You can use an oxyacetylene torch to warm the nut, which causes it to expand and break its bond with the stud or bolt. You can only use this technique on nuts where it is possible to heat without the danger of fire or around components that are not made of soft materials like aluminum.

Removing a Damaged Nut

A good technician knows how to remove damaged nuts. Cars that have accumulated many years and high mileage and have been repaired many times often have damaged fasteners. For example, hex nuts may have rounded off corners due to the use of an incorrect wrench size (Figure 3-29), or a nut may have rusted to a bolt or stud like the one shown in Figure 3-30.

A nut that is rounded off at the corners often will not allow the correct size wrench to fit tightly enough to remove the nut. These usually result when a technician uses the wrong size wrench or when a nut is stuck or "seized" to the bolt or stud.

First try to use a file and remove the displaced metal. Then try the largest adjustable, open-end wrench that will correctly fit on the nut. Be careful to push on the wrench with the hand positioned properly in case it slips off the nut. If this procedure does not work, try large vise grips. Adjust these to fit tightly on the nut. Try to turn the nut with the vise grips. Be careful to protect the hand in case the grips slip off the nut. If neither of these steps work, the nut will have to be split as described later.

Another problem is with nuts that are stuck on a bolt or stud. A nut may be stuck because it was overtightened and its threads are jammed or distorted into the threads of the stud or bolt. Corrosion and rust build up between the two fasteners and lock them together. This is called **condition seizing**. In either case, the nut may be impossible to loosen, and pulling on it with too much force can cause the bolt or stud to break. This makes it even more difficult to remove the broken stud or bolt.

If a stuck nut is encountered, first apply a **penetrating oil** as shown in Figure 3-31. Penetrating oil is made to seep into the threads between stuck fasteners and remove the corrosion. Follow the directions on the oil container. Typically, the penetrating oil must be applied several times, and a long period of time is needed for it to work.

When penetrating oil fails to remove the nut or when the nut is too damaged to be removed, the last resort is to split the nut. Use a sharp flat chisel to make a complete cut between the nut and the stud or bolt as shown in Figure 3-32. Try not to contact the

FIGURE 3-29 An example of a nut with rounded corners.

FIGURE 3-31 Penetrating oil can free a rusted nut.

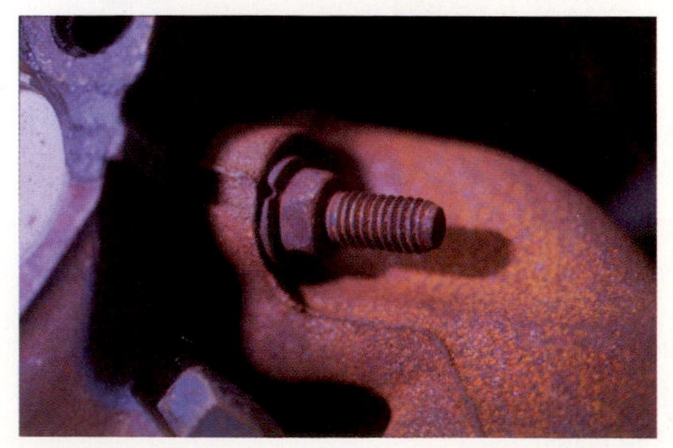

FIGURE 3-30 An example of a nut that has rusted on a stud.

FIGURE 3-32 Splitting a nut with a flat chisel.

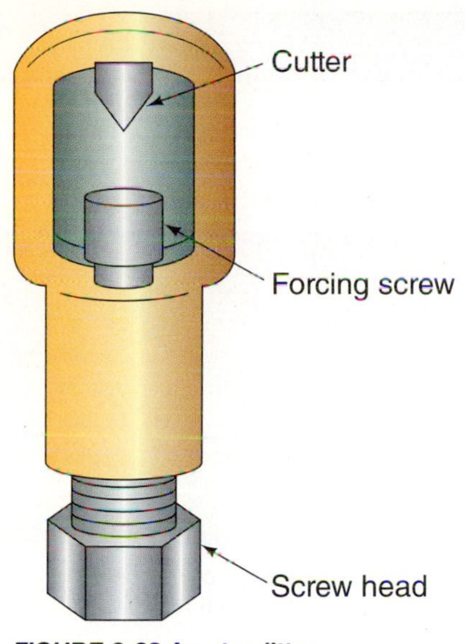

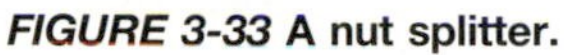

FIGURE 3-33 **A nut splitter.**

FIGURE 3-34 **Antiseize compound prevents nuts from sticking to bolts and studs.**

threads of the bolt or stud. Usually one cut will free the nut. Sometimes you will have to cut it on opposite sides to get it off.

A tool called a **nut splitter** is available to split a nut (Figure 3-33). It is positioned over the stuck nut. A forcing screw driven by a wrench causes a cutter to penetrate and split the nut.

Fasteners used in areas of the car subject to heat and corrosion often seize over time. An example is the fasteners used under the car on the exhaust system. Seizing is also a problem when steel fasteners are used on aluminum parts. A good prevention technique is to coat these fasteners with an antiseize compound when they are assembled. This will make your job much easier if you have to remove the parts later. An example of an antiseize compound is shown in Figure 3-34.

CAUTION:
Oxygen and acetylene cylinders must be secured in an upright position. Never allow a cylinder to fall over. If a cylinder were to fall over and break off its valving, it could become a rocket. When cylinders are being stored, they must have a steel protecting cap over the valve area and the regulators must be closed.

AUTHOR'S NOTE: When a little extra time is taken to do a professional job, like using a torque wrench or coating fasteners with antiseize compound, the repair job will last longer. A technician's reputation as a craftsperson will pay off in the long run when customers want her or him to work on their cars.

Classroom Manual
page 71

SETTING UP AND USING AN OXYACETYLENE TORCH FOR HEATING

The oxyacetylene welding and cutting outfit is used in the shop for welding, cutting, or heating. Photo Sequence 5 shows a typical procedure for setting up and lighting an oxyacetylene torch for heating. Welding and cutting are skills many technicians develop in order to do many advanced repair procedures. A good first step to mastering these skills is to learn how to set up the system for heating. Heating parts to aid in their removal is an important use of the acetylene torch.

SPECIAL TOOLS
Welding goggles
Welding gloves
Oxyacetylene welding and cutting outfit
Striker

WARNING: The installation of welding regulators and hoses on the oxygen and acetylene cylinders should be done by a technician with special training. We describe the setup of the system from the torch handle only.

PHOTO SEQUENCE 5

Typical Procedure for Setting Up and Lighting an Oxyacetylene Torch for Heating

P5-1 Connect the green or black oxygen hose to the torch handle connection marked "OX" by turning it clockwise (right-hand threads).

P5-2 Connect the red acetylene hose to the torch handle connection marked "AC" by turning it counterclockwise (left-hand threads).

P5-3 Connect the correct size torch tip to the end of the torch handle.

P5-4 Close both torch handle valves by turning them clockwise.

P5-5 Turn the main cylinder valve on the oxygen cylinder valve on by turning it counterclockwise.

P5-6 Turn the main cylinder valve on the acetylene cylinder valve on by turning it counterclockwise.

P5-7 Turn the oxygen regulator valve clockwise until the working gauge reads 10 psi.

P5-8 Turn the acetylene regulator valve clockwise until the working pressure gauge reads 5 psi.

P5-9 Turn the torch handle acetylene valve on slightly.

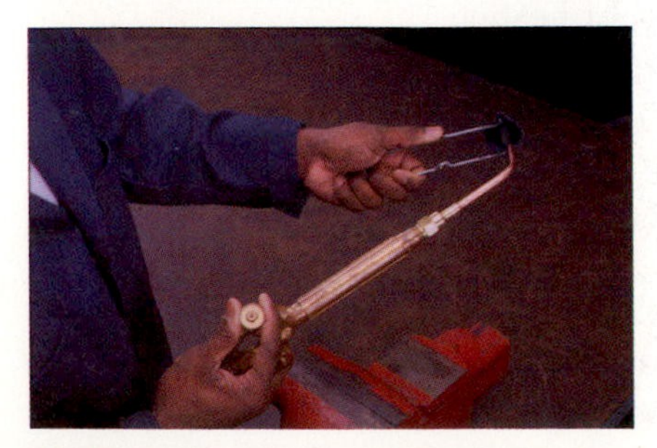

P5-10 Use a striker to ignite the acetylene gas at the tip of the torch.

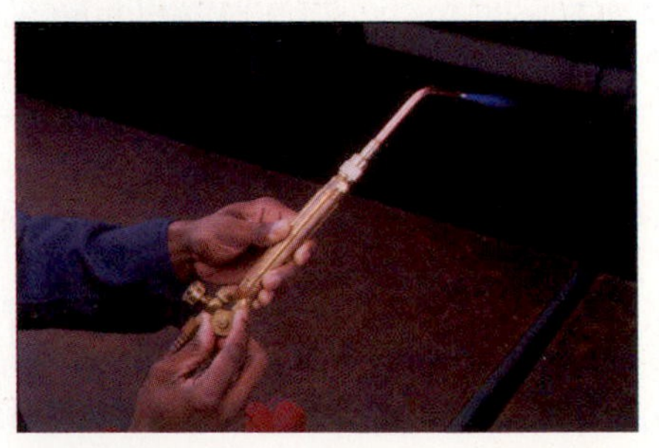

P5-11 Open the oxygen valve at the torch handle and add oxygen to achieve a neutral flame.

FIGURE 3-35 **Turning off the main cylinder valve.**

Before beginning any setup of the oxyacetylene outfit, make sure the gas flow from both the oxygen and acetylene cylinders is off. Look at the hose pressure and cylinder pressure gauge on top of each cylinder (Figure 3-35). Both gauges should be at zero. If not, turn the cylinder valve on top of the cylinder clockwise to close. Repeat this procedure for the other cylinder.

If the valve was open and there is a pressure reading on either gauge, bleed the gas out of the system before going any further. Make sure the cylinder valve is closed. Open the oxygen valve at the torch handle. Turn the oxygen regulator handle clockwise (in). This will allow oxygen in the line to flow through the hose and out of the torch handle. Both gauges will drop to zero. Turn the regulator handle counterclockwise (out). Repeat the procedure with the acetylene cylinder.

Now you are ready to install the torch handle (Figure 3-36) onto the hoses. The handle used for heating is similar to that used for welding. It has two connections: one is for the

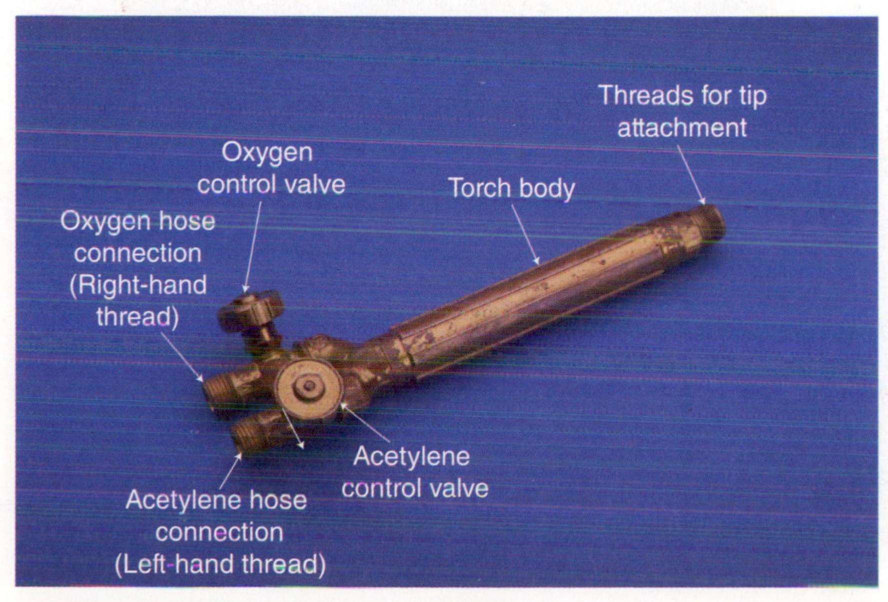

FIGURE 3-36 **Parts of oxyacetylene torch handle.**

A **welding tip** is the part attached to an oxyacetylene torch handle where the gases are mixed to create the flame.

CAUTION:
The oxyacetylene torch develops dangerous ultraviolet light rays. Always wear shaded welding goggles when using this equipment. The flame at the end of the torch is very hot. Wear gloves and be careful not to direct the flame at another person or anything that could ignite.

oxygen hose and the other is for the acetylene hose. The torch handle connections are marked "AC" for acetylene and "OX" for oxygen. To ensure that the correct hose is connected to the proper connection, the hoses are color coded (red for gas, green or black for oxygen) and have left- and right-handed threads. The oxygen hose is green or black and its connection has standard right-hand threads. The acetylene hose is red and its connection has left-hand threads.

Connect the red hose to the AC connection on the torch handle, observing the left-hand threads. Connect the green or black hose to the "OX" connection on the torch handle. Use a wrench to gently tighten the hose connections as shown in Figure 3-37.

The torch handle can be equipped with a number of different size **welding tips**. Welding tips are stamped with a number. The larger the number, the larger the tip and the more heat that can be developed. Common sizes are 1 to 4 as shown in Figure 3-38. Most heating is done with a small tip such as a number 1. Select a size 1 welding tip and screw it onto the end of the torch handle as shown in Figure 3-39.

Now adjust the gas pressure for heating. There are two valves on the torch handle. The one next to the oxygen hose controls the flow of oxygen to the tip. The one next to the acetylene hose controls the flow of acetylene to the tip. Close both these valves by turning them clockwise.

Turn the main valve on top of each cylinder counterclockwise to open the valve. The needle on the cylinder pressure gauge will rise to show the pressure in the cylinder. Turn the handle on the oxygen regulator clockwise (in) until the needle registers 10 psi

FIGURE 3-37 **Installing hoses to the torch handle.**

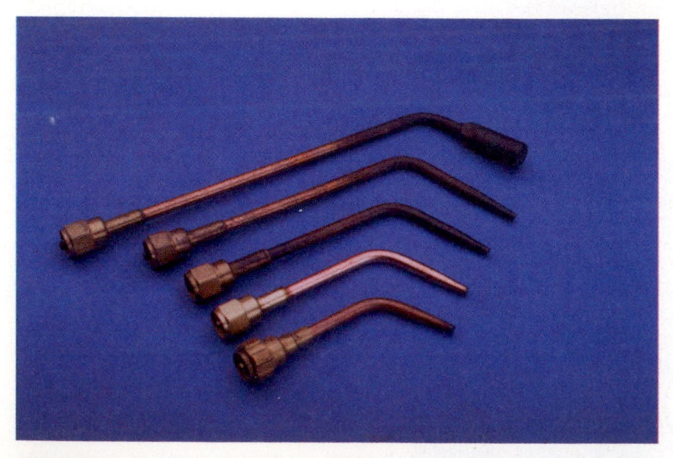

FIGURE 3-38 **The larger the welding tip size, the more heat it can develop.**

FIGURE 3-39 **Installing the welding tip on the torch handle.**

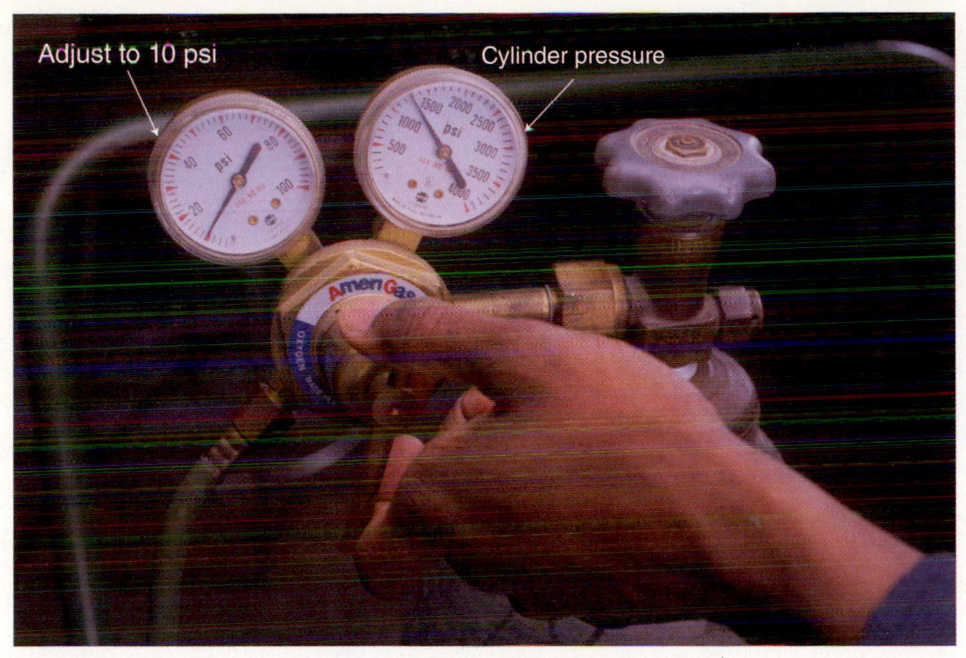

FIGURE 3-40 **Adjusting the working pressure with the regulator handle.**

(Figure 3-40). Turn the handle on the acetylene regulator the same way until the needle points to 5 psi. This will be the working pressure for heating.

You are now ready to ignite the torch. Put on your welding goggles and welding gloves. The torch is ignited with a tool called a **striker**. The striker (Figure 3-41) has a *striker bar* and a *flint* that cause a spark when rubbed together. The spark is used to ignite the torch flame. There is a cup on the striker that protects the operator from burns during start-up.

Hold the torch handle in your hand with the tip pointing downward, away from your body and away from the welding cylinders. Hold the striker by the handle and position it near the tip. Turn the acetylene valve on the torch handle just slightly open. Rub the striker flint over the striker bar to create a spark (Figure 3-42). The spark will ignite the

The **striker** has a flint that creates a spark to ignite an oxyacetylene flame.

CAUTION:
Always use a striker to ignite a torch. Never use matches or a cigarette lighter because these would put your hand next to the flame. Never allow the flame to point at any part of the welding outfit. When you heat something, always mark it "hot" with chalk so a coworker will not get burned picking it up.

FIGURE 3-41 **The striker is used to create a spark to ignite the torch flame.**

FIGURE 3-42 **Igniting the torch flame with a striker.**

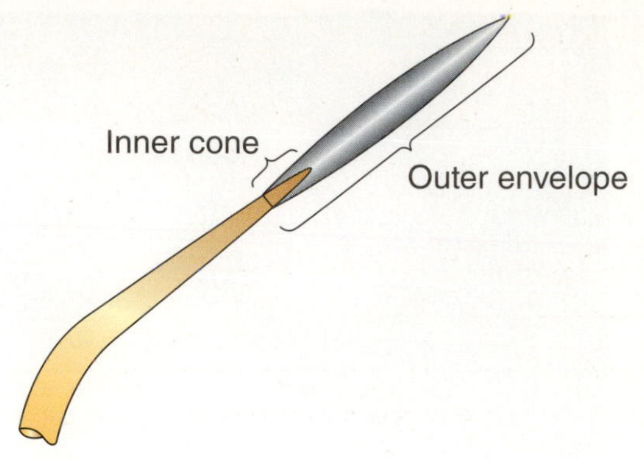

FIGURE 3-43 **A neutral flame has a small, sharp inner cone.**

acetylene gas coming out of the tip. Open the acetylene valve slowly until the sooty smoke starts to disappear. Then slowly open the oxygen valve on the torch handle.

As you open the oxygen valve, you will see the color of the flame change at the tip. The yellow flame of the acetylene will change to blue as you add oxygen. Continue to slowly add oxygen until you see a small, sharp blue inner cone in the middle of the flame envelope as shown in Figure 3-43. This is called a **neutral flame** and is the type of flame wanted for heating.

A **neutral flame** is an oxyacetylene flame with a small, sharp inner cone used for heating and welding.

After heating, it is important to know how to shut the equipment down. First, turn off the acetylene valve on the torch handle. This will extinguish the flame. Then turn off the oxygen valve on the torch handle. Turn the main cylinder valve clockwise on top of both cylinders. Open the valves on the torch handle to bleed the system. Turn the oxygen and acetylene regulator handles counterclockwise until they become loose. Close both valves on the torch handle. Move the cylinders to their storage area.

Jump Starting and Charging a Dead Battery

To **charge** a battery means to pass an electric current through the battery in an opposite direction than during discharge. If the battery needs to be recharged, the safest method is to remove it from the vehicle. If the battery is to be charged while still installed in the vehicle, it is important to remove the negative battery cable to prevent damage to the computer.

Depending on the time available, the battery can be charged at either a slow or fast rate. A slow charge rate is 3–15 amperes. A fast charge is more than that. Slow charging has an advantage in that it will not overheat a battery during charging. If a battery overheats during charging, it will become fully charged due to high resistance in the cells.

Charge rate is the rate of current that a battery can be charged at.

The **charge rate** is the speed at which a battery can be safely charged. Several factors will depict the maximum charging rate. Refer to Figure 3-44 for a reference chart for charging a battery. If you are recharging a gel cell battery, you will have to use a special charger. The high voltage battery on a Hybrid Electric Vehicle can only be recharged by special chargers that are sold by the manufacturer and are available at the dealership.

Jump starting a battery uses another battery source to quickly charge a dead battery.

Jump starting a battery will likely occur frequently in your first couple of years as a technician. In colder climates it is more frequent, but may still occur in warmer climates, especially in hot temperatures. Consistent hot temperatures can shorten the life of a battery. But it is in colder temperatures that the power of a battery is really put to the test.

The most important safety factor when jump starting a battery is to check to make sure that the battery is not frozen or low on fluid. You can open the caps on the top of the battery to visually check this. If a battery is sealed, you will have to measure the temperature of the battery, but it is always best to just remove it from the vehicle and bring it in to warm up and thaw out instead of jump starting it. A frozen battery, or one with very low fluid, can explode when jump starting or charging it.

Open Circuit Voltage	Battery Specific Gravity	State of Charge	Charging time of Full Charge at 80° F (267° C)					
			at 60 amps	at 50 amps	at 40 amps	at 30 amps	at 20 amps	at 10 amps
12.6	1.265	100%	Full Charge					
12.4	1.225	75%	15 min.	20 min.	27 min.	35 min.	48 min.	90 min.
12.2	1.190	50%	35 min.	45 min.	55 min.	75 min.	95 min.	180 min.
12.0	1.155	25%	50 min.	65 min.	85 min.	115 min.	145 min.	280 min.
11.8	1.120	0%	65 min.	85 min.	110 min.	150 min.	150 min.	370 min.

FIGURE 3-44 **Charging rate chart showing charge times based on battery voltage and charging ampere.**

You should refer to Figure 3-45 and read the following steps to safely jump start most vehicles:

1. Make sure the two vehicles are not touching.
2. Place the transmission in park on both vehicles (or set the emergency brake and put in neutral if it is a manual transmission).
3. Turn off the ignition and all accessories on both vehicles.
4. Attach one end of the positive jumper cable to the disabled battery's positive terminal.
5. Connect the other end of the positive jumper cable to the booster battery's negative terminal.
6. Attach the other end of the negative jumper cable to the booster battery's negative terminal.
7. Attach the last end of the negative jumper cable to an engine ground on the disabled vehicle.
8. Attempt to start the disabled vehicle. If it doesn't start right away, wait a few minutes and run the booster vehicle's engine to increase the charge rate.
9. Once the vehicle starts, disconnect the ground-connected negative jumper cable from its engine block.
10. Disconnect the negative jumper cable from the booster battery, then from the other battery.

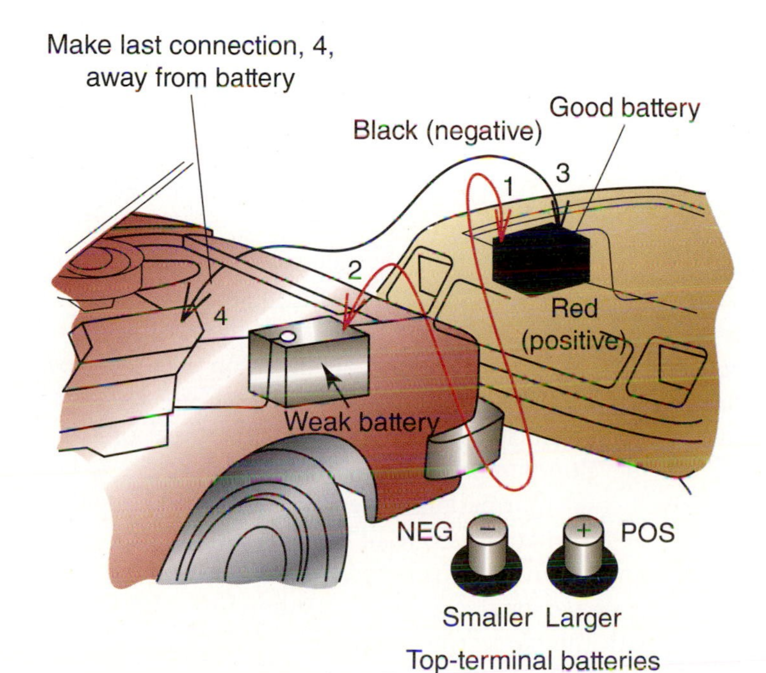

FIGURE 3-45 **Proper jumper cable locations and sequence. Note the size of the battery terminals.**

Understanding and Using Service Information

Classroom Manual
pages 75, 76

Most service data is easily obtained using paper or computerized service manuals. However, there are times when the data is not located in an expected area of the manual. Torque specifications for a certain component like cylinder head bolts may be listed in the specification table in one manual while another manufacturer places it in the section concerning engine assembly or in the cylinder head installation section.

The best method is to determine the correct term for the part and the system in which it is located. Most manufacturers use the same terms for the majority of automotive parts and components. They may, however, list a term as part of a subsection. For instance, data on the cooling system may be in a separate section on cooling or in a subsection under engine performance or diagnosis. Experience in the use of manufacturer service manuals is the best method to determine how each is set up.

Data banks are basically a computerized version of a paper manual. The two largest automotive repair data systems, Mitchell and AllData, use manufacturer data for the information, but both have some differences in the way their data bank is set up.

One of the best methods to pinpoint service data is to use the vehicle identification number (VIN) and identify tags or markings on various major components to determine exactly the type of component used. For instance, in a single manufacturer product line, there may be different versions of basically the same automatic transmission. Externally, the transmission may appear to be exactly the same; however, internally, different gear ratios or a different shift strategy will possibly require different valves or gear arrangement. Somewhere on the external casing will be a tag or etched markings that clarify exactly which transmission version is installed. This information is critical for a proper repair of that transmission and identifying the correct replacement parts.

For the purpose of this discussion, let's use a customer noise complaint to determine the initial steps for repairing the problem. The customer says his or her vehicle has a bumping or jerk as the steering wheel is turned to either far right or far left and it sometimes happens during a slight turn over a speed bump. This fault is probably in the steering and suspension system. Assume that the vehicle is a 2004 GMC 1,500 pickup. During the test drive and initial diagnosis procedure, it was determined that the left lower ball joint is excessively worn. Using this information, the VIN, and the service manuals, computerized or paper, locate the repair procedures for replacing the ball joint. If there are computerized and paper manuals available, use both and compare your findings between the two because there may be a mistake in moving the paper version to a data bank. Because the author does not know what is available at the reader's site, collect the service information and take it to the instructor or an experienced technician for review. Once the repair procedures are gathered, go to the labor guide and determine the labor hours for this job. Don't forget to add the diagnostic time. If available, consult the parts guide and determine the cost of replacement parts.

Was there anything in the procedures, warnings, or cautions that could cause injury or damage? Are special tools or equipment required? What is the labor time and cost of the parts? Based on this research, what is the total cost of this repair based on an $85.00-per-hour labor rate? Does the instructor or advisor agree with the repair estimate?

It is suggested that other vehicles and other complaints, real or simulated, be used to collect information and develop a repair estimate. If available, use a blank service repair order to record each vehicle's information. One thing that will become obvious during a technician's career is that information is one of the best avenues to gaining a great salary. This information may come from the customer, another technician, service manuals, the Internet, or the technician's personal experience. Either way, the route to a quick, first-time fix is much easier with the correct information.

TERMS TO KNOW

Charge
Charge rate
Condition seizing
Female coupling
Jump starting
Lug nuts
Male coupling
Neutral flame

CASE STUDY

Older technicians have been heard to say "I have an educated arm, I don't need a torque wrench." They think they can feel the correct tightness for the fasteners they are tightening and do not need a torque wrench.

A group of technicians got a chance to prove this at a recent automotive tools trade show. A torque wrench manufacturer had an engine on a stand with the cylinder head bolts ready for tightening. Technicians were offered a chance to tighten a head bolt to the specified 85 ft.-lb without a torque wrench. If they could get to within 10 ft.-lb of the specification, they would win a prize. Technician after technician took a turn. After each tightening, the torque on the head bolt was measured. The trade show lasted three days and not one prize was given. So much for the educated arm theory!

TERMS TO KNOW
(continued)

Nut splitter
Penetrating oil
Preloaded
Quick-disconnect coupling
Spring-loaded switch
Striker
Race
Welding tip

ASE-STYLE REVIEW QUESTIONS

1. The use of an air impact wrench is being discussed.
   Technician A says the air impact wrench works well for tightening wheel lug nuts.
   Technician B says the air impact wrench should be used primarily for disassembly work.
   Who is correct?
   A. A only
   B. B only
   C. Both A and B
   D. Neither A nor B
2. The use of an air impact wrench is being discussed.
   Technician A says standard 12-point sockets can be used on the impact wrench.
   Technician B says 6-point impact sockets must be used on an impact wrench.
   Who is correct?
   A. A only
   B. B only
   C. Both A and B
   D. Neither A nor B
3. The use of an air impact wrench is being discussed.
   Technician A says to check the direction of the reverse switch before using the wrench.
   Technician B says eye protection must be worn when using an air impact wrench.
   Who is correct?
   A. A only
   B. B only
   C. Both A and B
   D. Neither A nor B
4. The use of a torque wrench to tighten lug nuts is being discussed.
   Technician A says torque specifications are found in the shop service manual.
   Technician B says the rule is to get the lug nuts as tight as possible.
   Who is correct?
   A. A only
   B. B only
   C. Both A and B
   D. Neither A nor B
5. The use of a torque wrench is being discussed.
   Technician A says wheel lug nut specifications are in newton-meters.
   Technician B says the specifications are in foot-pounds.
   Who is correct?
   A. A only
   B. B only
   C. Both A and B
   D. Neither A nor B
6. The use of a torque wrench is being discussed.
   Technician A says fasteners must be tightened to the correct torque.
   Technician B says fasteners must be tightened in the correct sequence.
   Who is correct?
   A. A only
   B. B only
   C. Both A and B
   D. Neither A nor B

7. The use of a gear puller is being discussed.
   Technician A says gear pullers can be used to pull parts of a shaft.
   Technician B says gear pullers can be used to pull parts out of a hole.
   Who is correct?
   A. A only
   B. B only
   C. Both A and B
   D. Neither A nor B
8. The removal of a broken stud is being discussed.
   Technician A says first try to remove the broken stud with a screw extractor.
   Technician B says first try to remove the broken stud with a chisel.
   Who is correct?
   A. A only
   B. B only
   C. Both A and B
   D. Neither A nor B
9. The removal of a rounded-off nut is being discussed.
   Technician A says it can be removed with a flat chisel.
   Technician B says it can be removed with a nut splitter.
   Who is correct?
   A. A only
   B. B only
   C. Both A and B
   D. Neither A nor B
10. A nut is rusted onto a stud.
   Technician A says soak it in antiseize compound.
   Technician B says soak it with penetrating oil.
   Who is correct?
   A. A only
   B. B only
   C. Both A and B
   D. Neither A nor B

ASE-STYLE CHALLENGE QUESTIONS

1. *Technician A* says the lug nut should be tightened in a circular sequence since the wheel's lug nut area is rounded.
   Technician B says tightening the lug nuts in a star sequence will cause uneven stress on the rim.
   Who is correct?
   A. A only
   B. B only
   C. Both A and B
   D. Neither A nor B
2. *Technician A* says to use chrome sockets when tightening chrome lug nuts because the chrome indicates a match of fastener to wrench.
   Technician B says if the chrome covering comes off the lug nut, then a smaller size socket may be needed.
   Who is correct?
   A. A only
   B. B only
   C. Both A and B
   D. Neither A nor B
3. Using gear pullers are being discussed.
   Technician A says the puller jaws should not pull on the outer circumference of the gears.
   Technician B says the center bolt of the puller should be aligned with or on the bearing journal.
   Who is correct?
   A. A only
   B. B only
   C. Both A and B
   D. Neither A nor B
4. *Technician A* says to remove any broken screw drill completely through the screw so the extractor can get a full-length grip.
   Technician B says a chisel may be used to remove a broken screw.
   Who is correct?
   A. A only
   B. B only
   C. Both A and B
   D. Neither A nor B
5. *Technician A* says the use of antiseize may reduce the chance of a fastener sticking or corroding to its mating threads.
   Technician B says a nut splitter can be used to remove a stud from its thread bore.
   Who is correct?
   A. A only
   B. B only
   C. Both A and B
   D. Neither A nor B

TABLE 3-1 GUIDELINES FOR SERVICE

Raise a vehicle with a lift.

Steps	Classroom Manual	Shop Manual
1. Position the vehicle between the posts.	69, 70, 71	36
2. Position the lift arms and pads under the vehicle.		36
3. Raise the vehicle until its wheels clear the floor. Check for stability.	69, 70, 71	37
4. Lift the vehicle to work height and set the locks.	69, 70, 71	37

TABLE 3-2 GUIDELINES FOR SERVICE

Lift a vehicle with a jack.

Steps	Classroom Manual	Shop Manual
1. Position the vehicle, place in park, and block wheels.	67, 68, 69	37, 39
2. Place jack lift pad under the vehicle's lift point.	67, 68, 69	37, 39
3. Unlock the handle and knob.	67, 68, 69	37, 39
4. Operate the handle to raise the jack.	67, 68, 69	38, 39
5. Place jack stands under the vehicle.	67, 68, 69	38, 39
6. Lower the vehicle to the jack stand.	67, 68, 69	38, 39

TABLE 3-3 GUIDELINES FOR SERVICE

Connect and use an air impact wrench to remove the lug nuts on a wheel.

Steps	Classroom Manual	Shop Manual
1. Raise car on a hoist or floor jack and support on safety stands.	57, 58	37
2. Remove wheel cover and mark lug to wheel position.	57, 58, 59	38
3. Connect air impact wrench to air source.	57, 58, 59	38, 39, 40, 41, 42, 43
4. Set impact wrench to reverse and install correct size socket.	57, 58	38, 39, 40, 41, 42, 43
5. Remove lug nuts.	57, 58	38, 39, 40, 41, 42, 43
6. Reinstall wheel and start lug nuts by hand.	57, 58	38, 39, 40, 41, 42, 43
7. Reverse switch on impact wrench and spin lug nuts into position to be torqued.	57, 58	38, 39, 40, 41, 42, 43

Safety
Use lift locks.
Wear eye protection.
Use impact socket.

TABLE 3-4 GUIDELINES FOR SERVICE

Look up the torque specifications for wheel lug nuts and torque wheel lug nuts to the correct torque.

Steps	Classroom Manual	Shop Manual
1. Look up torque setting for lug nuts and tightening sequence in service or owner's manual.	75, 76, 77	45
2. Connect correct size socket to torque wrench.	57, 58, 59	42
3. Tighten the lug nuts in stages to correct torque.	57, 58, 59	41
4. Tighten lug nuts in the correct sequence.	57, 58, 59	41
5. Replace wheel covers.	—	41
6. Lower car from jack or hoist.	57, 58, 59	37, 38

Safety
Wear eye protection.

TABLE 3-5 GUIDELINES FOR SERVICE

Select, install, and use a gear puller to correctly remove a pulley from a shaft.

Steps	Classroom Manual	Shop Manual
1. Remove the retaining nut from the pulley.	65, 66	48, 49
2. Select the correct size and type of gear puller.	37, 38, 39	48, 49
3. Adjust the cross arm and jaws to install the puller on the pulley.	67, 68, 69	48, 49
4. Drive the forcing screw in to force the pulley off the shaft.	67, 68, 69	48, 49

Safety
Wear eye protection.

TABLE 3-6 GUIDELINES FOR SERVICE

Use a screw extractor to remove a broken stud or screw.

Steps	Classroom Manual	Shop Manual
1. Select correct size screw extractor and drill for broken fastener.	64, 65	51
2. Drill correct size hole in center of broken fastener.	64, 65	51
3. Install screw extractor in drilled hole.	64, 65	52
4. Install correct wrench or driver on screw extractor and rotate it to remove broken fastener.	64, 65	52

Safety
Wear eye protection.

TABLE 3-7 **GUIDELINES FOR SERVICE**
Use a flat chisel to split and remove a stuck nut from a stud or bolt.

Steps	Classroom Manual	Shop Manual
Soak stuck nut with penetrating fluid.	59, 60	52, 53
Cut the nut completely along one side with a cold chisel.	59, 60	52, 53
Remove the nut with a wrench or pliers.	59, 60	52, 53
Safety		
Wear eye protection.		

JOB SHEET 2

Name ______________________ Date ____________

Raising a Vehicle

Objective

Upon completion and review of this job sheet, you should be able to raise a vehicle using a double-post lift.

Tools and Materials Needed

None

Procedures

Task Completed

1. **A.** Identify the vehicle's lift points. ☐
   Make ____________ Model ____________ Year ____________
2. Center the vehicle between the two posts. The front door handle should be about even with the posts. ☐
3. Rotate and extend lift arms under the vehicle. ☐
4. Align lift pads with lift points. ☐
5. Raise lift arms until pad contacts points by holding the switch in the ON position. ☐
6. Check contact between pads and lift points. ☐
7. Raise the vehicle until all wheels clear the floor. Check the vehicle by shaking to ensure it is securely on the lift. ☐
8. Continue to raise the vehicle until the center of the wheel assembly is at eye level. ☐
9. Lower the vehicle until the locks engage the lift arms. ☐
10. Raise the vehicle to remove weight from the locks. ☐
11. Place the locks in the unlocked position. ☐
12. Lower the vehicle to the floor and the arms to their lowest travel. ☐
13. Swing the arms from under the vehicle and move the vehicle from the bay. ☐

Instructor's Response ____

JOB SHEET 3

Name ______________________________ Date ______________

Using an Impact Wrench and a Torque Wrench

Objective

Upon completion and review of this job sheet, you should be able to remove and install a tire assembly using an impact wrench and a torque wrench.

Tools and Materials Needed

½-inch-drive impact wrench
Set of ½-inch-drive impact sockets
½-inch-drive torque wrench (50–250 ft.-lb)
Hubcap hammer or long, flat screwdriver
Lift or jack and jack stands

Procedures

Task Completed

1. Locate the following vehicle service information. ☐
   Lug nut wrench size ______________________________
   Lug nut torque ______________________________
2. Lift the vehicle to a good working height. Use Job Sheet 2 for instructions. ☐
3. Remove the hubcap by using a hubcap hammer or a flat tip screwdriver. ☐
4. Select the correct impact socket and install on impact wrench. ☐
5. Connect the impact wrench to the air supply. ☐
6. Set the impact wrench to reverse and set the socket over the first lug nut. ☐
7. Use one hand to support the wrench and squeeze the impact wrench's trigger. ☐
8. Reduce trigger pressure as soon as the nut breaks loose from the stud. ☐
9. Repeat steps 6 through 8 to remove other lug nuts. If necessary, use one hand to keep the tire assembly on the hub while the last nut is removed. ☐
10. Lay the wrench aside and grasp the tire assembly at the 4 o'clock and 8 o'clock positions. (12 is straight up) ☐
11. Keeping the back straight and bending the knees, remove the tire assembly from the hub and place on floor. ☐

Installation of the tire assembly is the reverse:

12. Grasp the tire at the 4 o'clock and 8 o'clock positions. ☐
13. Keep the back straight and use the knees to lift the tire assembly to the hub. ☐
14. Align the wheel with the studs and slide the wheel assembly onto the studs. ☐
15. Use one hand to keep the assembly on the studs while the first lug nut is started. ☐

Task Completed

☐ **16.** Start each lug nut by hand.

☐ **17.** Inspect the impact wrench's controls. If there is an adjustment, set the wrench to its lowest forward setting. If there is no adjustment, screw the nuts on as far as they go by hand.

☐ **18.** Use the impact wrench to screw the nuts into contact with the wheel.

☐ **19.** Lower the vehicle until the wheels just touch the floor. A second option is to have someone apply the brakes while the nuts are being torqued.

☐ **20.** Set the torque wrench to the specified torque and transfer the socket from the impact wrench to the torque wrench.

☐ **21.** With the wheels held, use the torque wrench to tighten the first lug nut. Once the torque is achieved, move to the next nut.

☐ **22.** The second nut should be across the wheel from the first one.

☐ **23.** Torque each lug nut following a cross pattern.

☐ **24.** Install the hubcap.

☐ **25.** Lower the vehicle and clean the area.

☐ **26.** Complete the repair order.

Instructor's Response ____

____

____

____

JOB SHEET 4

Name ______________________________ Date ______________

Removing a Harmonic Balancer

NATEF Correlation

This job sheet addresses the following NATEF task:

A.1.C. Inspect or replace crankshaft vibration damper (harmonic balancer).

Objective

Upon completion and review of this job sheet, you should be able to use a harmonic balancer puller to remove a harmonic balancer from a crankshaft.

Note: The harmonic balancer was selected to illustrate the procedures for using a bolt-type puller. Other equipment could be used in lieu of the balancer.

Tools and Materials Needed

A crankshaft with harmonic balancer installed, in or out of vehicle

Vise if crankshaft is out of vehicle

Harmonic balancer puller set

Hand-tool set

Procedures

Task Completed

1. If crankshaft is out of vehicle, place it in a vise or other holding device. ☐
2. Locate the harmonic balancer puller set. ☐
3. Select the two (or three) bolts that thread easily into the outer threaded holes on the balancer. ☐
4. Select a center cone that fits the best into the large threaded hole in the crankshaft. ☐
5. Screw the large puller bolt into the puller body and position the bolt point into the center cone. ☐
6. Slide one of two (or three) previously selected pulling bolts through the puller body and screw into the balancer hole. Repeat for the other bolt (or two). ☐
7. Tighten each of the two (three) bolts until all penetrate the balancer. Ensure each protrude an equal length through the puller body. Ensure the puller body is parallel to the balancer face. ☐
8. Use a wrench to tighten the center puller bolt. It may be necessary to adjust the shaft anchor as the pulling operation continues. ☐
9. Continue to screw in the large bolt until the balancer is loose and can be removed by hand. ☐
10. Capture and store the woodruff key that pins the harmonic balance to the crankshaft. ☐
11. Remove the puller components from the balancer and store in case. ☐

Task Completed

☐ **12.** Record any problems encountered during this pulling operation.

____

____

Instructor's Response ____

____

____

____

JOB SHEET 5

Name ______________________________ Date ______________

Removing Pittman Arm from Steering Gear Box

NATEF Correlation

This job sheet addresses the following NATEF task:

IV.B.17. Inspect and replace pitman arm, relay (centerlink/intermediate) rod, idler arm and mountings, and steering linkage damper.

Objective

Upon completion and review of this job sheet, you should be able to use a two-jaw puller to remove a pitman arm from a steering gear box.

Note: A pitman arm was selected to illustrate the use of a two-jaw puller. Other equipment could be used in lieu of the pitman arm, but a two- or three-jaw should be used. It is unlikely a three-jaw puller could be used on a pitman, but it would be preferable if a gear or a shaft is being used instead of the pitman arm.

Tools and Materials Needed

Recirculating ball steering gear, in or out of vehicle

Vise if steering gear is out of vehicle

Lift if steering gear is in vehicle

Two-jaw gear and bearing puller

Hand-tool set

Procedures

Task Completed

1. If steering gear is out of vehicle, place it in a vise or other holding device. ☐
2. Loosen the large retaining nut holding the pitman arm. Screw it nearly off but do not remove. This will catch the pitman arm and puller when the arm is pulled free of its shaft. ☐
3. Locate a two-jaw puller. ☐
4. Unscrew the puller's large center bolt until the jaws slip over the pitman arm. The center bolt should be centered on the end of the steering shaft. ☐
5. Tighten the center bolt until the puller holds in place. ☐
6. Inspect the position of the jaws over the edges and back side of the arm. Ensure the contact surfaces of the jaws are flush against the back side of the pitman arm. If not positioned correctly, either reposition or select a different puller. ☐
7. With puller positioned correctly, use a wrench to screw the center bolt inward. This should begin to pull the pitman arm off. ☐
8. Continue to tighten the center bolt until the pitman arm and puller drop loose of the steering gear. ☐

Task Completed

☐ **9.** Remove the puller from the pitman arm if necessary and store in case. Secure and store the pitman arm.

☐ **10.** Record any problems encountered during the pulling operation.

WARNING: Once the arm is loose it will drop off. If the retaining nut had to be completely removed for use of the puller, be prepared to catch the arm and puller. Failure to catch it could result in an injury.

____

____

Instructor's Response ____

____

____

____

JOB SHEET 6

Name ______________________________ Date ________________

Collecting Service Information

NATEF Correlation

This job sheet addresses the following NATEF task:

A.1. Complete work order to include customer information, vehicle identifying information, customer concern, related service history, cause, and correction.

Objective

Upon completion and review of this job sheet, you should be able to collect service information based on a customer complaint.

Tools and Materials Needed

Service manuals, computerized or paper

Labor and parts guide

Blank repair order if available

Procedures

Task Completed

This job sheet is based on the following information provided by the customer or from the vehicle.

Customer complaint: The vehicle pulls to the right when the brakes are applied. It has no unusual noise and other than the brake pull, it operates fine. He stated the vehicle had the brakes repaired at an out-of-town shop about 3 weeks ago.

VIN: 1B46P54R7TB3XXXXXX. The vehicle has antilock brakes on rear wheels only.

NOTE: Labor fee for this shop is $65.00 per hour.

Note: Information gathered during this job sheet may be recorded in the spaces provided, but the use of a blank repair order is preferred.

1. Use the VIN to determine the year, make, model, and engine for this vehicle. ☐

   ____

   ____

2. Locate and record any recall or TSBs that pertain to the customer's complaint. If none, state none. ☐

   ____

   ____

3. Record the vehicle's repair history as stated by the customer or shop records. ☐

   ____

   ____

4. Locate and record a brief outline of the diagnostic procedures for the vehicle fault as stated by the customer. ☐

   ____

   ____

Task Completed

☐ 5. Locate and record a brief outline of the repair procedures for this problem. Note any cautions or warnings.

____

____

☐ 6. Use the parts guide to locate and price the parts needed to complete the repair. Record both the suggested retail price and the shop's cost.

____

____

☐ 7. Use the labor guide to calculate the number of hours needed to complete a total repair.

____

____

☐ 8. Calculate the total cost of this repair. Tax is 7 percent on parts; no tax on labor.

____

____

☐ 9. Were any problems encountered during the completion of this job sheet? What type of service manual was used? If you used a different source (e.g., paper instead of computer) before this job sheet, which appears to be the easiest to use?

____

____

☐ 10. Complete the repair order if used and present it or this job sheet to the instructor.

____

____

Instructor's Response ____

____

____

____

Chapter 4

Precision Measurements

Upon completion and review of this chapter, you should understand and be able to describe:

- How to use a feeler gauge for measuring spark plug gap.
- How to use an outside micrometer to measure shaft diameter.
- How to use an outside micrometer and telescopic gauges to measure a bore.
- How to measure a brake disc for thickness with an outside micrometer.
- How to measure bores with inside micrometer.
- How to use a dial indicator to measure runout and gear backlash.
- How to use a dial indicator to measure shaft endplay.
- How to measure and read a pressure gauge.
- How to measure and read a vacuum gauge.

Introduction

This chapter will guide technicians through methods of making measurements using tools studied in the Classroom Manual. The tasks discussed will be typical of the types of measurement jobs performed in a shop.

SPECIAL TOOLS

Round wire feeler gauge set or tapered plug gauge

Nonmagnetic feeler gauge set

Service information

Classroom Manual

pages 86, 87, 88

Measuring Spark Plug Gap

Even with the best electronics, all gasoline-fueled engines use spark plugs to fire the air/fuel mixture. While many vehicles require spark plug replacement only every 100,000 miles, the new plugs will have to be correctly gapped (Figure 4-1). If the gap is too wide, the spark may be weak in strength and short in duration because of the high voltage needed to bridge the gap. With a too-narrow gap, the spark will have a low voltage and not much heat. In both cases, the spark will have difficulty igniting the air/fuel mixture, resulting in a poorly operating engine.

There are or will be scrapped spark plugs in almost every shop or technical school. Select five or six of them and a round wire feeler gauge (Figure 4-2). If necessary, clean the plugs before gapping. Assume that the correct gap for the first one is 0.035 inch.

Author's Note: When possible, identify the vehicle from which the plugs were removed. The student can use the vehicle service data to locate the correct gap for the plug.

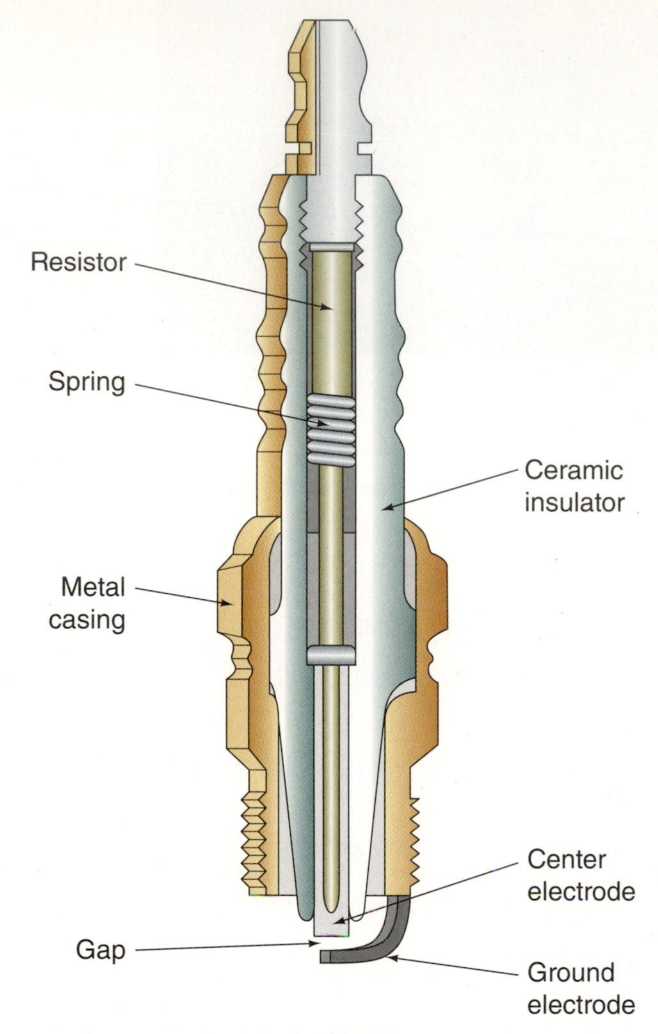

FIGURE 4-1 **The electricity bridges the gap and creates a spark to ignite the air/fuel mixture.**

FIGURE 4-2 **A typical round wire feeler gauge used for measuring spark plug gap.**

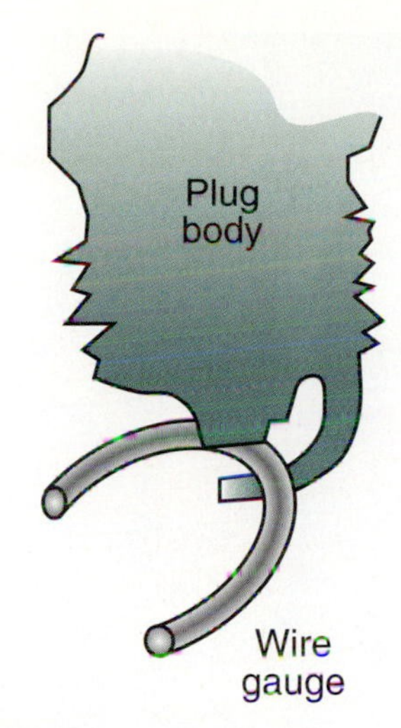

FIGURE 4-3 **The gauge wire should have a slight drag as it is moved between the two electrodes.**

Locate the 0.035-inch wire from within the gauge set. Attempt to slide the wire between the plug's two electrodes (Figure 4-3). If the gap is too small or too wide, use the gauge set's small bending tool to move the outer electrode. Check the gap again and correct any further space by using the bending tool on the gauge set. With one plug correct, select different gaps for the other plugs and set them. Most of the current vehicles use spark plug gaps between 0.035 inch and 0.046 inch. Some older General Motors vehicles use 0.080-inch gaps.

Measuring Air Gap

AUTHOR'S NOTE: One of the two easiest means to practice the measuring or setting of an air gap is using the wheel speed sensor for an antilock brake system (ABS) or the crankshaft position sensor on some engines.

For the purpose of this discussion, we will use an ABS wheel speed sensor on a front-wheel-drive vehicle. Some vehicles have ABS only on the rear wheels and will not have the sensor at the front wheels. Use the service data for the vehicle being used and retrieve the necessary instructions and specifications for the wheel speed sensor air gap. Each speed sensor will have a **toothed wheel** or **toothed ring** that triggers the sensor (Figure 4-4). The ring may be on the axle, rotor hub, or on a gear inside the differential. In this instance, we'll assume the vehicle is a front-wheel drive with the tooth ring on the brake rotor hub. Ensure the ignition is in the off position and the transaxle is in neutral. Lift the vehicle and remove one of the front wheels to gain access to the speed sensor. Locate the nonmagnetic feeler gauges and select the correct blade based on the air gap specification. Rotate the hub until a tooth on the ring is aligned perfectly with the center of the sensor. While holding the hub in place, slide the nonmagnetic feeler gauge between the tooth and the sensor (Figure 4-5). If it does not go, the air gap is too small. If it slides in and out with too little drag, the air gap is too large. Consult the service information to determine the proper method of adjusting the air gap. Adjust as needed and measure to ensure the air gap is correct.

The **toothed wheel** or **ring** is a metal ring or plate with teeth at specific distance apart. As each tooth passes the sensor, it will cause the sensor to generate a voltage.

FIGURE 4-4 **This is a version of a toothed ring. This ring is mounted to the drive near the outer CV joint. The raised ridges on the ring create a voltage signal as they pass the sensor.**

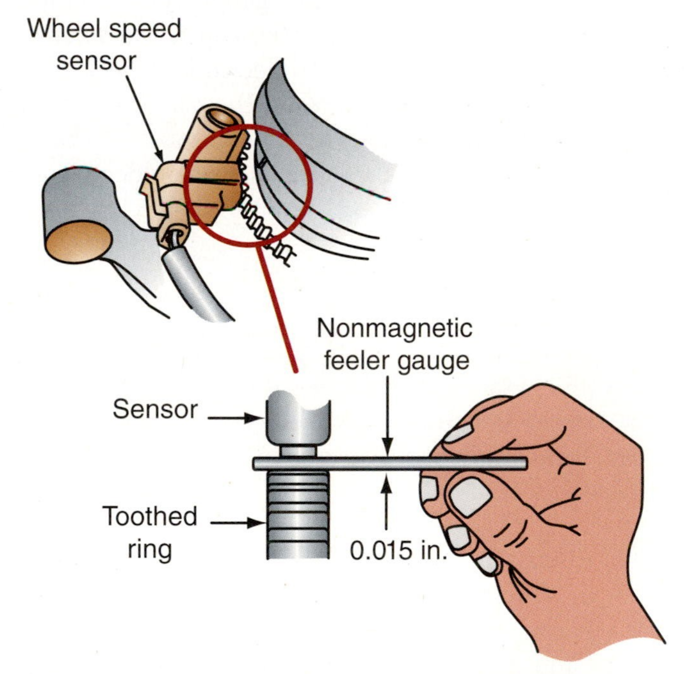

FIGURE 4-5 **Measuring the air gap between the sensor and a tooth or trigger. A nonmagnetic (usually brass) feeler gauge is used to measure the air gap.**

Measuring Cylinder Head Warpage with a Feeler Gauge and Straight Edge

A feeler gauge set is a very versatile measuring tool. It has to be kept clean and free of rust. Many parts of the engine have to seal together. Small imperfections in the metal surface will make it difficult to seal the area where two metal objects are bolted together. For the example of a cylinder-head-to-cylinder-block mating surface, a gasket is used in between. The gasket seals the cylinders because it forms to the surfaces that are irregular.

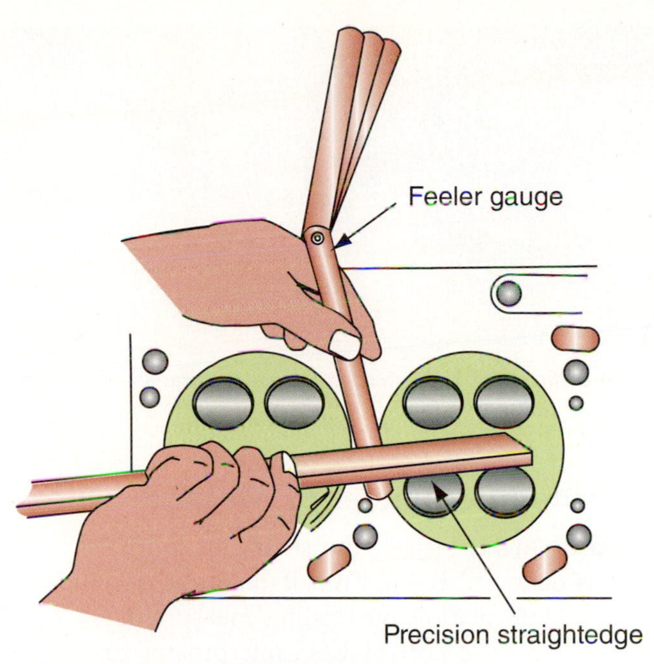

FIGURE 4-6 **Use a straightedge and a feeler gauge to determine warpage of the cylinder head.**

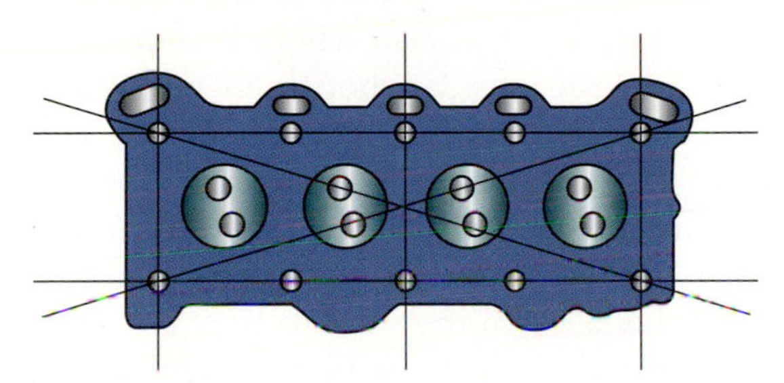

FIGURE 4-7 **A cylinder head, like other gasket mating surfaces, must be checked in several areas for straightness.**

When replacing any gasket in the engine, including the cylinder head gasket, you must check to make sure that the surface it is adhering to is straight. In order to do this, you need a **straight edge** and a feeler gauge. Place the straight edge against the surface to be measured and then use the feeler gauge to "feel" the gap that exists between the unknown edge and the straight edge. For a cylinder head, just like any other gasket mating surface, you must do this in several locations to ensure that it is straight (Figures 4-6 and 4-7).

A **straight edge** is a metal bar that has a machined surface and is very straight and smooth.

Measuring Shaft Diameter and Wear

There will be times when a technician is required to measure various components to determine if they are serviceable or need machining or replacement. Many shops do not rebuild engines in-house but may still have to determine if a replacement is needed. The same applies to transmissions and transaxles. Sometimes the noise made by a bad engine does not mean that the engine is only good for scrap metal. The amount, loudness, or lack of noise does not accurately indicate the amount of damage.

Classroom Manual
pages 92, 93, 94, 95, 96

Select a shaft that has one or more precision-machined areas on it (Figure 4-8). An output shaft from a scrapped transmission, a crankshaft, or a camshaft works well for this exercise. Select a set of United States Customary (USC) and metric micrometers that will measure the various areas of the shaft.

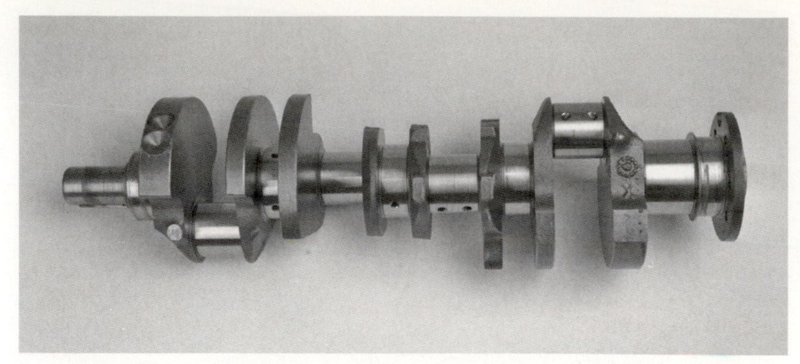

FIGURE 4-8 **The shiny, highlighted areas are the crankshaft bearing journals.**

SPECIAL TOOLS

Outside USC micrometers

Shafts

Using a USC Outside Micrometer to Measure a Bearing Journal

Before starting the measurement, inspect the shaft and journals. It is a waste of time to measure a journal if it is damaged beyond repair unless it is being done for practice. We will review the classroom information on reading measurements as we proceed through the rest of this chapter. Select the correct USC micrometer for the first area to be measured. The area selected should be machined (Figure 4-9). Measuring the rough or unmachined area is not needed in most instances. If possible, locate some data for the shaft from the instructor or the service manual, specifically data on the correct diameter and the amount of wear that may exist before the shaft has to be replaced is needed.

Move the spindle away from the anvil far enough for the micrometer to be slipped over the shaft (Figure 4-10). Rotating the thimble counterclockwise opens the gap between the two measuring faces. With the two measuring faces, anvil and spindle, 180 degrees apart and placed over the shaft, rotate the thimble clockwise until it contacts the shaft. Keep the micrometer in place as the ratchet is turned clockwise until it slips. Lock the spindle and check the alignment of the faces on the shaft. If they are positioned correctly, remove the micrometer so the measurement can be easily read. Unlock the micrometer and reposition it if the faces are not in place correctly.

Begin reading the measurement by noting the size of the total width of the micrometer (Figure 4-11). For instance, if this is a 1-inch to 2-inch micrometer, the first number will be 1 and will be to the left of the decimal point.

The second number and the first to the right of the decimal point will be read from the index line on the sleeve (Figure 4-12). This will be the highest visible number along the index line. We will assume for this exercise that the number is 3.

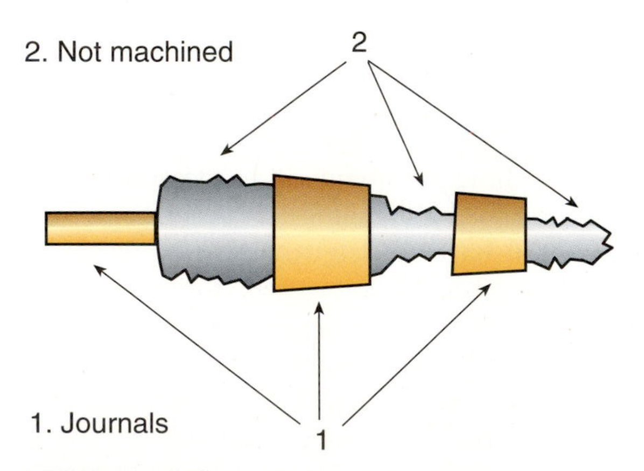

FIGURE 4-9 **Most shafts have several bearing journals to support the shaft evenly along its length.**

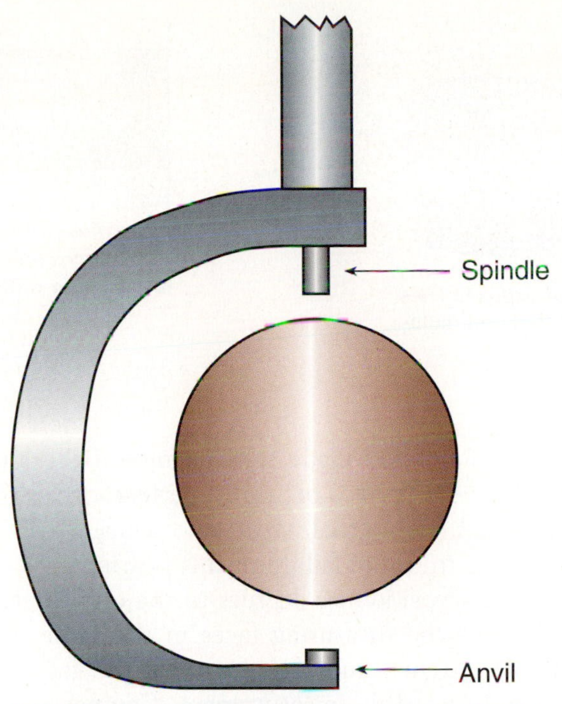

FIGURE 4-10 **Open the micrometer and slide it over the component being measured.**

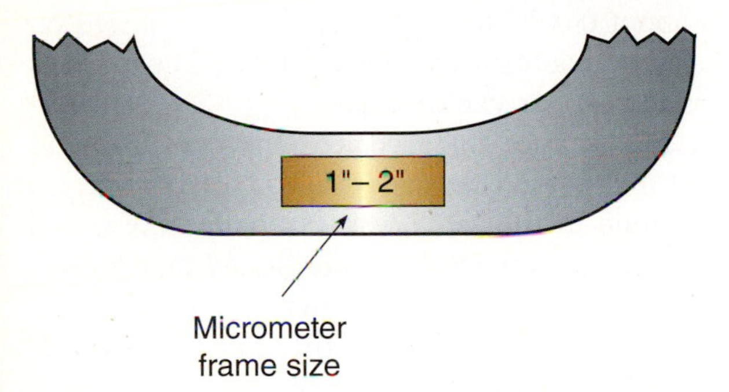

FIGURE 4-11 **The micrometer frame size is the digit to the left of the decimal point.**

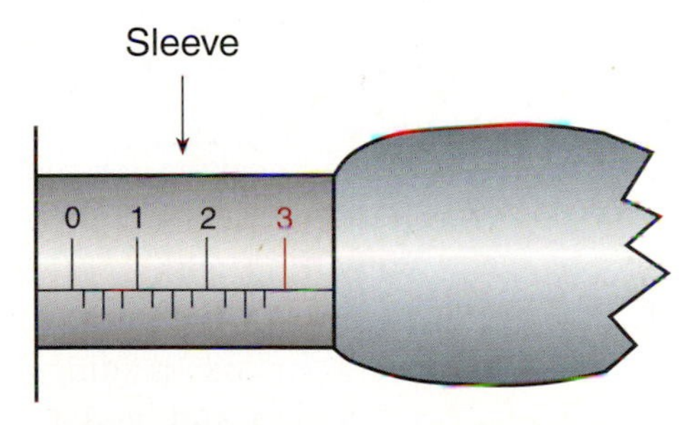

FIGURE 4-12 **The highest exposed sleeve number is the first digit to the right of the decimal point.**

The second number to the right of the decimal comes from the graduations under the index line on the sleeve (Figure 4-13). Each mark is worth 0.025 inch. If the second graduation mark is the last one showing, then the number will be 0.050 inch. Using the values from the last three paragraphs, we can say the shaft is at least 1.350 inches thick. The numbers are found as follows:

1 inch from the micrometer frame plus
0.3 inch from the index line plus
0.050 inch from below the index line for a total of 1.350 inches

The last number comes from nearest graduation mark on the thimble that most nearly aligns with the index line on the sleeve. There may be some room for interpreting the reading at this point. If the index line is near the center between two thimble marks, two technicians may come up with two different readings (Figure 4-14). This 0.001-inch difference between the two is not of concern on this type of shaft. If the two readings are at the limits of the wear tolerance, the shaft should probably be replaced anyway. The last number derived from the thimble is added to the previous readings. Assume that the number is 0.004 inch and it is added to the readings above, then the shaft is 1.354 inches thick at this **journal**.

A **journal** is a finely machined portion of a shaft that is fitted through a bearing or housing. This machined surface reduces the friction and heat as the shaft operates.

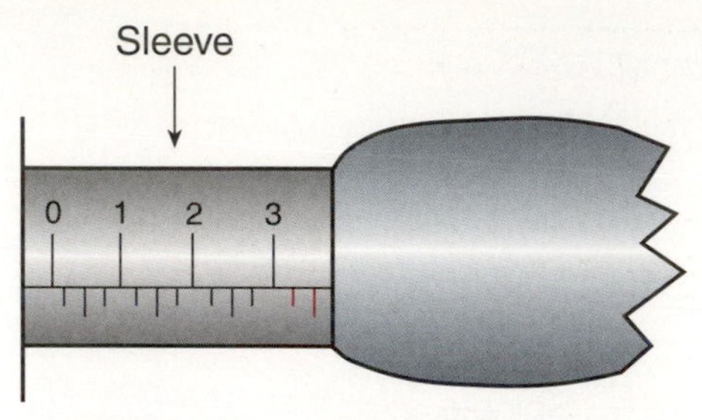

FIGURE 4-13 **The number of exposed lines past the top number is counted and multiplied by 0.025.**

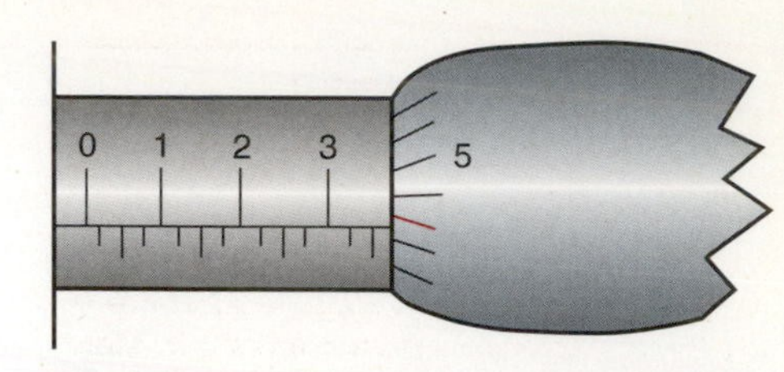

FIGURE 4-14 **The closest thimble graduation to the index line is always selected as the last number of the measurement.**

The term **out-of-round** only applies to a journal or shaft that is supposed to be perfectly round.

A second measurement should be made to determine if the journal and shaft is **out-of-round**. If the journal is not a perfect or nearly perfect circle, the journal and bearing will make contact. The life of the journal, shaft, and bearing will be severely shortened. Other damage may also result from the failure of this bearing.

Move the micrometer to a position 90 degrees to the one where the last measurement was made (Figure 4-15). Place the measuring faces in the same position relative to each other. Operate the thimble, ratchet, and lock to move the faces to the journal surface. Remove the micrometer and record the measurement. Compare the two measurements to each other and the specifications for the journal.

If one of the two measurements is smaller then the other, the journal is out-of-round and has a slight oval shape. A difference of about 0.0005 to 0.001 inch may be allowed on some shafts. Consult the service data for the allowable tolerance (Figure 4-16). The journals on some shafts can be machined and undersize bearings can be used.

While checking for out-of-round data, also check for the specified minimum diameter of the journal. Even a perfectly round journal that is too thin may result in a failure.

Measuring with a metric micrometer is done in the same way except for the actual compilation of the measurement. Once the micrometer has been positioned and locked, remove it for easy reading. The numbers above the index line are graduated one millimeter apart. The last and largest exposed number on the index line will be the digit to the left of the decimal point, just the opposite of the USC micrometer (Figure 4-17). If the journal is more than 25 mm, this first number must be added to the frame size.

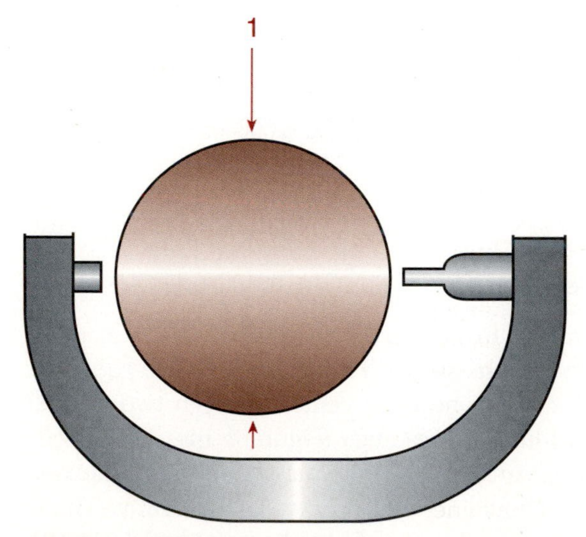

FIGURE 4-15 **The second shaft measurement is at a right angle to the first (arrows).**

Mark 0	54.998–55.003 mm (2.1653–2.1655 in.)
Mark 1	54.993–54998 mm (2.1651–2.1653 in.)
Mark 2	54.998–54.993 mm (2.1649–2.1651 in.)

FIGURE 4-16 **There is only 0.005 mm (0.0002 inch) of wear allowed on a crankshaft. Mark 1 main journal (54.998 − 54.993 = 0.005).**

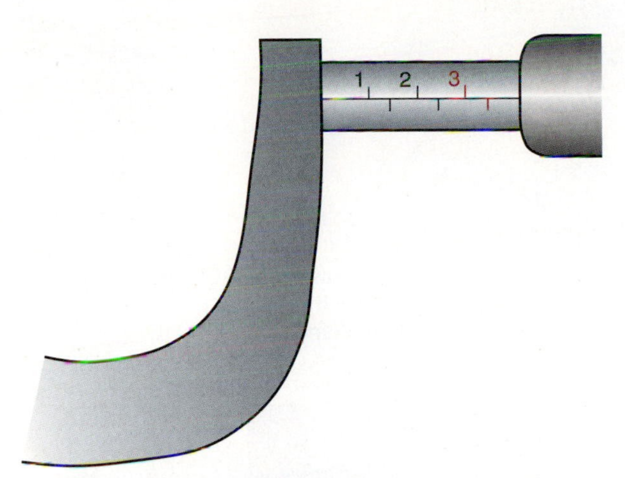

FIGURE 4-17 **On a metric micrometer, the last exposed number on the index line is the digit to the left of the decimal point.**

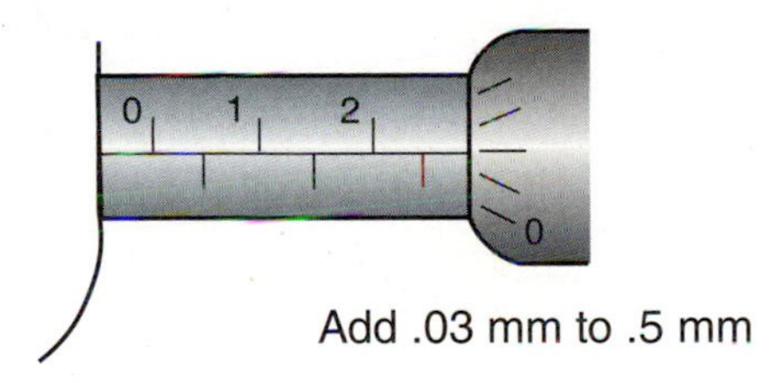

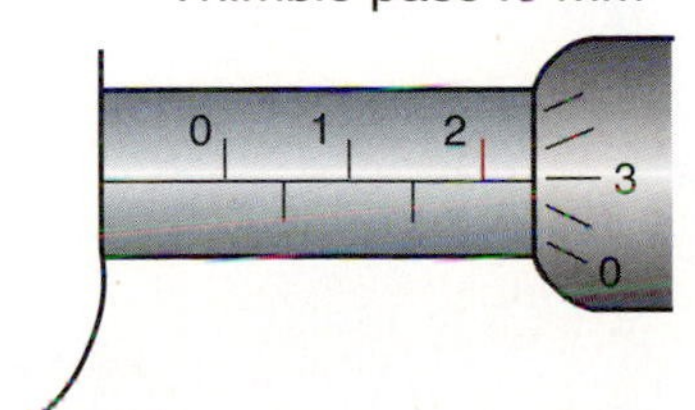

FIGURE 4-18 **Ensure that the below-the-line reading is correct. A 0.5-mm error could be made.**

Digits to the right of the decimal are a combination of the graduations below the index line and the nearest line on the thimble. See if the 0.5 mm line below the index line is exposed (Figure 4-18). Determine which line of the thimble most closely aligns with the index line and add that number to the 0.0 or 0.5 mm line found one step back in Figure 4-18. The digits to the left of the decimal point come from above the index line

PHOTO SEQUENCE 6

Typical Procedure for Reading a Micrometer

P6-1 **The tools required to perform this task are a micrometer set and a clean shop towel.**

P6-2 **Select the correct micrometer. If the component measurement is less than 1 inch, use the 0- to 1-inch micrometer.**

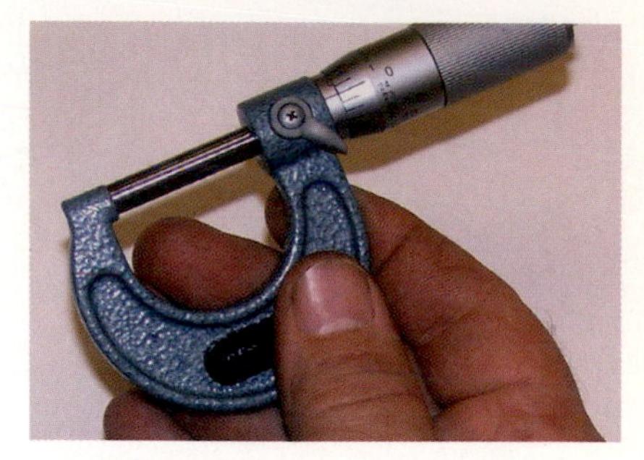

P6-3 **Check the calibration of the micrometer.**

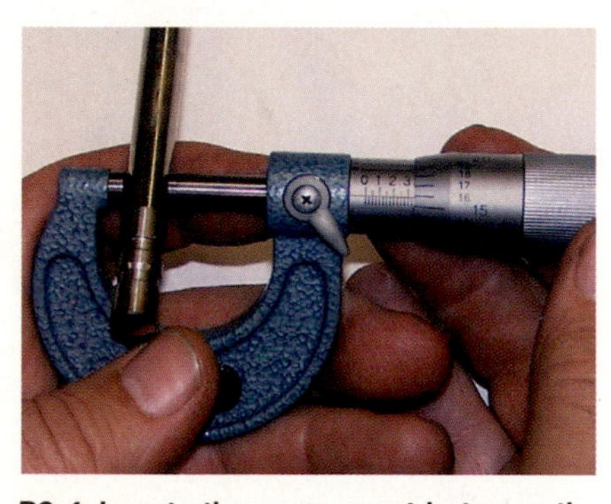

P6-4 **Locate the component between the anvil and spindle of the micrometer, and rotate the thimble to slowly close the micrometer around the component. Tighten the thimble until a slight drag is felt when passing the component in and out of the micrometer. If the micrometer is equipped with a ratchet, it can be used to assist in maintaining proper tension.**

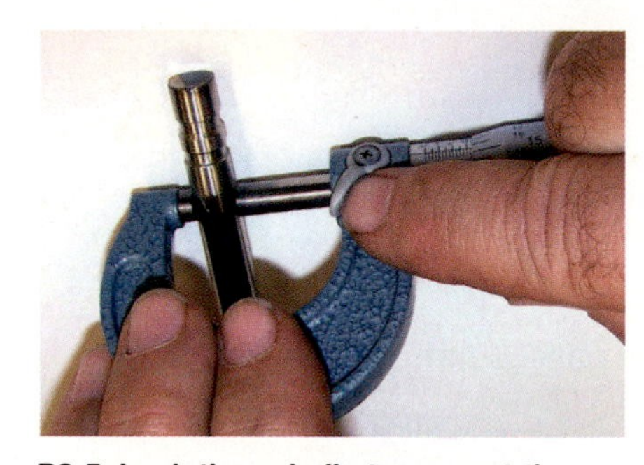

P6-5 **Lock the spindle to prevent the reading from changing.**

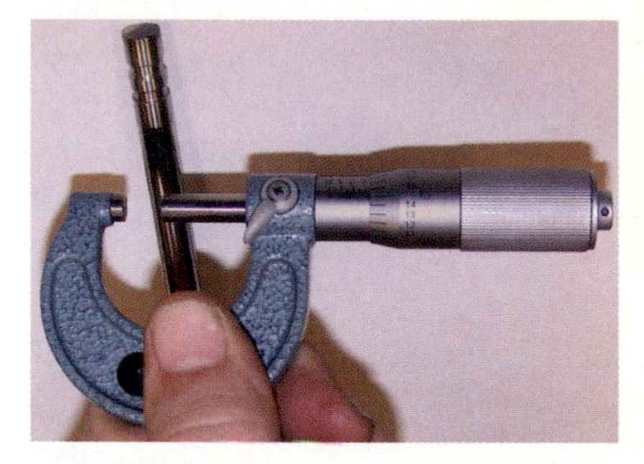

P6-6 **Remove the micrometer from the component.**

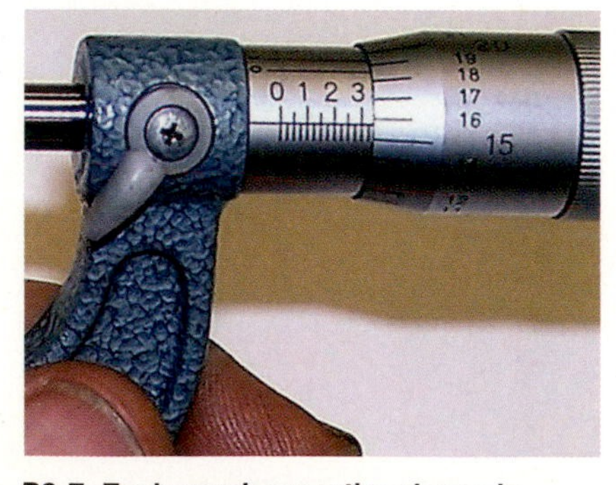

P6-7 **Each number on the sleeve is 0.100 inch, and each gradation represents 0.025 inch. To read a measurement, count the visible lines.**

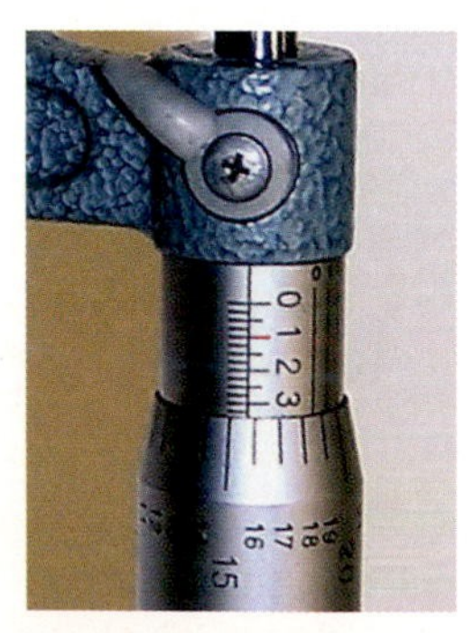

P6-8 **The graduations on the thimble define the area between the lines on the sleeve in 0.001 increments. To read this measurement, use the graduation mark that aligns with the horizontal line on the sleeve.**

P6-9 **Add the reading obtained from the thimble to the reading on the sleeve to get the total measurement.**

plus the frame size. All the numbers below the line and the thimble go to the right of the decimal. See Photo Sequence 6 for the use of an outside micrometer.

Another measuring device is the *dial caliper*. The dial caliper is a machined bar with two measuring jaws, one stationary and one moveable. The bar is graduated in either USC or metric and mounts the stationary jaw. The moveable jaw slides along the bar and mounts a dial indictor. The graduations on the bar are used to measure gross dimensions (whole inches, centimeters) while the dial indicator provides more precise dimensions (0.001 inch, 0.1 mm). There is an electronic version that is much easier to use and is inexpensive. See Photo Sequence 7 for details on using a standard dial indicator.

Classroom Manual
pages 94, 95, 97, 98

SPECIAL TOOLS

Outside UCS and metric micrometers

Telescopic gauges

A component with machined bores

MEASURING BORE WITH AN OUTSIDE MICROMETER AND TELESCOPING GAUGE

The diameters of machined **bores** in vehicle components have to be just as perfect as the various bearing journals. In fact, some of the bores may actually act as bearing journals and all are a very important part of vehicle operation. For instance, the bores or cylinders in an engine tend to wear more on one side. This is **thrust wear** and results in an out-of-round cylinder (Figure 4-19). The bore walls also wear throughout the piston **ring travel area** (Figure 4-20). This is normal wear for an engine, but the bore should be corrected when the engine is being rebuilt. Bores used to house bearings may also wear and allow the bearings to move, thereby causing damage.

Thrust wear is caused by expanding gases in the combustion chamber that pushes the piston toward the cylinder wall.

As with the previous journal, inspect the bore for any damage. The component operating in the bore requires a smooth finish to reduce friction, heat, and wear.

With the inspection complete, select a telescoping gauge that fits into the bore and will extend far enough to reach the walls. Select an outside micrometer to match the gauge. Insert the gauge just below the top of ring travel and release the spindles. Move the gauge until it is square in the bore. This means the gauge is at right angles (90 degrees) to the vertical centerline of the bore (Figure 4-21). It must also touch the opposite walls at the widest point. If an engine cylinder bore is being measured, the spindles should be parallel to the length of the block or at a right angle to the length of the block. When the gauge is properly positioned, lock the spindles by turning the lock thimble clockwise.

Ring travel is the area where the piston's sealing rings move up and down in the cylinder.

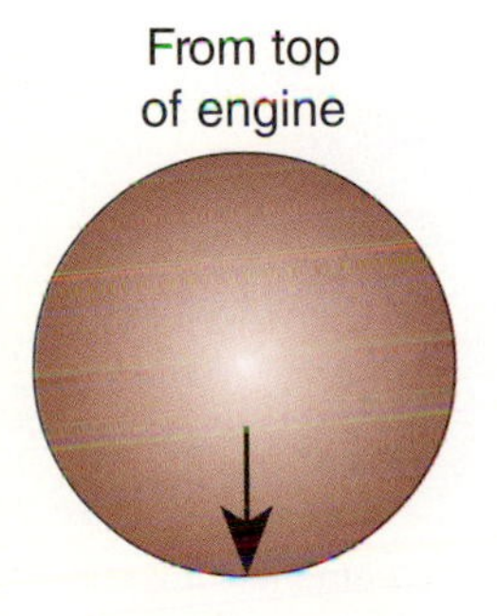

FIGURE 4-19 **Thrust wear will cause a bore to elongate or become somewhat egg-shaped.**

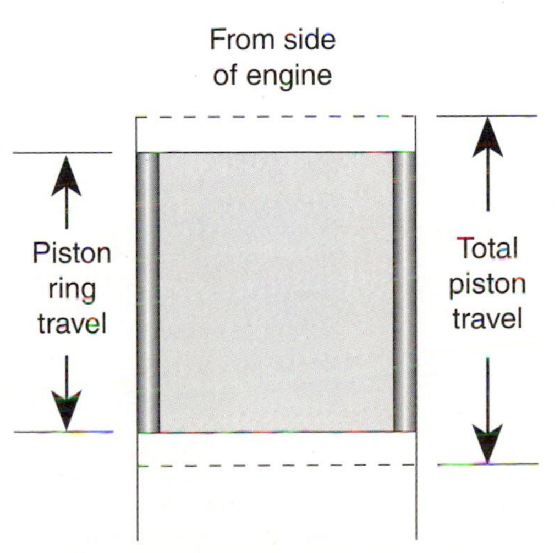

FIGURE 4-20 **The piston's rings are mounted just below the top of the piston, making a smaller wear pattern than the amount of piston travel would seem to indicate.**

PHOTO SEQUENCE 7

Typical Procedure for Reading a Dial Caliper

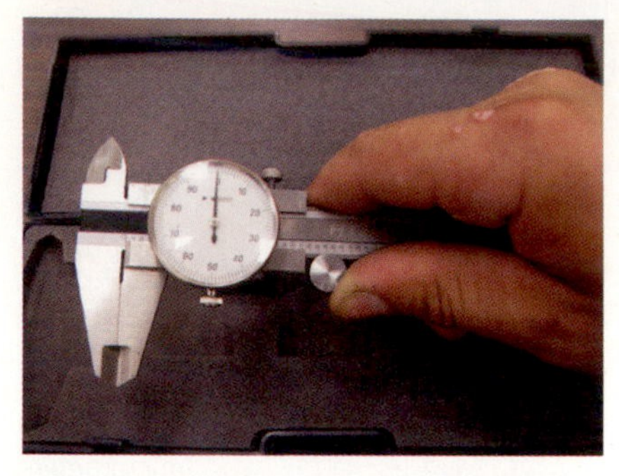

P7-1 **Set the dial caliper in your right hand so that the gauge can be easily read and your thumb is on the wheel.**

P7-2 **Make sure the jaws are clean and close them all the way. Zero the gauge again if needed.**

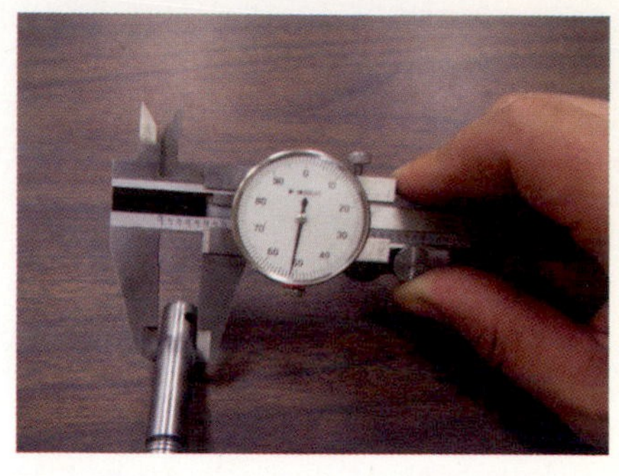

P7-3 **Using your thumb to slide the jaws of the caliper, move it into the position desired to measure.**

P7-4 **Make sure that you are using the beveled edge of the caliper jaws as the contact point.**

P7-5 **Once you have taken your measurement, lock it in place and remove it from the part. This makes reading of the measurement easier.**

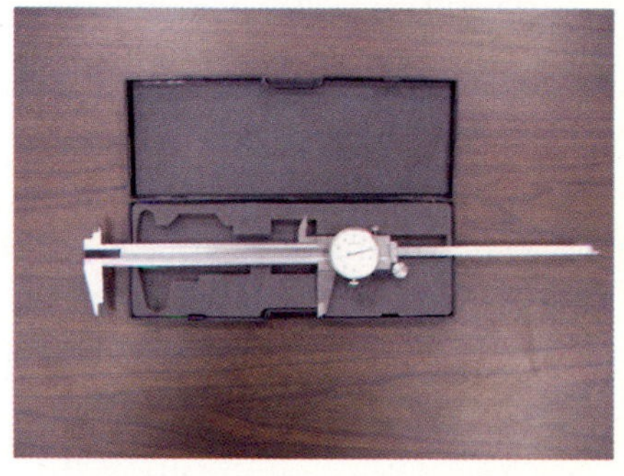

P7-6 **The dial caliper being used in this example has a range from 0 to 6 inches. Each division between the inches is 1/10th of an inch (0.1 inch).**

P7-7 **View and record the measurement on the sliding ruler. In this example, the caliper reads 0.4 inch.**

P7-8 **After reading the inches and first decimal place, you can read the dial. In this example, each mark on the dial measures 1/1000th of an inch (0.001 inch). This will give you the second and third decimal places. In this example, the dial reads 55. That means that 0.055 inch is added to the first number.**

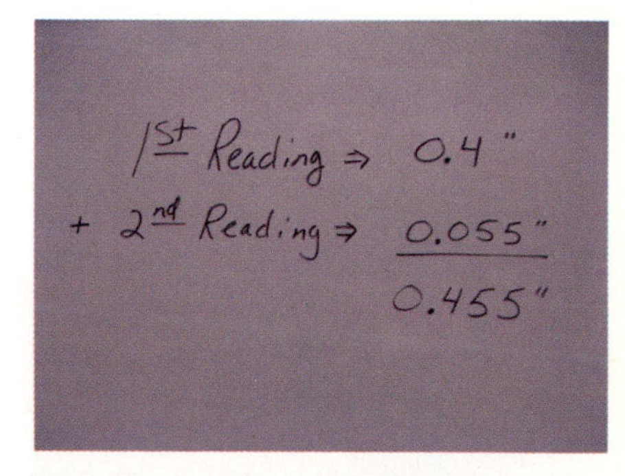

P7-9 **The total measurement is the sum of all the recorded numbers.**

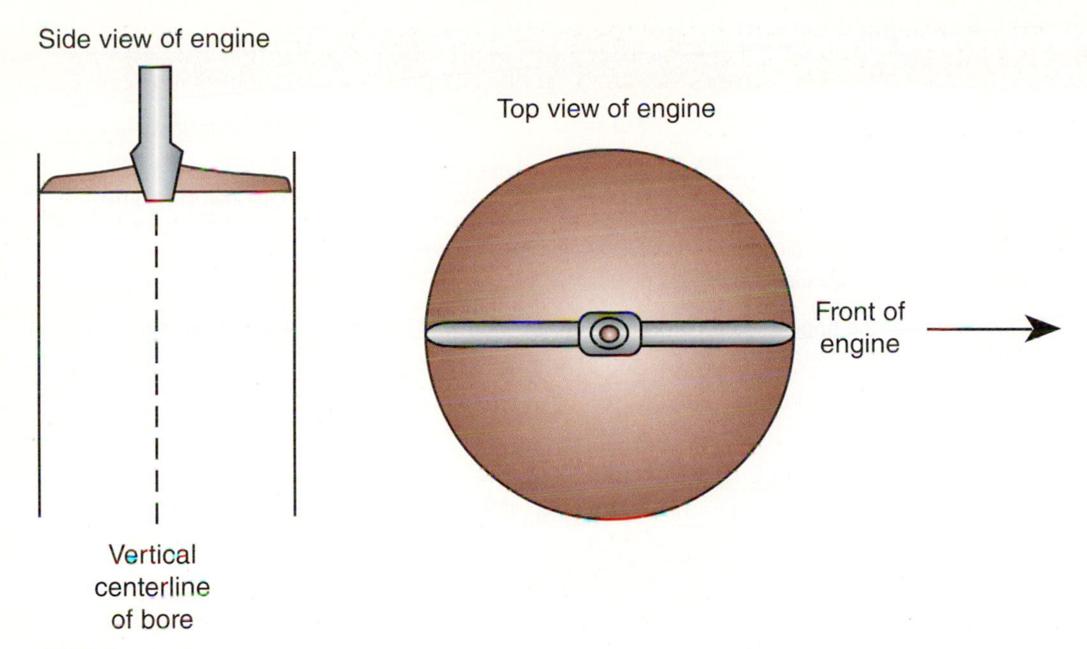

FIGURE 4-21 **The gauge must be square in the bore for an accurate measurement.**

Remove the gauge from the bore and place it between the measuring faces of the micrometer (Figure 4-22). Rotate the thimble and ratchet until the faces are set against the ends of the spindles. Lock the micrometer, remove the telescoping gauge, and read and record the measurement. A second measurement must be made to determine if the bore is out-of-round.

The same setup will be used for the second measurement except that the gauge is turned 90 degrees from the first measurement position within the bore (Figure 4-23). The measurement should be at the same height as the first measurement. Position the telescoping gauge, lock it, and remove it from the bore. Measure the distance between the spindles' ends with the micrometer and record the measurement.

If one of the two measurements is smaller than the other, the bore, like the journal, is out-of-round and has a slight oval shape. Most bores have a little more

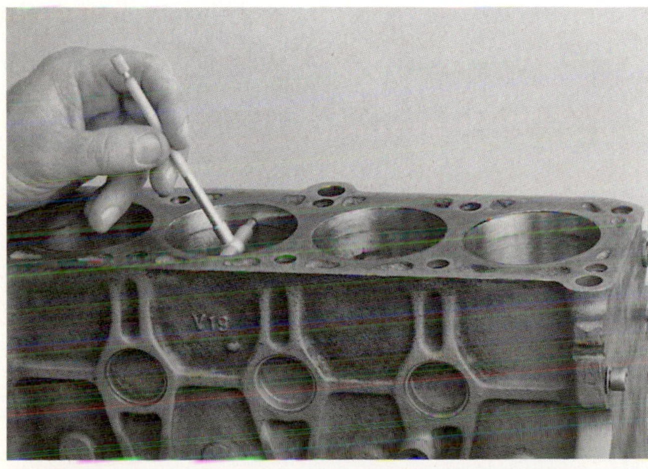

FIGURE 4-22 **Use an outside micrometer to measure the extended telescoping gauge. Note the placement of the gauge within the bore in the center of the figure.**

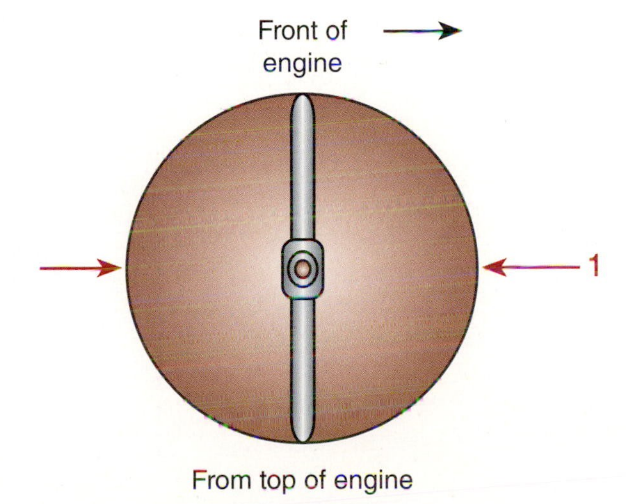

FIGURE 4-23 **Like the shaft, the second measurement of the bore is at a right angle to the first.**

Cylinder Bore Specifications

Engine	Standard	Maximum
3.9 L	3.775 inch (98.9 mm)	3.780 inch (96.0 mm)*
5.2 L	3.998 inch (101.5 mm)	4.003 inch (101.7 mm)*
5.7 L	4.010 inch (101.9 mm)	4.015 inch (102.0 mm)*

*Maximum out-of-round on cylinder bores is (1/2) 0.003 inch (0.007 mm) of standard.

Note: Chart does not indicate accurate specifications.

FIGURE 4-24 **Each of the engines listed allows only 0.005-inch (0.01 mm) wear and a maximum 0.003-inch out-of-round.**

SERVICE TIP: There is an old saying in carpentry: "Measure twice, cut once." The same saying applies to automotive measuring.

SPECIAL TOOLS

Inside micrometer set

A component with machined bores

Classroom Manual

pages 98, 99

out-of-round tolerance than some shafts. Consult the service data for the allowable tolerance (Figure 4-24). Most bores can be machined and oversize components can be used.

While checking for out-of-round data, also check the specified maximum diameter of the bore. A bore that is too large will have thin walls. High compression or stress could cause the walls to crack causing serious damage.

Any shaft or bore can be measured with an outside micrometer and telescoping gauges if space and the correct measuring procedures are used. The two primary factors ensuring a true measure are the placement of the device's measuring faces and the proper reading of the scales. Costly mistakes can be made if either of the two is not done correctly.

Measuring Bore with an Inside Micrometer

Measuring a bore with an inside micrometer is a little easier and can be more accurate than the procedures mentioned above. This is because the micrometer itself is placed inside the bore and makes a direct measurement.

If the bore is more than 1 inch in diameter, select an extension spindle to fit the micrometer. If necessary, review the Classroom Manual on how to select an extension. Fit the extension to the micrometer and rotate the thimble to bring the micrometer to zero. Place the micrometer inside the bore in a position similar to the placement of a telescopic gauge (Figure 4-25). Rotate the thimble until the end of the extension and the end of the micrometer are square in the bore. Once positioned and extended to the widest diameter of the bore, lock the micrometer and remove it from the bore.

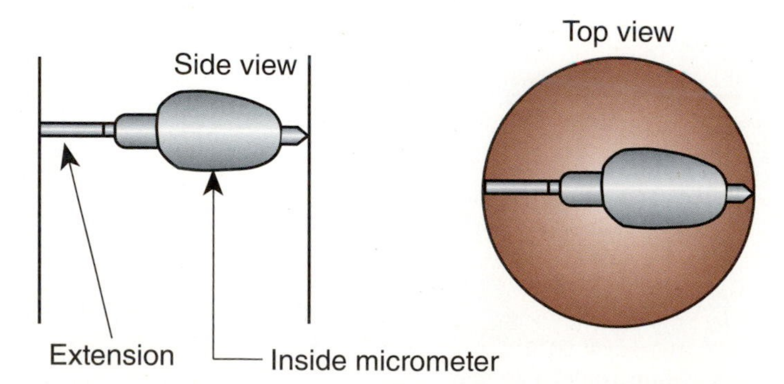

FIGURE 4-25 **An inside micrometer is placed in a position similar to the placement of a telescoping gauge.**

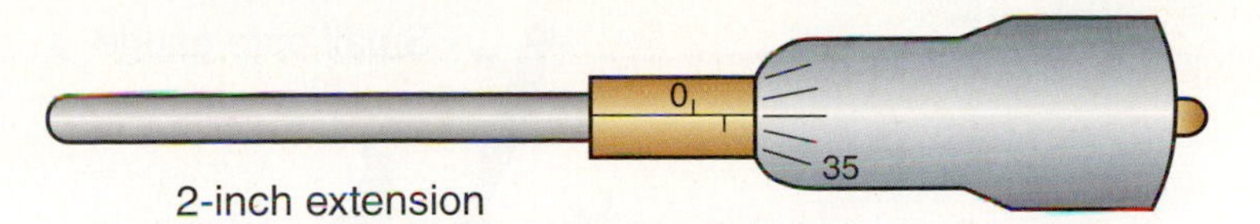

FIGURE 4-26 **Remember to add the length of the extension spindle to the reading on the micrometer.**

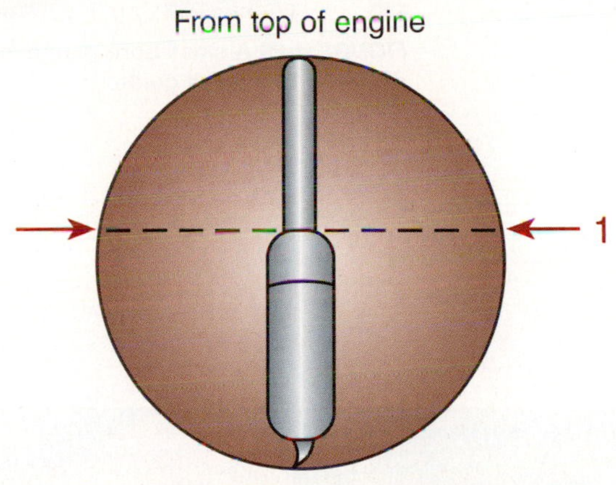

FIGURE 4-27 **A second measurement is at a right angle to the first (arrows).**

Read the measurement on the micrometer in the same manner as an outside micrometer. Remember to add the size of the extension to the left side of the decimal (Figure 4-26). For instance, if the micrometer reading is 0.037 inch and the extension is 2 inches long, then the bore is 2.037 inches in diameter.

Place the micrometer back in the bore at a right angle (90 degrees) to the first position (Figure 4-27). Position and extend the micrometer and extension to the widest point of the bore. Lock, remove, and read the measurement of the second measurement.

Like previous measurements, compare the two readings and determine if the cylinder is out-of-round. Also make a decision on any service that may be needed on the bore. Data on wear limits and guidance on repair or replacement can be found in the service manual.

Measuring Valve Guides with a Small Bore Gauge

Some objects are too small to fit an inside micrometer into. A valve guide in a cylinder is a good example of this. In order to measure a valve guide, you must place a small bore gauge into the guide and then pull it out and measure the gauge with a micrometer (Figure 4-28). Like all other procedures, this sounds simple, but in practice you will find that it is challenging.

When you place the small bore gauge into the guide, you have to tighten the bore gauge. It has a screw in it that pulls a cone shape against two metal shells. These shells spread out and open. You will tighten the gauge with your fingers only. When the screw stops moving, you have to pull it. A technician must measure the valve guide in three different places in order to get a good reading of the amount of guide wear. These measurements are the inside diameter of the guide (Figure 4-29). You also need to measure the outside diameter of the valve itself at three places (Figure 4-30) and then

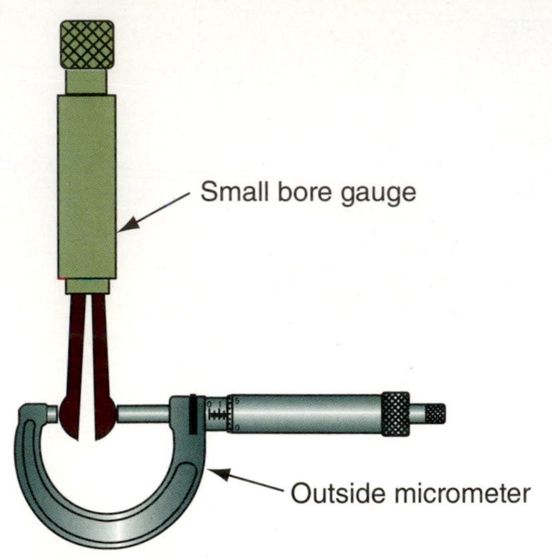

FIGURE 4-28 **Use an outside micrometer to measure the size of the small bore gauge.**

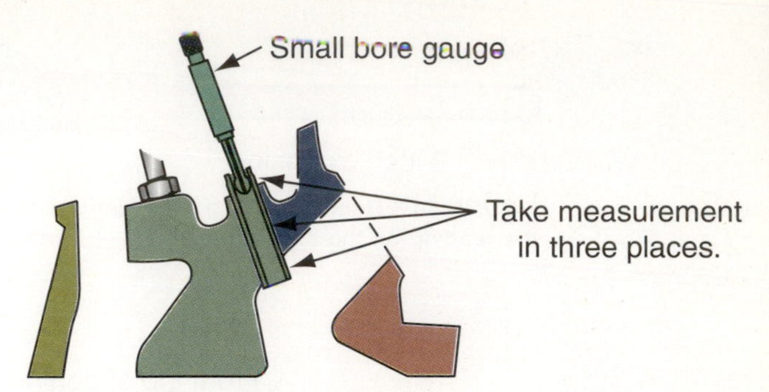

FIGURE 4-29 **A small bore gauge is used to measure the inside diameter of a valve guide.**

FIGURE 4-30 **Using an outside micrometer, measure the outside diameter of the valve stem.**

compare the outside diameter of the valve stem to the inside diameter of the guide (Figure 4-29). This measurement is known as valve stem-to-guide clearance.

SPECIAL TOOLS

Dial indicator set

A vehicle with the disc caliper and wheel assembly removed

Classroom Manual

pages 94, 95, 96, 97

Measuring Runout on a Disc

Runout is the term used to determine if there is a wobble in a wheel or disc (Figure 4-31). Generally, a disc or wheel should spin without any side-to-side movement. Movements of this type cause a vibration or shake and if serious enough could cause damage. It can be frustrating to passengers when the vibration is transmitted throughout the vehicle. In most cases, technicians do not specifically measure runout on a brake disc, but it is a good training device.

For this job, we will use a brake system disc mounted on the vehicle, which allows an easy mounting area for the disc and the dial indicator. The dial indicator requires a mounting bracket that can be attached to a solid portion of the engine block. It may be a *clamp type* or a *magnetic type* (Figures 4-32 and 4-33). For this setup, a clamp should be used. Sometimes the shape and position of the mounting area may be such that only one type of mount can be used. Regardless of the mount type, ensure that it is securely fastened so the dial indicator can be moved and positioned properly against the component being measured. The dial indicator should be the balanced type, but a continuous type can be used.

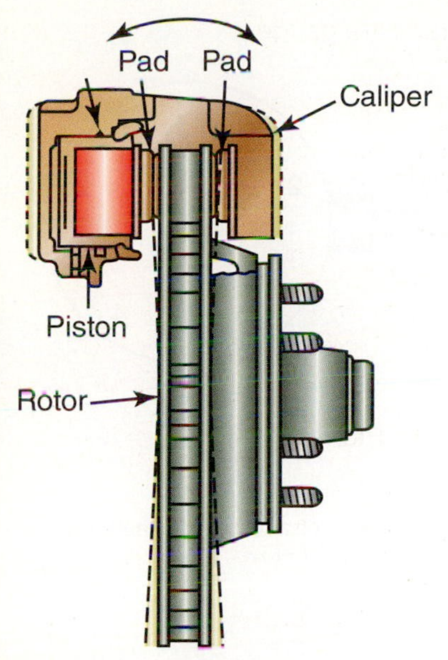

FIGURE 4-31 **The side-to-side movement of the brake rotor will cause a brake pedal pulsation. This is known as "runout."**

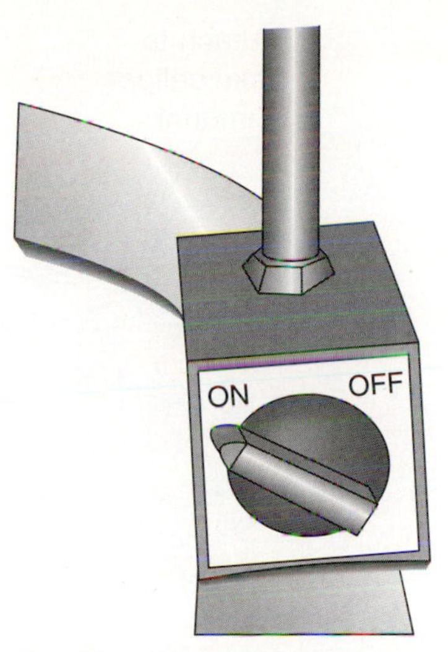

FIGURE 4-32 **A dial indicator with a magnetic mount. The magnetism in this mount can be switched on or off for easier setup and can be fitted to a round component as shown in the figure.**

FIGURE 4-33 **The dial indicator can be mounted with an adjustable locking pliers as well.**

Set the disc on the hub and install at least half of the lug nuts. The nuts only need to be tight enough to hold the disc firmly in place. Once the disc is positioned, mount the dial indicator bracket to the brake caliper mount. Assemble the various swivels and rods so the links reach over the disc (Figure 4-34). Before mounting the dial indicator to the mount, release the indicator's plunger completely. Install the indicator on the mounting rod and shift the linkage until the plunger is resting against the disc about midway from the machined area (Figure 4-35).

Move the dial indicator toward the disc until about half of the plunger is retracted into the indicator. At this point, tighten all of the mount's swivel or links to hold the dial indicator in place. Hold the dial indicator with one hand and adjust the scale until the zero mark is directly under the needle (Figure 4-36). It may take a few tries to get it correct.

When the dial indicator is firmly mounted, the plunger is partially retracted, and the scale is set to zero, the measurement can be taken. Using a screwdriver between the wheel lugs or by hand, rotate the disc in one direction slowly and evenly. Sharp, sudden movements should be avoided. Instead, try to make a smooth, continuous rotation to

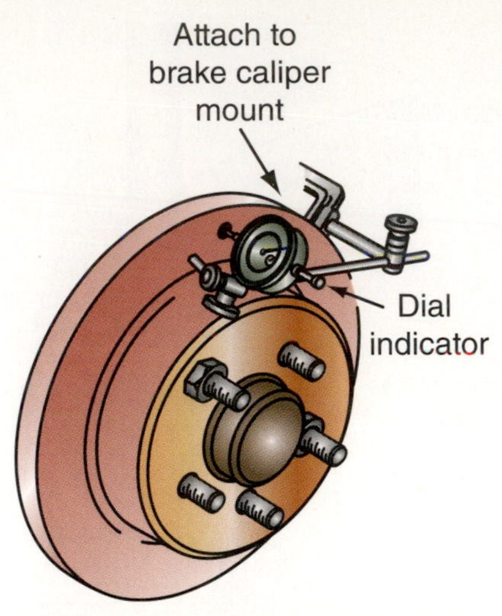

FIGURE 4-34 **The dial indicator brackets allow the indicator to be positioned for an accurate measurement. Other types of mountings are also available.**

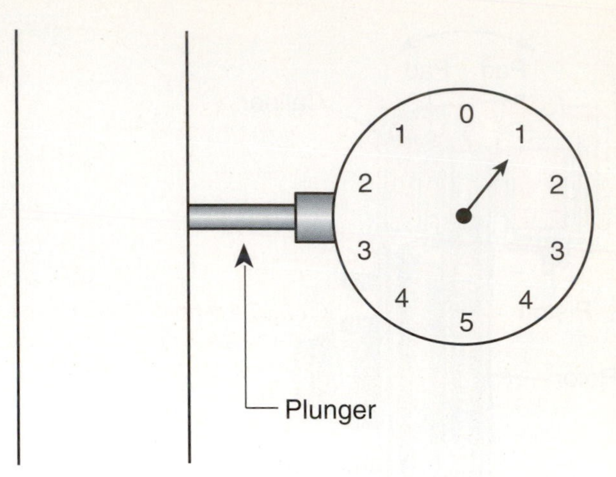

FIGURE 4-35 **With the plunger retracted about halfway, the needle will probably not be at zero.**

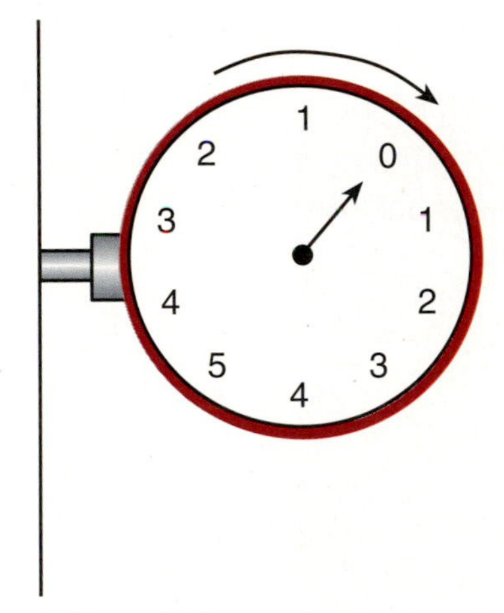

FIGURE 4-36 **Rotate the knurled ring to turn the scale so that zero is under the needle. The indicator will have to be held in place.**

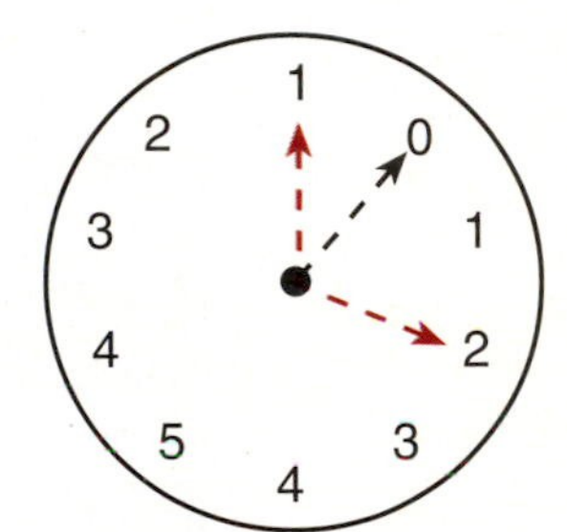

FIGURE 4-37 **The needle may move to the left and right of zero as the disc is rotated.**

prevent moving the dial indicator. As the disc is rotated, read the dial indicator's scale for movement of the needle (Figure 4-37).

The needle will probably move to the left and right of zero (Figure 4-38). Generally, any movement exceeding 0.002 inch indicates the disc needs to be machined or its mountings checked. On a balanced scale, the reading can be read from the numbers directly under the needle. If a continuous dial indicator is used and the disc's rotation caused the needle to move counterclockwise, the marks will have to be counted. Start at zero and count counterclockwise until the mark directly under the needle is reached. Check the size or calibration of the scale. It will be marked on the face as either 1/1,000 or 1/10,000 inch. Each mark on the scale represents one of the calibration units. Usually, it is best to make the measurement two or three times to ensure an accurate reading.

There are several flat, disc-type components, including the brake disc and flywheel, which should be measured for runout. A warped component that is supposed to be flat will cause its mated component to vibrate and shake, thereby resulting in driver discomfort and possible damage.

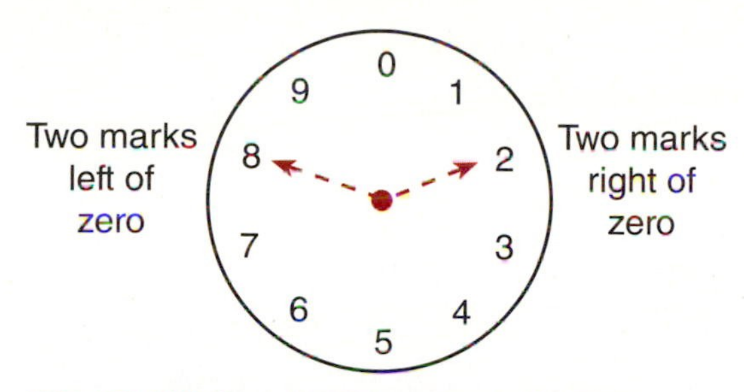

FIGURE 4-38 **A continuous scale dial indicator requires the technician to count the number of graduations the needle moved to the left of zero.**

SPECIAL TOOLS

Brake disc

Brake caliper (micrometer)

Using Brake Calipers to Measure the Thickness of Brake Discs

Classroom Manual
pages 97, 98

During brake repair, the disc (or **rotor**) should be checked for excessive wear and warpage. The procedure discussed above will give the technician an idea of the flatness of one side of the rotor, but a more common measurement is the thickness and parallelism of the two opposing friction areas. Thickness can be measured with a regular outside micrometer or a special brake disc micrometer (Figure 4-39). The primary difference between these micrometers is quick thimble action and larger numbers on the brake micrometer. This makes it easier and quicker to use. Warpage or parallelism between the two opposing friction areas can also be checked with a micrometer or a disc parallel gauge.

Brake disc thickness and parallelism can be checked on the car, a table, or a brake lathe. Find the minimum thickness specification for the disc. Many times it is stamped on the inside or outside of the hub portion of the disc and most are listed as metric measurements (Figure 4-40). If the brake disc micrometer is USC, convert the specification to USC.

Place the micrometer's measuring faces on each side of the disc midway to the wear area (Figure 4-41). Rotate the thimble until the spindle and anvil faces are against the disc. Read and record the measurement. In almost all cases, the measurement will be less than 1 inch. Repeat the measurement at a minimum of 12 different points around the disc (Figure 4-42). Remember to measure at the midpoint of the friction areas at each of the measuring points. Compare your readings to the specified minimum thickness. Any measurement that is less than specified means the disc is not to be reused and should be discarded. Differences over 0.002 inch between any of the 12 points mean the disc's friction areas are not parallel. If the disc is not too thin, it can be machined and reused.

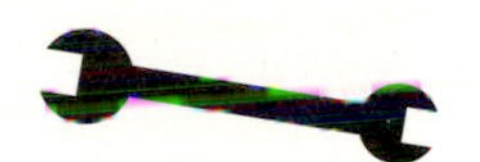

SERVICE TIP:
In the last few years, brake rotors have an additional minimum specification: *minimum refinishing thickness.* If the rotor is at that thickness or thinner, then the rotor is replaced. The minimum refinishing limit is usually greater than the minimum or discard thickness. The refinishing specification is found in the service manual and usually not stamped on the rotor.

FIGURE 4-39 **An electronic brake (rotor) micrometer.**

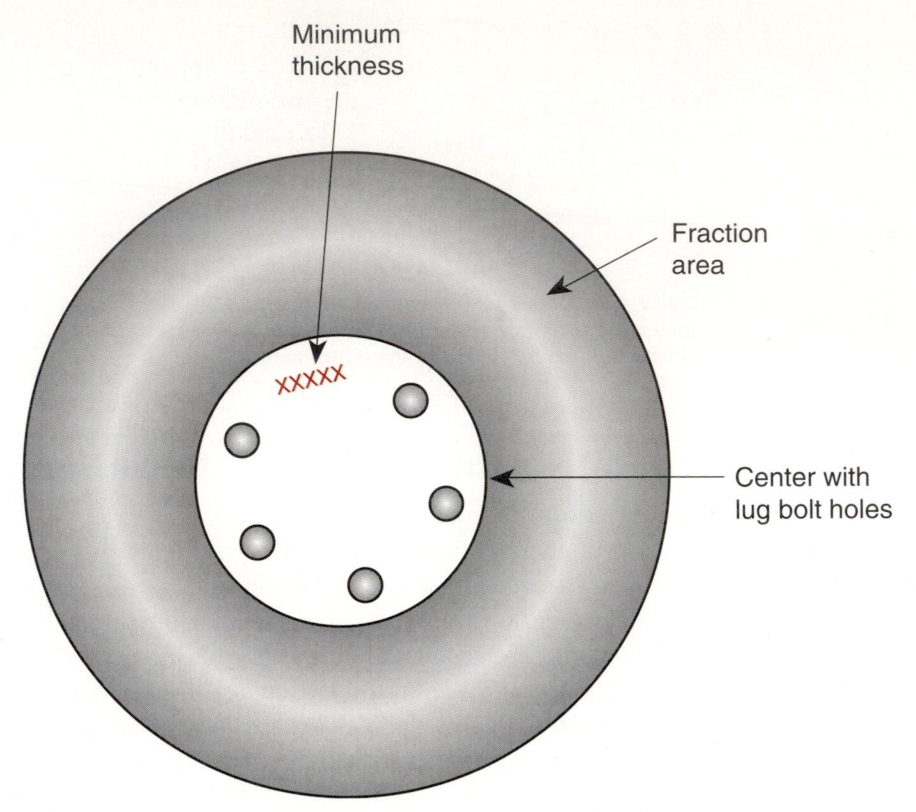

FIGURE 4-40 **The minimum thickness is usually on the disc, but it can also be found in the service manual.**

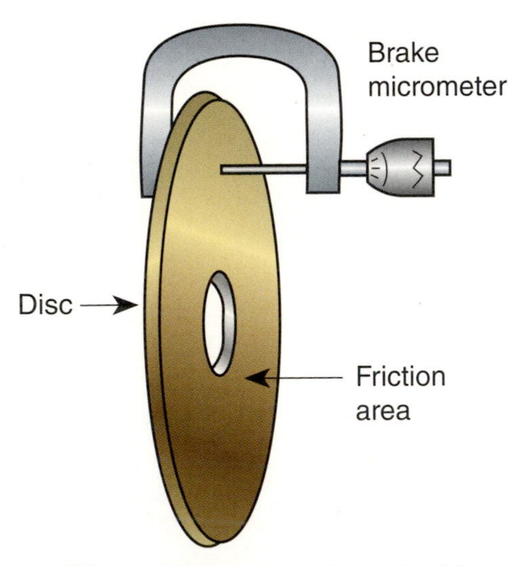

FIGURE 4-41 **The measuring faces should always be at the halfway point of the friction area. The outer edges may be thinner or have a ridge.**

FIGURE 4-42 **Check for thickness variation by measuring 8 to 12 points around the rotor.**

Measuring Shaft Endplay

> **AUTHOR'S NOTE: The easiest component to use for this exercise is either a manual or an automatic transmission/transaxle input (turbine) shaft.**

FIGURE 4-43 **This dial indicator has been mounted on a transaxle prior to measuring shaft endplay.**

SPECIAL TOOLS

Dial indicator with clamp

Assembled manual transaxle

Transaxle holder or table

SPECIAL TOOLS

Tire pressure gauge

Vehicle

Classroom Manual
pages 100, 101, 102

For the purpose of this section, an automatic transmission will be used and the input shaft endplay will be measured. Determine the type of transmission either from a transmission ID tag or by researching the type of vehicle from which it was removed. Use the service information available and locate the specification for the shaft's maximum endplay. This measurement is one of the very first steps in disassembling any transmission or transaxle. Place the transmission in the holder or solidly on a rigid table. If using a table, it is suggested that blocks of wood or an assistant keep the transaxle from shifting during the measurement. Since most transaxle cases (housings) are made from aluminum, it will be necessary to use a dial indicator with a clamping mount. Before mounting the dial indicator, either push or pull the input shaft until it stops. It does not make any difference whether the shaft is all the way in or all the way out. However, experience seems to suggest that pulling the shaft out makes the final movement easier.

Using the clamp mount over the lip of the transaxle bell housing, position the dial indicator plunger over the center of the outward end of the shaft (Figure 4-43). Be careful not to disturb the position of the shaft. Clamp the mount to the housing while keeping the plunger in the general area of the shaft's end. Flex the dial indicator support arm (bracket) until the plunger is directly on the shaft end and approximately halfway in. Tighten all fasteners or clamps to secure the dial indicator in place.

CAUTION:
Never add air to a large truck or equipment tire that is low on air. The design of some large wheel rims can allow the tire or locking equipment to become dislodged and can erupt in a dangerous explosion. Very serious injury or death can occur. The wheel assembly must be removed and the tire broken loose from the rim before inflating. The wheel assembly must also be positioned in a safety cage before inflating.

With the dial indicator locked in place, loosen the dial lock and rotate the graduated scale until the zero number is directly under the dial's needle (Figure 4-44). Release and recheck the position of the zero to the needle. If correct, lock the scale in place. Have an assistant hold the transaxle steady while the shaft is pushed (pulled) to its opposite stop. Read the measure from the dial indicator scale. If it is set up properly and the shaft started at its outmost position, the needle should have moved clockwise and the measurement can be read directly (Figure 4-45). If the shaft started at its innermost position, it may be necessary to count the graduations. Compare the measures and determine if the endplay is within specified limits. Photo Sequence 8 shows the steps in this shaft measurement.

Measuring Pressure

Pressure on an automobile is normally measured on the fuel, oil, air conditioning, power steering, brake, and automatic transmission systems. With antilock brake systems being used on most vehicles, special gauges are used to check brake pressure. Pressure tests on

Air conditioning systems use pressure to force the refrigerant to change its physical state from liquid to vapor and then back to liquid again.

Tire pressure directly affects the fuel economy, handling characteristics, and riding comfort of the vehicle.

CAUTION:
Never exceed the tire's maximum pressure rating. An explosion could occur if the tire splits open and injury or death could occur. The pressure rating can be found on the side of the tire.

CAUTION:
Never stand directly to the side of the wheel assembly. A tire, new or old, may have a weak spot that will burst with a highly explosive effect, thereby causing injury or death. Technicians should stand beside the fenders and attempt to check or adjust pressure with only their hands near the tires. Many tire pressure gauges have long metal reaches, thus moving the technician away from the danger.

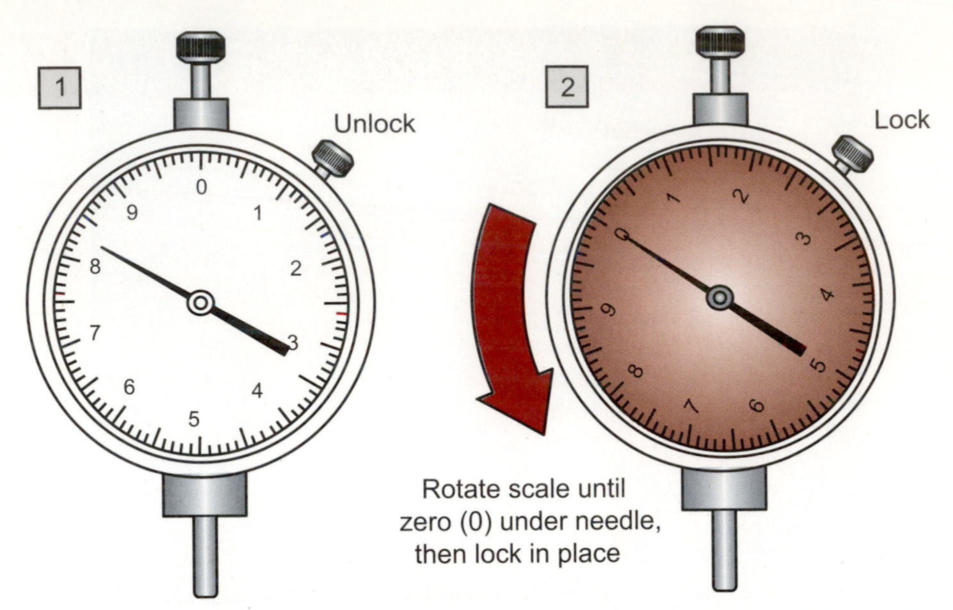

FIGURE 4-44 **The scale must be rotated so that the zero mark is aligned with the needle.**

FIGURE 4-45 **After moving the shaft to the opposite position, the needle will move clockwise. The total endplay can now be read.**

these systems are used to detect malfunctions in fluid delivery. Tapping into the fluid flow in some way is the method to perform any pressure test (Figure 4-46). Gauges and taps must be selected depending on the system being tested. Gauges may be either a pressure/vacuum type, measuring only up to 10 psi, or a pressure type that measures up to 200 psi or more. Normally, the pressure or vacuum gauges are used only on carburetor fuel systems. Almost all other systems require gauges that can read high pressures. Only an experienced technician should take pressure measurements, especially on the high-pressure systems. For our discussion, we will measure the pressure of a tire.

To obtain a correct tire pressure measurement, the tire must not have been driven for more than 2 or 3 miles. In other words, it must be "cold." Driving heats the air in the tire, thereby increasing pressure. With the vehicle on the ground or floor, remove the cap from the tire's valve stem.

PHOTO SEQUENCE 8

Measuring a Shaft's Endplay

P8-1 **Position the transmission so the input shaft is facing up.**

P8-2 **Mount a dial indicator to the transmission case so the indicator's plunger can move with the input shaft as the shaft is pulled up and pushed down.**

P8-3 **Lightly pull up on the shaft, then push it down until it stops. Then zero the indicator.**

P8-4 **Pull the shaft up until it stops and then read the indicator. This is the total amount of the endplay.**

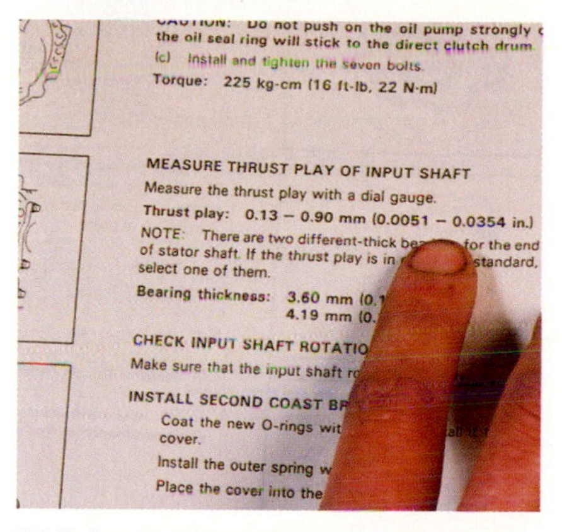

the oil seal ring will stick to the direct clutch drum.
(c) Install and tighten the seven bolts.
Torque: 225 kg-cm (16 ft-lb, 22 N·m)

MEASURE THRUST PLAY OF INPUT SHAFT
Measure the thrust play with a dial gauge.
Thrust play: 0.13 – 0.90 mm (0.0051 – 0.0354 in.)
NOTE: There are two different-thick be[...] for the end of stator shaft. If the thrust play is in [...] standard, select one of them.
Bearing thickness: 3.60 mm (0.[...]
4.19 mm (0.[...]

CHECK INPUT SHAFT ROTATIO[...]
Make sure that the input shaft r[...]

INSTALL SECOND COAST BR[...]
Coat the new O-rings wit[...]
cover.
Install the outer spring w[...]
Place the cover into the[...]

P8-5 **Locate the endplay specifications in the service manual.**

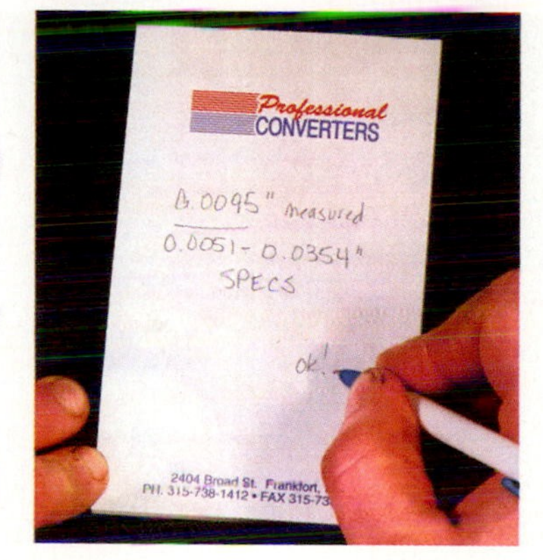

P8-6 **Compare your measurements to the specifications.**

Place the connection end of the gauge firmly over and down the end of the valve stem (Figure 4-47). There should be no noise of escaping air around the gauge. Read the pressure with the gauge held in place. A typical passenger car should have pressure between 32 and 35 psi. If the reading is high, remove the gauge and release some air from the tire. If too low, add air through the valve stem. Do not exceed the tire's maximum allowable pressure.

Measuring Engine Vacuum

Measuring vacuum on an automobile engine can tell the technician a great deal about the internal condition of that engine. The downward movement of the pistons within their cylinders creates engine vacuum. Anything that causes an engine's vacuum to drop will result in lower performance. The causes and results of improper engine vacuum will be covered in Chapters 7 and 8.

SPECIAL TOOLS

Vacuum gauge or pressure and vacuum gauge

Vehicle

WARNING: An experienced technician or instructor should guide the student or new technician through this section the first time. Disruption to the engine's performance or injury to the technician could occur if skill, attention to detail, and safety procedures are not used.

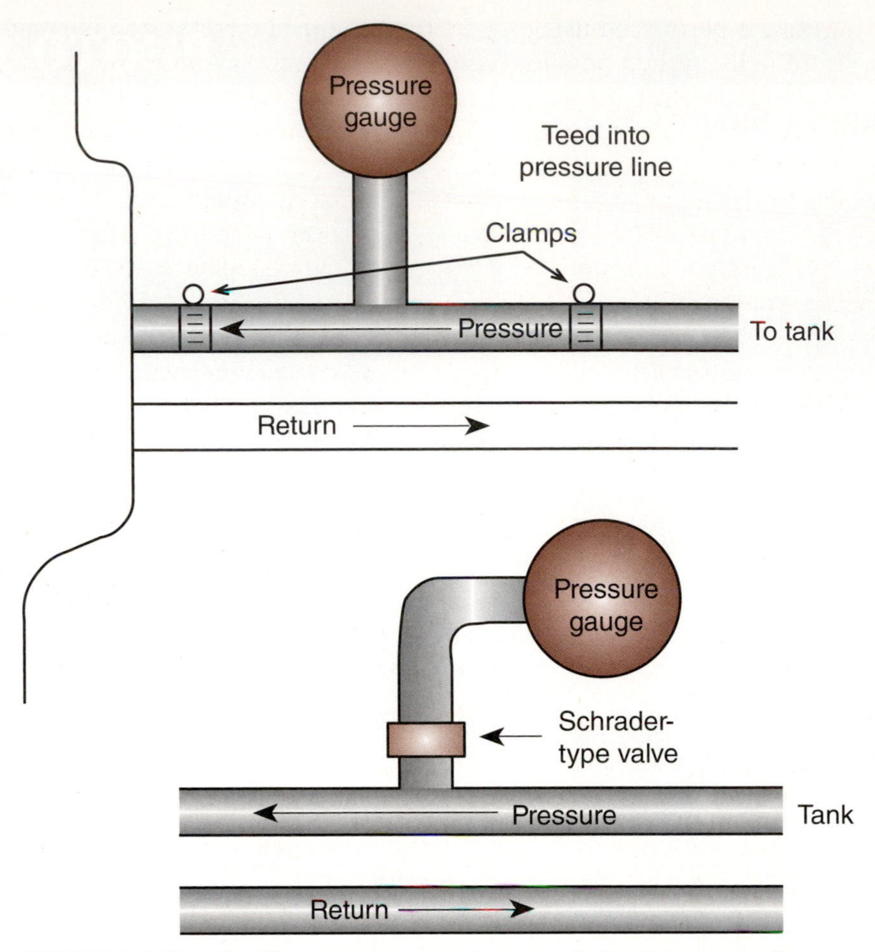

FIGURE 4-46 **Note that the pressure gauge is connected to the pressure line. This is true in almost every measurement taken on any hydraulic system.**

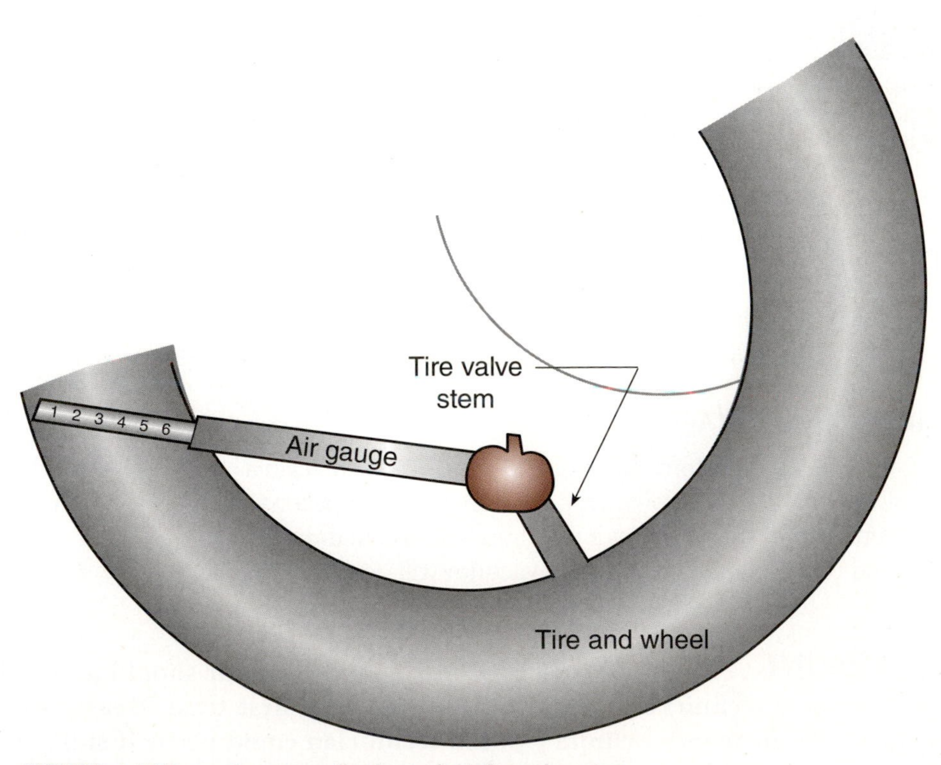

FIGURE 4-47 **Press the gauge firmly against the valve stem.**

A vacuum test is performed using a port that is tapped directly into the engine's intake manifold (Figure 4-48). When possible, a multiple connection, sometimes known as a *vacuum tree,* should be used for the gauge's connection (Figure 4-49). If an empty port is not available, the gauge will have to be teed into one of the vacuum lines from the intake manifold. If a vacuum line is disconnected and a gauge is put in its place, the device operated by the vacuum line may cause a malfunction in engine operations and cause errors in the measurements. This can lead an inattentive technician in diagnostic circles.

Classroom Manual
pages 102, 103

With the vacuum gauge connected, start the engine and allow it to warm. Observe the vacuum measurements at idle and high idle engine speeds (Figures 4-50 and 4-51). Readings between 17 and 20 in. Hg at idle are normal. Note that there will be a drastic drop in vacuum as the engine is accelerated. A steady, below-specified vacuum or a fluctuating needle normally denotes an engine needing repairs. Most scan tools have the ability to display the engine vacuum by measuring it on the engine's manifold absolute pressure (MAP) sensor (Figure 4-52).

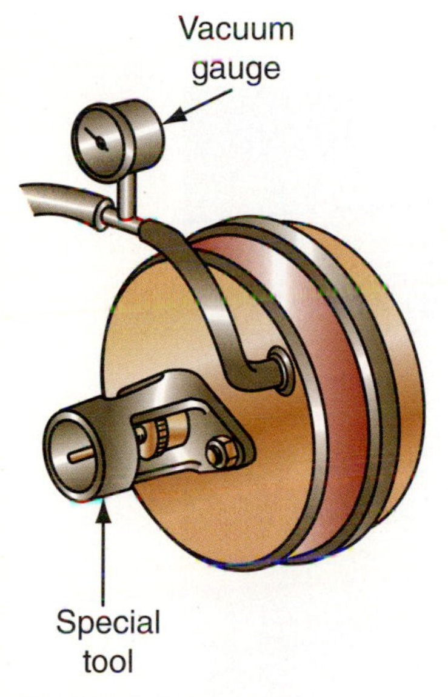

FIGURE 4-48 **This vacuum gauge is connected to the brake booster and can measure vacuum supply or engine vacuum (depending on the setup).**

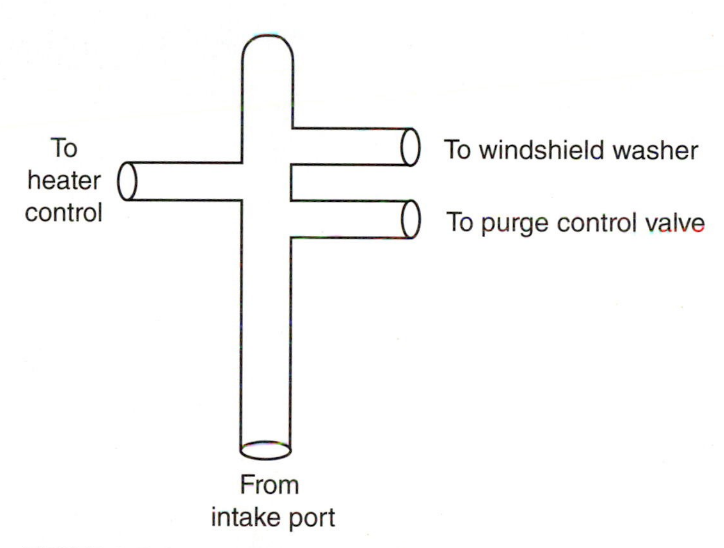

FIGURE 4-49 **A *vacuum tree* is a small plastic component with connection ports for vacuum hoses. It is usually found on the firewall.**

FIGURE 4-50 **This engine has a healthy 18 in. HG at idle.**

FIGURE 4-51 **Vacuum gauge readings can indicate an engine's running condition.**

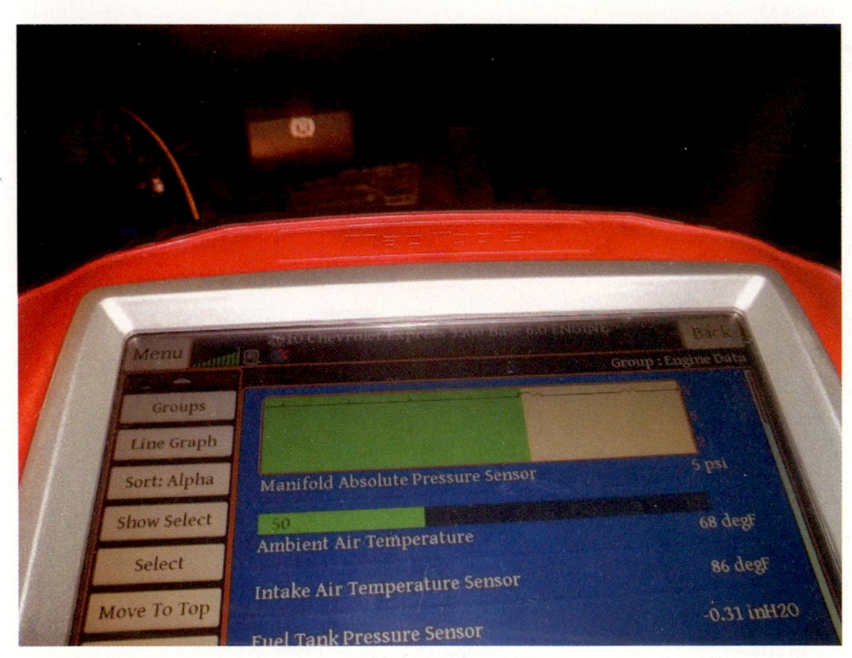

FIGURE 4-52 **Reading engine vacuum on a scan tool.**

TERMS TO KNOW

Bore
Dial indictor
Journal
Out-of-round
Ring travel area
Rotor
Straight edge
Thrust wear
Toothed ring
Toothed wheel

CASE STUDY

An engine was checked for a poor idle condition. One of the tests was a measurement of the engine's vacuum. A vacuum gauge was connected to an intake port for the adjustment. As soon as the vacuum line was disconnected and the gauge connected, the engine began to run smoothly. A closer inspection found that the hose was connected to a small vacuum motor in the heating system and was disconnected. This allowed a substantial idle leak and leaned the air/fuel mixture. However, it was not enough of a leak to cause a noticeable difference at cruising speeds. The disconnected hose was under the dash, and road noise and the radio muffled the sound of the leak.

ASE-STYLE REVIEW QUESTIONS

1. Pressure and vacuum measurement is being discussed. *Technician A* says a pressure gauge is used to check some brake systems.

   Technician B says an engine vacuum test may indicate the problems with a poorly performing engine.

   Who is correct?

   A. A only C. Both A and B
   B. B only D. Neither A nor B

2. *Technician A* says pressure may be used within the steering system.

   Technician B says tire pressure must be checked when the tire is cold to obtain an accurate measurement.

   Who is correct?

   A. A only C. Both A and B
   B. B only D. Neither A nor B

3. *Technician A* says any out-of-round wear in a cylinder requires the replacement of the component with the cylinder.

   Technician B says an out-of-round measurement can be made with just a telescoping gauge.

   Who is correct?

   A. A only C. Both A and B
   B. B only D. Neither A nor B

4. Measurement systems are being discussed.
   Technician A says a metric micrometer may be used to measure a USC crankshaft journal.
   Technician B says a conversion factor is used to convert measurements between measuring systems.

   Who is correct?

   A. A only C. Both A and B
   B. B only D. Neither A nor B

5. Cylinder measurements are being discussed.

   Technician A says a larger-than-specified measurement may indicate an out-of-round cylinder.

   Technician B says only one measurement is made to determine if a cylinder has an out-of-round condition.

   Who is correct?

   A. A only C. Both A and B
   B. B only D. Neither A nor B

6. All of the following will require the technician to machine an engine block EXCEPT:

   A. Cylinder wall has minor scratches
   B. Cylinder is 0.003 inch out-of-round
   C. Cylinder has over 0.006 inch of wear
   D. Cylinder is within out-of-round and wear specifications

7. *Technician A* says a bearing journal is measured twice.

   Technician B says the journal measurements are performed 180 degrees apart.

   Who is correct?

   A. A only C. Both A and B
   B. B only D. Neither A nor B

8. Shaft measurements are being discussed.

   Technician A says a measurement smaller than specified requires the shaft to be machined or replaced.

   Technician B says diameter measurements larger than those specified usually indicate wrong data or a poor measurement.

   Who is correct?

   A. A only C. Both A and B
   B. B only D. Neither A nor B

9. Dial indicators are being discussed.

   Technician A says a dial indicator may have a scale that is numbered from zero and goes up in two different directions.

   Technician B says a brake disc caliper is a modified dial indicator.

   Who is correct?

   A. A only C. Both A and B
   B. B only D. Neither A nor B

10. Micrometers and dial indicators are being discussed.

    Technician A says a dial indicator can be used without a mount.

    Technician B says the use of a micrometer requires a clamping mount.

    Who is correct?

    A. A only C. Both A and B
    B. B only D. Neither A nor B

Name ______________________ Date ____________

MEASURING DEPTH WITH A DEPTH MICROMETER

AUTHOR'S NOTE: The author uses two blocks of scrap metal grooved by the machine tool technology class. Grooved and planed blocks of wood or an engine with cylinder heads removed will work just as well with minor changes in the procedures.

Objective

Upon completion and review of this job sheet, you should be able to measure the depth of a groove cut into a block or the depth of an object within a bore.

Tools and Materials Needed

Component or material with grooves
OR
Engine with cylinder head removed and pistons installed
Depth micrometer

Procedures

Task Completed

1. Secure the material to be measured so it will not move as the measurement is made. ☐
2. Select a depth micrometer and, if necessary, extension spindles. ☐
3. If there is specification data available, record it below. Space is also available to record the actual measurement. ☐

First Groove Specification	Actual
____________	____________
____________	____________
____________	____________

Second Groove Specification	Actual
____________	____________
____________	____________
____________	____________

Third Groove Specification	Actual
____________	____________
____________	____________
____________	____________

4. Fit the extension to the micrometer as needed. ☐
5. Retract the micrometer until it will not reach the bottom of the groove when it is positioned. ☐

Task Completed

☐ 6. Ensure that the surface where the micrometer is to be set down is flat and smooth.

☐ 7. Place the micrometer over the hole with the spindle pointing into the hole.

☐ 8. Position the base of the micrometer so it extends over two edges of the groove.

☐ 9. Holding the micrometer firmly in place by hand, rotate the thimble until the spindle contacts the bottom of the groove.

☐ 10. Lock the micrometer and remove it from the material for easy reading.

☐ 11. Record the measurement.

☐ 12. Perform any other measurements needed to complete this exercise and record all readings next to the specifications.

☐ 13. If specifications were available and recorded, compare them to the actual measurements and determine any needed repairs.

Instructor's Response ______________________________

____

____

____

JOB SHEET 8

Name ______________________________ Date ______________

MEASURING WITH AN OUTSIDE MICROMETER

AUTHOR'S NOTE: For this task the author uses four different shafts from transmissions. Each shaft has three to five machined areas that can be measured. Each machined area is labeled so the answers can be tracked and graded. The NATEF task is from the engine repair section.

NATEF Correlation

This job sheet addresses the following task:

Inspect crankshaft for endplay, straightness, journal damage, keyway damage, thrust flange and sealing surface condition, and visual surface cracks; check oil passage condition; measure journal wear; check crankshaft sensor reluctor ring (where applicable); determine necessary action.

Objective

Upon completion and review of this job sheet, you should be able to measure a shaft with an outside micrometer.

Tools and Materials Needed

Shafts to be measured

Outside micrometer (at least one should be metric)

Procedures

Task Completed

1. Select the correct size micrometer for the areas to be measured. ☐
2. Open the micrometer enough to slide over the area to be measured. ☐
3. Close the measuring faces until they make contact with the shaft. ☐
4. Position the measuring faces 180 degrees apart at a right angle to the shaft. ☐
5. Rotate the thimble with the ratchet until it slips. ☐
6. Slide the micrometer back and forth over shaft to ensure the micrometer is closed and squared correctly on the shaft. ☐
7. Flip the thumb lock on and remove the micrometer from the shaft. ☐
8. Read the measurement in the following manner:
   - **A.** Record the size of the micrometer, that is, 0 inch, 1 inch, 2 inches. ______________ ☐
   - **B.** Record the last number on the sleeve uncovered by the thimble. ______________ ☐
   - **C.** Record the number of thousandths of an inch (or millimeters) between the uncovered sleeve number and the edge of the thimble. ☐

   ____

Task Completed

☐ **D.** Read and add the thimble graduation aligned with the index line to the number arrived at in item C above.

____

☐ **E.** Record the total measurement in the appropriate blank at the end of this job sheet. Note that the number of blanks may not match the number of measurements. ______________________________

☐ **F.** Repeat the steps above for the other measurements.

Measurements

1. Shaft 1

   Area 1 ________ + ________ + ________ = ________ inches/millimeters
   size Sleeve Thimble

   Area 2 ________ Area 3 ________ Area 4 ________

2. Shaft 2

   Area 1 ________ Area 2 ________ Area 3 ________

3. Shaft 3

   Area 1 ________ Area 2 ________ Area 3 ________

   Use the blanks below for additional measurements.

   ________ ________ ________ ________ ________
   ________ ________ ________ ________ ________
   ________ ________ ________ ________ ________
   ________ ________ ________ ________ ________

Instructor's Response ______________________________

____

____

____

JOB SHEET 9

Name ______________________ **Date** ______________

MEASURING A SHAFT'S ENDPLAY

AUTHOR'S NOTE: For this task, the author used a Chrysler 727 automatic transmission because it was readily available and this measurement is easily done on this transmission. Any other transmission or a crankshaft mounted in an engine can be used because the procedure will be the same.

NATEF Correlation

This job sheet addresses the following NATEF task:
Measure endplay or preload; determine necessary action.

Objective

Upon completion and review of this job sheet, you should be able to measure the endplay of an automatic transmission turbine (input) shaft.

Tools and Materials Needed

Assigned automatic transaxle or transmission
Dial indicator
Transaxle holder or table
Service manual

Procedures

Task Completed

1. Mount the transaxle in a holder or secure to a table. ☐
2. Identify the transaxle type by using an on-transaxle ID tag or by the vehicle. Record the information. ☐

   ____

   ____

3. Locate the transaxle shaft endplay specification and record. ☐

   ____

   ____

4. Secure a dial indicator with a clamp-type mount and mount the indicator to the bell housing of the transaxle. ☐
5. Flex the dial indicator's cable or linkage until the dial is near the end of the shaft. ☐
6. Push the shaft in until it stops and then pull it out as far as it will go. This procedure may be reversed if desired. ☐
7. Position the dial indicator so the plunger rests against the center of the shaft's end with the plunger extended out about halfway of its full travel. Lock indicator in place. ☐

Task Completed

☐ **8.** If necessary, unlock the indicator's scale and rotate the scale until the zero graduation is directly under the needle. Lock the scale in place and recheck the position of the zero and needle alignment.

☐ **9.** Push (pull) the shaft in (out) until it stops.

☐ **10.** Read and record the measurement.

____

____

☐ **11.** Perform steps 5 through 10 again to recheck the measurement. Record the results.

____

____

☐ **12.** Do the two measurements match? ________ If not, remeasure starting at step 4. Record this measurement and determine if it matches the previous reading.

____

____

☐ **13.** Compare the measurement with the specifications and answer the following questions. Is the shaft endplay within limits? If not, what repairs are suggested by the service manual?

____

____

Instructor's Response ____

____

____

____

JOB SHEET 10

Name ______________________ Date ______________

Measuring Cylinder Diameter

NATEF Correlation

This job sheet addresses the following NATEF task:

Inspect main and connecting rod bearing for damage and wear; determine necessary action.

Objective

Upon completion and review of this job sheet, you should be able to measure a crankshaft main bearing journal for wear.

Tools and Materials Needed

Assigned crankshaft removed from engine

Outside micrometer, either 1 to 2 inches or 25 to 50 mm (or micrometer appropriate to bearing diameter)

Three V-blocks to hold crankshaft

Service manual

Procedures

Task Completed

1. Mount the crankshaft in the three V-blocks. Ensure one main bearing is accessible. ☐
2. Based on the engine that housed the crankshaft, use the service manual to determine the standard size of the journal, the maximum wear, and amount of allowable out-of-round (if any) specifications. Record this information and convert the data if necessary. ☐

   ____

   ____

3. Secure an outside micrometer, USC or metric (both if available). ☐
4. Place the micrometer in position around the journal for the first measurement. ☐
5. Tighten the measuring faces to the shaft, engage the lock, and remove the micrometer from the shaft. Read and record the measurement. ☐

   ____

   ____

6. Place the micrometer measuring faces, anvil and spindle, at a position 180 degrees to the location of the first measurement. ☐
7. Lock and remove the micrometer from the shaft. Read and record the measurement. ☐

   ____

   ____

Task Completed

☐ 8. Compare the two measurements and answer the following questions. Are the two measurements exactly the same? If not, what is the difference? Does this difference indicate the journal has excessive wear? If yes, what actions are suggested by the service manual?

____

____

Instructor's Response ____

____

____

____

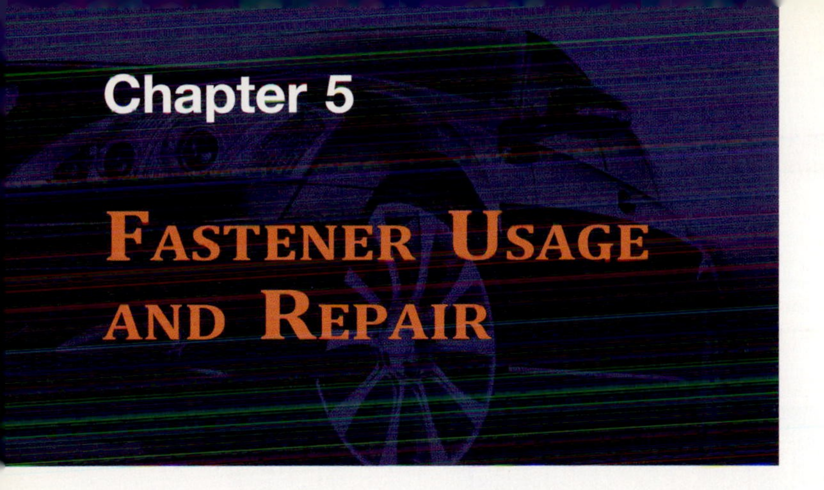

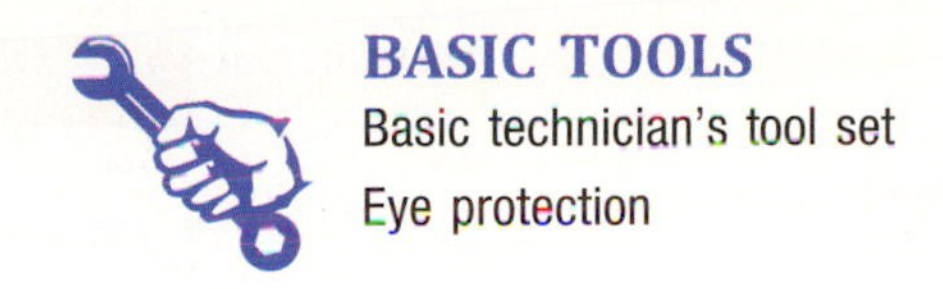

Upon completion and review of this chapter, you should understand and be able to describe:

- How to use a pitch gauge to identify U.S. (English) and metric thread sizes.
- How to install a torque-to-yield bolt.
- How to use a tap to repair damaged internal threads in an automotive component.
- How to remove and replace a stud with a stud remover.
- How to remove and replace a stud with a jam and drive nut.
- How to strip wire and install a solderless terminal and solderless butt connector.
- How to use a soldering gun or soldering iron and rosin-core solder to solder an electrical terminal to a wire.
- How to use a soldering gun or soldering iron and rosin-core solder to splice two wires together.
- How to repair connectors and terminals.

Introduction

Almost every job a technician does involves the use of fasteners. In the Classroom Manual we described the types and uses of fasteners. This chapter describes several tasks involving fasteners. It describes how to use a pitch gauge to identify threads, how to use a tap to repair internal threads, how to remove and replace a damaged stud, how to install solderless electrical connectors, and how to solder wires and electrical connectors.

Fastener Torque

One of the most critical procedures that must be followed is applying the correct amount of torque to a fastener so the fastener can do the job for which it was designed. Overtightening can strip the threads, break the fastener, or damage the components. Undertightening can cause the fastener to loosen and come apart. The bottom line is to torque to specifications.

Classroom Manual
pages 107, 108, 109, 110, 111, 112

Torque specifications do not necessarily match the size of the fastener. A physically large fastener may require less torque than expected. The torque is based on the clamping force required and the tensile strength of the fastener. In this section, we will discuss the proper method of collecting torque specifications and installing a bolt into a threaded hole.

Standard Bolts

AUTHOR'S NOTE: Two pieces of scrap metal can be used for this exercise. One piece should have a threaded hole.

SPECIAL TOOLS

Face shield
Bottom or tapered tap
Tap wrench
Hinge torque wrench
Angle indicator
Proper-sized socket
OSHA approved air blow gun
Light lubricant

CAUTION:
A face shield is the appropriate protection. Bits of metal or other material will be blown out of the hole with some force. Injuries to the face and eyes could result.

SERVICE TIP:
A common practice among senior technicians is to tighten the fastener in stages, particularly when multiple fasteners are used. For instance, with a final torque of 45 ft.-lb., each fastener is tightened to 20 ft.-lb., then to 35 ft.-lb., and finally to the specified 45 ft.-lb. This brings the two components together evenly, helping to prevent cocking of one part and a possible bad seal.

Item		ft.-lb. (N·m)
1st step	in sequence	25 (33.9)
2nd step	in sequence	35 (47.5)
3rd step	in sequence	An additional 90 degrees

FIGURE 5-1 **Many torque specifications require that a specific torque be applied in steps and that an additional amount of rotation be measured in degrees.**

FIGURE 5-2 **Typical cylinder head torque sequence.**

The first step in the installation of a bolt-type fastener is to clean the internal threads of the hole in which the bolt is to be installed. This can be done with a tap, either a bottom or tapered tap, depending on the hole. Once the tap has reached the bottom of the hole (or through the hole), use compressed air to blow the debris out.

With the threads cleaned, it is time to collect the torque specification. In a typical torque chart, a multiple step torque sequence may be listed. Figure 5-1 shows an example of a torque setting chart for a cylinder head. It is also important to note that cylinder head bolts have to be tightened in a specific sequence because of the multiple setting. Refer to Figure 5-2 to see a typical sequence for a cylinder head. For the purpose of this section, assume the following information is provided:

Intake manifold to cylinder head 35 in.-lb. (3.95 N·mm)

Note that this torque is in inch-pounds of force. This indicates a heavy clamping force is not needed, and a 1/4-inch or 3/8-inch drive torque wrench would be appropriate. A 1/4-inch drive torque wrench may be easier to set and easier to "feel" when the hinge breaks over. Thirty-five inch-pounds on a 3/8-inch drive torque wrench is a very low setting for a wrench of this size. When choosing a ratchet drive size, a good rule of thumb is to look at the torque range on the torque wrench. For example, if the range is from 25 to 125 ft.-lb., then it is safe to assume that it will be difficult to "feel" the setting when using this torque wrench on a fastener with a torque specification of 25 ft.-lb.

Apply a light coat of lubricant to the bolt threads. The lubricant should be a thin, lightweight penetrating oil such as WD-40. Select the desired wrench and set the scale to 35 in.-lb. Install the correct-size socket onto the wrench and place over the bolt head. Hold the wrench at the end or handle. This will apply the proper amount of force. Move the end of the wrench in a clockwise direction until the wrench "breaks over" or clicks. The fastener is now tightened to the proper torque. Remove and store the tools.

Torque-to-Yield

AUTHOR'S NOTE: Old torque-to-yield bolts work well in this exercise without the additional costs of new bolts.

Torque-to-yield bolts are generally tightened in the manner just described except for the final step concerning the stretch of the fastener. For this exercise, follow the steps listed here:

1st step	25 ft.-lb. (2.82 N·mm)
2nd step	35 ft.-lb. (3.95 N·mm)
3rd step	Turn an additional 30 degrees

An **angle indicator** will be needed for step 3 in addition to the socket and torque wrench (Figure 5-3). The angle indicator is used to measure the exact degree of wrench rotation. Also, a 3/8-inch drive torque wrench will probably be the best wrench for the torque required.

Experience is the best method to learn how to set up and use an **angle indicator**. The manual one is awkward to use in tight places.

Select and install the correct socket onto the torque wrench. Set the torque wrench to 25 ft.-lb. Tighten the bolt until the wrench clicks. Remove the wrench from the fastener and set it to 35 ft.-lb. Now tighten the bolt until the wrench clicks at 35 ft.-lb. Remove the socket from the wrench and install the angle indicator onto the wrench. Connect the socket to the drive end of the angle indicator.

The angle indicator has an anchor to hold its scale in place as the wrench is moved. Place the socket over the bolt head and hold in place until the indicator's anchor is positioned and locked (Figure 5-4). While still holding the wrench in place, hand rotate the indicator needle (or scale, depending on the model) until it aligns with the zero on the scale. Ensure the anchor is still in place and locked. Hold the wrench end and turn in a clockwise direction until the needle aligns with the 30-degree mark on the scale. The torque-to-yield fastener is now at its specified torque. Remove and store the tools.

SPECIAL TOOLS

U.S. (English) pitch gauge

Metric pitch gauge

Selection of U.S. (English) and metric fasteners

Note that the torque wrench was not set to a higher torque when the final 30-degree turn was made. This increased setting is not necessary and the click when the torque is applied acts as a "reminder" of what is intended. If desired, the setting may be adjusted to a higher setting without affecting the final results as long as the fastener is not turned past the 30-degree mark.

FIGURE 5-3 **An electronic angle indicator is used to measure the degree of torque.**

FIGURE 5-4 **Note the anchor (upper left) set between the two rocker arms to prevent the scale from moving. The socket is under the scale.**

Using a Pitch Gauge

Replacing damaged or lost fasteners is a common problem when reassembling an automotive component. Whenever a fastener must be replaced, be sure to get a replacement that is the exact quality and thread size as the original. One of the easiest ways to select the correct fastener is to use an original bolt or nut to determine the correct replacement size. For example, if a replacement bolt is needed, take the nut and find the bolt that fits. Make sure the replacement bolt threads into the nut. Be cautious because some metric nuts/bolts may appear to screw onto USC bolts/nuts, and vice versa. However, they will not effectively hold.

Classroom Manual
page 118

There are times when matching bolts and nuts is impossible. In these cases you may need to measure the thread on a fastener to determine its size. The tool used to do this is the pitch gauge. A pitch gauge (Figure 5-5) has a series of blades, each with teeth on one side. Each blade is marked with a thread size. The teeth on the blades are used to match up with threads on a fastener. If they match up, determine the thread size. There are pitch gauges for both U.S. (English) and metric threads.

The best way to learn how to use a pitch gauge is to practice with a bolt or hex head cap screw with a known size. This way you can get to see how the blades of the pitch gauge should match the threads on a fastener.

Begin with a common U.S. (English) thread size like the 1/4 bolt or hex head cap screw shown in Figure 5-6. Pull each blade out of the pitch gauge tool and set the teeth on the fastener as shown in Figure 5-7. Keep trying the blades until a set of teeth fits perfectly into the threads. When a match is found, thread pitch can be identified. The next step is to simply read the thread pitch off the pitch gauge blade. In our example, the pitch gauge blade reads 16 threads per inch.

SPECIAL TOOLS

U.S. (English) or metric tap and die set

Machinist square

Now the technician is ready to try a fastener with an unknown thread size. Choose a metric fastener with an unknown thread size like the one shown in Figure 5-8. This time, use a metric pitch gauge. Pull out each blade and set it on the threads as done before. When a perfect match is found, determine the thread pitch. The thread pitch of the fastener is 1.5, as shown on the pitch gauge blade in Figure 5-9.

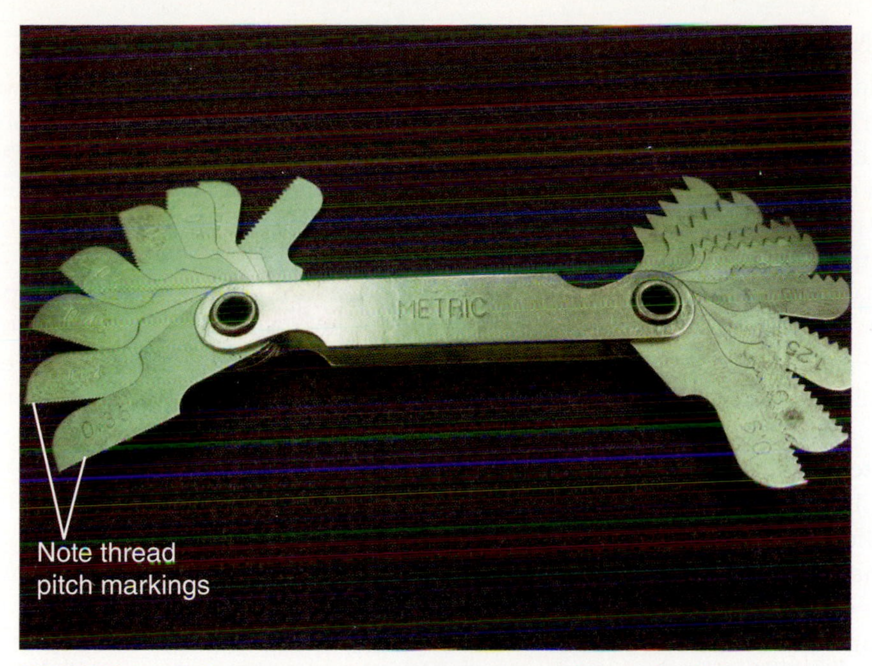

FIGURE 5-5 **A pitch gauge is used to measure the pitch or angle of the threads.**

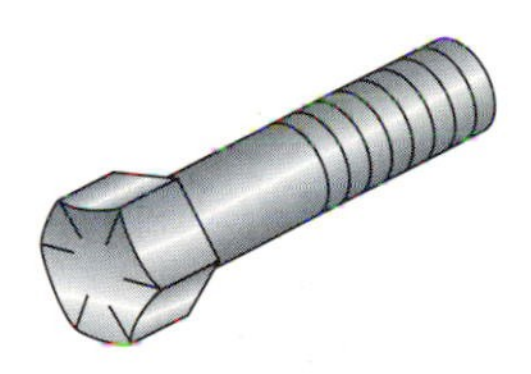

FIGURE 5-6 **A class 8 1/4-inch bolt.**

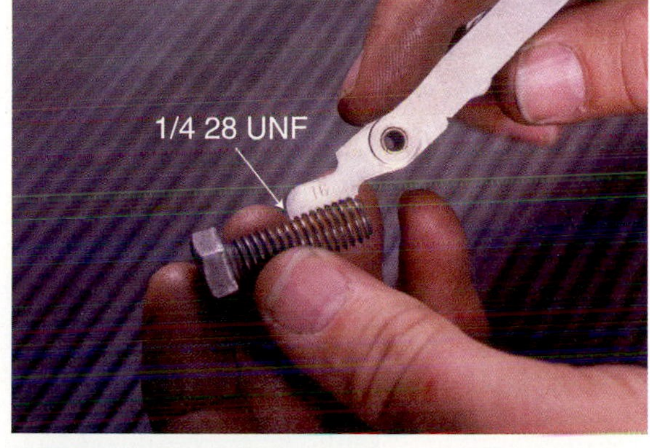

FIGURE 5-7 **The pitch gauge blade that matches the threads reads 16 threads per inch.**

FIGURE 5-8 **A metric bolt with an unknown thread size.**

FIGURE 5-9 **The thread pitch gauge blade shows that this fastener has a thread pitch of 1.5.**

Repairing Internal Threads

Classroom Manual
pages 119, 120, 121

Technicians often find fasteners that have been damaged. Photo Sequence 9 shows a typical procedure for repairing damaged threads with a tap. A common problem is fasteners that have not been started properly by hand. When a fastener is started into a threaded hole crooked, and then a wrench is used to force the fastener into the threaded hole, the mismatched threads are often damaged. This damage (Figure 5-10) is often called cross threading or **stripped threads**.

The damaged internal threads must be repaired before the automotive component can be reassembled. The tools made to repair internal threads come in a set called a tap and die set. There are many different types and sizes of tap and die sets. There are sets for both metric and U.S. (English) thread sizes. Most shops have a set that contains the most common sizes used for the work done in the shop.

The tool used to repair or make new internal threads is called a *tap*. The tap (Figure 5-11) is a hardened cutting tool. It has a square drive end that is used to rotate it for thread cutting. The shank is the round part of the tap and is where the size of the tap is found. The actual thread cutting is done with sharp cutting threads spaced between flutes where the cut metal can collect. The end of the tap is chamfered or tapered so that it can fit easily into the hole to be tapped.

Before using a tap to repair the threads, make sure to select the correct size and type. This often means you will have to determine the thread size of the screw that fits into the threaded hole you are repairing. This can be done with the pitch gauge, as explained earlier.

If a technician were to measure the thread on the screw that fits into a damaged internal thread as 1/4 28 unified fine threads UNF, then a tap with the size 1/4 28 UNF on the shank would be selected, as shown in Figure 5-12.

FIGURE 5-10 **A damaged or stripped internal thread.**

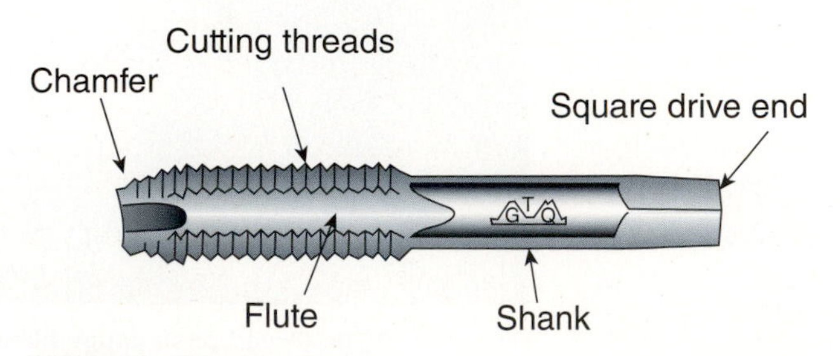

FIGURE 5-11 **Parts of a tap.**

PHOTO SEQUENCE 9

Typical Procedure for Repairing Damaged Threads with a Tap

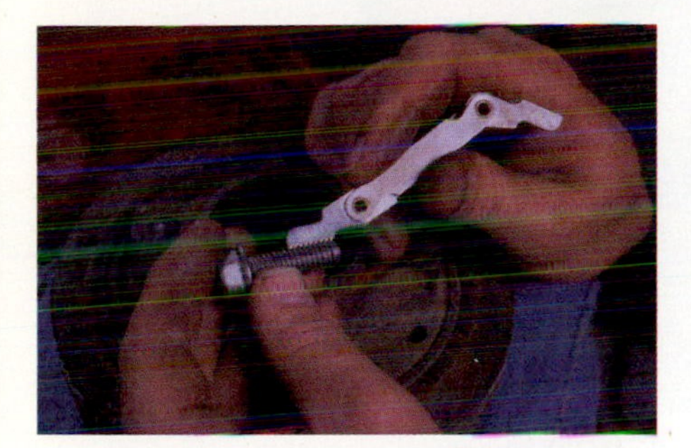

P9-1 **Determine thread size of fastener that fits into damaged internal threads by matching fastener to pitch gauge.**

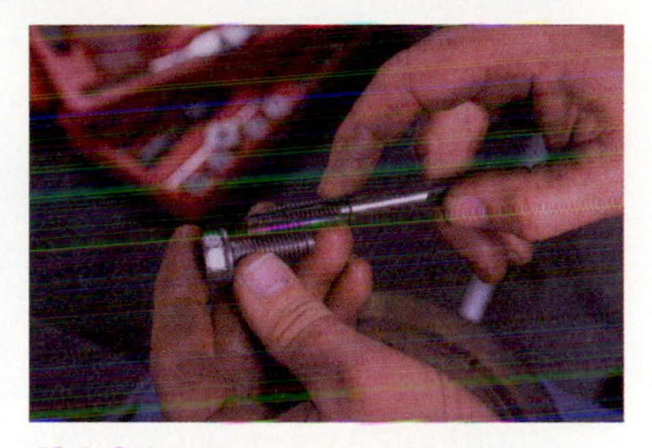

P9-2 **Select the correct size and type of tap for the threads to be repaired.**

P9-3 **Install the tap into a tap wrench.**

P9-4 **Start the tap squarely in the threaded hole using a machinist square as a guide.**

P9-5 **Drive the tap clockwise down the threaded hole the complete length of the threads.**

P9-6 **Drive the tap back out of the hole by turning it counterclockwise.**

P9-7 **Clean the metal chips left by the tap out of the hole.**

P9-8 **Inspect the threads left by the tap to be sure they are acceptable.**

P9-9 **Test the threads by threading the correct-size fastener into the threaded hole.**

A technician must also determine if she or he is repairing a thread that goes all the way through a part, called a **through hole**. The other type of hole does not go all the way through and is called a **blind hole**. Refer to the illustration (Figure 5-13) of a blind and a through hole.

There are different types of taps used for blind and through holes. A taper tap is used to repair threads in a through hole. The taper tap has several incomplete threads at the end to help it start in the hole. A bottoming tap is used to repair threads in a blind hole. The bottoming tap has only two incomplete threads at the bottom, so it can tap

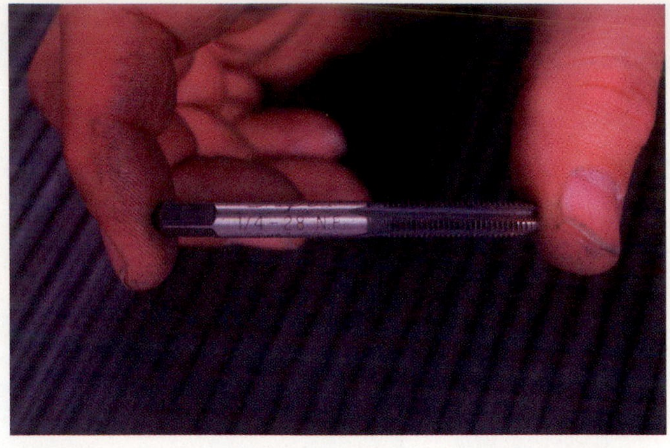

FIGURE 5-12 **A tap with the size 1/4 28 UNF on the shank.**

SERVICE TIP:
The tap should screw into the hole without much effort. If the tap becomes hard to turn, back it off a few turns to allow the cut metal to drop out of the flutes. If it is still hard to turn, or seems to bottom out, then the tap is either cross-threaded or the hole threads may be damaged beyond repair. Do not attempt to turn the tap if you have to strain. The tap is brittle and will break in the hole. Removing a broken tap is very time consuming and difficult.

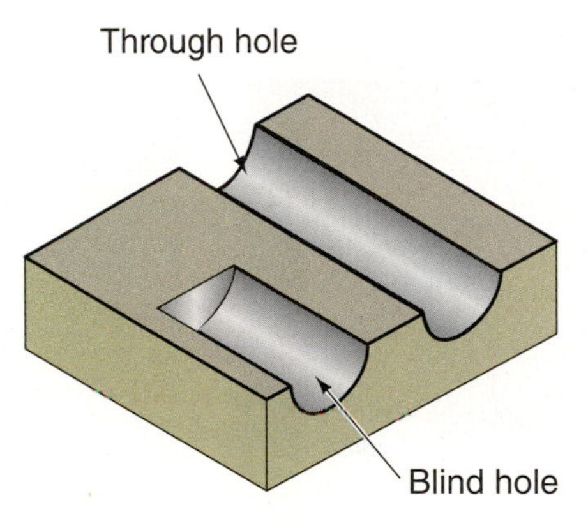

FIGURE 5-13 **A through hole and a blind hole.**

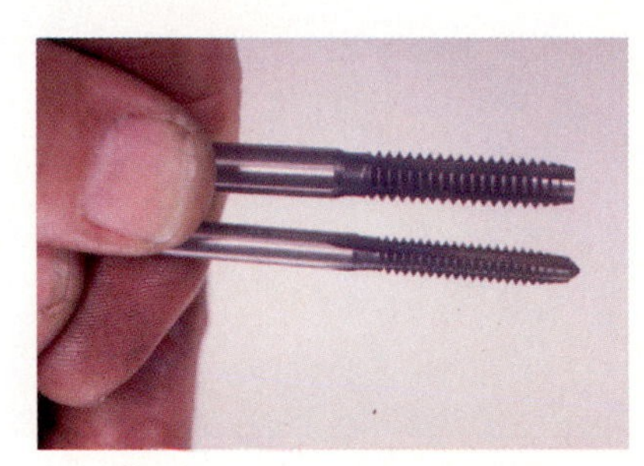

FIGURE 5-14 **A taper tap is used for through holes and a bottoming tap is used for blind holes.**

threads all the way to the bottom of a hole. A taper and a bottoming tap are shown in Figure 5-14.

WARNING: Practice using a tap on a scrap automotive component before attempting this job on a good part. Using a tap incorrectly can cause further damage to threads.

After selecting the correct type and size of tap, attach the tap to a tap wrench. There are two common types of wrenches used to drive a tap: a T-handle and a hand-held tap wrench (Figure 5-15). The T-handle has an adjustable chuck on the end that is opened and tightened over the square drive of the tap. The hand-held tap wrench also has an adjustable chuck that is opened and closed by rotating one of the handles. The T-handle is usually the best choice for repair work, especially on smaller threads like the 1/4 28 UNF.

Slide the square drive end of the tap into the chuck of the T-handle tap wrench. Tighten the chuck to hold the tap securely in the wrench. Set the tap in the hole to be repaired and rotate the tap carefully one or two turns. Make sure the tap enters the hole square and not at an angle. A good way to check this is to place a machinist square next to the tap as shown in Figure 5-16.

Carefully and slowly rotate the tap through the threaded hole. When the bottom is reached, slowly rotate the tap counterclockwise up and out of the thread. Clean up all

CAUTION:
Always wear eye protection when using a tap. Taps are hardened tools and may shatter if they break.

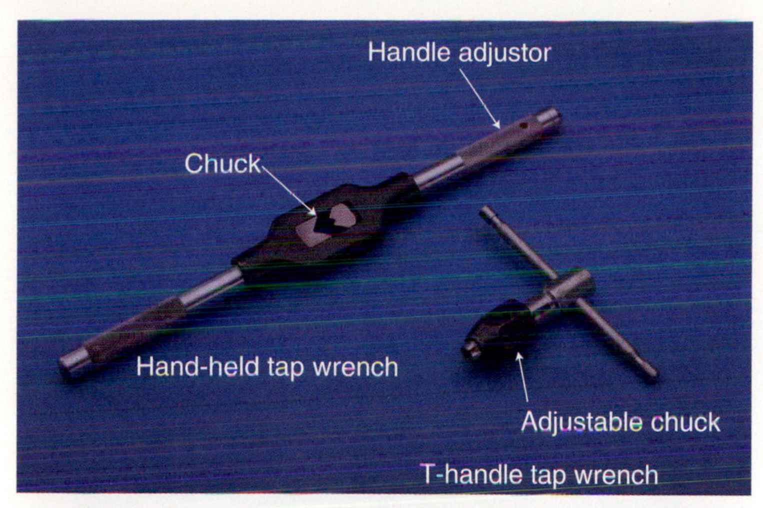

FIGURE 5-15 **Two common types of tap wrenches.**

FIGURE 5-16 **The tap must be started square in the hole.**

SERVICE TIP: Sometimes threads are too damaged to repair. Badly damaged threads can be repaired with the installation of a part called a *helicoil.* A special helicoil tap is used to form a special thread in the hole. The helicoil is a metal coil that is threaded into the hole using a special tool. The coil forms new threads that are of the same size as the damaged ones.

SPECIAL TOOLS

Helicoil
Helicoil installation tool
Drill and drill bit
Tap and tap wrench
Lubricant

metal filings that result from cutting out damaged threads. Test the repaired thread with the correct-size screw.

Installing a Helicoil

A helicoil is used to replace internal threads. The coil is a steel tread that looks like a miniature coil spring (Figure 5-17). The outside of the coil has small barbs that will grip the inside of the hole when a bolt is installed. Several steps must be performed before the coil is actually installed into the hole.

The first step is to determine which size of coil to use. Use a pitch gauge to determine the thread pitch and then measure the diameter of the bolt's shank. This will provide the information for selecting the helicoil and the tools needed for its installation. For this discussion, assume the bolt is M10 × 1.25 mm × 25 mm. The length of the shank is not needed for this operation.

A helicoil is also called a *thread insert.*

Find the correct helicoil and the instructions that go with it. The following instructions are general. If the helicoil's instructions are different, follow them.

There is usually a special-sized tap for the helicoil. Ensure that the tap is used or the helicoil may not fit correctly. The required drill bit size is usually etched into the head of the tap. Install the correct drill bit into an air- or electric-powered drill. While drilling the

SERVICE TIP:
General Motors and others use specially designed and engineered helicoil sets for their vehicles. Aftermarket helicoils can be used in their place except for critical areas like the cylinder head, engine block, and transmission. Some manufacturers do not recommend the use of helicoils on their vehicles. As always consult the service manual.

FIGURE 5-17 **Shown is a set of one-size helicoil. This set does not include the tap and drill bit.**

CAUTION:
Wear a face shield during the drilling and cleaning steps. Injuries to the eyes and face could be caused by flying metal and debris.

damaged threads out of the hole, ensure the drill bit is exactly aligned to the hole. Once the drill bit reaches the bottom of the hole, remove it and blow out the hole with compressed air.

Using the tap specified by the helicoil instructions, install the internal threads into the hole. After reaching the bottom (passing through) of the hole, remove the tap and blow out the hole again. The hole is now ready to accept the helicoil. Install the helicoil onto its installation tool (Figure 5-18). The installation tool is used to screw the helicoil into the hole. Screw the helicoil in until its upper edge is even with the top of the hole. The helicoil must not extend above the hole or be lower than the edge. Remove the tool and use either a small screwdriver or punch to break off the installation tab at the bottom of the helicoil. Blow out the hole for the last time. Spread a light lubricant on the bolt threads and screw the bolt into the helicoil. If installed properly, the bolt will screw in with a minimum of force. Remove the bolt and store the tools.

SPECIAL TOOLS
Six-inch scale
Stud remover
Penetrating fluid
Thread sealant
Antiseize compound

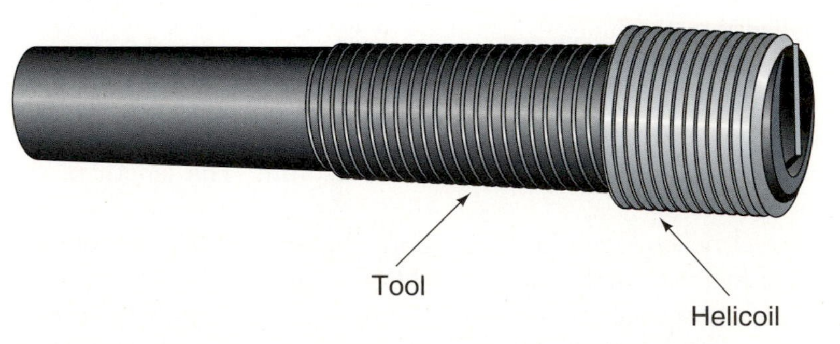

FIGURE 5-18 **The helicoil is placed (screwed) onto the tool and then screwed into the newly tapped hole.**

Removing and Replacing a Damaged Stud

A technician will often find damaged external threads on screws, bolts, and studs. These can be repaired with a die much like we used a tap to repair internal threads. However, because the screws, bolts, and studs are easily replaced with new fasteners, we rarely perform this job. All technicians need to master the replacement of studs.

Classroom Manual
pages 110, 111, 112

Studs often get damaged from overtightening or from cross threading a nut on the threads (Figure 5-19). A damaged stud is replaced by unscrewing it from the part and installing a new one of exactly the same size.

Stud replacement can be difficult because the stud has often been in place for a long period of time. There has been constant heating and cooling, corrosion, and rust buildup between the stud threads and the internal threads of the part. These factors combine to make some studs very difficult to remove.

The first step in stud removal is to use penetrating fluid to remove the corrosion to free the stud from its mating threads. Soak the area of the threads with penetrating fluid, as shown in Figure 5-20. Allowing the fluid to soak into the threads overnight will make it easier to remove the stud.

Before removing the old stud, measure the distance it sticks up from the surface. This measurement will be needed later when installing the new stud. Use a 6-inch scale to measure from the part surface to the top of the stud. Write the measurement down so it can be referred to later.

A **stud remover** is used to remove the studs (Figure 5-21). It is installed over the stud. The jaws on the stud remover grip the outside of the stud. A wrench fits on the stud remover and allows a technician to rotate the stud in a counterclockwise direction to remove the stud, as shown in Figure 5-22.

If a stud remover is not available, a stud can be removed with two nuts. Locate two nuts that are of the correct thread size to thread onto the stud. Start one nut and thread it all the way down to the bottom of the stud. This nut will be the drive nut. Start another nut and thread it down until it contacts the first nut. This is called the jam nut.

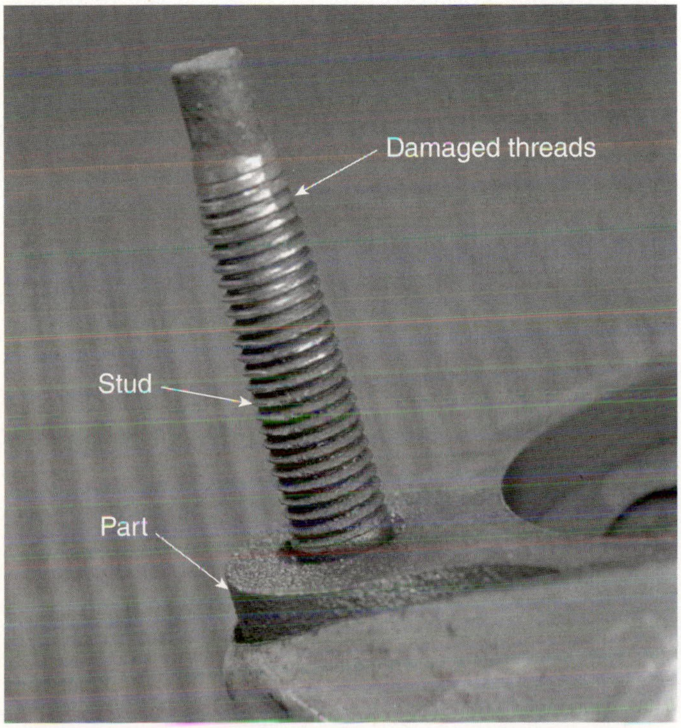

Damaged threads
Stud
Part

FIGURE 5-19 **A stud with damaged threads must be replaced.**

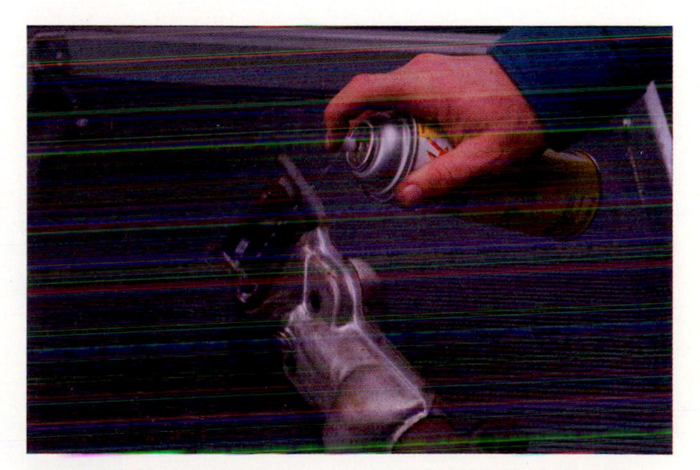

FIGURE 5-20 **Use penetrating fluid around the stud threads before removing them.**

FIGURE 5-21 **A stud remover is used to unscrew a stud. Some studs have ends that can be driven with a wrench.**

FIGURE 5-22 **Using a stud remover to remove a stud.**

Put a wrench on the bottom drive nut and hold it in place. Put another wrench on the jam nut and tighten, or "jam," it against the drive nut. The jam nut will now hold the drive nut in position on the stud (Figure 5-23).

Now put an open-end wrench on the bottom drive nut. Turn the nut in a counter-clockwise direction. Turning the nut in this direction causes it to want to unscrew up the stud (Figure 5-24). The jam nut, however, prevents the drive nut from moving up the stud. Instead the forces cause the stud to unscrew.

When the old stud is out, inspect the internal thread. If it appears rusty or damaged, clean up the thread by running the correct size tap through the threads as previously explained. Compare the new stud with the old one (Figure 5-25). The studs should be of exactly the same thread size and the same length.

CHEMICALS USED WITH FASTENER THREADS

SERVICE TIP:
When using **thread-locking** compound or **thread-sealing** compound, make sure you are using the right chemical. They come in similar containers and one will not do the job of the other.

Check the shop service manual for the car being worked on to determine if the threads of the new stud or bolt should be coated. If the stud should be locked in place and not easily removed, the technician may be instructed to use a **thread-locking compound** or **thread-sealing compound** (Figure 5-26). Thread-locking compounds are used on studs and other fasteners when vibration or impact might cause them to unscrew. Thread sealants are used when a stud extends into an area where liquids, such as oil or coolant, could get on the

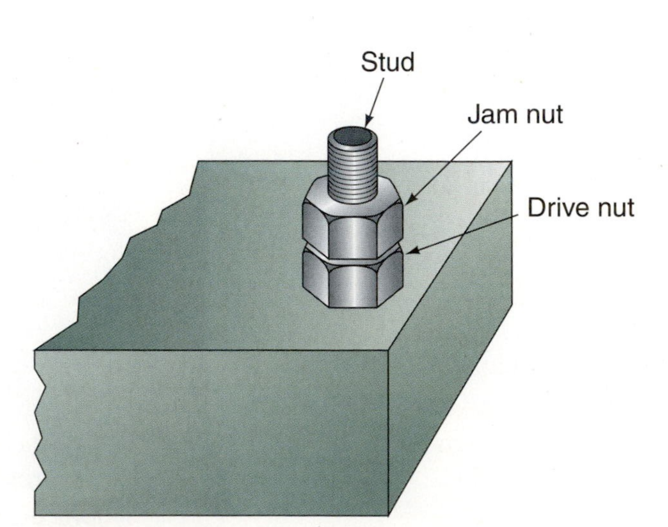

FIGURE 5-23 **A drive nut and jam nut installed on a stud.**

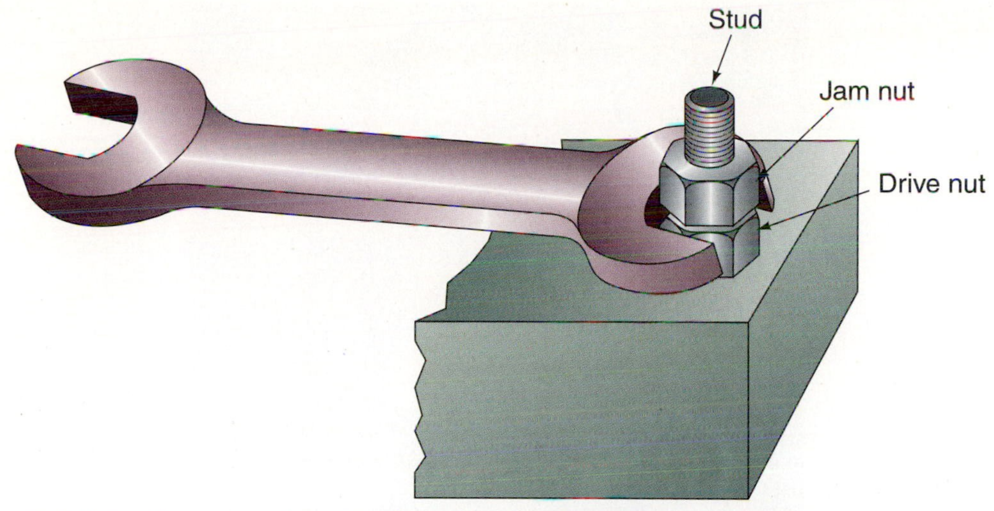

FIGURE 5-24 **Use a wrench on the drive nut to unscrew the stud.**

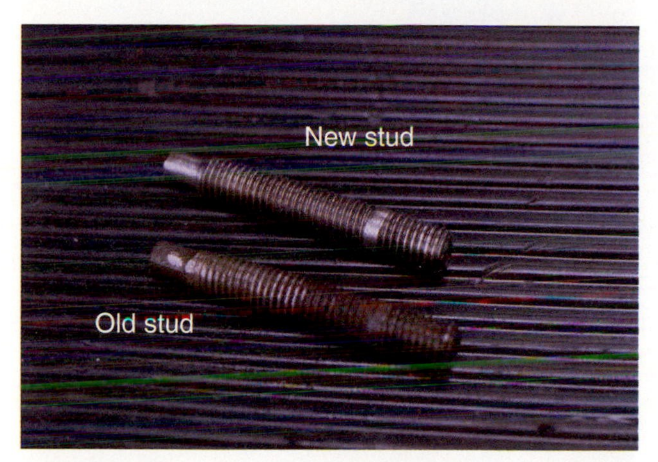

FIGURE 5-25 **The replacement stud must be the same size as the old stud.**

Anaerobic means without air. When you drop liquid sealant onto the threads of a bolt, it will remain liquid until you install it. After installing it and removing the air, it starts to harden and become plastic.

fastener. Both of these compounds are known as **anaerobic** sealants. This means they start the curing process in the absence of air.

Antiseize compound (Figure 5-27) is used on the stud threads to prevent the stud from reacting with the metal of the internal threads. If this happens, the stud could stick or seize. Antiseize compound prevents this reaction and makes the stud easier to remove next time.

After the new stud is properly coated, it can be installed. Start the stud by hand, making sure it enters the threads squarely. Turn the stud in as far as possible by hand before using any tools. Then use two nuts as described earlier to drive the stud into the part. Use the depth measurement made on the old stud to be sure it is driven to the correct depth (Figure 5-28).

SPECIAL TOOLS

Wire stripping and crimping pliers

Solderless connectors

Installing Solderless Electrical Connectors

Replacing burned or damaged wires is a common electrical service procedure. When installing new wires in an electrical system, a technician will need to be able to install the connecting hardware on the wires. There are two general types of wire connectors: solderless and soldered. This section describes how to use the solderless type. The next section shows how to solder on wire connectors.

Classroom Manual
pages 127, 128, 129

FIGURE 5-26 Threadlocker is a chemical that will assist in keeping a fastener tight. A small amount is placed on the threads of a fastener, usually the bolt.

FIGURE 5-27 Antiseize is a chemical that will assist in keeping a fastener from seizing in its counterpart during its life. It is most commonly applied to the threads of a spark plug.

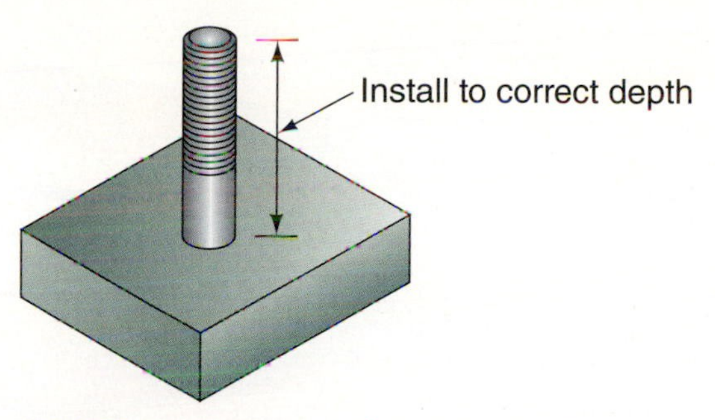

FIGURE 5-28 **The new stud must be installed to the correct depth.**

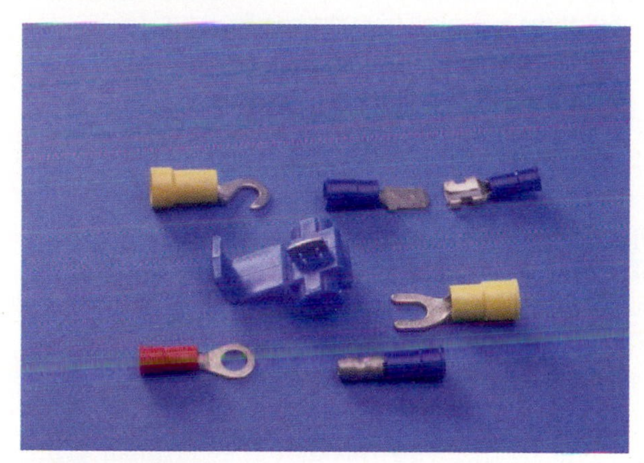

FIGURE 5-29 **Parts of a solderless connector.**

WARNING: Some manufacturers do not recommend the use of solderless butt connectors for making wiring repairs. They recommend soldering the two wires together or complete wire or wiring harness replacement.

There are many types and styles of solderless connectors, as discussed in the Classroom Manual. The two basic groups are butt connectors used to connect two wires together and terminal connectors used to connect a wire to a terminal. Both types have the same basic parts shown in Figure 5-29. The connector is made in two parts. The part that connects to the wire or terminal is made from metal that is a good electrical conductor. A plastic insulator fits over the parts that must not be allowed to touch another conductor. There are crimping tabs on the end of a terminal connector and on each end of a butt connector. The tabs are used to attach the wire. The crimping tabs are formed in a circular shape so that wire can be inserted.

Connectors are classified according to the wire size they fit. A 16-gauge (1.0 mm^2) connector has a crimping tab hole that will fit a 16-gauge (1.0 mm^2) wire. The wire size for solderless connectors is printed on the package.

Crimping is a procedure used to join a wire to a solderless connector.

WARNING: Always use the correct-size connector for the wire you are installing. Wire that is too small or large for the connector will not be retained properly in the connector.

The basic tool used to install solderless connectors is the stripping and crimping pliers shown in Figure 5-30. The jaw area of the tool is used to crimp or squeeze the crimping

Stripping and crimping pliers are used to strip insulation from wire and crimp a solderless terminal on wire.

Stripping is the procedure used to remove insulation from an electrical wire.

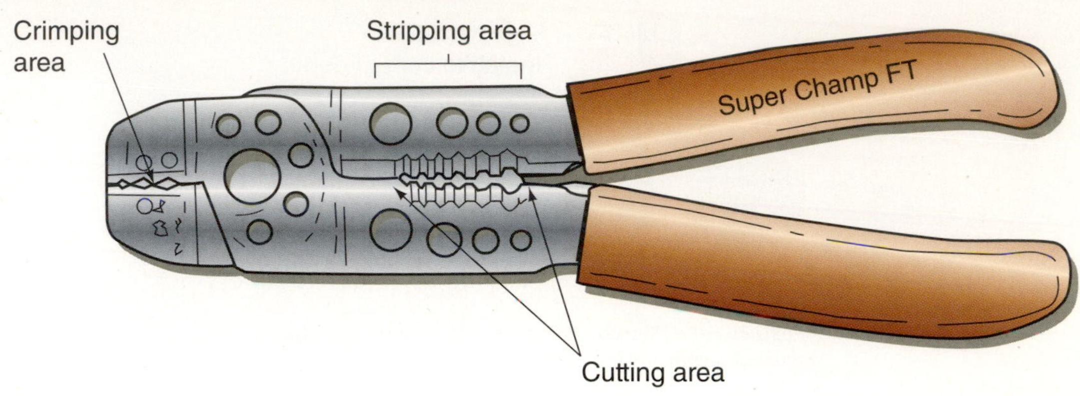

FIGURE 5-30 **A wire crimper is used to strip the wire and crimp on a solderless connection or terminal. Some have the cutting area at the end away from the handle.**

SERVICE TIP: Many manufacturers and most experienced technicians recommend the stripped end of the wire be "tinned" before installing it into a solderless terminal. This seals the strands together and makes a better electrical connection. **Tinning** is done by heating the stripped wire and applying a light coat of solder. When making a solder connection, tinning should also be used. (See Figure 5-39.)

tabs of the connector onto the wire. The area behind the jaws has a set of cutters used to strip away the insulation on the wire.

The first step in installing the connector is to strip the insulation off the end of the wire. The wire stripper has several sets of cutters in the stripping area. Each is labeled with a wire gauge size. When **stripping** a 16-gauge (1.0 mm^2) wire, insert approximately 1/4 inch of the end of wire between the stripper cutters. Squeeze down on the tool handles. This action cuts the insulation but not the wire. Pull the wire out of the stripping area while holding the handles together. The insulation should be removed from the end of the wire as shown in Figure 5-31.

Insert the stripped end of the wire through the terminal crimping tabs as shown in Figure 5-32. When installing a butt connector, install a wire at both ends as shown in Figure 5-33. Be sure the stripped wire fits completely into the crimping area.

The crimping jaws on the end of the pliers are used to squeeze the crimping tabs down on the wire. Most stripping and crimping pliers have several sets of jaws that are marked with different wire gauge sizes. Use the one that is correct for your solderless terminal size.

Place the terminal and wire in the tool jaws as shown in Figure 5-34. Squeeze on the tool handles to crimp the tabs against the wire. The two parts of the tool jaw will flatten part of the terminal and turn in the crimping tabs on the wire as shown in Figure 5-35.

Test the connection by gripping the wire in one hand and the terminal in the other. Gently try to pull the two apart. A properly installed connector will not have any movement back and forth.

SPECIAL TOOLS

Soldering gun or soldering iron

Rosin-core solder

Stripping and crimping pliers

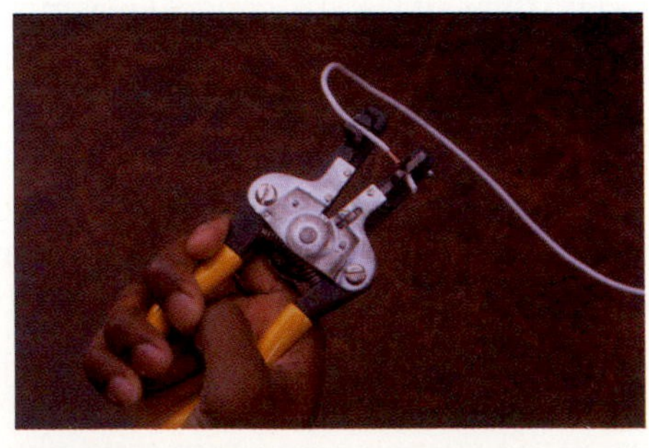

FIGURE 5-31 **Stripping the insulation from a wire.**

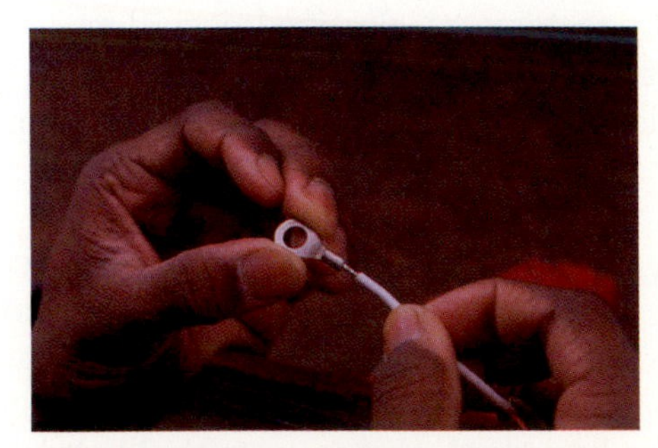

FIGURE 5-32 **The stripped end of the wire is inserted into the crimping taps.**

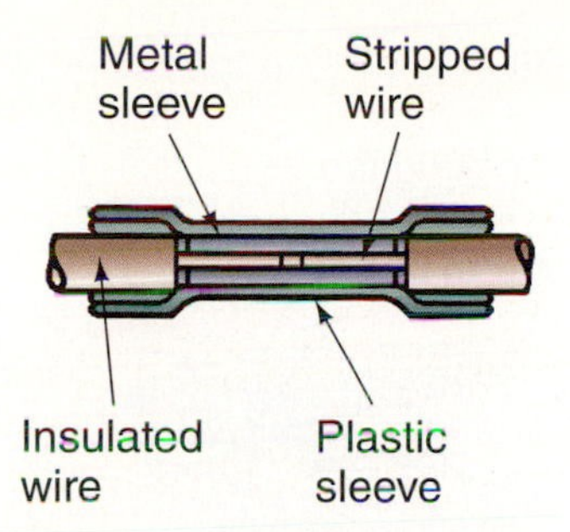

FIGURE 5-33 **How two wires fit in a butt connector.**

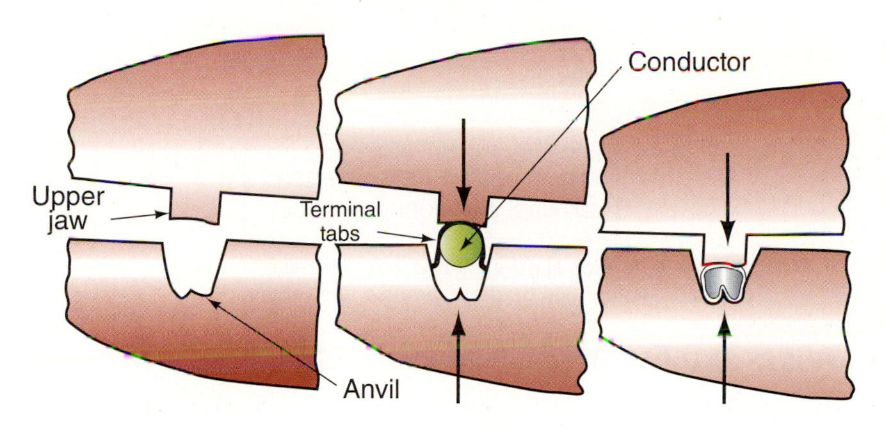

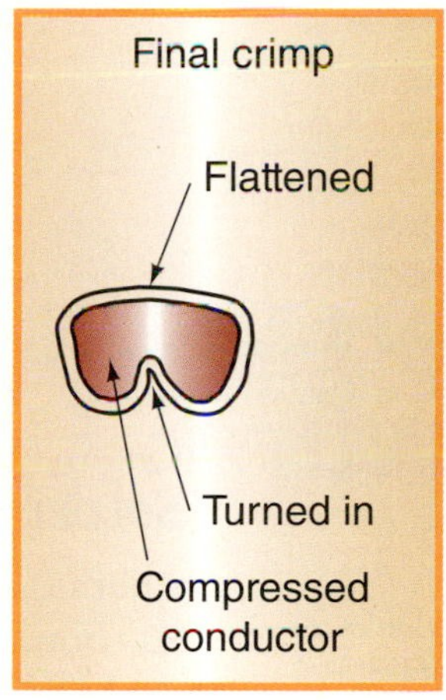

FIGURE 5-34 **Crimping a solderless connector in crimping jaws.**

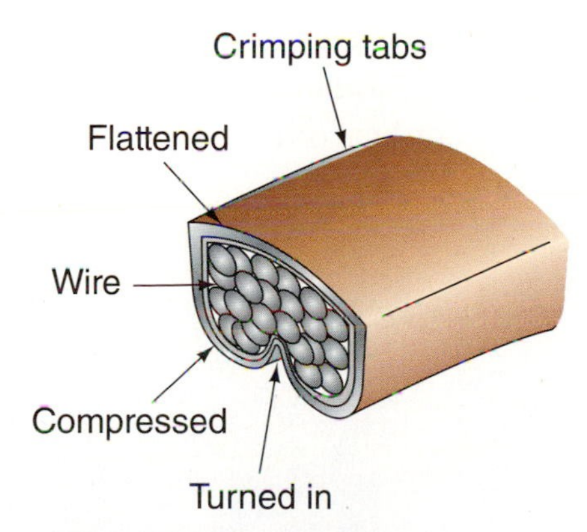

FIGURE 5-35 **The crimping tabs are crimped against the wire.**

Repairing Weather Pack Connectors

Hard-shell and weather pack connectors usually provide a means of disassembling the terminal for repair. On some occasions there may be a Technical Service Bulletin that outlines a problem with a connector. On other occasions you may find that wiggling the connector is when the problem occurs.

You can remove these connectors using special tools to release a locking tang inside the connector (Figure 5-36). Once disassembled, check the connector for corrosion, worn parts, and a good solder joint. Sometimes you can rebuild the connector, and sometimes you have to get a new connector. If you have to purchase a new connector, it will usually come with extra wire already connected to it. You then have to solder the wire into the circuit.

SERVICE TIP:
A good way to practice the wire stripping and soldering techniques presented in this chapter is to make a set of jumper leads for circuit testing. Cut a 24-inch length of 16-gauge (1.0 mm^2) wire. Strip both ends of the wire. Buy two alligator clips from the local auto parts or electronics store. Solder these on each end of the wire. The jumper is an excellent test tool when used properly.

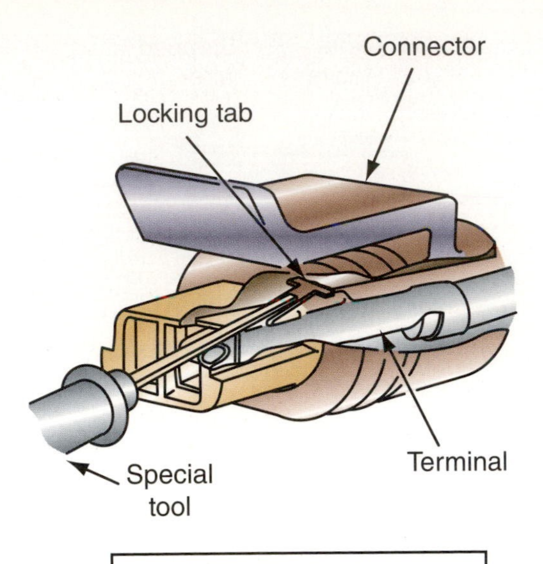

FIGURE 5-36 **Hard-shell and weather-pack connectors can be disassembled and rebuilt.**

Soldering Wires and Electrical Connectors

Classroom Manual
page 127

Another way to join electrical wires or terminals to wires is by **soldering**. Photo Sequence 10 shows soldering wire and electrical connectors. This is a method of joining wires and connectors using a metal called solder. Solder is a metal combination, or alloy, of tin and lead that melts at a low temperature. It can be melted and then allowed to cool back to a solid to join electrical components. Solder has excellent electrical conductivity.

In order to solder parts together, the solder must be heated to its melting point. This is done with either a **soldering gun** or a **soldering iron**. An electric soldering iron is shown in Figure 5-37. The unit has a power cord that is plugged into an electrical outlet.

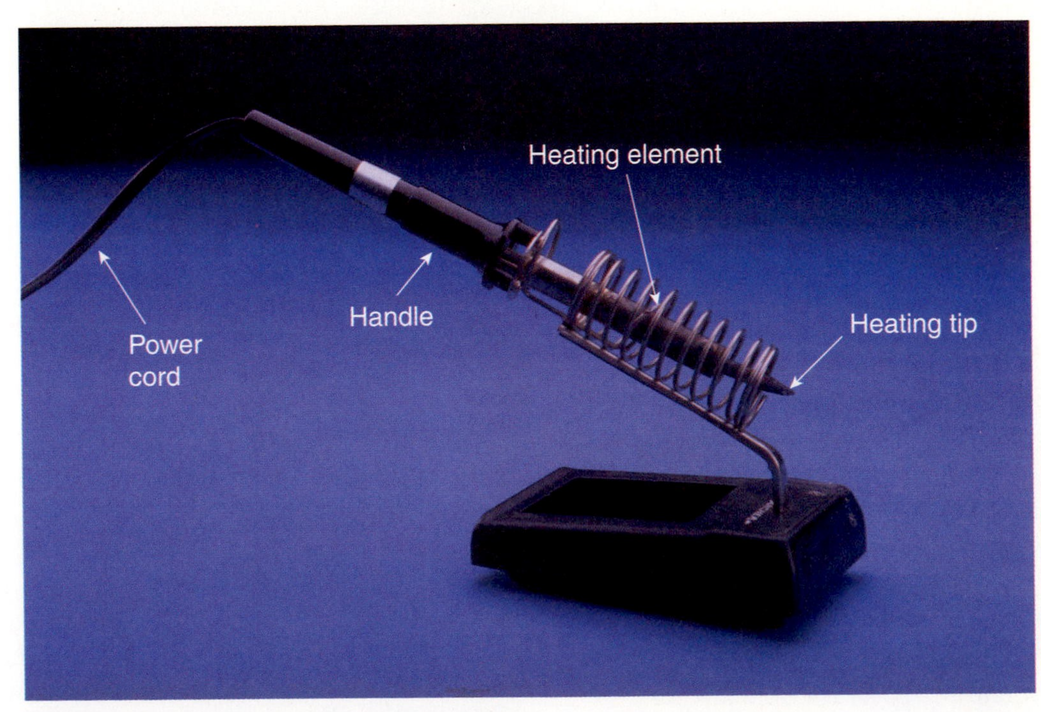

FIGURE 5-37 **A typical electrical soldering iron.**

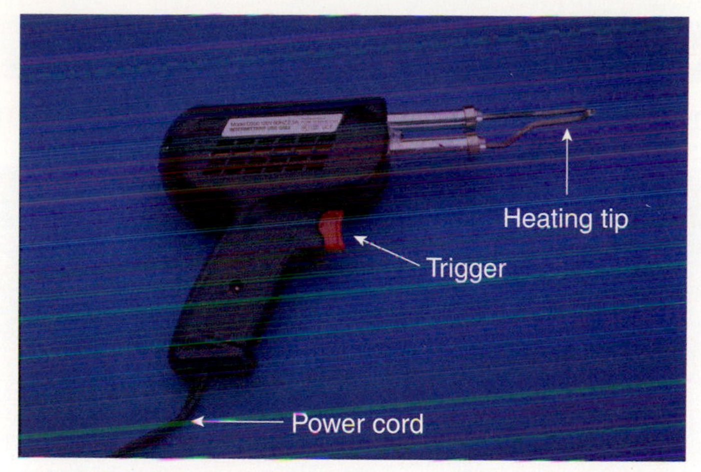

FIGURE 5-38 **A typical soldering gun.**

Electricity flows through a heating element inside the unit and heats the tip hot enough to melt solder. An insulated handle allows you to hold the handle safely. Inexpensive soldering irons do not have on-off switches. Once plugged in, they take a short time to heat up to get the tip hot enough to melt solder.

The soldering gun gets its name from the fact that it looks a little like a gun and has a trigger, as shown in Figure 5-38. The power cord of the soldering gun is plugged into an electrical outlet. The tip is heated instantly by a heating element inside the gun whenever the trigger is pulled. When the trigger is released, the tip cools down. Some units have a light that illuminates when the tip is hot enough to melt solder. Both the soldering gun and soldering iron are used the same way to melt solder.

CAUTION: The heating tip of the soldering gun and soldering arm gets to a temperature in excess of 300°F. Be careful not to touch the tip or it will cause burns.

There are many types of solders to do many different joining jobs. Solder used for electrical work is sold in spools (Figure 5-39). It is shaped in the form of a wire. Many different diameters are available. A solder wire of 0.062-inch diameter is commonly used for general electrical work. In addition, many different alloys of tin and lead are available. Most electrical work is done with 40/60 alloy. The diameter and alloy are marked on the solder spool.

Soldering must be done with a material called soldering **flux**. Soldering flux is used to remove oxides from the wires and terminals. Oxides are formed when metal is exposed to air. The flux cleans the wires and terminals. It also helps the solder flow more easily. Although flux is available separately, solder for electrical work typically has the flux inside the core of the solder wire, as shown in Figure 5-40.

There are two general types of flux: *acid* and *rosin*. Acid-type flux is used for joining sheet metal. It is too corrosive for electrical work. Electrical soldering is done with a noncorrosive flux called resin. The solder spool will indicate the type of flux in the solder core. Choose solder that is marked "resin core."

FIGURE 5-39 **Spool of 40/60 rosin-core solder.**

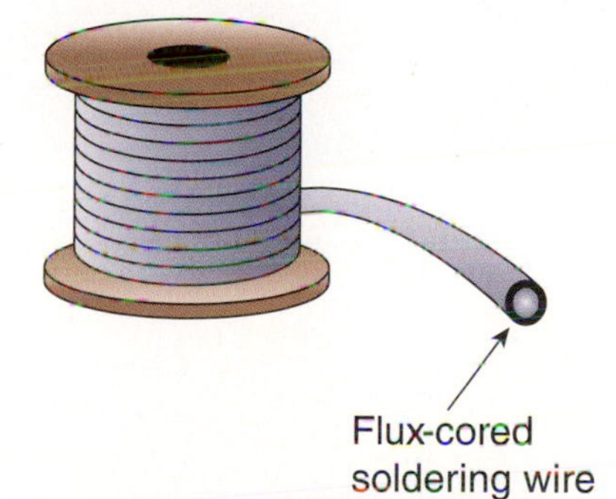

FIGURE 5-40 **The center of the solder has a core of rosin flux.**

PHOTO SEQUENCE 10

Typical Procedure for Soldering Electrical Terminals and Joining Wires

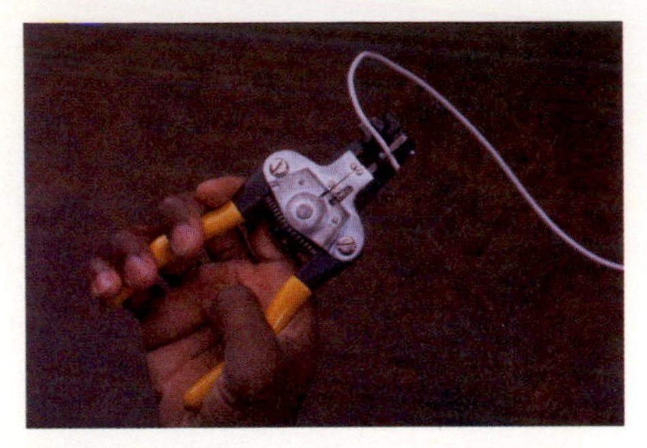

P10-1 To solder a terminal to a wire, first place the end of the wire in the correct size cutter on the stripping and crimping pliers.

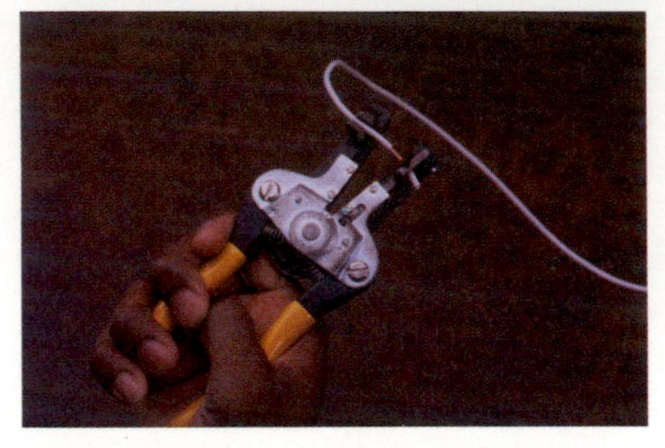

P10-2 Squeeze the pliers' handles and pull the insulation off the end of the wire.

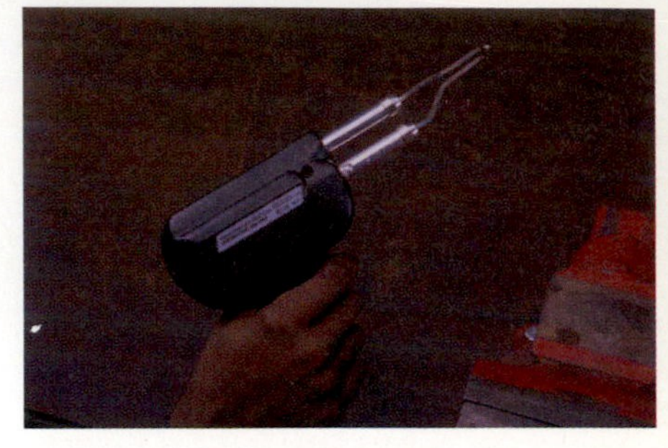

P10-3 Pull the trigger on the soldering gun to heat the end of the gun.

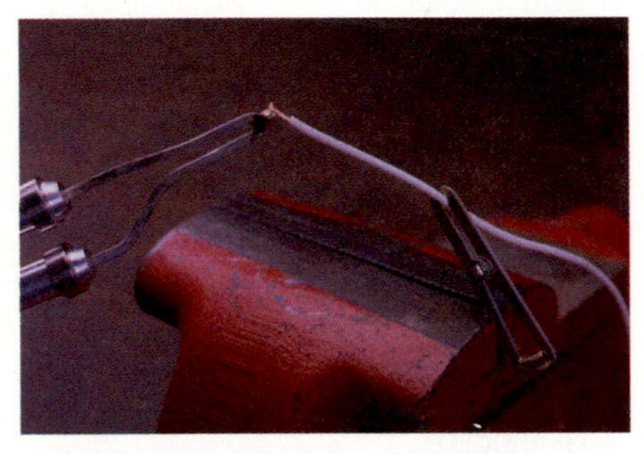

P10-4 Touch the heated end of the soldering gun to the stripped end of the wire to heat the wire.

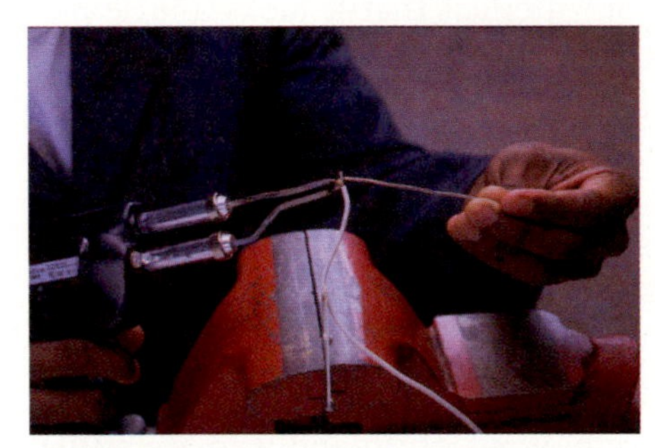

P10-5 Melt a small amount of rosin-core solder over the surface of the wire to "tin" the wire.

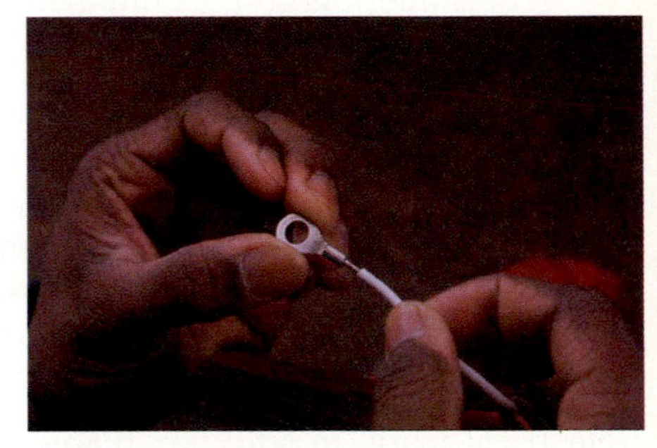

P10-6 Install the end of the wire into the correct-size electrical terminal.

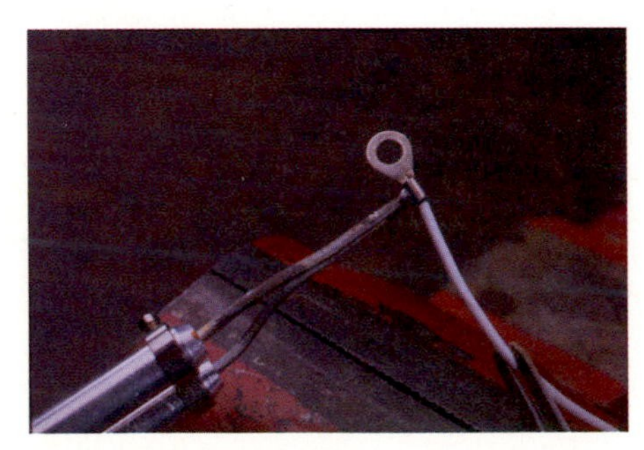

P10-7 Heat the electrical connector and wire with the soldering gun.

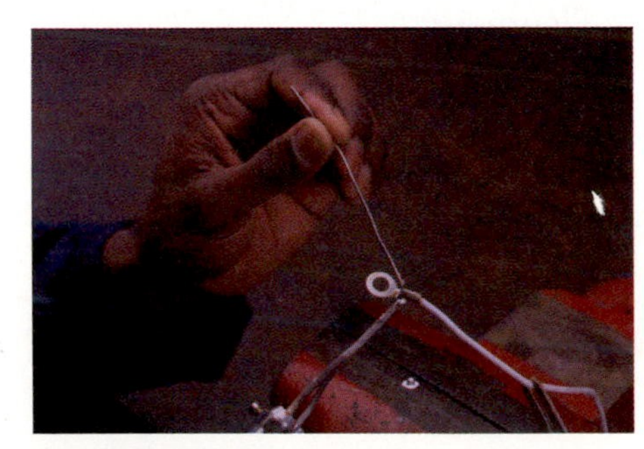

P10-8 Melt solder into the joint between the wire and connector and allow to cool.

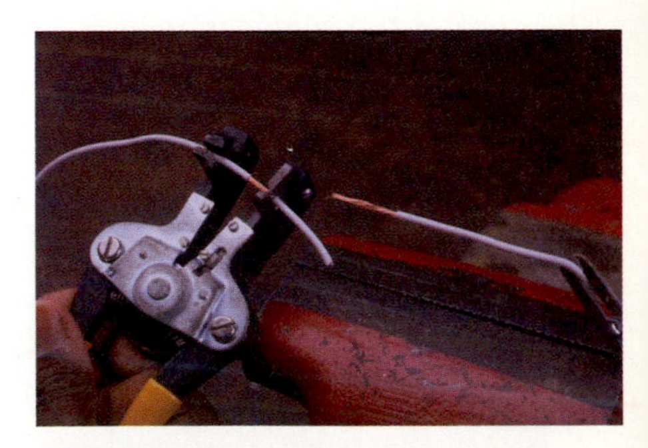

P10-9 To join two wires, first use the stripping and crimping pliers to strip the ends of both wires.

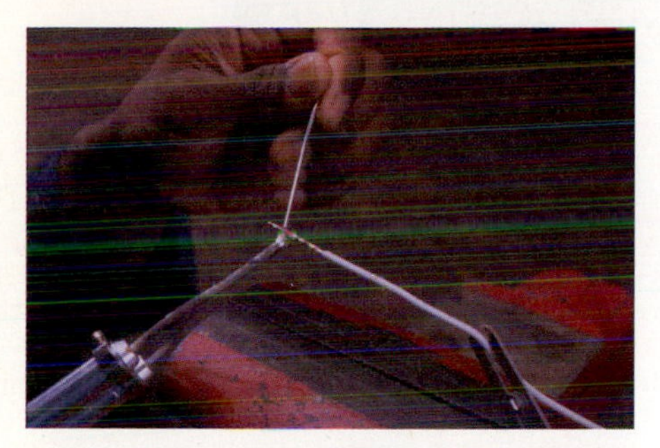

P10-10 Slip a piece of heat shrink tubing over one wire and slide it out of the way. Then use the soldering gun to heat each wire end and melt solder on the wire to tin the wire.

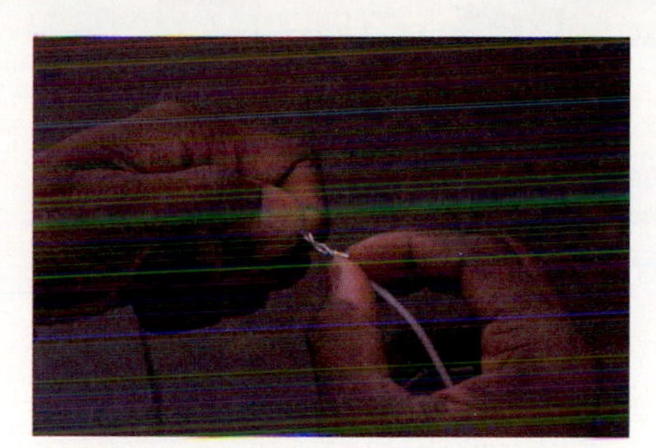

P10-11 Wrap the two ends of the wire together to form a tight connection.

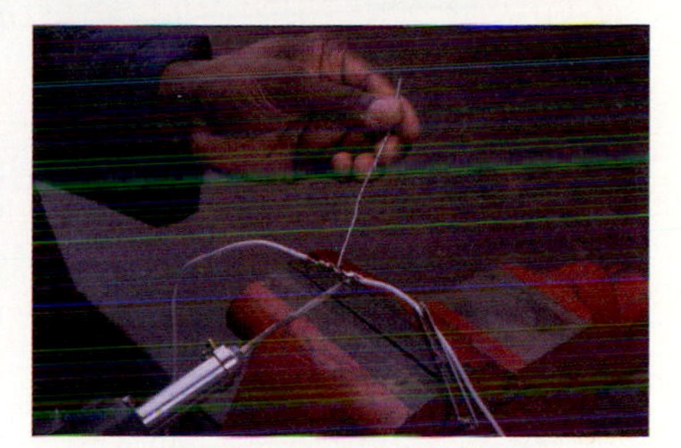

P10-12 Melt solder between the two wires to form a connection.

P10-13 Cover the exposed connection with the piece of heat shrink. Use a heat gun to shrink the covering.

WARNING: Never use acid-core solder on electrical components. It can cause corrosion that can interrupt electrical flow.

CAUTION:
Both the flux and the solder cause fumes that are dangerous to breathe for a long period of time. Always solder in a well-ventilated area. Always wear eye protection when soldering.

With the correct solder and the soldering gun or iron heated, try soldering a terminal to a wire. Just as in solderless connectors, choose wire and terminals that are of the same gauge size.

Use the stripping and crimping pliers to strip the insulation off approximately 1/4 inch of the end of the wire, as shown in Figure 5-41. Tin the wire before it is joined to the terminal. Tinning improves the final joining of the terminal and wire. Touch the tip of the soldering gun or iron to the wire and allow it to heat a moment. Then touch the solder to the heated wire and allow the solder to flow completely over the wire. Tinning the wire is shown in Figure 5-42.

Place the tinned wire into the terminal. Heat the terminal with the soldering iron tip. Give it a moment to heat up, then apply solder into the terminal as shown in Figure 5-43. Allow the solder to solidify and cool. Then test the connection by holding onto the terminal and pulling on the wire. A defective solder joint will break under pressure.

Soldering can also be used to join, or **splice,** two or more wires together. The common methods of joining wires are shown in Figure 5-44. In all the methods, the wire ends are stripped and tinned, then they are soldered as shown in Figure 5-45. The exposed area of

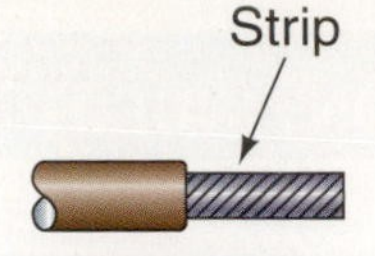

FIGURE 5-41 **Stripping the insulation off the wire.**

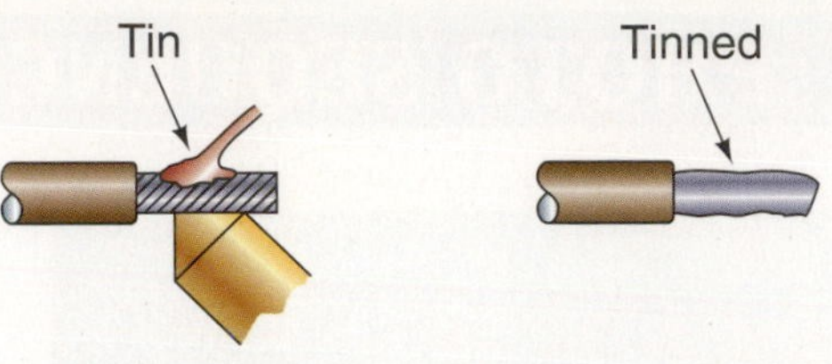

FIGURE 5-42 **Tinning the end of the wire with melted solder.**

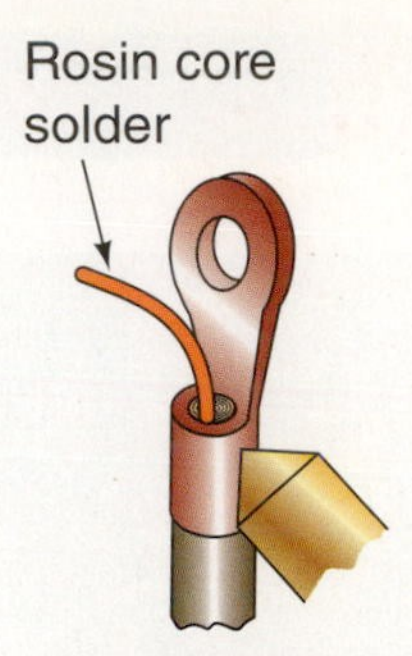

FIGURE 5-43 **Soldering the terminal to the wire.**

SERVICE TIP: Replacement wires can be covered with a product called *heat shrink*. Heat shrink is a thin-wall plastic tubing. The tubing is available in many sizes to cover a single wire or a number of wires to form a loom. Heat the tubing after the wires are installed through the tubing. The tubing shrinks into a tight fit around the wires. The tubing must be installed onto one section of the wire before connection is made.

the wires and solder will then have to be covered with the correct type of electrical tape or *heat shrink*.

When you have to solder a twisted pair wire you must make sure to solder it and then not leave much straight wire. Always try to return the wiring to the way the manufacturer designed and produced it.

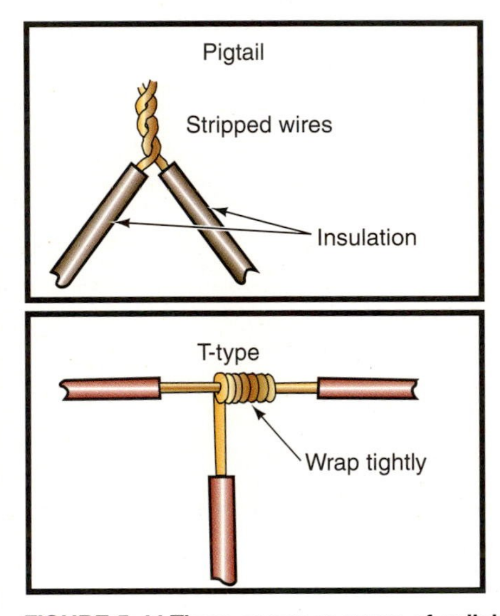

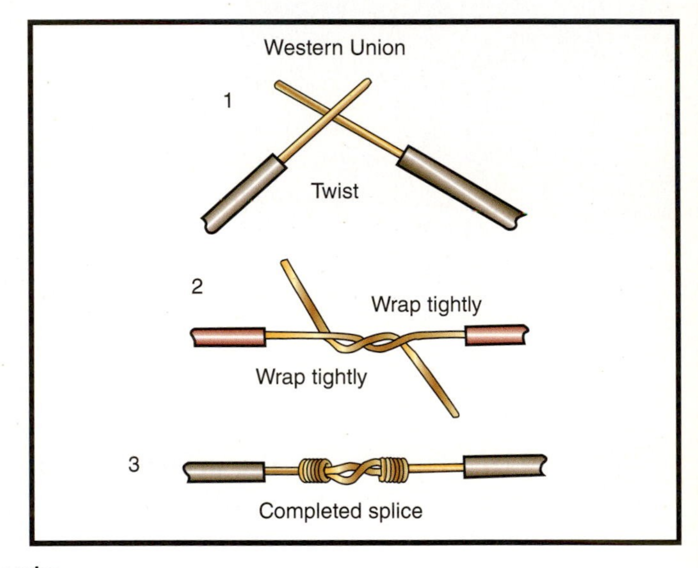

FIGURE 5-44 **Three common ways of splicing wire.**

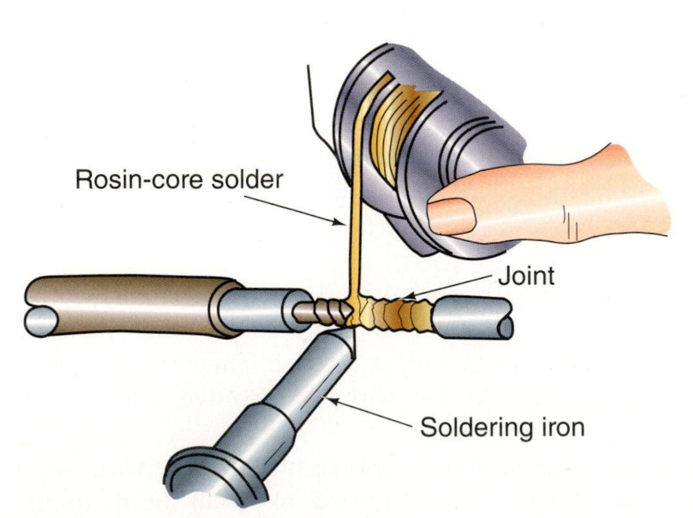

FIGURE 5-45 **Soldering the wire joint.**

TERMS TO KNOW

Anaerobic
Angle indicator
Antiseize
Blind hole
Flux
Soldering
Soldering gun

CASE STUDY

Using the wrong type of soldering flux can have disastrous results. A friend related a problem several years ago. He was doing a ground-up restoration on an old 356 Porsche. All the old wiring had been pulled out of the car in order to put in a complete new wiring harness. After pricing commercially available wiring harnesses, he decided to make his own. He bought and cut hundreds of feet of new wire. He wanted to do a good job, so he soldered each terminal and connector in his new loom.

The newly restored car had been on the road a couple of years when he started having electrical problems. Light switches, turn signals, and accessories stopped working for no apparent reason. He began to notice that corrosion was building up on his electrical connections. Further investigation showed that the corrosion was caused by his use of an acid-core solder. He was quite unhappy when he found out that all of the hundreds of connections had to be redone with the new terminals and the correct type of rosin-core solder.

TERMS TO KNOW (continued)

Soldering iron
Splice
Stripped threads
Stripping
Stud remover
Thread-locking compound
Thread-sealing compound
Through hole
Tinning

ASE-STYLE REVIEW QUESTIONS

1. The use of a pitch gauge is being discussed.
   Technician A says the pitch gauge can be used to identify U.S. (English) threads.
   Technician B says the pitch gauge can be used to identify metric threads.
   Who is correct?
   A. A only
   B. B only
   C. Both A and B
   D. Neither A nor B

2. The use of a tap to repair internal threads is being discussed.
   Technician A says the tap must have the correct thread size on the shank.
   Technician B says one tap fits many different thread sizes.
   Who is correct?
   A. A only
   B. B only
   C. Both A and B
   D. Neither A nor B

3. The use of a tap to repair internal threads is being discussed.
   Technician A says a taper tap is used for a through hole.
   Technician B says a taper tap is used for a blind hole.
   Who is correct?
   A. A only
   B. B only
   C. Both A and B
   D. Neither A nor B

4. The use of a tap to repair internal threads is being discussed.
   Technician A says a taper tap has numerous incomplete threads at its bottom.
   Technician B says a bottoming tap has numerous incomplete threads at its bottom.
   Who is correct?
   A. A only
   B. B only
   C. Both A and B
   D. Neither A nor B

5. The removing and replacing of a stud is being discussed.
   Technician A says a stud remover can be used to remove the stud.
   Technician B says two nuts can be used to remove the stud.
   Who is correct?
   A. A only
   B. B only
   C. Both A and B
   D. Neither A nor B

6. The removing and replacing of a stud where vibration is a problem is being discussed.
   Technician A says to coat the stud with antiseize compound.
   Technician B says to use a softer grade of bolt so it will grip better.
   Who is correct?
   A. A only
   B. B only
   C. Both A and B
   D. Neither A nor B

7. The removing and replacing of a stud that goes into an area with engine coolant is being discussed.

   Technician A says to coat the stud with antiseize compound.

   Technician B says to coat the stud with sealing compound.

   Who is correct?

   A. A only
   B. B only
   C. Both A and B
   D. Neither A nor B

8. The installation of a solderless electrical connector is being discussed.

   Technician A says the connector and wire must have the same wire gauge size.

   Technician B says the connector should be crimped to the wire before soldering.

   Who is correct?

   A. A only
   B. B only
   C. Both A and B
   D. Neither A nor B

9. The soldering of a terminal on a wire is being discussed.

   Technician A says to use acid-core solder so the wire is cleaned of contaminants.

   Technician B says to use rosin-core solder so the connection resists corrosion.

   Who is correct?

   A. A only
   B. B only
   C. Both A and B
   D. Neither A nor B

10. The soldering of a terminal on a wire is being discussed.

    Technician A says to tin the stripped wire before installing it in the connector.

    Technician B says to tin both the wire and connector after soldering.

    Who is correct?

    A. A only
    B. B only
    C. Both A and B
    D. Neither A nor B

JOB SHEET 11

Name ______________________________ Date ______________

Restoring Damaged External Threads

NATEF Correlation

This job sheet addresses the following task:

Perform common fastener and thread repair to include: remove broken bolt, restore internal and external threads, and repair internal threads with thread inserts.

Objective

Upon completion and review of this job sheet, you should be able to restore damaged external threads on a fastener.

Tools and Materials Needed

Thread restoring file
Thread restoring tool
Vise
Basic tool set
Gloves

Procedures

Task Completed

1. Determine the type of restoring device desired. ☐
   Restoring file—Go to step 2.
   Restoring tool—Go to step 10.
2. Match the fastener threads to a set of teeth on the file. ☐
3. Locate a nut that fits the fastener. ☐
4. Mount the fastener firmly in a holding device. A table-mounted vise is usually sufficient. ☐
5. Put on the gloves. ☐
6. Align the selected file teeth with the fastener threads. Allow the teeth to overlap onto undamaged threads if possible. This will act as a guide. ☐
7. Work the file over the damaged area. The fastener may have to be repositioned within the holding device for easy access to all damage. ☐
8. Use the nut to make periodic checks of the repairs during the filing. ☐
9. When the nut screws easily onto the fastener, the task is complete. ☐
10. When using a restoring tool, select the tool insert that fits the damaged threads. ☐
11. Select a nut that fits the fastener. ☐

Task Completed

- ☐ **12.** Install the insert into the tool handle with teeth facing the clamping section of the tool handle.
- ☐ **13.** Mount the fastener into a holding device. A table-mounted vise is usually sufficient.
- ☐ **14.** Clamp the tool around an undamaged thread area.
- ☐ **15.** Turn the tool so the insert runs over and through the damaged threads.
- ☐ **16.** Move the tool back and forth over the damaged area until it turns smoothly.
- ☐ **17.** Tighten the clamp slightly and move it over the damaged area several more times.
- ☐ **18.** Remove the tool and test the repair with the nut.
- ☐ **19.** Repeat steps 14 through 18 until all repairs are complete.

Instructor's Response ____

____

____

____

JOB SHEET 12

Name ______________________________ Date ______________

Install a Helicoil

NATEF Correlation

This job sheet addresses the following NATEF task:

Perform common fastener and thread repair including: remove broken bolt, restore internal and external threads, and repair internal threads with thread inserts.

Objective

Upon completion and review of this job sheet, you should be able to install a helicoil to repair or replace internal threads.

Tools and Materials Needed

Helicoil and installation tool
Drill and drill bit
Pitch gauge
Tap and tap wrench
Face shield
OSHA-approved air blow gun

Procedures

Task Completed

1. Collect the following information concerning this procedure. ☐
   Thread pitch ______________________________
   Bolt diameter ______________________________
   Drill bit required ______________________________
   Tap diameter and pitch ______________________________
   Bottoming or tapered tap required ______________________________
2. Secure the material to be drilled in a vise or similar holding tool. ☐
3. Install the correct bit into the drill. Drill the hole to the bottom or completely through, whichever is appropriate for this job. ☐
4. Clean the bore or hole with compressed air. ☐
5. Select and install the helicoil onto the installation tool. ☐
6. Use the installation tool to screw (install) the helicoil into the bore or hole. Stop when the helicoil is even with the top of the bore or hole. ☐
7. Use a punch or small screwdriver to remove the installation tap at the bottom of the helicoil. ☐
8. Clean the bore or hole with compressed air. ☐

Task Completed

☐ **9.** Screw the bolt into the bore or hole as a test. Did it screw in smoothly and easily? If not, what would now be the proper repair method of this internal thread?

____

____

____

Instructor's Response ______________________________

____

____

____

JOB SHEET 13

Name ________________________________ Date ________________

Soldering a Terminal to a Wire

NATEF Correlation

This job sheet addresses the following NATEF task:
Repair connectors and terminal ends.

Objective

Upon completion and review of this job sheet, you should be able to solder a wire to an electrical terminal.

Tools and Materials Needed

Resin-core solder
Soldering iron or gun
6-inch section of 14-gauge wire or metric equivalent
Crimper pliers
14-gauge (or metric equivalent) terminal
Wire holder (small vise grip pliers will work)
Heat shrink
Heat gun

Procedures

	Task Completed
1. Use the crimpers to strip about 1/4 inch of insulation from the wire.	☐
2. Place the wire (or the terminal) into a holder. If using vise grip pliers, apply just enough force to hold in place.	☐
3. Cut an 8- to 12-inch section of resin-core solder.	☐
4. Plug in the soldering gun (iron) and allow to heat.	☐
5. Slide a sufficient length of heat shrink over the wire.	☐
6. Use the gun (iron) to heat the wire and apply the solder as a tinning agent.	☐
7. Slide the tinned end of the wire into (onto) the terminal.	☐
8. Apply heat to the terminal until solder will melt when touched to the wire and terminal. Do not apply direct heat to the solder.	☐
9. Allow heated solder to flow over and into the wire and terminal. Remove the solder and then the heat when sufficient solder has been applied.	☐
10. Unplug the soldering gun (iron) and set aside to cool.	☐

Task Completed

☐ **11.** Slide the heat shrink over the connection. Turn on the heat gun and apply heat to the shrink until it seals over the connection tightly.

WARNING: Do not apply flame heat to the heat shrink. The shrink will not contract correctly and the seal will not be tight.

☐ **12.** Allow the heat gun and soldering gun (iron) to cool completely before storing.

Instructor's Response ______________________________

____

____

____

JOB SHEET 14

Name ______________________________ Date ______________

Connecting Two Wires with Solder

NATEF Correlation

This job sheet addresses the following NATEF task:

Perform solder repair of electrical wiring.

Objective

Upon completion and review of this job sheet, you should be able to solder two wires to form an electrical splice.

Tools and Materials Needed

Resin-core solder
Soldering iron or gun
Two 6-inch sections of 14-gauge wire or metric equivalent
Crimper pliers
Wire holder (small vise grip pliers will work)
Heat shrink
Heat gun

Procedures

Task Completed

1. Use the crimpers to strip about 3/8 inch of insulation from one end of each section of wire. ☐
2. Position the bare wires so the two meet about midpoint of the stripped area. Twist the two ends around each other in a pigtail wrap. Make the wrap as flat as possible with no protruding strands. ☐
3. Cut an 8- to 12-inch section of rosin-core solder. ☐
4. Plug in the soldering gun (iron) and allow to heat. ☐
5. Slide a sufficient length of heat shrink over the wire. ☐
6. Use the gun (iron) to heat the twisted wires. ☐
7. Apply heat to the wrapped wires until the solder will melt when touched to the wires. Do not apply direct heat to the solder. ☐
8. Allow heated solder to flow over and into the wires. Remove the solder and then the heat when sufficient solder has been applied. ☐
9. Unplug the soldering gun (iron) and set aside to cool. ☐

Task Completed

☐ **10.** Slide the heat shrink over the connection. Turn on the heat gun and apply heat to the shrink until it seals over the connection tightly.

WARNING: Do not apply flame heat to the heat shrink. The shrink will not contract correctly and the seal will not be tight.

☐ **11.** Allow the heat gun and soldering gun (iron) to cool completely before storing.

Instructor's Response ______________________________

____

____

____

Chapter 6

Automotive Bearings, Gaskets, Seals, and Sealant Service

Upon completion and review of this chapter, you should understand and be able to describe:

- How to remove and install a tapered roller bearing and wheel seal.
- How to inspect wheel bearings.
- How to clean and repack wheel bearings.
- How to remove and install an outer bearing race in a hub.
- How to remove and install a sealed hub bearing assembly.
- How to install a lip seal.
- How to remove an engine cover and reinstall it with a new gasket or chemical sealant.

Introduction

Replacing and servicing bearings are typical tasks for automotive technicians. Reassembling automotive components requires the installation of different types of gaskets, seals, and chemical sealants. This chapter will give you a short explanation of these tasks and some items that need to be checked as the repairs or servicing are completed.

SERVICE TIP: When you diagnose a vehicle as having one bad wheel bearing, remember that the other side likely has the same amount of wear. It isn't necessary to replace both bearings at the same time. If the other side is not making noise and is not loose, then it is not bad. Notify the customer that the other wheel bearing has the same amount of service on it. Sometimes he or she will want you to install two wheel bearings during the same visit.

Inspecting Wheel Bearings Before Removal

Inspection of a wheel bearing is usually part of a periodic inspection. There are three methods of inspecting a wheel bearing. Raising the vehicle and shaking the tire while everything is still installed and looking for movement as a result of a loose bearing is the most common method. Removing the bearing and disassembling it is a less preferred method and is only possible with a non-sealed bearing. Test driving the vehicle to listen for a bad bearing is also common.

A sealed bearing assembly can be disassembled for inspection, but you will have to purchase a new bearing assembly after disassembly. Older style non-sealed wheel bearings are easy to inspect but must still be disassembled to inspect, and you must install a new seal to put it back together.

To inspect a wheel bearing during a test drive, you must drive the vehicle in an open parking lot. Always use caution when test driving a vehicle. Turn the vehicle at low speeds and listen carefully for a growling or humming noise that occurs only when turning. After the test drive return to the shop where you can lift it up and inspect the bearings.

When wheel bearings and hubs are a sealed (integral) assembly, the bearing endplay should be measured with a dial indicator stem mounted against the hub. If the endplay exceeds manufacture specifications, the hub bearing assembly must be replaced. A typical specification is 0.005 inch (0.127 mm). Some manufactures require replacement when there is any movement. Photo Sequence 11 shows a typical procedure for measuring front wheel hub bearing endplay.

PHOTO SEQUENCE 11

Typical Procedure for Measuring Front Wheel Hub Endplay—Integral, Sealed Wheel Bearing Hub Assemblies

P11-1 **The vehicle should be properly positioned on a lift and raised to a comfortable working height for performing this measurement.**

P11-2 **Remove the wheel cover and dust cap.**

P11-3 **Attach a magnetic dial indicator base securely to the inside of the fender at the lower edge of the wheel opening. Position the dial indicator stem against the vertical wheel surface as close as possible to the top wheel stud, and preload the dial indicator stem.**

P11-4 **Zero the dial indicator pointer.**

P11-5 **Grasp the top of the tire with both hands. Push and pull on the top of the tire without rotating the tire, and note the dial indicator readings with the tire pushed inward and the tire pulled outward. The difference between the two readings is the wheel hub endplay. Repeat this procedure twice to verify the endplay reading.**

P11-6 **Maximum wheel bearing endplay should be 0.005 inch (0.127 mm). If the endplay measurement is not correct, wheel bearing hub replacement is necessary.**

P11-7 **Remove the dial indicator and install the dust cap and wheel cover.**

Removing a Non-Sealed Wheel Bearing and Seal

We will discuss the removal of a wheel bearing and seal from a non-driving wheel in this section. (Later, we discuss their installation.) Lift the vehicle and remove the wheel. Use a screwdriver or the appropriate special tool to remove the dust cover (Figure 6-1). Under the dust cover, a nut with a **cotter key** will be visible. Remove the key and nut and place to the side. A flat, keyed washer is removed next. Many times the outer bearing will slide out with the washer. Shaking the hub will dislodge the bearing and washer for an easier grip.

With the outer bearing, washer, nut, and cotter key removed, the hub can be removed. However, read the next short section on how the inner bearing and seal can be removed quickly.

Thread the nut back onto the spindle about two or three threads. Grasp the edges of the hub and pull outward and downward (Figure 6-2). The inner bearing will hang on the nut and force the seal out. Force is not needed to perform this task. If the nut will not fit through the hub, or the seal and bearing do not come out, remove the hub and place it on the table with the seal facing upward.

A seal remover or large flat-tipped screwdriver can then be used to remove the seal. Place the tool or screwdriver across the hole in the seal (Figure 6-3). Hook the end of the

SERVICE TIP:

(*continued*)
But if the customer does not want to remove a part that isn't bad yet, then at least he or she will be prepared and aware when the other side does need replacement.

CAUTION:

Before lifting the vehicle, ensure that the lift pads are correctly placed under the vehicle's lift points. Consult the service manual for the location of the lift points. Failure to properly place the hoist lift pads and arms could result in damage and injury.

FIGURE 6-1 **These tools can remove most automotive seals and dust covers.**

SPECIAL TOOLS

Dust cover removal tool
Cotter pin remover

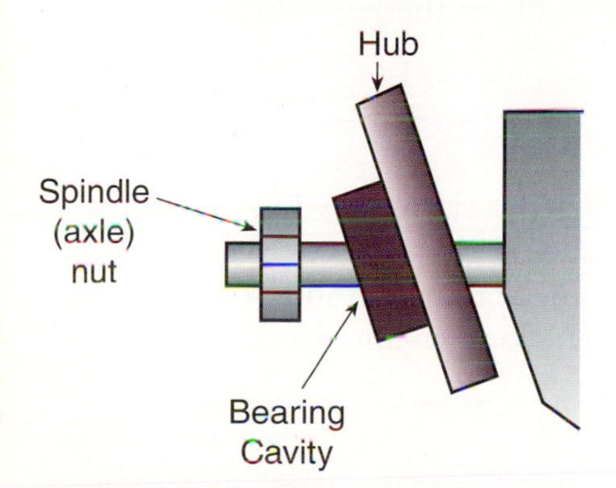

FIGURE 6-2 **Screw the nut slightly onto the spindle and pull outward and downward on the hub assembly.**

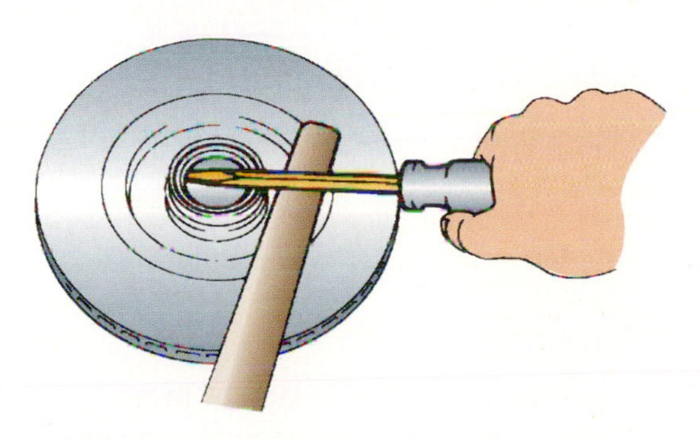

FIGURE 6-3 **A screwdriver can be used to remove many automotive seals.**

Classroom Manual
pages 138, 145

A **cotter key** is a piece of metal doubled back on itself. One use is to keep a nut from loosening.

SERVICE TIP:
The terms *inner* and *outer* do not necessarily refer to the position of the races to the bearing. Inner, with regard to bearings, races, and seals, generally means the part that is closer to the center line of the vehicle. Outer refers to the part furthest from the centerline. For example, an outer bearing, seal, and race will be visible after the wheel and the hub's dust (or grease) cap is removed. The inner ones will not be visible until the hub is removed from the vehicle.

CAUTION:
Always use protective gloves and face protection when cleaning bearings. Skin and eyes can be injured by the cleaning solution.

SPECIAL TOOLS
Seal remover
Blow gun
Parts washer

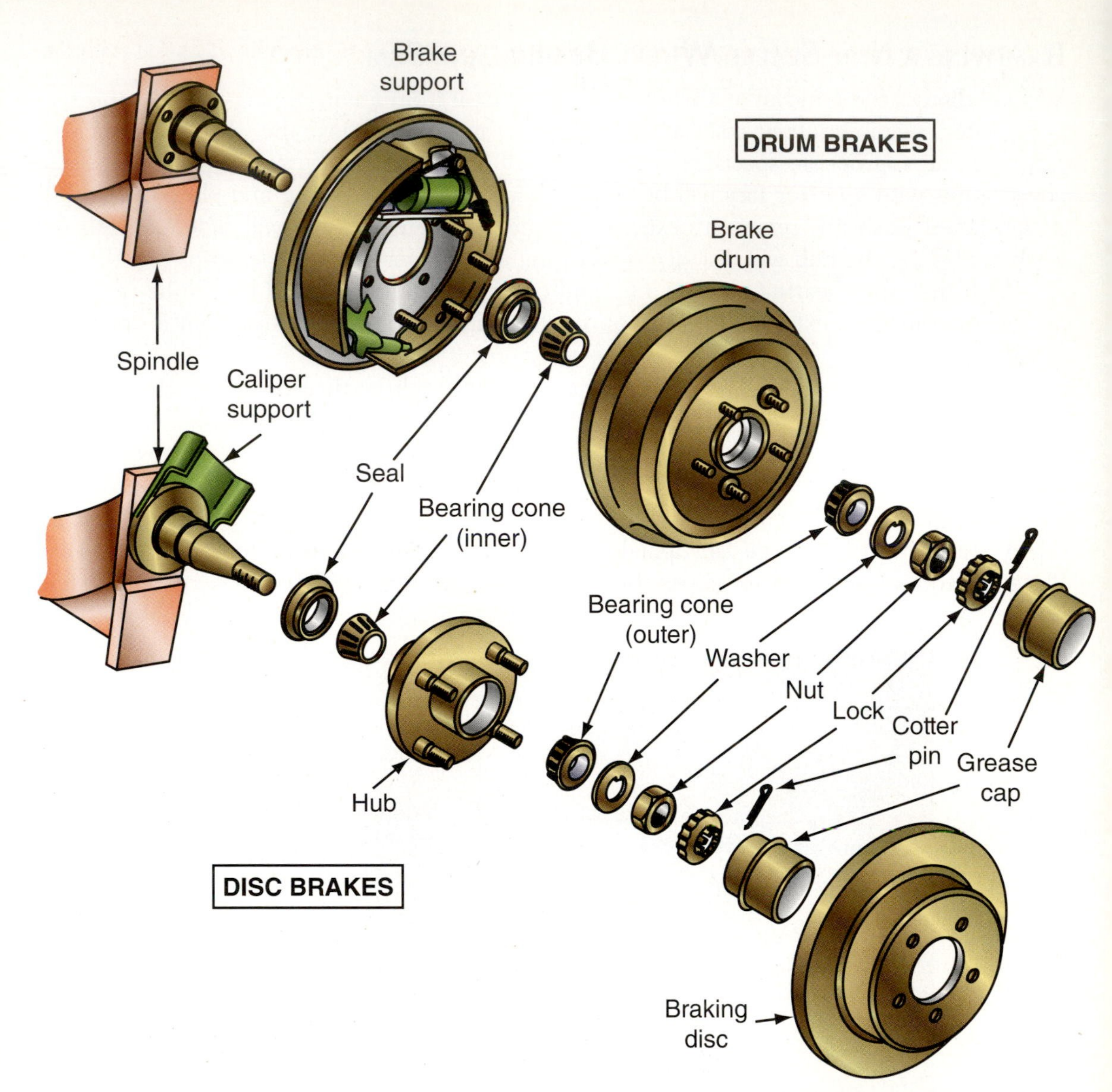

FIGURE 6-4 **Components of a non-sealed wheel bearing (this setup is found on both brake drums and rotors).**

tool or screwdriver under the inner edge of the seal and use the hub as a pivot to apply force to the seal. The seal should pop out fairly easy. The bearing can now be removed. A picture of a typical non-sealed bearing setup and related components is outlined in Figure 6-4.

Before beginning work on the bearings, clean the inside of the hub completely. Dry it with compressed air and set it aside.

Inspecting and Repacking a Wheel Bearing

During a typical brake job, it is usual for the technician to service the **wheel bearings** on non-driving wheels. During the service, the bearings should be inspected. The inspection discussed next applies to every bearing but directly highlights ball and roller bearings. Engine bearing failures are usually made obvious by sound and engine operation.

Inspection

Bearings fail because of a lack of proper lubrication, overloading, or improper installation. In most cases, the design of the total component prevents overloading. Few bearing failures result directly from improper installation, thereby leaving lubrication as the main cause of failure.

Most shops have a standard fee for replacing the brakes on one axle. The fee includes repacking the wheel bearing and a new seal.

Bearing failure is indicated by a noise from the bearing area. At times, it is almost impossible to isolate the problem bearing. Axle bearings can be confused with wheel or differential bearings. A slow test drive around the lot may find the end of the vehicle from which the noise originates. Turning in tight circles may pinpoint the side of the car with the most noise. An experienced technician should perform the test drive if one is needed.

Bearing failure causes two types of damage. When pieces break off the bearing, the damage is called **spalling** (Figure 6-5). Dents in the bearing or race are signs of **brinelling** (Figure 6-6). Either can cause more damage if not replaced. A dry bearing can weld itself to the races and the components by friction-produced heat.

Classroom Manual
pages 148, 149, 150, 151, 152, 156, 157

Spalling comes from the German and it means "to break or split off."

Brinelling, from the French, is a type of bearing failure where dents appear due to the roller "hammering" against the race.

CAUTION:
Do not let the bearing spin while drying it with compressed air. The bearing may fly apart, thereby causing injury, or it could catch skin between the bearing and cage or race.

FIGURE 6-5 **Note how the metal is flaking from this race (left) and bearing.**

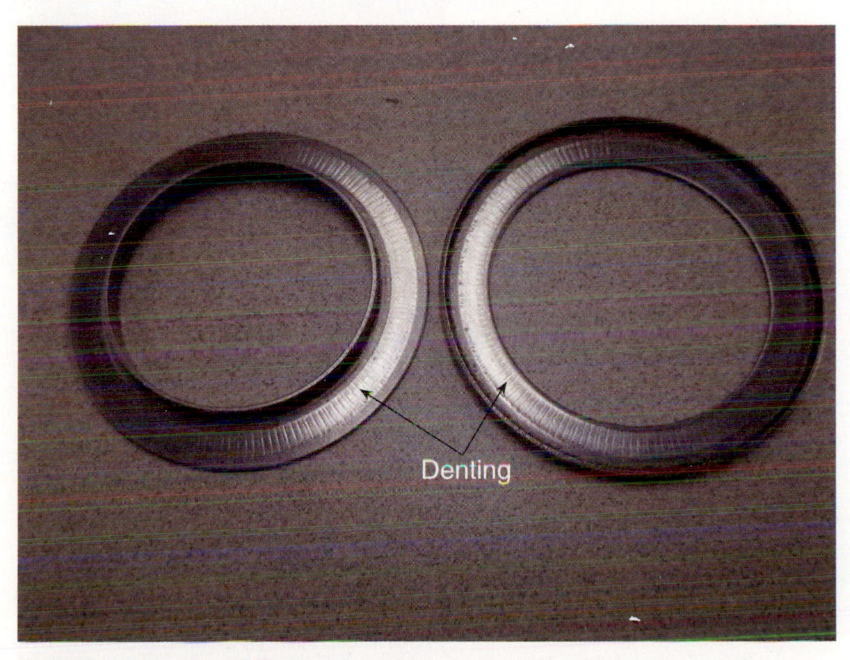

FIGURE 6-6 **Denting of the bearing metal is caused by the bearing moving up and down within its races. This may be a result of improper installation.**

SPECIAL TOOLS

Parts washer

Blow gun

Catch pan or vat

A **burned area** will be bluish in color.

The **shaft** with the type bearings discussed here is called a *spindle*. The shaft (or spindle) is part of a steering knuckle, which is replaced as a unit.

SPECIAL TOOL

Bearing packer

Remove the suspected bearing and clean it thoroughly with a parts cleaner. Ensure that all of the old lubricant is removed. The bearing can be air dried with compressed air. This will help remove the cleaning solution and any particles trapped in the bearing. However, when using compressed air to dry a bearing, there are some critical things that must be done. Hand protection must be used to protect the skin from the cleaner, compressed air, and any particles dislodged by the air (Figure 6-7). Eye and face protection should always be worn during the cleaning process. The bearing must be held in a manner to prevent any of the rollers or balls from spinning. A free-spinning bearing can disintegrate, thereby causing injuries. If at all possible, use a regulator to reduce air pressure. This will lessen the possibility of injury due to spinning bearings and loose particles within the bearing.

Dry the bearing over a vat or pan to catch the cleaner and particles blown off by the air (Figure 6-8). The waste will probably have to be treated as hazardous waste and stored accordingly. With the bearing cleaned, clean the races. Most can be cleaned while still installed in the component.

With all of the bearing components clean and dry, inspect each ball or roller for dents, nicks, **burned areas,** or any other damage to the machined surfaces (Figure 6-9). Inspect the cage for any bending or breakage. Inspect the races as much as possible. Most races can be inspected while installing a component. Also, inspect the component into which the bearing and races are fitted. If a race has been removed from the hub or **shaft,** carefully inspect the area where the race fits (Figure 6-10). This area must be as clean and undamaged as other parts of the bearing. Any type of damage to any part requires replacement of the complete bearing, including the races.

If the shaft or hub is damaged, consult the owner. Replacing or machining some shafts and hubs can be an expensive job. While there is not much choice concerning the repairs needed in this situation, the owner is the one paying and has the final word based on the technician's recommendation.

FIGURE 6-7 **Protective gloves can be used by automotive technicians, and they can still retain a good sense of touch.**

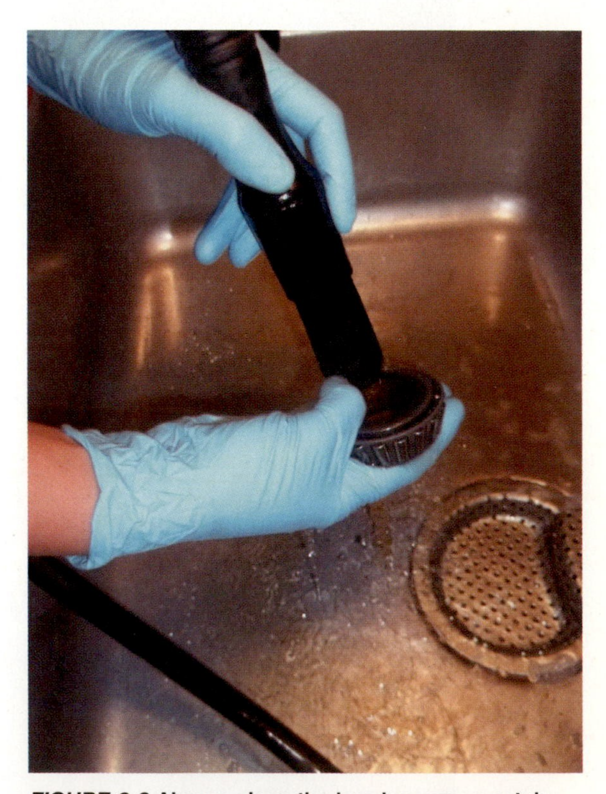

FIGURE 6-8 **Always clean the bearing over a catch container or parts washer to capture the old grease and debris you have removed.**

Tapered Roller Bearing Diagnosis

Consider the following factors when diagnosing bearing condition:

1. General condition of all parts during disassembly and inspection.
2. Classify the failure with the aid of the illustrations.
3. Determine the cause.
4. Make all repairs following recommended procedures.

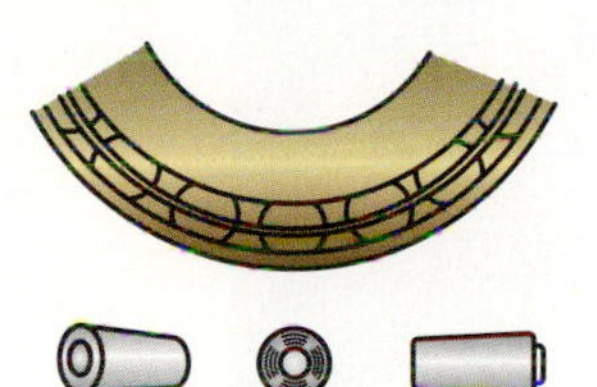

Abrasive Step Wear

Pattern on roller ends caused by fine abrasives.
Clean all parts and housing, check seals and bearings and replace if leaking, rough, or noisy.

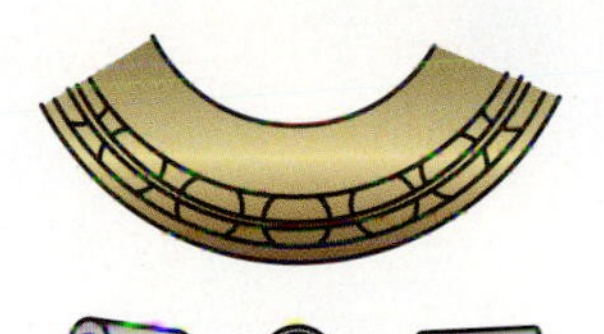

Galling

Metal smears on roller ends due to overheating, lubricant failure, or overload. Replace bearing, check seals and check for proper lubrication.

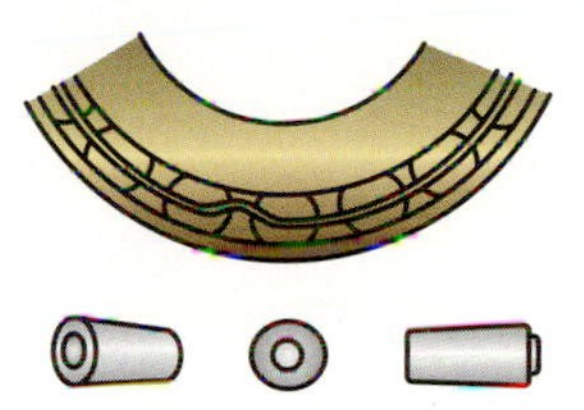

Bent Cage

Cage damaged due to improper handling or tool usage.
Replace bearing.

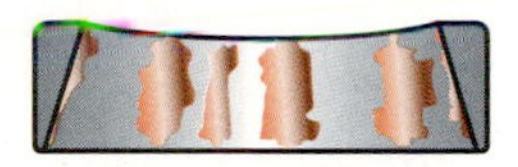

Abrasive Roller Wear

Pattern on races and rollers caused by fine abrasives.
Clean all parts and housings, check seals and bearings and replace if leaking, rough, or noisy.

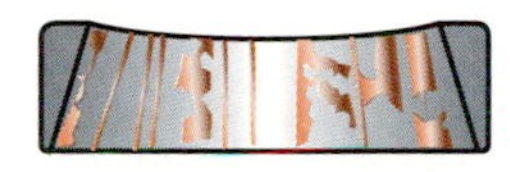

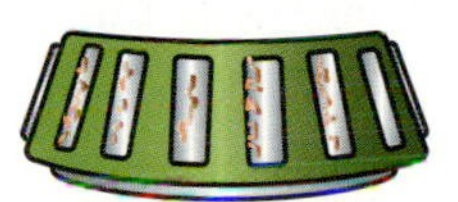

Etching

Bearing surfaces appear gray or grayish black in color with related etching away of material usually at roller spacing.
Replace bearings, check seals, and check for proper lubrication.

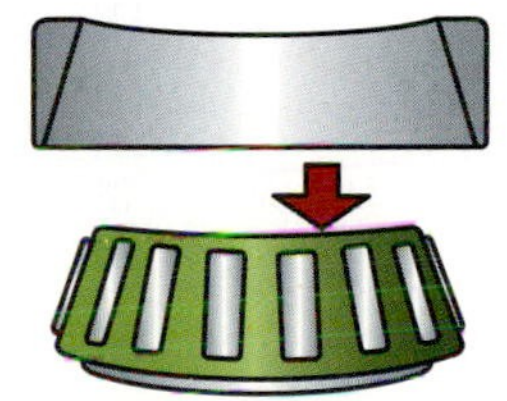

Bent Cage

Cage damaged due to improper handling or tool usage.
Replace bearing.

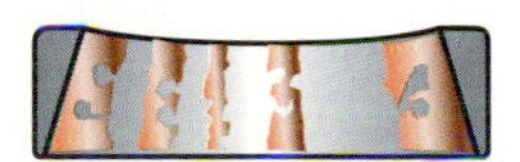

Indentations

Surface depressions on race and rollers caused by hard particles of foreign material.
Clean all parts and housings. Check seals and replace bearings if rough or noisy.

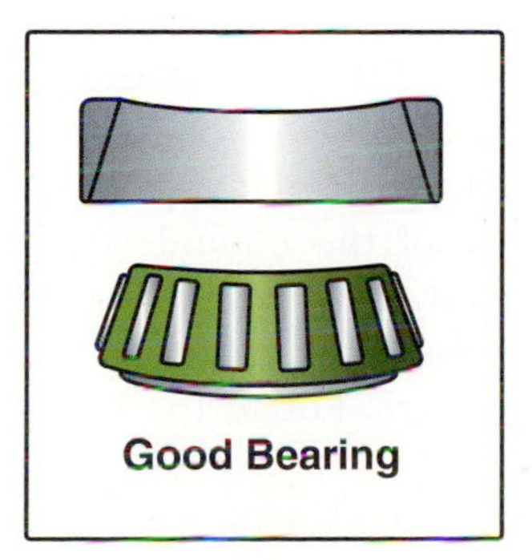

Good Bearing

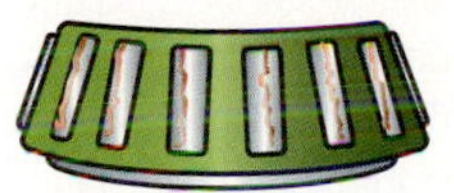

Misalignment

Outer race misalignment due to foreign object.
Clean related parts and replace bearing. Make sure races are properly sealed.

FIGURE 6-9 **Typical bearing and race problems and their probable causes.**

FIGURE 6-10 **The spindle or shaft must be as clean and as clear of damage as the bearings and races.**

Repacking the Bearing

Classroom Manual
pages 139, 140, 141

For a bearing to be completely lubricated, each roller or ball must be greased before installation on the vehicle. *Repacking* is usually the term used to indicate the procedure for prelubricating the bearing.

Several types of tools can be used to repack bearings. Some operate by using compressed air whereas others are hand operated (Figure 6-11). The tools that have been around the longest and are sometimes the fastest and most accurate are, quite simply, the technician's hands. That is the procedure we will discuss here.

Before attempting to repack the bearing, ensure that it is clean and dry. Parts cleaner left on the bearing will break down the grease and lubrication will be reduced. Also ensure that the correct grease is being used. It should be multipurpose or a G classification. After repacking, make sure there is a clean strip or ribbon of grease extending upward and past the ends of each roller (Figure 6-12). Even with the best cleaning, a little dirty grease may sometimes be trapped within the bearing. The new grease will force it out.

A rough cloth will remove most of the **grease** from a technician's hands after the bearing has been repacked.

With the grease on hand and the bearing clean and serviceable, scoop some **grease** and place it in the palm of one hand. The amount of grease depends on the technician's hand, of course, but it should be about the size of half a Mounds candy bar. Holding the bearing with the smaller cone (end) up, press a portion of the outer edge of the bearing down into the grease from the top of the mound (Figure 6-13). When the bearing edge touches the palm, press just slightly harder and drag that edge across the palm toward the wrist. It does not have to be dragged far. Repeat this step until clean grease ribbons are visible at the top end of the rollers. Rotate the bearing until the next rollers are aligned to be placed in the grease. Repeat the pressing, dragging, and rotating until all of the bearing rollers have clean grease between them. More grease may be needed to replenish what is in the palm. Before placing the bearing on a clean cloth or rag, spread a light coat of grease around the outside of the bearing. This will provide lubrication during the first several rotations of the bearing within its supported component. Repeat the procedures for any other roller bearing.

FIGURE 6-11 **Typical hand-operated bearing packer.**

FIGURE 6-12 **Until ribbons of grease appear above the rollers, the bearing is not properly packed.**

Ball bearings can be repacked in a similar manner, but it may take a little longer. Most automotive ball bearing assemblies are not used in areas where they have to be repacked. Many ball bearing assemblies are prelubed and sealed at the factory. Some ball bearings are exposed to the lubricant within the component. However, new ball bearings used in these applications should be prelubed before installation. The best way is to drip the bearing in the same lubricant used in the component. A new bearing will run dry and be damaged if not prelubed in some manner. If the technician works on large vehicles,

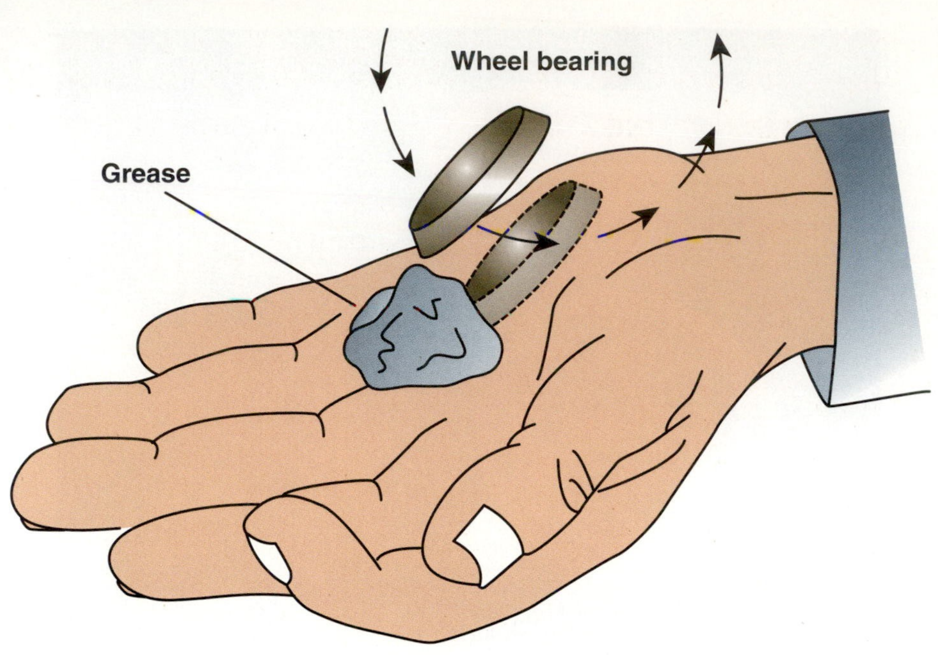

FIGURE 6-13 **The bearing is moved down through the grease and rotated toward the heel of the palm.**

SPECIAL TOOLS

8-ounce ball peen hammer

6-inch punch

Spacer material

Classroom Manual
page 133

CAUTION:
Always wear gloves and use a chisel or punch holder when removing or installing races and seals with a driving tool. A misaimed swing could result in a serious hand or finger injury.

it is common practice to prelube the large bearings even if it is an old one being reinstalled.

Repacking a bearing with packing tools is fairly easy and much cleaner because each tool does what can be done by hand. The bearing is placed in a cone-shaped holder and another cone fitted over it (refer to Figure 6-11). The grease is forced through the bearing rollers. Ensure clean ribbons of grease protrude past the rollers and some grease is applied to the outside by hand.

Removing and Installing a Race in a Hub

A race is removed from a hub or shaft with a hammer, punch, or a puller. On the hub we are working with in this chapter, the races can be removed with a hammer and pin punch. Set the hub on a solid surface such as a metal workbench. If the race for the inner bearing is being removed, some type of spacers may have to be placed under the hub so the race has room to slide out. Two short pieces of 2 × 4 lumber work fine without damaging the hub. The punch will enter the hub from the opposite side of the race being removed.

A typical hub with two bearings and one seal will be used for this purpose. Photo Sequence 12 shows a typical procedure for removing a bearing race. The procedures can be followed on other applications except that some bearings may need to be pressed out and in. This is best done by an experienced technician.

With the hub cleaned and in position, observe downward through the opening. A very small edge of the bottom race or two notches in the hub directly behind the race will be visible (Figure 6-14). In many cases, both are visible. The notches provide a place to rest the end of the punch against the race.

Set the punch in place and strike it sharply with the hammer (Figure 6-15). Move to the other notch and repeat. Do not strike the punch very hard unless absolutely necessary. Most races will pop out without too much trouble. It will be necessary to move the punch from notch to notch to keep the race fairly even as it is forced out. This is a simple task, and there is no need to rush. Rushing usually causes a mistake with the hammer and results in a sore hand or finger. Repeat the procedure for the other bearing race.

Removing and Installing a Bearing Race on a Non-Sealed Wheel Bearing

P12-1 Before attempting to remove the bearing race, clean the hub thoroughly.

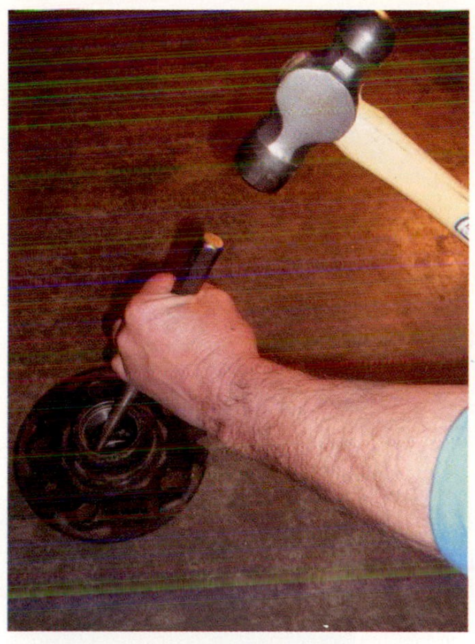

P12-2 Flip the hub so the race to be removed is down. It may be necessary to place short 2 x 4 strips under the hub for clearance.

P12-3 With the driving end of the punch held against the inner edge of the race, use the hammer to drive the bearing race out.

P12-4 The space where the bearing was installed is plainly visible.

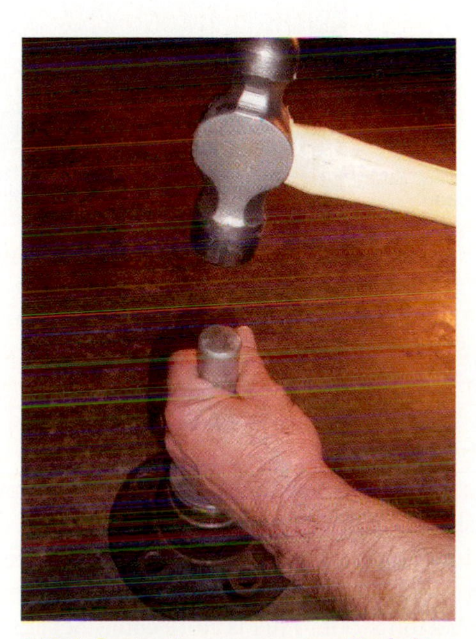

P12-5 Center the new race over the hub's cavity and fit the driver into the race.

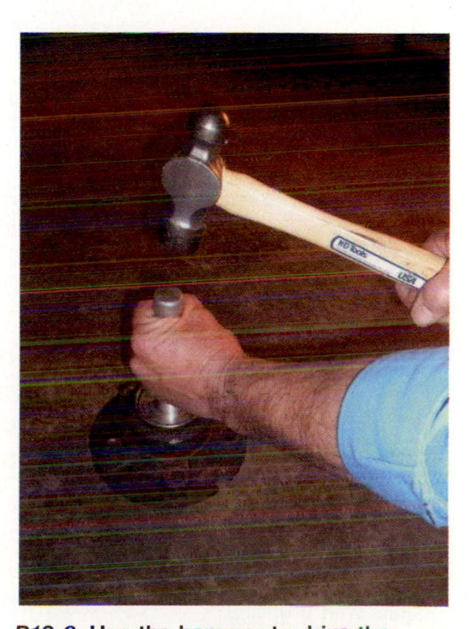

P12-6 Use the hammer to drive the bearing race into the hub. Continue to drive until the race bottoms against its stop inside the hub.

PHOTO SEQUENCE 12 (CONTINUED)

P12-7 **Typically when a bearing race is fully installed, there will be a space between the outer edge of the race and the outer edge of the hub.**

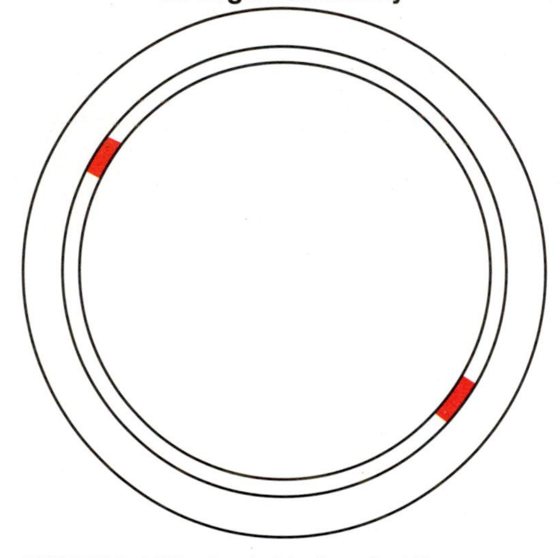

FIGURE 6-14 **The backside (inner) of the race can be seen through the two notches at the lower edge of the hub's opening.**

FIGURE 6-15 **A hammer and pin punch are used to drive the race from the hub.**

CAUTION:
Always wear gloves and use a chisel or punch holder when removing or installing races and seals with a driving tool. A misaimed swing could result in a serious hand or finger injury.

Installing the race is pretty much the opposite of the removal. A bearing and seal driver should always be used to install a race. The driver set comes with different sized drivers (Figure 6-16). Select a driver that fits snugly in the bearing side of the race and will slide inside the hub opening.

Set the race over the opening with the larger opening upward. Place the driver into the race and tap slightly with a hammer (Figure 6-17). The race and driver must be completely aligned with the opening. If the race is off just a little, it will not enter the opening or will be cocked in the cavity. Once the race is started, harder strikes can be made with the hammer until the race bottoms out. This is noticeable because of the sound made when the race hits the bottom. The sound will be dull compared to the sounds made when the race is being driven in. Remove the tool and inspect the race for nicks or scratches. There should be no damage if the proper driver is used. Flip the hub over and install the other race.

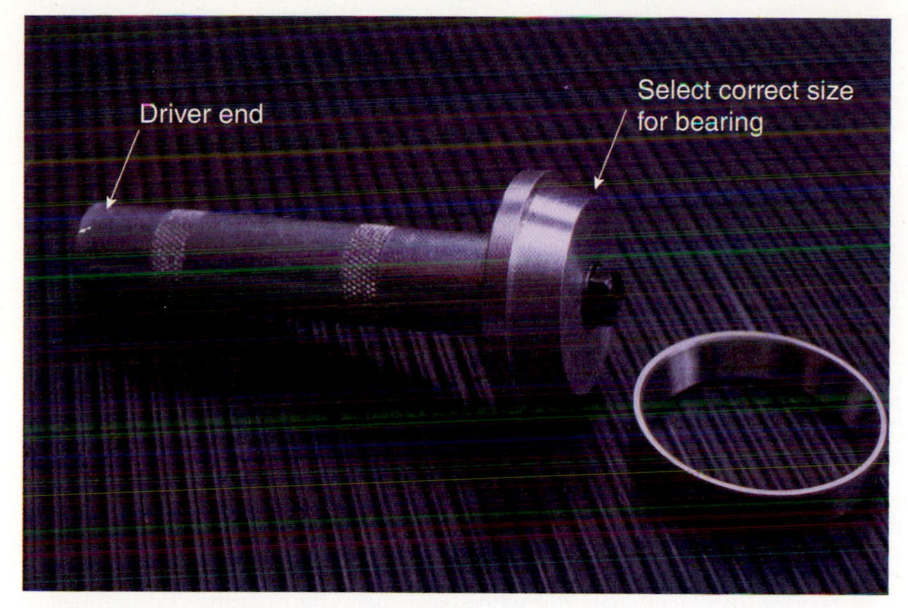

FIGURE 6-16 A bearing and seal driver set.

SPECIAL TOOLS

8-ounce hammer

Bearing and seal driver

Classroom Manual

pages 148, 149, 150, 156, 157

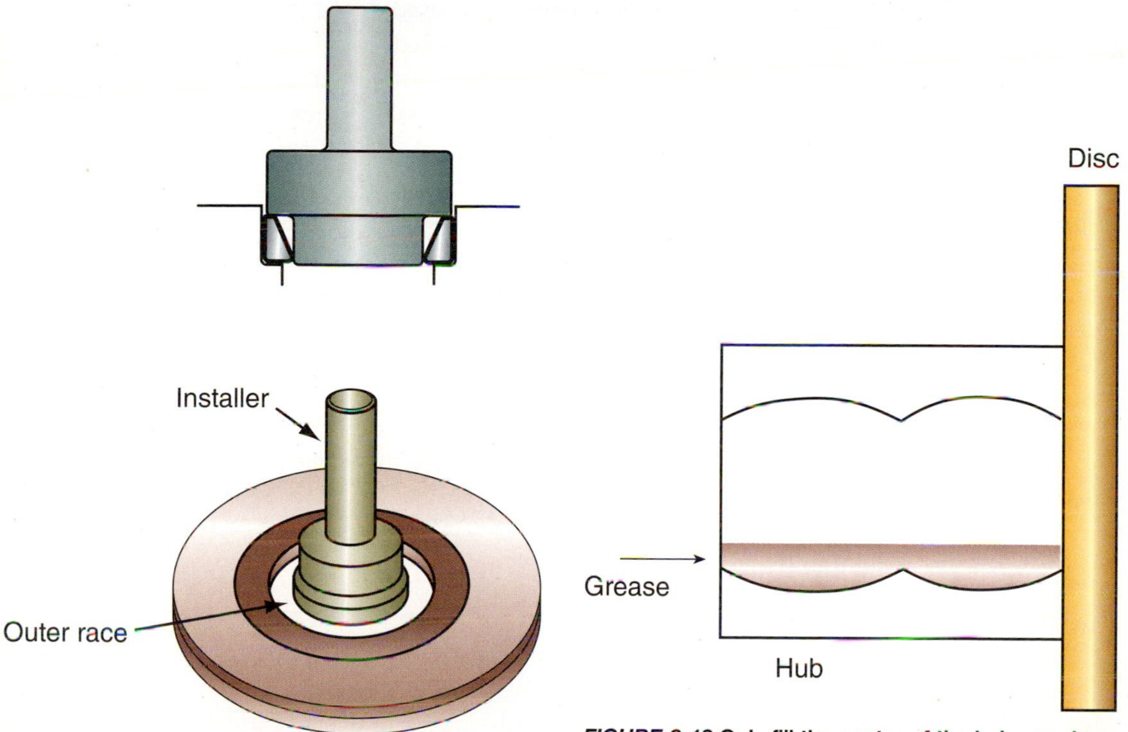

FIGURE 6-17 Proper selection and positioning of the correct driver prevents damage to the race.

FIGURE 6-18 Only fill the center of the hub anywhere from a third to half full of grease. Too much could ruin the bearing.

SERVICE TIP:
In the past, the hub grease cap was packed with grease. Many technicians, and most vehicle owners, figured that if a little bit is good, then a lot must be *very* good. Before you fill that hub cavity with grease, think about what happens inside that hub after several hours of driving at interstate speeds (actually it doesn't take that long). The grease dissolves into a near liquid, increasing pressure within the hub. The seal is not designed to retain that pressure nor the semiliquid; grease is forced out of the hub. The problem starts when the vehicle stops for a period of time and what remains of the grease solidifies.

Installing a Wheel Bearing and Seal

With the race installation completed, the inner bearing and seal can be installed. Put some loose grease into the hub. It should fill the inner portion of the hub's opening but not overfill it (Figure 6-18). Excessive grease here will cause heat buildup, which will melt the grease on the bearing, thereby causing damage to the bearings and seal. Place the hub with the inner bearing race facing up. This is normally the larger race of the two.

Set the greased inner bearing into its race. It should drop into the race far enough to be almost flush with the race. If it is not flush, it is probably in backward. The bearing should not block the seal.

SERVICE TIP:
(*continued*)
At the next start-up, the now-dry bearings will have little or no lubrication. In a short period of driving, the damage is done.

FIGURE 6-19 **Installing a seal with a seal driver.**

With the bearing in place, align the wheel seal over the opening. Remember to place the seal's lip inward or toward the lubricant (Figure 6-19). The same driver used to install the race can be used to install the seal, but normally a hammer is used. Hold the seal in place by hand and tap slightly around the edges of the seal until it is started evenly. Use the hammer to tap the seal the rest of the way into the cavity. If the seal has a flange that overlaps the hub, the sound will change as the flange mates with the hub. Many seals do not have a flange. They are driven in until the outer edge of the seal is even or flush with the outer edge of the hub.

There are many ways used to adjust the wheel bearing tension. The only right way is to follow service manual instructions. The other ways rely on touch, feel, and guessing.

SPECIAL TOOL
Torque wrench

Reinstall the hub and slide the outer bearing over the spindle and into the hub. Screw the nut on while turning the hub. This helps center the bearings on the spindle and in the hub. Consult the service manual for the torquing specification on the nut. Many times the nut is tightened to a torque and then backed off a certain amount. Overtightening may result in bearing failure because of excessive friction and heat. Undertightening will also damage the bearing and cause the wheel assembly to be loose on the spindle.

With the nut torqued, select a new cotter key and install it. Place some grease into the dust cup, about 1/3 full, and install it into the hub.

Replacing a Sealed Hub Bearing Assembly

Classroom Manual
pages 138, 139

Sealed wheel bearing removal and replacement will vary from model to model. Most removal procedures are similar, or follow a similar pattern. In this section, you will learn the typical procedures needed to remove and replace a bolted-in sealed hub bearing assembly. The procedures for replacement of a pressed in sealed bearing are similar, with the exception of a press needed to move the bearing out. In practice, always remember to follow the manufacturer's guidelines for replacement.

If the bearing is on a driving axle, the axle shaft (or CV shaft) must be removed from the hub unit first. This usually includes using an air tool to remove the axle nut and a hammer and brass punch to move the shaft inward. In some cases, a special hub puller can be used (Figure 6-20). In most cases, you will only need to remove the outer axle from the hub. Some vehicles will require removal of the strut-to-knuckle bolts (Figure 6-21), the ball joint (Figure 6-22), and the outer tie rod end in order to remove the shaft. After removing the shaft, remove the bolts of the hub assembly from the knuckle (Figure 6-23).

At this point, the hub bearing assembly should just come off, but it won't because it is a close and tight fit and there is usually corrosion. You will have to use a chisel and hammer, and sometimes heat, to get the assembly out. Once it is out, clean the hub surface of

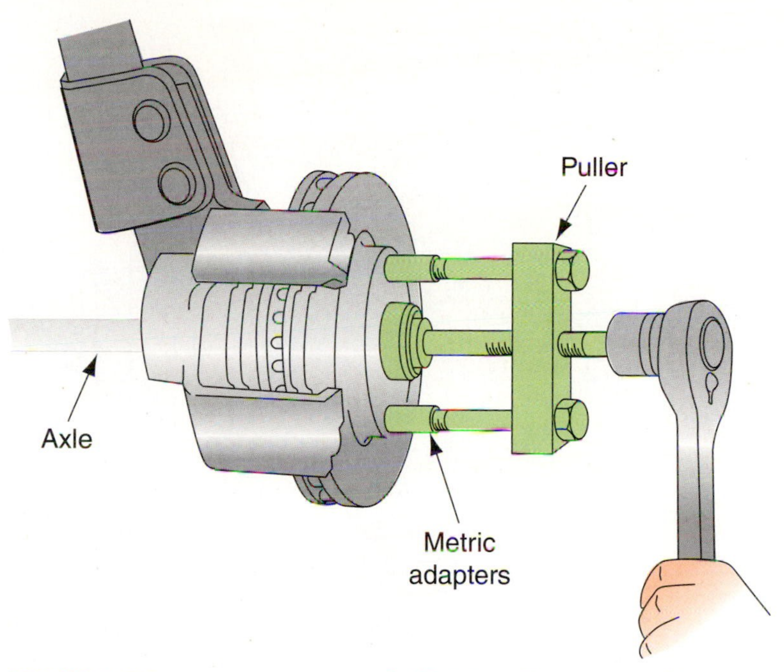

FIGURE 6-20 **Removal of the outer axle joint from the front wheel hub.**

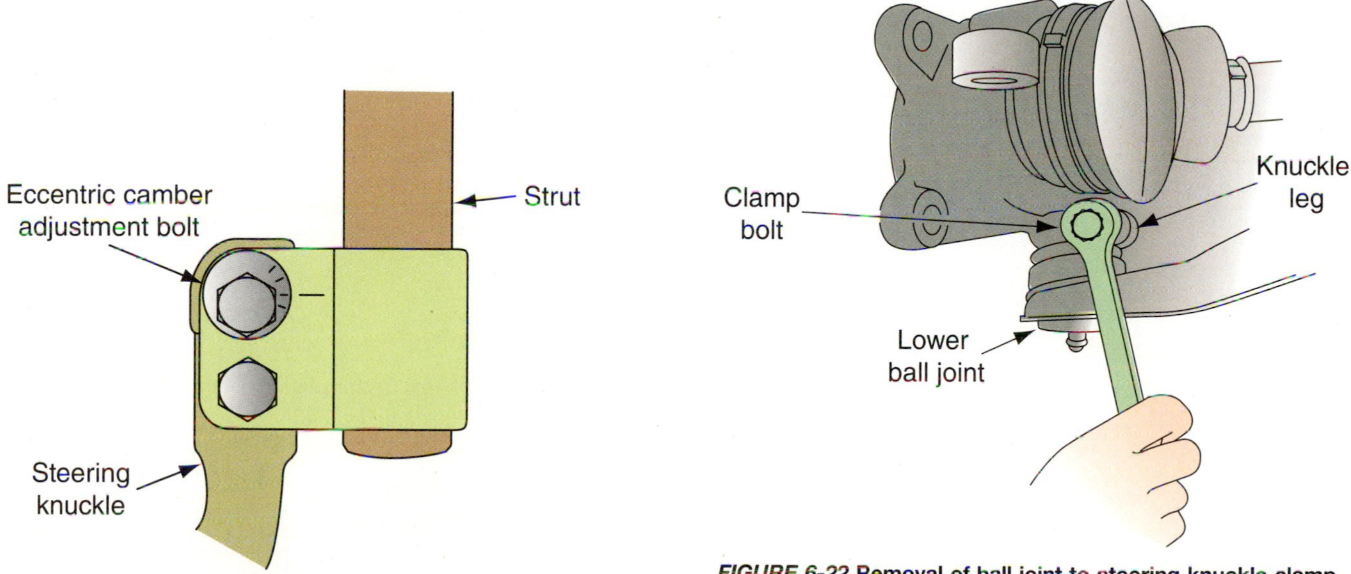

FIGURE 6-21 **Mark eccentric strut bolt before removal.**

FIGURE 6-22 **Removal of ball joint to steering knuckle clamp bolt.**

any rust or corrosion before reinstalling the new one. Reverse the procedure to install. Make sure to torque all fasteners and the wheel nut to specifications (Figure 6-24). It helps to have someone holding the brake or wedging the tire against the ground (just like when torquing lug nuts, do not let the vehicle all the way down to the ground when torquing the nut). Make sure to reinstall the cotter pin (Figure 6-25) and tighten the lug nuts in the proper order (Figure 6-26).

A sealed bearing that is a press fit must be removed with the knuckle (Figure 6-27). This is a more advanced technique that you will learn later. You can reference the book *Today's Technician: Automotive Suspension and Steering*, 5th edition, for more information.

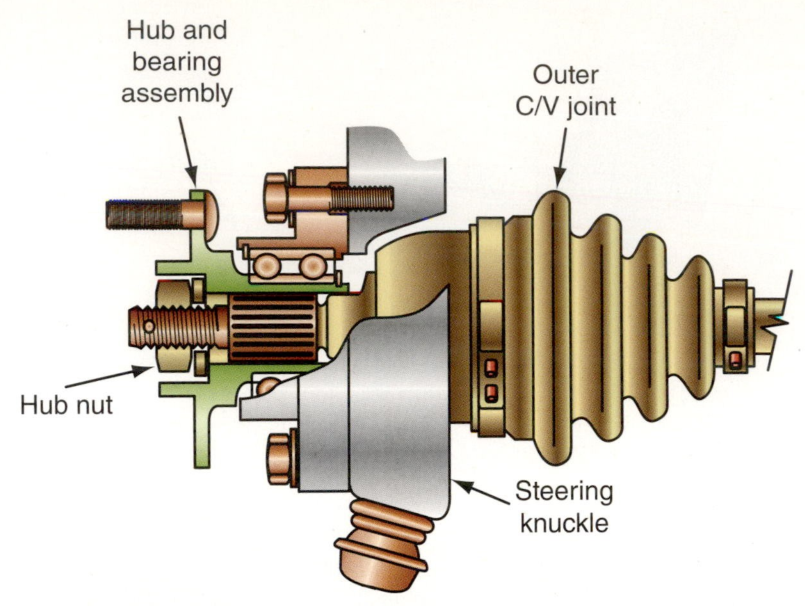

FIGURE 6-23 **Remove the bolts attaching the hub to the steering knuckle.**

FIGURE 6-24 **Hub nut torquing.**

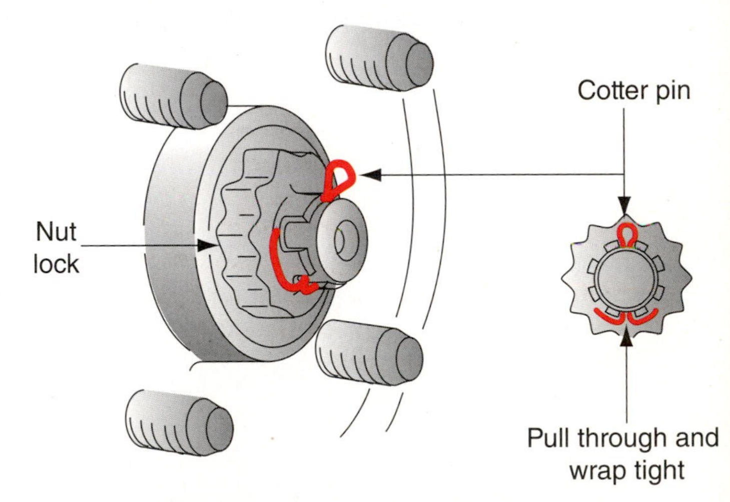

FIGURE 6-25 **Nut lock and cotter pin installation.**

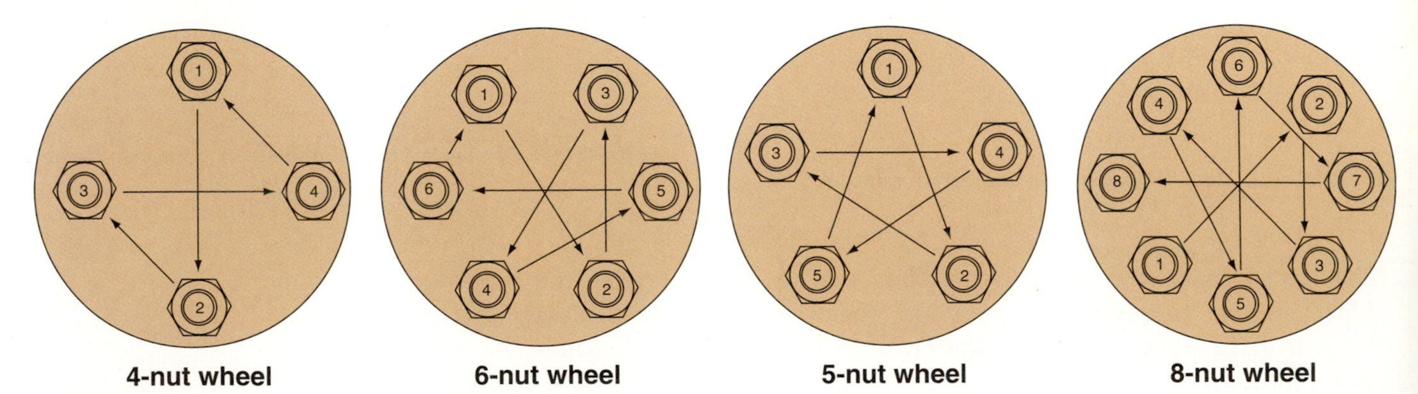

FIGURE 6-26 **Wheel nut tightening sequence.**

Classroom Manual
pages 144, 145

Using a Chemical Sealant as a Gasket

This section is intended as a practice exercise. If cost is a factor, a tube of cheap toothpaste can be used instead of Room Temperature Vulcanizing (RTV). Find a scrap engine, transmission, differential cover, or a plain piece of flat metal. If a flat piece of metal is used,

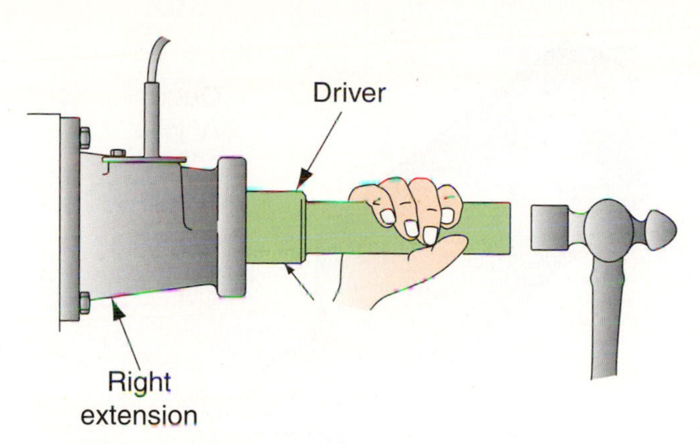

FIGURE 6-27 **Removal of left inner drive axle joint in Ford automatics transaxles.**

SERVICE TIP: Cutting the tube at an angle allows only a small portion of the tube to rest on the cover. This helps prevent spreading the RTV.

SERVICE TIP: Do not place RTV on a gasket of any material unless directed to do so by the gasket installation instructions. The RTV forms a rubber-like bead and will not stick to cork and most other types of gaskets, thus forming a leak between the RTV and gasket. Typically, the only time RTV and gaskets are used together is when RTV is used to seal the pieces of a multipiece gasket.

a marker can be used to track and simulate bolt holes. Any of the items can be cleaned after the practice is completed. The technician will have to use RTV in many applications.

Remove the cover from the RTV tube and cut it at an angle until its opening is about 1/8 inch in diameter (Figure 6-28). Once the bead is started, do not break it. If it is necessary to move a hand, keep the tube end in place and begin to squeeze the tube again from that point. Also, when going by a bolt hole, lay the bead to the inside (lubricant side) of the hole (Figure 6-29). Do not run the bead over the hole.

Place the end of the tube at a convenient starting point. The tube should be angled at 45 degrees to the cover (Figure 6-30). Squeeze from the bottom and continue squeezing as the tube is moved. The tube should be moved with an even spread. The RTV should come out in a smooth, even flow without humps or dips in the bead. Continue until the bead is laid completely around the cover. Finish the bead by rolling the tube tip into the starting point of the bead. This will mesh the two ends together and provide a leak-resistant seal.

Before reinstalling the cover or pan or applying sealant to a cover/pan, it should be checked for straightness. It is not unusual for a thin stamped piece of metal like the typical cover or pan to become deformed around the clamping flange. This may be caused by overtightening on the last installation or the stresses of operation. Either way a warped cover or pan will not properly seal. Any time a pan or cover is removed it should be checked for straightness along its flange. For the purpose of this discussion, the term *pan* will be used because it is the most accepted. An oil pan is used for this discussion.

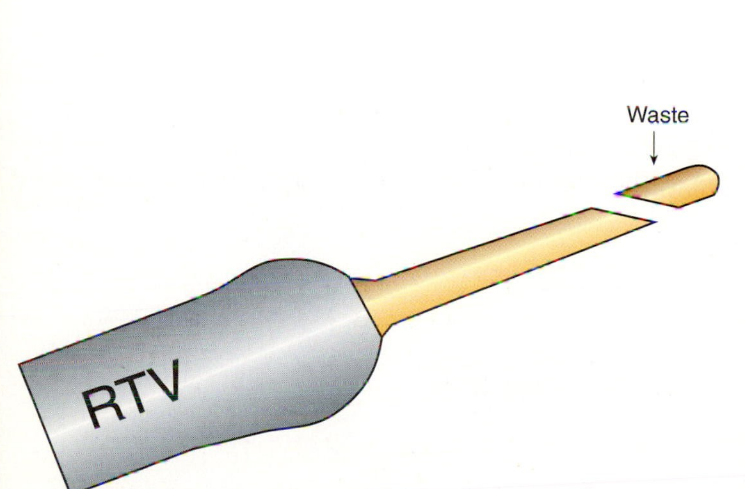

FIGURE 6-28 **Cut the RTV tube at an angle.**

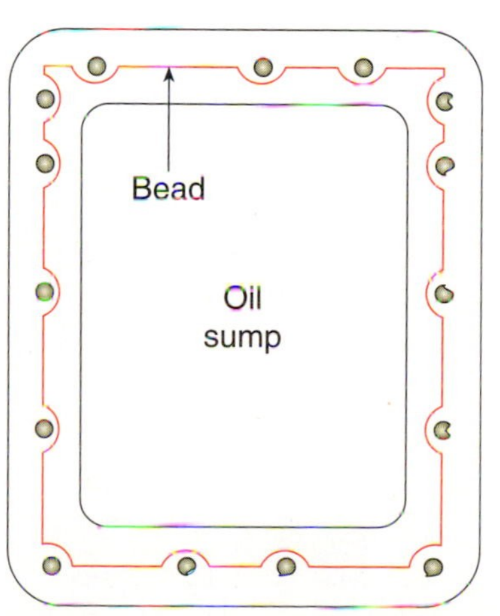

FIGURE 6-29 **Lay the RTV bead in the manner shown. Keep the bead to the lubricant side of the bolt holes.**

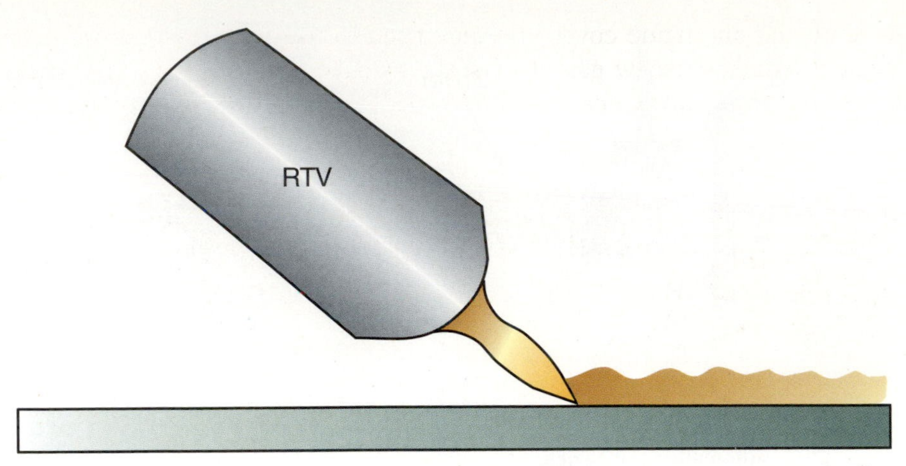

FIGURE 6-30 **Keep the point of the tube to the metal at about a 45-degree angle. The angled cut will allow for a smooth bead.**

There are several means for straightening the clamping flange of a pan but all require a solid, flat surface such as a heavy steel work table. The one discussed here is probably the most common technique. Place the oil pan right side up on the table and use a straightedge to locate high and low spots along the flange. Emphasis should be directed along the line between the bolt holes. Use some type of marker to mark the start and end of each low or high spot. Compare these spots to determine if the flange is warped downward here or if the spot is caused by an area that has been warped upward at another spot. Any high-warped areas can be flattened more easily than trying to raise warped low areas.

Place the oil pan upside down on the table with the worst side most easily accessible. Now the low areas are the high areas. Locate or make a 3- to 4-inch strip of metal or hard wood (such as ash or oak) block with a width equal to that between the oil pan body and the outer edge of the flange and several inches tall. It should be of strong enough metal or wood to transfer hammer blows or clamping force to the flange without deforming. It depends on the thickness of the flange and its metal composition.

SERVICE TIP:
Check with a machine shop. Sometimes they will have a metal jib that can do the straightening job much easier—or at the least they could machine a block for use in this type of repair. Most vehicle pans have about the same width of flange.

Secure the pan or have an assistant hold it in place. Place the block onto the first area to be straightened. The block should be laid lengthwise along the flange (Figure 6-31). Use a ball peen hammer to strike the block at different places along its length. Do not use excessive force with the hammer. Repetitive midforce blows will do a better job than hard blows. A couple of hard blows may cause even more warping. Continue to work the area until it is flat for the length of the working block. Check it with the straightedge before continuing. With one area straightened, move the block along the flange, taking care to have the block overlap the previously straightened area. Continue in this fashion completely around the pan or as necessary. The corners will provide some difficultly. However, the use of a shorter piece of metal or a flat nose punch with a diameter approximately the width of the block can be used at those points. Skill in this type of job is gained with experience.

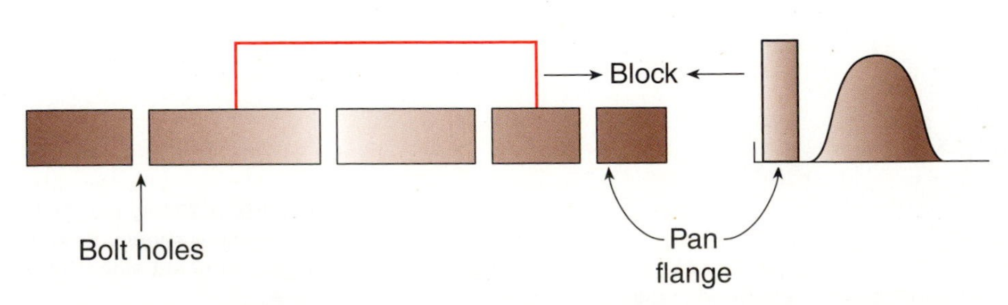

FIGURE 6-31 **Lay the block so the hammering force is transmitted over a larger area.**

When removing an engine cover you may need to use RTV. It's use is becoming very popular. When installing a new gasket or sealant, make sure to follow the manufacturer's service directions. Photo Sequence 13 outlines the typical procedures for replacing a valve cover gasket.

PHOTO SEQUENCE 13

Removing and Installing an Engine Valve Cover with a New Gasket

P13-1 **Consult the service information for instructions on how to replace the valve cover gasket properly.**

P13-2 **Remove the air distribution ductwork to gain access to the valve cover.**

P13-3 **Remove and label connectors and vacuum lines as needed.**

P13-4 **With the intake ducts out of the way, remove the spark plug wire loom and wires that will obstruct removal.**

P13-5 **Loosen the valve cover bolts using the proper sequence if one is specified.**

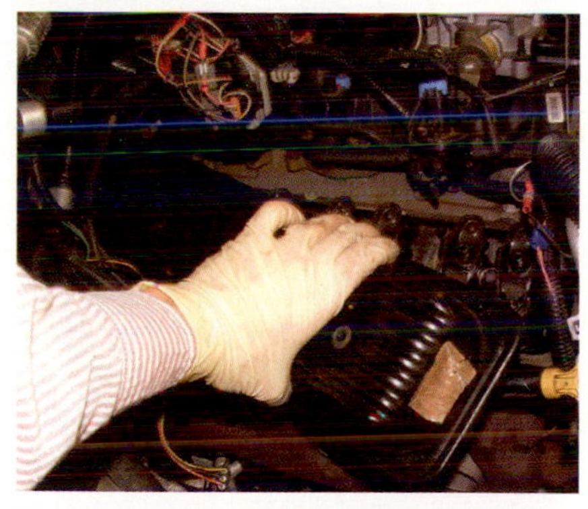

P13-6 **Maneuver the valve cover off the cylinder head.**

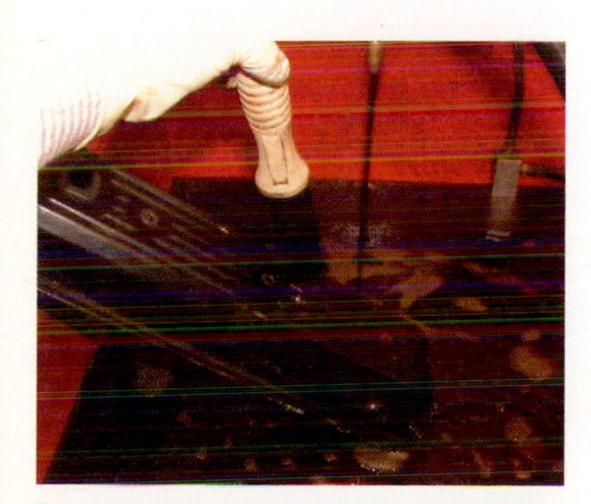

P13-7 **Remove the old gasket material, and thoroughly clean and dry the valve cover.**

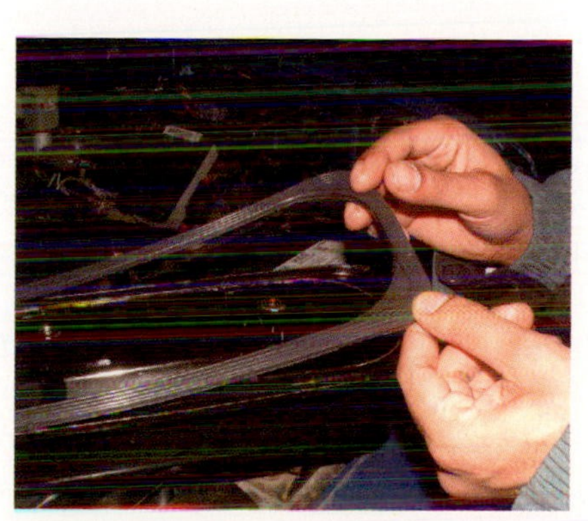

P13-8 **Install the new valve cover gasket securely. Use sealant only if it is specified.**

P13-9 **Always properly torque the valve cover to the specification to ensure a quality repair.**

PHOTO SEQUENCE 13 (CONTINUED)

P13-10 **Reinstall the ductwork securely.**

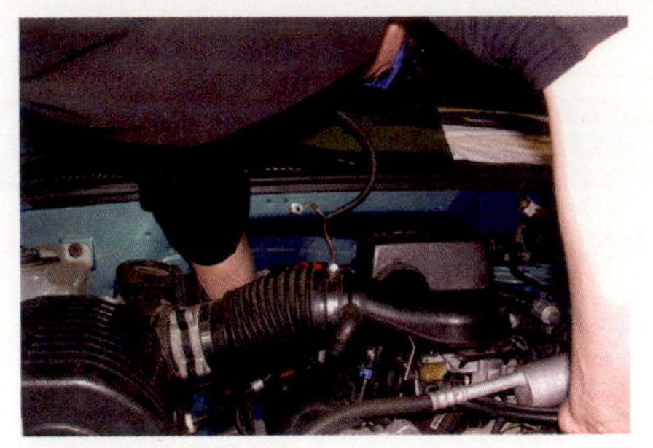

P13-11 **Reconnect all wiring and vacuum lines. Wipe down the valve cover and surrounding components. Finally, start the vehicle and check to be sure it runs properly and has no leaks.**

CASE STUDY

A technician performed a routine job of replacing the differential fluid. She sealed the cover with RTV as recommended by the service manual. The customer returned some weeks later with a small but unsightly leak around the cover. The technician removed, cleaned, and installed the cover. RTV was used. Within a few days, the customer returned with the same problem except the leak was now worse. A careful inspection of the cover and housing revealed no visible problem. As a matter of curiosity, the technician screwed in some of the cover bolts. They bottomed out with about half of the bolt threads exposed. Digging into the holes with a pick revealed that every hole was filled about halfway full with RTV. The technician had laid the bead straight over the hole on the cover, and the installed bolt pushed the RTV into the hole. The bolts were tightened with a quarter-inch-drive impact wrench, so the technician did not have a feel for the bolt as it bottomed on the RTV. Cleaning the holes completely, installing RTV properly, and torquing the fasteners properly corrected the problem. Rushing the "simple" tasks on a routine job can result in a loss of profits and customers.

TERMS TO KNOW

Brinelling
Burned areas
Cotter key
Grease
Shaft
Spalling
Wheel bearings

ASE-STYLE REVIEW QUESTIONS

1. Repacking a bearing is being discussed.
   Technician A says the grease should be an L classification.
   Technician B says the grease must extend through the rollers.
   Who is correct?
   A. A only
   B. B only
   C. Both A and B
   D. Neither A nor B

2. Laying an RTV bead is discussed.
   Technician A says RTV should cover the entire sealing surface.
   Technician B says to remove the tube from the work until the complete bead is laid.
   Who is correct?
   A. A only
   B. B only
   C. Both A and B
   D. Neither A nor B

3. All of the following are used to clean a bearing EXCEPT:
   A. compressed air.
   B. hand, eye, and face protection.
   C. gasoline or kerosene.
   D. parts washer.

4. Removing a wheel bearing is being discussed.
   Technician A says the seal must be removed before the outer bearing can be removed.
   Technician B says the nut or a screwdriver can be used to gain access to the outer bearing.
   Who is correct?
   A. A only
   B. B only
   C. Both A and B
   D. Neither A nor B

5. Installing a wheel bearing is being discussed.
   Technician A says the driving tool can be used to install the seal behind the inner bearing.
   Technician B says the outer bearing race may be installed with the same tool used to install the inner race.
   Who is correct?
   A. A only
   B. B only
   C. Both A and B
   D. Neither A nor B

6. Inspecting bearing components is being discussed.
   Technician A says the shaft is part of the bearing assembly.
   Technician B says the shaft should be inspected for the same problems as the bearing.
   Who is correct?
   A. A only
   B. B only
   C. Both A and B
   D. Neither A nor B

7. *Technician A* says spalling is when the bearing or race has dents in it.
   Technician B says brinelling causes pieces of the bearing to flake off.
   Who is correct?
   A. A only
   B. B only
   C. Both A and B
   D. Neither A nor B

8. *Technician A* says a special tool is used to remove the outer bearing.
   Technician B says a nut can be used to help remove the seal.
   Who is correct?
   A. A only
   B. B only
   C. Both A and B
   D. Neither A nor B

9. *Technician A* says the outer side of the bearing should receive a light coat of grease before it is installed in the hub.
   Technician B says bearings that are lubricated by the component's lubricant should be prelubed.
   Who is correct?
   A. A only
   B. B only
   C. Both A and B
   D. Neither A nor B

10. *Technician A* says the dust cap should be filled about three-quarters full of grease.
    Technician B says the hub cavity should be full of grease before installing the bearings and seal.
    Who is correct?
    A. A only
    B. B only
    C. Both A and B
    D. Neither A nor B

JOB SHEET 15

Name ______________________________ Date ______________

Removing, Repacking, and Installing Wheel Bearings

NATEF Correlation

This job sheet addresses the following NATEF tasks:

Remove, inspect, and service or replace front and rear wheel bearings.

Remove, clean, inspect, repack, and install wheel bearings and replace seals; install hub and adjust wheel bearings.

CAUTION: Use eye, face, and hand protection during parts cleaning. Injury or skin irritation may result if protective clothing is not worn.

Objective

Upon completion and review of this job sheet, you should be able to remove, repack, and install a wheel bearing and seal.

Tools and Materials Needed

Seal remover

Bearing packer

8-ounce ball peen hammer

Seal driver

Hub with inner bearing and seal installed

Blocking materials

Catch basin

CAUTION: Do not allow the bearing to spin when using compressed air as a drying agent. Serious injury could result if the bearing comes apart.

Procedures

Task Completed

1. Place the hub with the outside facing upward. If necessary, use blocks so that clearance is available between the hub and bench. ☐
2. Use the seal remover to force the seal from its cavity. ☐
3. Remove the inner bearing. ☐
4. Wash the bearing and hub cavity using a parts washer. ☐
5. Use reduced-pressure-compressed air to dry the bearing. Use a catch basin to collect the cleaner and waste. ☐
6. Place the bearing, small end down, into the bearing packer. ☐
7. Force the grease into the bearing until clean ribbons of grease are visible at the top of the rollers. ☐
8. Smear a light coat of grease on the outer sides of the rollers. ☐
9. Install the bearing, small end first, into the cleaned and dried hub cavity. 
10. Place the grease seal, lip first, over the cavity. ☐

Task Completed

☐ **11.** Hold the seal in place with the seal driver while using the hammer to drive the seal in place.

☐ **12.** Drive the seal in until it is flush with the edge of the hub.

☐ **13.** Use some fingers to rotate the bearing within the hub. If it moves freely, the task is complete.

Instructor's Response ____

____

____

____

Chapter 7

Diagnosing by Theories and Inspection

Upon completion and review of this chapter, you should understand and be able to describe:

- How to use the five senses aid in diagnosis.
- How understanding theories of operation can help a technician diagnose a fault.
- How the air/fuel mixture can affect thermodynamics and engine performance.
- The location of components from under the hood.
- How to inspect basic components from under the hood.
- How to inspect the top of the engine for leaks.
- The location of components under the vehicle.
- How to inspect under the vehicle for fluid leaks.
- How to inspect the exhaust system for leaks.
- How to inspect vehicle lighting and wipers for operation.

Introduction

One important step in diagnosing a problem is to look at the simple things first. Understanding the theories of operation will assist the technician in determining the fault. In addition, an inspection of the vehicle will give the technician insight into how the vehicle was operated and maintained. The two should give the technician an idea of what to expect as the diagnosis and repairs are being performed.

Inspection Routine Using the Five Senses

In Chapter 2 you learned how customer service plays an important role. Some of your best tools to do the job right are not always found in the tool box. In this section, you will learn how to use those customer service skills along with your senses as a human to diagnose a problem.

Five of the best tools available for technicians are the five senses. Taste, touch, vision, smell, and hearing can speed up the diagnosis and repair process. Granted, the sense of taste will not apply in most cases. But consider that sometimes the atmospheric environment in the area will actually cause a bad (or good) taste on the tongue, so don't ignore taste entirely. A solid inspection gives a good indication of the vehicle overall and some systems specifically. In addition to the diagnosis inspection, most shops require their technicians to perform a safety check. This check has two results in mind: notifying the customer of potential problems and increasing business. Remember, if a problem is found and presented to the customer, he or she has the right to refuse that service. But, on the other hand, most customers appreciate the "heads up," and the repair may be done that day or scheduled later.

SERVICE TIP: One of the simplest rules to remember is: Keep it simple, stupid, or KISS. If a starter doesn't work, check the battery first. A bad (dead) battery or corrosion on the terminals will cause very poor starter operation. Many times the simple things will cost time and labor because they were not checked before replacing parts.

Performance or operation checks are those simple tasks to see if something is working. To determine if a radio will receive a broadcast, turn it on. If sound comes out in some form, then it probably works. In many cases, the same scenario can be applied to other systems. Coupling a good inspection with performance checks determines quickly if the system should be working and whether it works correctly. For the discussion in this section, these procedures will be performed with the vehicle in the shop. To performance check the engine, transmissions, and other "moving vehicle" systems, it would be necessary to conduct a road test. Road tests will be discussed in later chapters.

Diagnostic Steps

There are seven general and acceptable steps to a proper diagnosis and repair:

1. Confirm customer's concern.
2. Collect vehicle data including maintenance history.
3. Collect vehicle Technical Service Bulletins (TSBs), recalls, or other information pertaining to this repair.
4. Test the system, including use of scan tools and other test instruments.
5. Determine the cause of the repair.
6. Complete the repair (correction).
7. Test the repair and verify that you took care of the original customer concern.

Each vehicle manufacturer outlines and suggests these steps in some form for the dealership technicians to follow. For this discussion, step 1, confirming customer's concern, is the one that most involves the inspection and performance check.

SERVICE TIP: A good trick to try when learning how to do a proper visual inspection is to put your hands in your pockets. People tend to touch what they see. Touching distracts from what you are supposed to be doing: looking and observing. Touching comes later after you have made your observations. Instead, concentrate on the situation in the engine compartment. Many times your eyes will tell you almost everything needed to make an initial diagnosis.

Sometimes there is a communication disconnect between the customer, the service writer, and the technician. A customer says the left headlight is out, but he or she was facing the vehicle from the outside and really the right headlight is out. It can be embarrassing to replace the wrong headlight because the customer's concern was not confirmed.

AUTHOR'S NOTE: One of the most embarrassing things I have ever done was when I was a young technician. I was installing a new belt on a Jeep because the old one was worn out—a fairly simple and regular maintenance item. It was a tight fit with not much room for my hands to work. My arm had to touch the water pump pulley when installing the new belt. The water pump pulley felt hot and warmer than any other water pump pulley I have ever felt before. A week later the customer's vehicle was towed in and the water pump had blown the seal, coolant sprayed everywhere, and the engine overheated. In my head I thought that if I would have just thought this through, I would have been able to prevent the vehicle breakdown by going further in my inspection and notifying the customer. That would have saved them from overheating the engine and towing it in.

For the purpose of this discussion, let's assume the customer's concern is engine overheating. Upon receiving the repair order, the technician approaches the vehicle. This is the first part of a good inspection. Observe the vehicle exterior. Assuming it has not been in a crash, damage to the body, tires, and overall appearance may point to a vehicle that is poorly maintained. The first clue to solving the overheating problem: When was the coolant system last serviced, and when was the drive belt last replaced?

Underhood Inspection

Raising the hood and taking a good look at the engine compartment also gives an idea of the vehicle's previous maintenance. If the engine is fairly clean and shows signs of recent repairs, such as new hoses or spark plug wires, then the operator at least tried to keep it running (Figure 7-1). The presence of new radiator hoses may also provide a key to the diagnosis. There could be a leak at one of the connections, possibly a thermostat was installed incorrectly, or the water pump was replaced or needs replacing. Since we have an overheating problem, check the drive belts for a nonelectric cooling fan and above all else check the level of coolant in the system. Many times this is checked in the recovery reservoir (Figure 7-2). Even if the reservoir is at the proper level, check the level in the radiator.

FIGURE 7-1 **This switch is used to deactivate or activate the passenger-side air bag on some pickup trucks.**

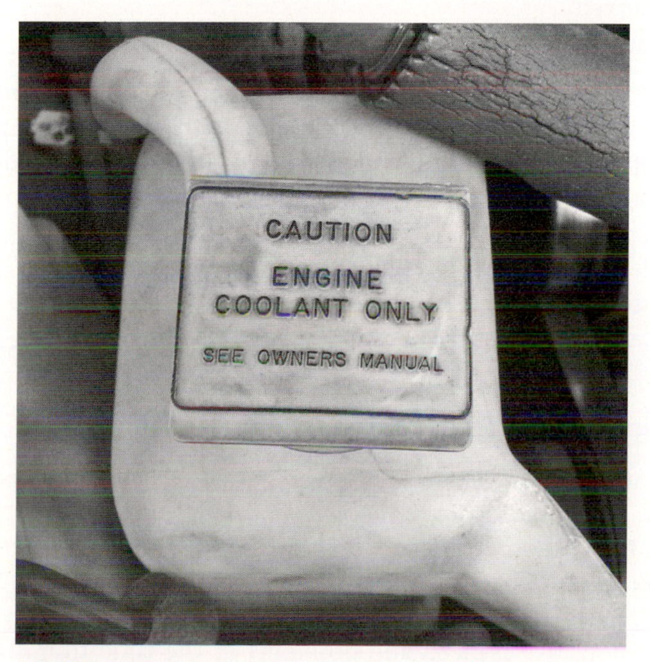

FIGURE 7-2 **This is for engine coolant only. There may be several other reservoirs nearby. Do not confuse reservoirs when topping them off.**

CAUTION:
Hybrid-powered vehicle have a high-voltage system. The positive cable for this system is colored orange. If this is your first hybrid or a different make or model than the ones you are used to repairing, *stop* and get someone with experience in this hybrid's electric power safety. The voltage on this cable can seriously injure or kill you.

CAUTION:
Never remove the cap on a hot radiator. Always allow the engine to cool before opening the radiator. The coolant temperature may be as high as 212°F (100°C) or higher. Serious burn injury could result.

CAUTION:
Use care when touching or moving anything under the hood. The engine and its components may be hot, or there will be moving belts and fans if the engine is running. Injuries could result from heat or moving components touching or grabbing fingers or hands.

Continue the underhood inspection by checking for leaks. If none are found and no other obvious signs of faults are found, a quick performance check can be done. If not already cool, allow the engine to cool to the point where the upper radiator hose can be touched without discomfort.

SERVICE TIP: If the coolant in the reservoir is below or above the *add* or *full* marks, it could point to a problem with the radiator cap. With the engine cooled to at least ambient temperature, remove the radiator cap and observe the coolant for discoloration and level. The coolant should cover the tubing and be near the bottom of the filler neck. It should be either green or reddish (sometimes an orange) in color. Clear, clean liquid points to the operator adding water and the antifreeze is diluted or missing. Sometimes plain water will cause certain engines under certain conditions to overheat to a small extent.

WARNING: Wear a face shield when checking coolant flow on an operating engine. A blown cylinder head gasket could cause the coolant to be ejected through the cap opening. Injury to the eyes, face, or body could occur.

This step requires the technician to pay attention and not allow the engine to operate for more than a minute once the upper radiator hose begins to warm. Start the engine and allow it to idle with the radiator cap off. Touch the upper hose repeatedly until heat is felt. It will become warmer closer to the engine first and then heat up rapidly to the radiator connection. This indicates the thermostat has opened. Use caution while observing the flow of coolant across the internal radiator tubes. Most times the flow is easily seen. Look for air bubbles in the coolant. They could mean outside air is in the system or could reveal a cylinder head problem. Shut down the engine and allow it to cool enough to install the radiator cap.

If the vehicle is equipped with an electric-operated radiator cooling fan, there is one test that may help in the diagnosis. With the engine operating, turn on the air conditioning. At least one cooling fan should switch on as the air-conditioning **compressor** engages. Though this doesn't completely eliminate the cooling fan system as the problem, it at least tells the technician that the fan itself is operating.

A **compressor** is used to pressurize a vapor.

Based on the visual observations and heat felt in the upper radiator hoses, most of the following can be determined to be correct or not:

a. Belts are present, tight, and serviceable.
b. Coolant is at proper level and appears to be the correct mixture of water and antifreeze.
c. Electric cooling fan is on with air conditioning on.
d. Thermostat is functioning.
e. No leaks are observed.
f. No air or combustion gases are present in coolant.

This diagnosis inspection may point directly to the problem or eliminate some components as the possible fault. But before continuing with the repair, make a good visual inspection of the other components under the hood. Items like corroded or old batteries, power-steering or transaxle leaks, damaged cables, and add-on electrical or mechanical components may reveal other possible or future problems. If the inspection finds other things that the technician honestly believes may be a problem, tell the customer. As stated before, it is the customer's vehicle and he or she has the right to refuse or accept the technician's advice.

Identifying the Area of Concern

There are many parts on the vehicle that wear out, break, fall off, make noise, get dirty, require adjustment, and need repair or replacement. It is your job as a technician to understand almost all of them and determine which area and component need repair.

To do this, you must be able to focus on one area of the vehicle. Often, components like the engine will have had problems for a period of time before you see the vehicle. The customer may not have noticed any problems at the time they started. When asking customers what is wrong with their vehicle, be sure that you understand them clearly. If language is an issue, get an assistant to help with communications.

Being able to communicate with a customer in technical terms proves to be the most challenging of all tasks. When trying to explain something to or communicating with a customer, you should avoid using technical terms and use clear, everyday terms. Avoid using very general terms, like "that thing," as these may make the customer feel very inferior or substandard. Some customers will understand the word "piston" and be able to relate it to an engine component, but they may not know what one looks like or what it does. You will have to be the judge of your actions when you are trying to understand what the customer needs.

When focusing on one area, try to ask as many questions as possible. Make sure to get a phone number where the customer can be reached quickly. Having a good working relationship helps here because you can feel comfortable talking with the customer. If you are still trying to establish the area to focus on with the customer, try test-driving the vehicle with the customer. Ask them to drive the vehicle and attempt to repeat the problem. Sometimes this may be difficult, but remember to be patient and calm (Figure 7-3).

Once you have determined the area to focus on, make an attempt to zone out other areas that are not related. This is tricky, especially because some other areas may be related to the problem. An example of trying to zone out other nonrelated area is when the customer has an engine noise, but it only makes noise when you accelerate hard from a dead stop. If the customer has a lot of papers and things lying on the dash or center console, those papers may make a lot of annoying noises when accelerating hard. You definitely don't want to clean it up or move it for the customer, but a phone call or simply asking the customer to move them so that you can hear the engine better is polite.

Sometimes using a **chassis ear** machine (Figure 7-4) works well when trying to figure out where the noises are coming from. You can attach the ends of the chassis ear to components on the engine and then bring the wires into the passenger compartment and listen to them while driving. The chassis ear is designed for listening to suspension, steering, and chassis problems but can be used for engines as well. Be careful when using this tool because you don't want to hear suspension or steering problems and assume they are engine problems. Make sure to fully read the directions before using the chassis ear.

A **chassis ear** machine uses several small microphones that can be placed under the hood and then routed to the passenger compartment where the driver can use a set of headphones.

FIGURE 7-3 **Sometimes a test drive with the customer is necessary.**

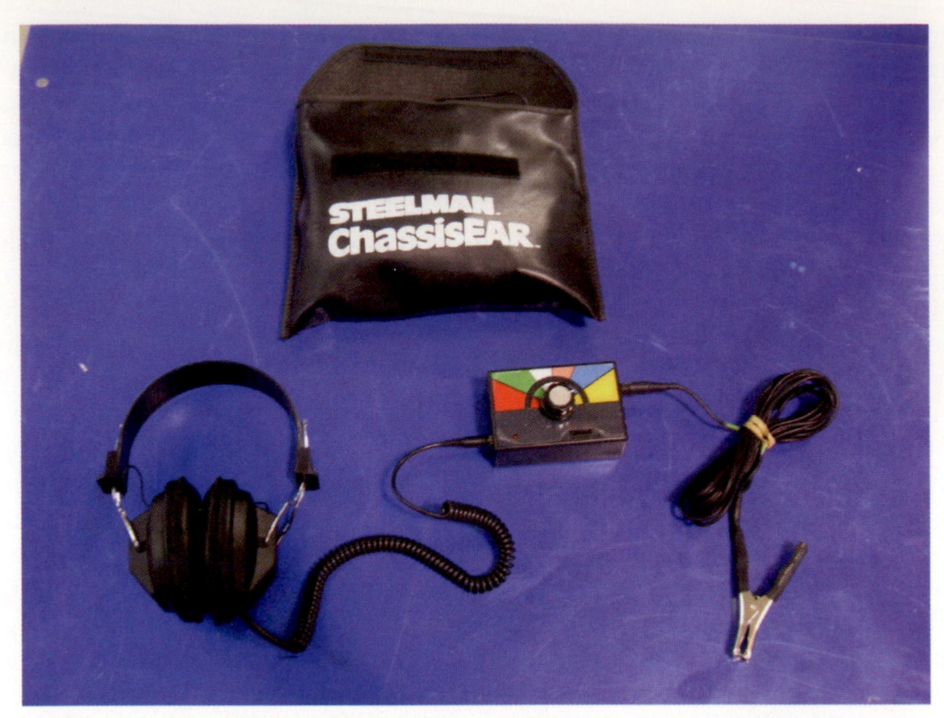

FIGURE 7-4 **A chassis ear machine can help you figure out where the noises are coming from.**

Once you have focused on the problem area, ask the customer about the vehicle's past service history. If they have had other major problems like a head gasket wear, overheating, or oil pressure problems in the past, ask what repairs have been completed. It is good practice to ask when and where the previous repairs have been done. They may have the old receipts and repair orders from other shops. These are all clues to the puzzle. If things are not adding up, try asking the owner if the vehicle has been in an accident or has a salvage or flood title. Although this seems strange, many flood and salvage vehicles can have preexisting damage from a previous accident that was never fully repaired. The body and exterior may be in great shape, but the engine may have suffered damage and have never received the proper repairs.

Under the Vehicle Inspection

WARNING: Before lifting a vehicle, ensure that the vehicle's lift points are identified and the lift pads are properly placed. Failure to do so may result in damage to the vehicle and the creation of a work hazard. If necessary, review Chapter 3 of the Shop Manual, "Tools and Equipment Usage," on using a lift.

CAUTION:
Never work under a vehicle that is not supported by jack stands or the locks of a lift. The vehicle could slip off a jack and cause serious injury.

Under-vehicle inspection and work can be done using a floor jack and jack stands. However, for this discussion, it is assumed that the vehicle will be raised on a double-post, above-ground lift.

Position the vehicle between the posts and align the lift's pads with the vehicle's lift points. Raise the vehicle slightly above the desired work height and then lower it until the lift's locks engage and support the vehicle's weight. Start at the front of the vehicle and work rearward observing for components and leaks not visible from the top of the engine. First, inspect the front of the engine for components not found or not readily visible from the top of the engine.

FIGURE 7-5 **The air-conditioning compressor is capable of producing up to 600 psi and draws horsepower from the engine during operation. Note the engine oil dipstick.**

Probably the most visible major component at this point is the air-conditioning (A/C) compressor, which is a type of pump (Figure 7-5). A/C compressors come in many different sizes but modern ones usually have a round body, which is slightly less in diameter than the length of the compressor. At the rear or on one side of the compressor will be two large, aluminum lines. This is the only component that will have this type of line. One end of the compressor will have the pulley and drive belt. Just behind the compressor pulley is a magnetic locking device with a small electrical plug. Make sure the plug is connected securely to the compressor and there are no broken or damaged wires visible.

On some vehicles it may be easier to see the power-steering pump from this viewpoint. This is not necessarily the position from which to remove the pump, but the technician may be able to find a hidden fastener that is invisible from the top. It may also be easier to see the bottom of the alternator from this view, but its removal will probably be from the top.

Other components visible or at least partially visible include those of the engine, transmission, steering, suspension, and brake systems. (Note the corrosion and grime in Figures 7-5 and 7-6. This is not unusual on older vehicles). Each system is discussed in later chapters. Directly under the lower portion of the engine is the oil pan. The oil filter is usually mounted low on the engine near the top edge of the pan. The filter may be found on either side or near either end of the engine. A plug will be located at the lowest part of the oil pan for easy removal of the engine oil (Figure 7-6).

Check around the oil filter, oil pan, and the engine in general for evidence of oil leaks. Engine oil will be black. Since liquids will flow to the lowest point before dropping off, do not assume that oil on the pan means the leak originated there. The oil could be coming from the top of the engine. Sometimes it is necessary to completely clean the entire engine compartment, test drive the vehicle, and then attempt to find the source of the leak.

A rear-wheel drive vehicle uses a transmission instead of the **transaxle.** They basically work in the same manner.

Behind the engine on the opposite end from the drive belts is the **transaxle** (Figure 7-7). The transaxle allows the driver to select gears for movement of the vehicle. There is a pan under this unit as well. It is usually flat and parallel to the floor, but it may be angled because of the space available for the transaxle. Usually, there is no place to drain the oil. The pan has

FIGURE 7-6 **This is a drain plug for the engine's oil pan. Some transaxles have a similar plug. Ensure that the correct one is removed and replacement oil is added to the correct component.**

FIGURE 7-7 **The transaxle is normally mounted to the rear of the engine and could be an area of leaks.**

to be removed for service. Most transaxles have the filter inside the pan. Inspect the transaxle at its pan and anywhere else that appears to be a connection between the two parts. Most of the external parts of a transaxle have some type of seal between its mating parts. Transaxle oil or fluid is usually red but may be very dark because of dirt and overheating. One fairly common leak area is the connection between the engine and the transaxle. Fluid found there might be engine oil or transaxle fluid, and both could be dark in color and difficult to identify. However, burned transaxle fluid will have a very burned smell.

Rear-wheel drive vehicles usually do not use a **subframe.**

The engine and transaxle are mounted on a **subframe** that reaches from the passenger floorboard to the front bumper (Figure 7-8). It also extends from side to side under the

FIGURE 7-8 **The subframe supports all the power components in a FWD vehicle.**

vehicle. Extending from the subframe to an area near the bottom of each wheel assembly is a lower **control arm** (Figure 7-9). This provides a pivot and support point for the wheel.

Looking upward past the arm and into the fender area, the lower part of the upper suspension components is visible. Also on both sides of this area is a long, slender rod extending from the center of the vehicle to an attachment near the wheel. These are the outer parts of the steering system (Figure 7-10). The braking devices for each wheel may also be visible. If this is a late-model Chrysler car, notice the fasteners that hold the inner panel for the left fender. The vehicle's battery is located behind that panel.

Extending from each side of the transaxle out to the wheels is a large, round bar with bellowed rubber boots at each end (Figure 7-11). These are the **drive axles**. They transfer power from the transaxle to the wheels.

Drive-axles transaxle the engine's power from the transaxle to the drive wheels.

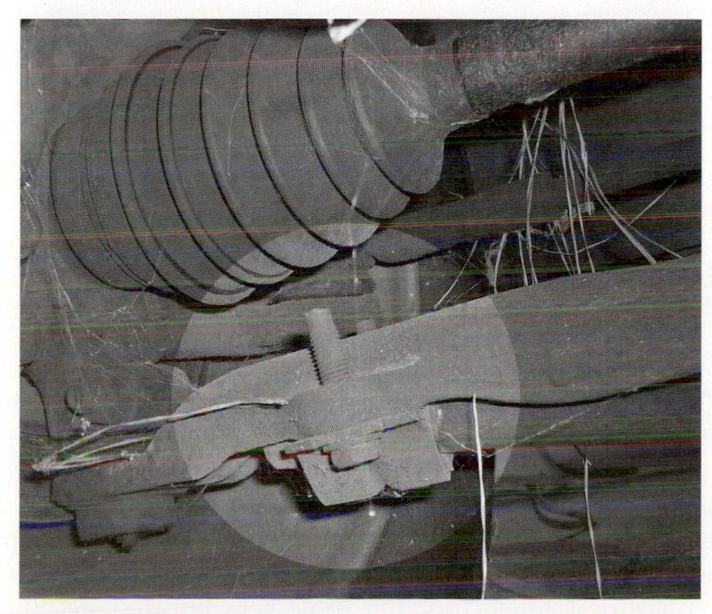

FIGURE 7-9 **The lower control arm provides the bottom mounting for the suspension components. There is one for each front wheel and on some rear wheels.**

FIGURE 7-10 **Shown is the outer tie rod end of one side of the steering linkage. It is often used to align the front wheels.**

FIGURE 7-11 **A FWD drive axle. One critical item to check during an inspection is the condition of the rubber boots at each end of the axle. Some RWD vehicles have similar axles at the rear.**

Classroom Manual
pages 156, 157, 158, 159, 160.

Seepage is seen as a damp spot with no visual of fluid drops. Seepage is considered to be normal. Many seals will not function correctly without some seepage.

Usually, no fluid leaks are found in these areas unless the brake system is leaking, but grease may be slung around the inside of the fender and any components mounted there. This is a result of a torn rubber boot. The boot is designed to hold special-purpose grease. A tear in the boot, combined with the high-speed rotation of the drive axle, slings the grease out in a vertical, circular pattern.

Moving rearward from the engine, the technician will find the floor pan or floorboard for the passenger compartment. Along one or both sides of the vehicle are the brake and fuel lines and the electrical conductors. On some vehicles, the lines and conductors are installed inside a tubular section of the frame and are not visible from this view. There should be no leaks or **seepage** along any of the lines.

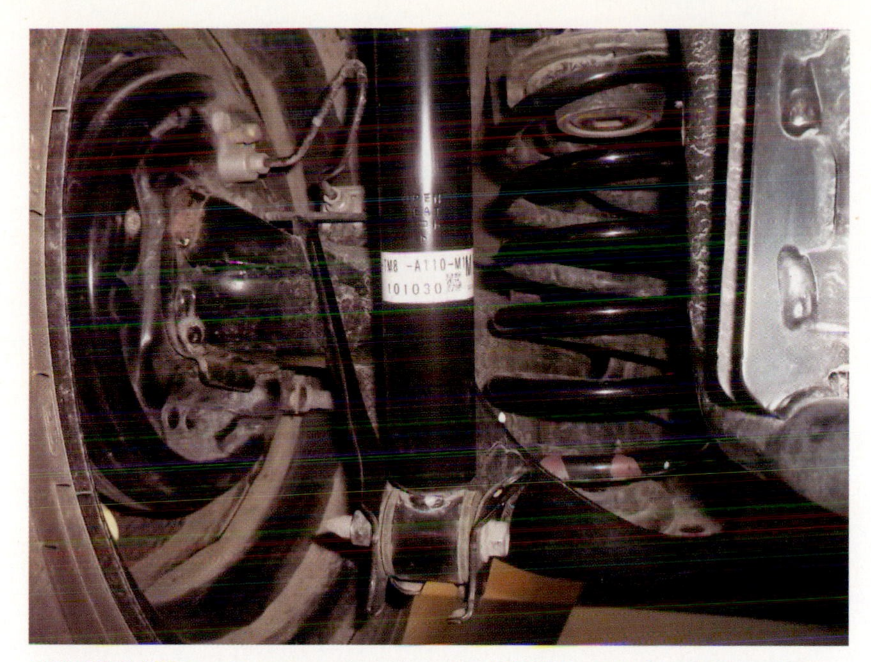

FIGURE 7-12 **The rear suspension allows the wheel to move in relation to the road surface and also keeps the wheel attached and in line with the body.**

At the rear of the vehicle is the most obvious component: the fuel tank. The suspension for each wheel is also visible. There can be several different components visible depending on the design of the suspension system. Almost everything visible between the body and the wheel is some part of either the suspension or the brake system (Figure 7-12). Each of the systems mentioned in this section is covered in detail in later chapters.

The inside of the wheel assemblies and the mounting components should be clean and free of any oil. Fluid or oil at this point usually indicates a leaking brake system. Naturally, there should be no leakage or seepage from the fuel tank or any of the lines and hoses under the vehicle. On the side of the vehicle where the fuel cap is located, check the area between the fuel tank and the fender or body for a large and small hose (Figure 7-13).

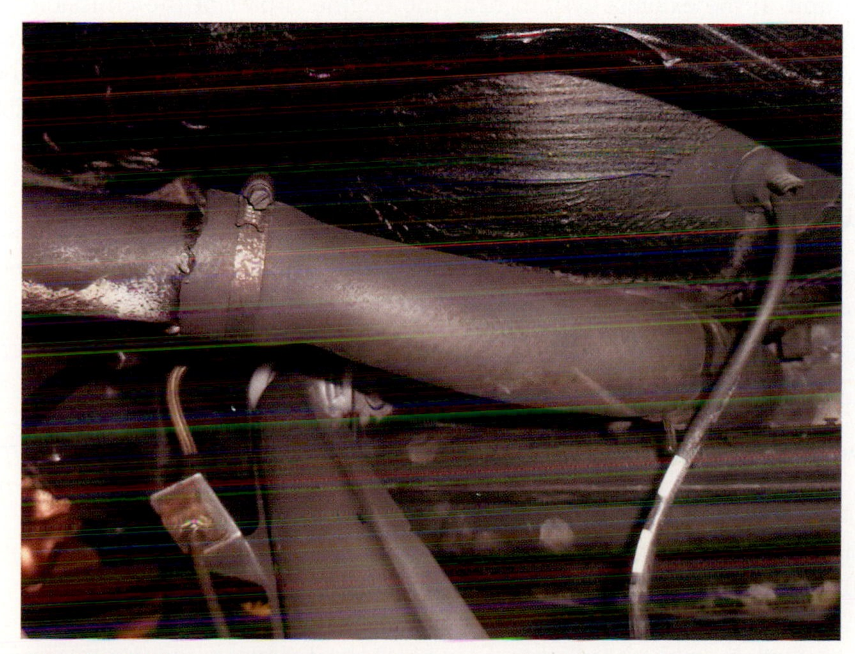

FIGURE 7-13 **There should never be leaks around the fuel filler hose, cap, or air breather. The rubber should be flexible without signs of deterioration.**

FIGURE 7-14 **There should be no exhaust leaks anywhere along the exhaust system. Leakage is commonly seen as soot.**

CAUTION:
Do not handle or touch any exhaust component until it is completely cool. Very serious burns could result.

Two lighting systems are considered emergency lighting. They are the brake lights and the hazard warning lights. They should operate in almost any condition as long as electrical power is available.

Classroom Manual
pages 160, 161, 162

The larger one is the fuel inlet hose. The second and smaller hose allows air to escape the tank during filling. Each is held in place with clamps. Like all fuel hoses, there should be no leaks or seepage.

Inspect the exhaust system that runs under the vehicle (Figure 7-14). There should be no soot at the connections. The larger parts of the exhaust are silencers called *mufflers* or *resonators*, and they may have small holes near one of the welded seams. Ideally, there should be no soot around the holes either, but they should be examined closely and corrective action taken as needed.

A large unit in the exhaust system near the engine is part of the emission control system. It operates at a temperature near or exceeding 1,600°F. It can be uncomfortable working near this system, so it should be allowed to cool before repairs are made. There should be no kinks, holes, or cracks anywhere else in the exhaust system. If such defects exist, the sections in which they appear must be repaired or replaced.

Inspecting Body-Mounted Components

The entry-level technician performs only basic repairs to body-mounted components. He or she can, however, inspect and test many body-mounted components.

Lights

Inspecting the lighting system is a relatively simple process, but it should be done every time the vehicle is brought in for servicing or repairs. An inoperative lamp may not be apparent to the operator until it is too late. Repairs to the lighting system are covered in Chapter 9, "Inspecting and Servicing the Electrical System."

Start your inspection with the lights that must be connected directly to the battery. In other words, the ignition switch does not need to be in the RUN position.

Turn on the parking lights and walk around the vehicle (Figure 7-15). All parking lights should be on. Switch on the headlights. Some imported vehicles require the ignition switch to be in the RUN position before the headlights will work.

FIGURE 7-15 **A typical headlight switch. Note the dimmer control.**

Even though light may be visible on the wall in front of the vehicle, do not assume everything is correct and that the lights are in working order. Move to the front of the vehicle and make sure both headlamps are operating. Check each headlight in its high- and low-beam position and remember to check the high-beam (bright) indicator to be sure it is working in the dash.

Check the hazard warning lights next. The headlights and parking lights may be left on or off, but ensure that the ignition switch is off. The hazard lights are on a circuit by themselves and must operate regardless of the position of any other switches. The hazard switch will be on the right side of the steering column near the ignition key slot, on the top center of the steering column just forward of the steering wheel, or on the instrument panel (Figure 7-16). It will be marked with a red marking in most cases.

Move the hazard switch to the ON position and observe the direction or turn signal indicators in the dash. Both should be flashing. The two front parking lights should be flashing, as should the brake lights in the rear. This system will work even if three lamps are burned out. The fourth light will still flash.

With the hazards working, apply the brakes. The front lights should still flash, but the rear lights should shine solidly without flashing. Releasing the brakes will allow the rear lights to flash again. Switch off the hazard lights and retest the brakes.

The ignition switch must be in the RUN position to test most turn or signal lights. Switch on the ignition and select the left turn signal. The turn signal indicator should blink. Select the right turn signal and observe the right indicator. With the right turn signal on, check the rear and front lights on the right side of the vehicle. Reselect the left turn signal and check the left side.

Leave the left turn signal on and apply the brakes. Depending on the design of the lighting circuits, one of several things should happen. If this is a typical two-lamp system,

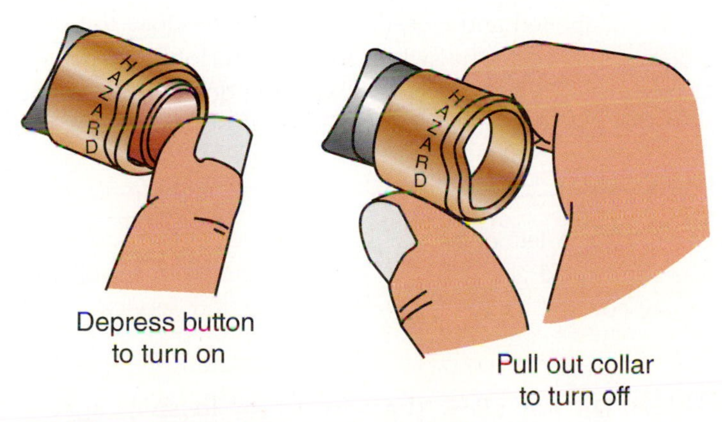

FIGURE 7-16 **This switch is used to flash all four (or more) marker (parking) lights in emergencies.**

FIGURE 7-17 **Some internal lights are switched on or off by pressing one end of the light lens. Other types may use push-button or flip-type switches.**

the rear left turn signal will flash and the right brake light will glow solidly. On a three-lamp system, the brake and turn signal are on separate lamps, so there will be brake lights on both sides in the rear and a flashing light for the left turn. After checking the brake and left turn signal, do the same test for the right turn signal.

With the ignition still in the RUN position, operate the headlights and hazard lights again in the same manner noted earlier. Headlights that require the ignition switch to be in the RUN position can be checked now. One other headlight system that requires checking is the daytime running lights. In this system, the headlights are automatically switched on when the ignition switch is on. The lights will be dimmer than normal because of the low voltage and current supplied. If in doubt, turn on the ignition switch and observe the headlight illumination. Switch on the headlight switch and the illumination should brighten considerably.

Check the lights in the dash while the ignition switch is on. There will be several red lights visible. The system at this time is operating in a test mode of the warning lights. The warning lights and gauges will be covered in Chapter 9, "Inspecting and Servicing the Electrical System."

There should be one or more interior lights that work when a door is opened. In addition, the interior lights usually have a switch that can be used for control. Observe the interior lights as each door is opened and closed. Remember to close the other doors as one is checked. Close all doors and locate the switch for the interior lights. Check the lights using this switch. Some interior lights will only work with the doors while others may have independent switches for each light (Figure 7-17). Check all of them.

Windshield Wipers

Windshield wipers tend to be left out of the maintenance process. A badly damaged wiper blade can scratch the windshield glass and will not sufficiently clear it of rain or debris. Repairs on the wiper system are covered in Chapter 9, "Inspecting and Servicing the Electrical System."

WARNING: Do not allow the wiper arm to snap against the glass. The glass may chip or crack, resulting in a high cost of replacement.

FIGURE 7-18 **Only the rubber blade portion of this wiper should touch the glass. The metal arm could scratch the glass.**

WARNING: Do not operate the windshield wipers on dry glass. Using the windshield wipers on dry glass could scratch the glass and/or damage the new wipers.

Inspect the rubber insert portion of the wiper that lies on the windshield (Figure 7-18). The rubber should not be torn or ripped. If so, replace the insert before proceeding. With good inserts on each side, turn the ignition to RUN. The wiper control is on a lever on the left or right side of the steering column or on the dash. Most vehicles have interval wipers, meaning that a position on the switch allows the wiper to sweep, pause for a set amount of time, and then make another sweep. Most wiper systems have two speed positions in addition to the interval. The system includes a washer position on the wiper lever or a separate switch for the washer. Before operating the wipers, use the washer to wet the glass.

Operate the wiper in each switch position and ensure that the glass is wet by using the washer. In the interval position, adjust the time by operating the interval switch (Figure 7-19). Some vehicles, however, do not have the option of adjusting the time. Also check the rear windshield. Some vehicles have a wiper and washer for this glass, and it should be checked in the same manner as the front.

Doors

While checking the interior lights with the door, observe the movement of the door. It should be smooth and quiet. A grinding or scraping noise indicates a lack of lubricant. The door should shut solidly and swing in an even arc through its entire range. A sagging door will not close quietly and may have to be slammed shut to latch properly. This usually indicates that the door hinges are worn, allowing the door to drop out of alignment. Repairs to the door should be performed by an experienced technician and in some cases by a body shop technician.

FIGURE 7-19 **A typical windshield wiper and washer switch mounted on the turn signal (multifunction) switch arm.**

Diagnosing by Theory

When a system fails, it is because some part of it did not fulfill the requirements of a theory. Often, a lamp does not work because the electricity did not flow from the power source, through the lamp, and back to the power source. A lamp may be "burned out," but the electricity failed to return to the power source, hence, no light. The same diagnostic procedure can be applied to any system on the vehicle. Every technician will encounter a fault that fails to respond to typical diagnostic tests. By thinking about the theory or theories that make the system work, the technician can track through the system and find the point where the theory breaks. In this area the fault will be found. It may be caused by the lack of electricity, fuel, or air. It may be a broken part or any other item that is needed to make the system work.

Classroom Manual
pages 149, 150, 151, 152, 153, 154

Engine Operations

In the Classroom Manual, we discussed the need for fuel, oxygen, and spark to be present in the combustion chamber to produce power. Two of those items, fuel and oxygen, require pressure and vacuum in order to be drawn into the chamber. The air and fuel must also be mixed in the correct **ratio**.

A **ratio** is the proportional amount of one element to another. An ideal air-to-fuel ratio is 14.7 parts of air to 1 part of fuel (14.7 to 1).

The laws of motion also apply to the movement of air and fuel. The engine is designed to make maximum use of physics to move, mix, and fire the fuel and transfer the produced power to the vehicle's driveline.

Understanding the theory of operation of vehicle systems will give the technician a head start on diagnosing. Let us discuss a possible situation. The technician does not have any high-tech diagnostic tools available but has a vehicle that will not start. As discussed, the engine must receive air and fuel in the correct mixture. The mixture is then fired with a spark. The easiest item to check is the airflow. Remove the air filter housing and try to start the vehicle. It is not often that an air intake is so plugged that no air can pass. But a dirty air filter could reduce the amount of air and cause a rich mixture or too much fuel for the air available. This could cause a no-start situation in the engine but more likely will allow it to start but run poorly.

A broken or missing vacuum line will cause the opposite condition or a lean mixture, which is too much air for the fuel available. Both conditions reduce the power produced and cause more harmful emissions. Most vacuum line connections are on the intake manifold or on the throttle body.

Let us consider a second situation concerning a poorly operating engine. The fact that it is running indicates that fuel, air, and spark are present and being used. Apparently, something is wrong with one of the three. Conduct a quick test to see if the problem is the same at idle and higher engine speeds. After allowing the engine to warm, speed up the engine until it reaches 1,200 to 1,500 rpm. If a meter (**tachometer**) is not available, hold the accelerator at about one-quarter throttle.

A **tachometer** is a meter used to measure the speed of a device in revolutions per minute (rpm).

If the engine smoothes out and runs evenly, then the problem exists only at idle. The engine intakes little air and fuel at idle. The air valve, sometimes called a throttle valve, is closed and air enters through an idle air control valve (Figure 7-20). Carbon buildup can partially block the valve and reduce the oxygen available to the engine. The air-to-fuel ratio is rich and combustion cannot be supported. Since the valve has no effect when the engine is off idle, the engine will run smoothly at speed.

CAUTION:
Never look into a carburetor when the engine is running. The fuel in a cylinder could ignite early and cause a flame to run back through the intake system and cause injury to the eyes and face.

A second problem that affects idle speed is a vacuum (air) leak around the intake manifold. As mentioned, a vacuum leak will cause the air-to-fuel ratio to be lean, but unless the leak is very large, the engine will run fairly smooth off idle. A fuel or spark problem will probably be present at idle and may increase at higher engine speeds. In this example, we have found that a small valve or a leak can upset the application of thermodynamic theory by changing the air-to-fuel ratio so combustion is not completed.

The fuel delivery cannot be checked easily on a fuel injection system. It is best left to technicians with the proper test equipment. On a **carburetor** system, operating the throttle once or twice without the engine running will check the fuel delivery. Observe the inside of the carburetor while moving the throttle. There should be a squirt of fuel sprayed into the carburetor each time the throttle is moved. Once or twice is sufficient.

A **carburetor** is an old air/fuel mixture device.

Remember that in order for the engine to work properly, the laws of physics must be followed. If one of the checks fails, then the faulty system has been found. From this point, the technician can inspect and test that system until the root cause is found.

FIGURE 7-20 **This throttle valve or air valve is controlled mechanically by the driver using the accelerator pedal. Newer vehicles may be driven by an electrical motor based on accelerator pedal movement input to a computer.**

When the first electronically controlled engines began to show up for repairs, many good sensors were replaced due to a lack of understanding about system and test equipment operation.

This was a simple diagnostic test using the theory of operation. Similar diagnosing occurs every day, except we now rely on electronic devices to do the initial testing for us. An electronic sensor "stuck on lean" indicates that there is not enough fuel or that there is too much air in the mixture. Fuel injection relies on constant pressure and volume. If a sensor indicates the mixture is lean, then the fuel is insufficient or the air volume may be too large. On most engines using a fuel injection system, the fuel delivery is the most common cause of a lean condition. The fuel may not be present in either pressure, volume, or both. A little common sense tells us that if there is no volume, then there can be no pressure, so in this instance we can assume there is insufficient fuel available. The most common cause of this problem is a partially clogged fuel filter. A $20 part could restore engine performance and ensure that thermodynamics is accomplished.

A point to remember as technicians advance in their careers is to avoid replacing a sensor until it is determined that the sensor is bad. In most cases, the sensor is doing exactly what it was designed to do. In the previous case, the oxygen sensor is telling the technician that the fuel flow is insufficient for the amount of air available.

Engine Noises and Smells

Much can be determined about an engine's operation based on its sound and the odor or color of the exhaust gases. If the engine sounds like metal is ripping apart during operation, it should be obvious that something is broken inside the engine. The only true way to determine what is wrong is to disassemble the engine. Most experienced technicians can give a fairly accurate diagnosis based on the sound(s) alone. Many internal mechanical failures produce sounds that point to the failed component. The only way to prevent further damage is engine shutdown as soon as the noise is first heard. Further operation will only increase the damage.

The smell of an engine, particularly when it is at operating temperature, can lead the technician to a possible diagnosis that can be confirmed with follow-up testing. The smell of burnt oil in and around the engine compartment usually means serious oil leaks. The smell of antifreeze may indicate a coolant leak. The smell of fuel in the same area requires immediate engine shut down and a nearby fire extinguisher. Leaking fuel on a hot surface can start a fire quickly. Another quick check can also narrow down some engine problems.

CAUTION:
Do not attempt to perform an inspection or repairs on hot engines. Serious burns may result or injury could occur if a temperature-sensitive cooling fan switches on.

Observe the color of the exhaust gases at the rear of the vehicle. Blue smoke indicates bad piston rings, allowing engine oil into the combustion chamber. Black smoke usually indicates excessive fuel or a rich air/fuel mixture. This could be caused by several things, but is most often caused by bad engine sensors or malfunctioning fuel injectors. With a diesel engine, black exhaust smoke is almost a sure sign that the air filter needs changing. Also, on a diesel-fueled engine, white smoke could be caused by a rich air/fuel mixture. Gray smoke that may be accompanied by drips of water from the tail pipe is a sign of cylinder head or head gasket failure.

Smell and feel senses can also be applied while observing the exhaust gases. A rotten egg smell is associated with a faulty catalytic converter, possibly caused by a long-term rich air/fuel mixture. If the eyes and nasal passage become irritated while behind an operating engine, it could be nitrogenous oxide or some version of smog in the exhaust gases. This could be caused by one of several subsystems, including the exhaust gas recirculating system, lean air/fuel mixture, or very high combustion chamber temperature. A clogged fuel filter will also contribute to this condition.

Component Location and Inspection

Classroom Manual
pages 160, 161, 162

The various components of the vehicle's systems are located in different areas of the body and frame. Belt-driven components are mounted on and driven by the engine. Steering, brake, and suspension systems have components mounted on the engine, body, and frame. It is also sometimes difficult to locate and view a component.

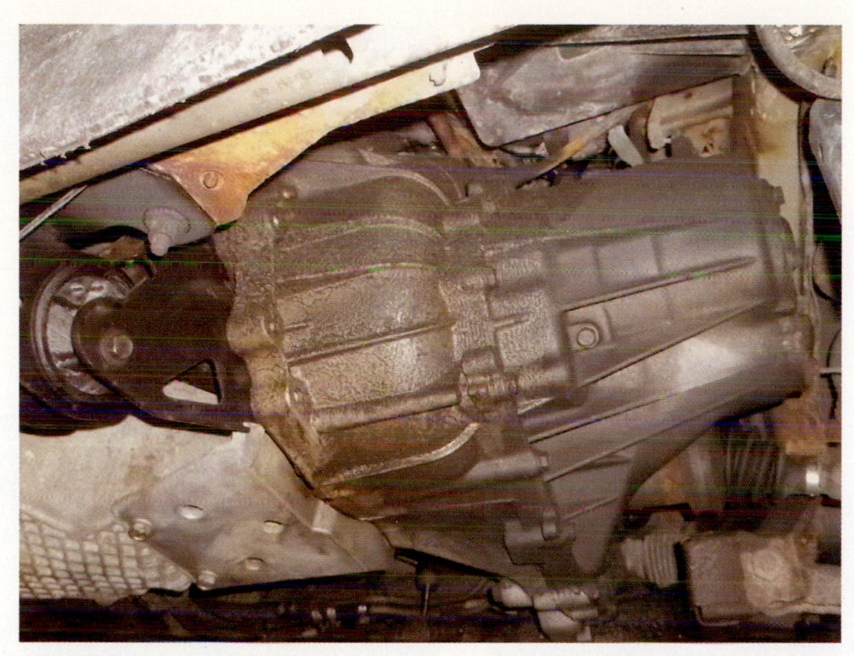

FIGURE 7-21 **FWD vehicles have the engine and transaxle mounted crossways (transverse) within the engine compartment. The front of this one is to the right of the vehicle (to left in figure); others may be reversed.**

Location and Identification

With front-wheel-drive (FWD) and **transverse** engines there is not much room under the hood. Most engines have the alternator, power-steering pumps, **water pump**, and air-conditioning compressor at the front of the engine where the drive belts are located (Figure 7-21). On an FWD vehicle, this may be located on the left or right side of the vehicle. When dealing with a vehicle's left or right side, always look at it from the vehicle's point of view. The left side is always the driver's side in most countries. One or all components may require the vehicle to be lifted, the engine mounts removed, or both for access and repair. In addition, the battery may be located within the wheel well or elsewhere in the vehicle and requires the removal of other components for access and service. In most cases, the battery has remote terminals mounted in the engine compartment for jump starting and electrical tests (Figure 7-22).

Transverse means the engine is mounted crossway to the length of the vehicle.

Some **water pumps** are driven by the timing belt positioned behind a cover on the front of the engine.

Rear-wheel-drive vehicles usually have sufficient room under the hood and elsewhere to repair or replace components (Figure 7-23). In addition, the major components are easy to identify. For these reasons, we will discuss component location and inspection using a typical FWD vehicle. Left and right directions apply to any vehicle.

Before starting the inspection, study the engine compartment and locate the alternator (Figure 7-24). It is usually at the top of the engine, but may be mounted low on the engine. It will usually have a single groove pulley with a fan behind it. The alternator is short, round, and will appear to be made of aluminum. It will have slots running across the rear cover next to the area where the wires are positioned (Figure 7-25). The end opposite the pulley will have a two-wire electrical connection and usually a heavy, red wire attached to another terminal. In most cases, the alternator will be the easiest belt-driven device to access.

While checking for the alternator, locate any other belt-driven components. The water pump may not be located on the outside of the engine, but there may be several pulleys, counting the crankshaft pulley at the bottom center of the engine. There may be up to eight pulleys depending on idler and tension pulleys used and the number of belt-driven components available.

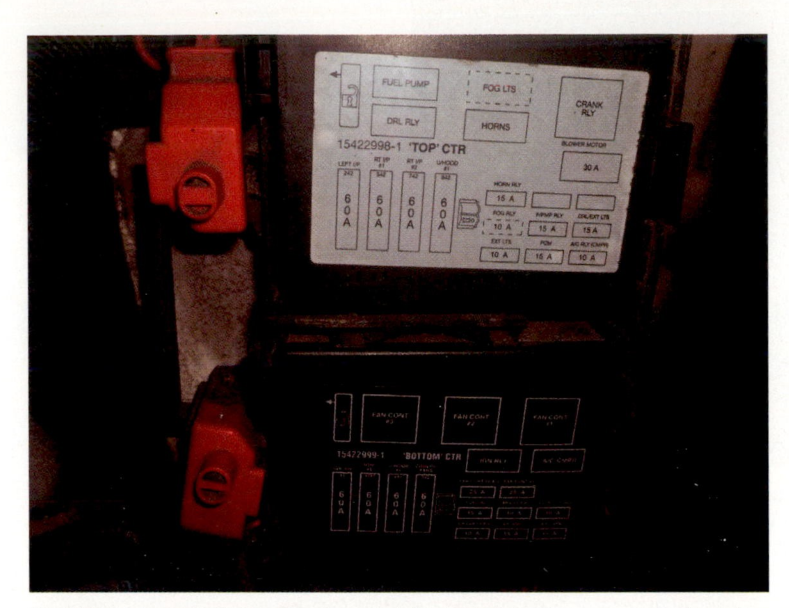

FIGURE 7-22 **This battery is out of sight and sometimes out of mind. Remote terminals provide a means to recharge, jump off the battery, and make electrical measurements.**

FIGURE 7-23 **The alternator may be mounted at the top or to one side of the engine. Note the drive belt.**

FIGURE 7-24 **This alternator is mounted at the top of the engine. Note the drive belt and pulley.**

FIGURE 7-25 **Slots at the rear of the alternator allow the airflow produced by the alternator fan to exhaust.**

The power-steering pump and reservoir are oddly shaped (Figure 7-26). The easiest way to identify these devices is to locate a drive pulley that has a small cap behind the metal bracket between the pulley and the pump. The cap is used for checking the fluid level. However, some vehicles have a remote-mounted reservoir with two hoses, one of which routes down to the pump. In either case, the filler cap will be marked in some way to indicate that this is the power-steering reservoir. Usually, the marking will indicate that only a certain type of power-steering fluid should be used in that system.

Many times, the power-steering pump is located to the lower left side of the engine in rear-wheel-drive vehicles and the upper right side of the engine in front-wheel drive. Again, remember that the directions are based on the vehicle or engine, *not* the position the technician is facing. The front of an engine is almost always the end with the drive belts, and the left and right of an engine may not be the left and right of the vehicle (Figure 7-27).

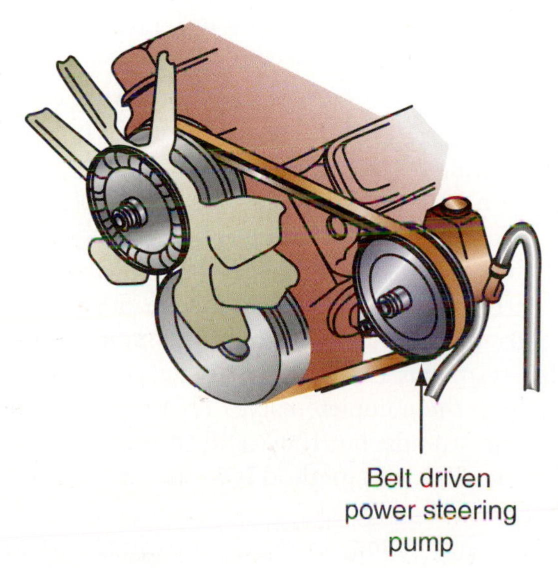

FIGURE 7-26 **Layout for a power-steering pump.**

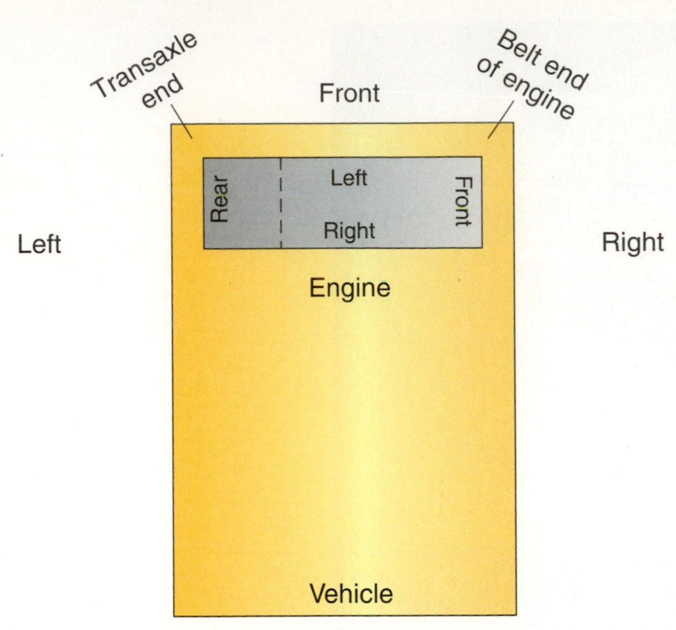

FIGURE 7-27 **Directions are based on the front of each major component.**

FIGURE 7-28 **Alternator electrical connectors should be clean and fit tightly to each other.**

Basic Wiring Inspection

CAUTION:
Hybrid-powered vehicles have a high-voltage system. The positive cable for this system is colored orange. If this is your first hybrid, or a make or model that you have not worked on before, then *stop* and consult someone with experience in this hybrid's electric power safety. The voltage on this cable can seriously injure or kill you.

WARNING: Before performing any electrical repairs, disconnect the negative cable from the battery. Damage to electronic components or injury to the technician may result.

Before we start discussing the electrical system, we need to quickly survey the electrical system of a hybrid vehicle. All of the systems normally associated with non-hybrid vehicles still operate on 12 VDC except for the alternator, drive wheels, and sometimes the starter. The hybrid drive batteries provide high voltage that is converted or stepped down to 14 volts, while high voltage is used to operate the vehicle's drive motor(s). The drive batteries are charged by a large alternating-current generator (alternator) usually mounted between the engine and transaxle or transmission.

There are two things that quickly damage electrical wiring and batteries: lack of maintenance and **corrosion**. If preventive maintenance is performed, corrosion will not be a problem. But along with antifreeze changes, battery preventive maintenance does not appear to be of great concern for some drivers until the vehicle quits working.

Corrosion can be caused by oxidation (oxygen acting on metal) or by acid or acid fumes.

Inspect the wiring at the back of the alternator for tightness and routing (Figure 7-28). The battery end of the cable is the most probable place for massive corrosion. This is partially due to the heavy current needed during cranking. Sometimes the corrosion can be removed, but if it extends under the cable's insulation it is best to replace the cable (Figure 7-29). The cable's terminals can also become a point of corrosion. Terminals can be replaced without changing the complete cable. There are temporary terminals that can be clamped to the bare cable strands, but they tend to increase resistance and lose voltage between the clamp and cable. The best method is to use the type of terminal that has to be soldered to the cable (Figure 7-30).

Other starting and charging system inspections include checking the security of the battery and corrosion on the battery tray or box. The tray should be cleaned and the

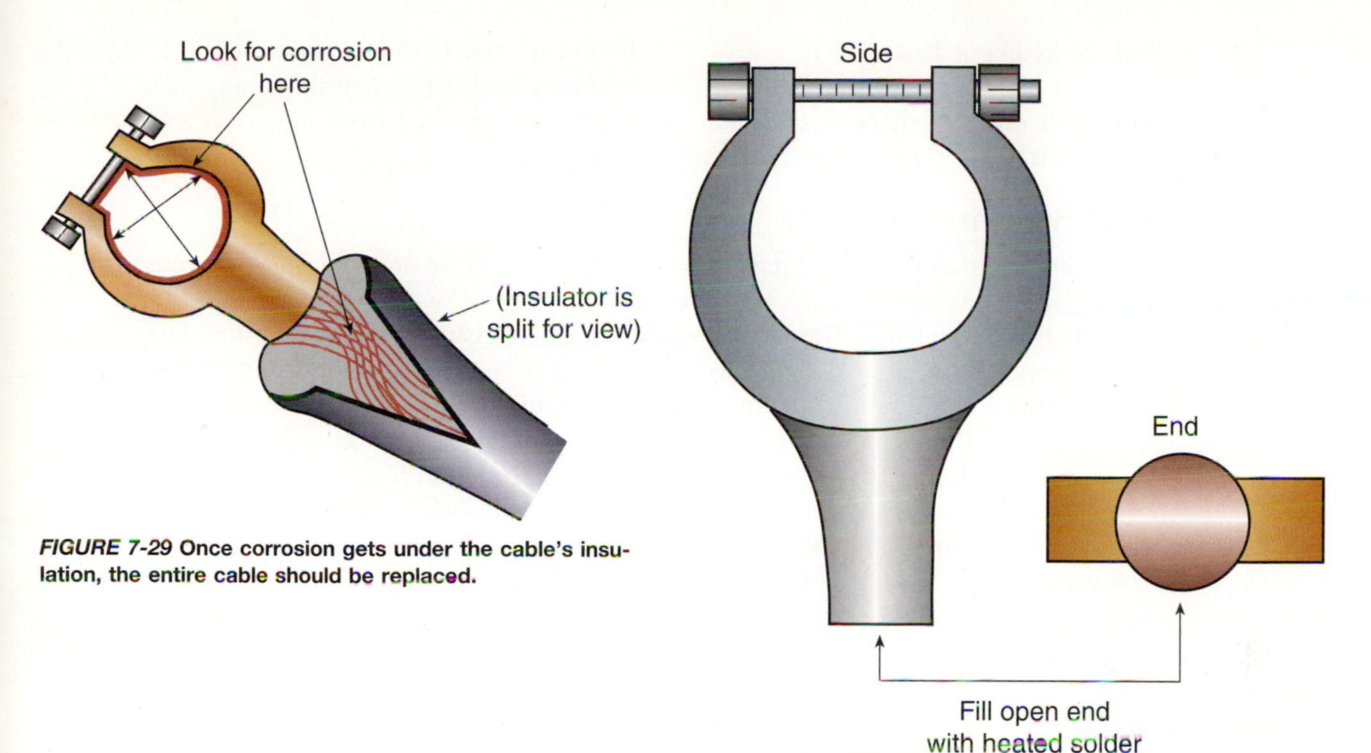

FIGURE 7-29 **Once corrosion gets under the cable's insulation, the entire cable should be replaced.**

FIGURE 7-30 **The permanent type of battery terminals are best suited for preventing corrosion.**

battery held in place by proper hold-downs. **Baking soda** and water work well to remove corrosion and electrolyte leakage. Use plenty of water to wash all of the soda from the vehicle. A final check includes the routing of all starting and charging system wires and harnesses. Ensure that there are no contacts between the wire and engine hot spots.

Baking soda is a common household product.

Checking for Leaks in the Engine Compartment

There are many places for leaks to occur under the hood. Among them are fuel clamps and hoses, air intake hoses, cover gaskets, and engine seals. The origin of some leaks may not be easy to locate, but they are evident based on the amount of liquid on the outside of the engine and engine mounted components. Smell and color can identify leaking fluids and their possible source.

Classroom Manual
pages 156, 157, 158, 159, 160

1. Pink or red—Power-steering fluid; automatic transmission fluid; some manual drivelines.
2. Green—Engine coolant, antifreeze; has distinct odor and very greasy to touch; newer fluids may be pinkish.
3. Black or very dark—Engine oil; manual drivelines; driveline fluid has distinct odor.
4. Clear—Brake fluid; some power-steering fluid; brake fluid is very slick compared to other fluids; fluid may look dirty.
5. Clear with distinct odor—Gasoline; will feel cold even in warm weather.
6. Dirty clear with distinct odor—Diesel fuel; oily to the touch.

The oil will be dirty and the color not readily apparent sometimes, but a little investigation by the technician can pinpoint the type and system. Automatic transmission fluid contains a lot of detergent, so if the leak is similarly colored and relatively clean, it may be transmission fluid. Engine oil tends to flow slowly and collect road dust and dirt, so if the leak looks like it has a lot of grease and dirt on it, it is likely an engine oil leak. Always remember that leaks will be conditioned by gravity and wind. A leak coming from a part of the valve cover where it is difficult to see may be first seen from under the vehicle and on the back side. Always trace the source of the leak.

Fuel System

A **throttle body** may mount one or two injectors and regulate the fuel pressure.

Fuel rails provide a common channel to maintain constant fuel pressure and volume at each injector in a multi-injector system.

The fuel system ends at the intake manifold. It may be just one fuel line connected to the carburetor or multiple lines to a **throttle body**.

Electronic fuel injection (EFI) systems use a pressure line and a return line. On an EFI system, there are two fuel lines—pressure and return. Both lines are close to each other. One leaves the fuel tank and goes to the **fuel rail**, the other leaves the fuel rail and goes back into the fuel tank (Figure 7-31). Some EFI systems are returnless and only have a pressure fuel link. EFI systems using individual injectors and fuel rails may have both lines close to each other or the lines may be at opposite ends of the fuel rail (Figure 7-32). The checks are the same on either of the systems. Visually check each connection in the lines and each connection to the fuel rail, throttle body, or carburetor. Feel under each connection with a finger.

WARNING: Use a flashlight for illumination if a fuel leak is suspected. An electric light may create an electrical spark when it is switched on and off, thereby causing a fire.

There must be no leaks or any type of seepage in any fuel system. In other words, if there is dampness at a fuel connection or anywhere on the line, find the problem before releasing the vehicle. This is especially true if fuel lines were disconnected during repair work. While checking the lines, also inspect the fuel rail connection to the injectors, the fuel injectors, and the carburetor or throttle body to the extent possible. Some of the fittings, hoses, lines, and components may not be visible. Older carburetors tend to seep at their various seals and gaskets. The carburetor should be rebuilt or replaced.

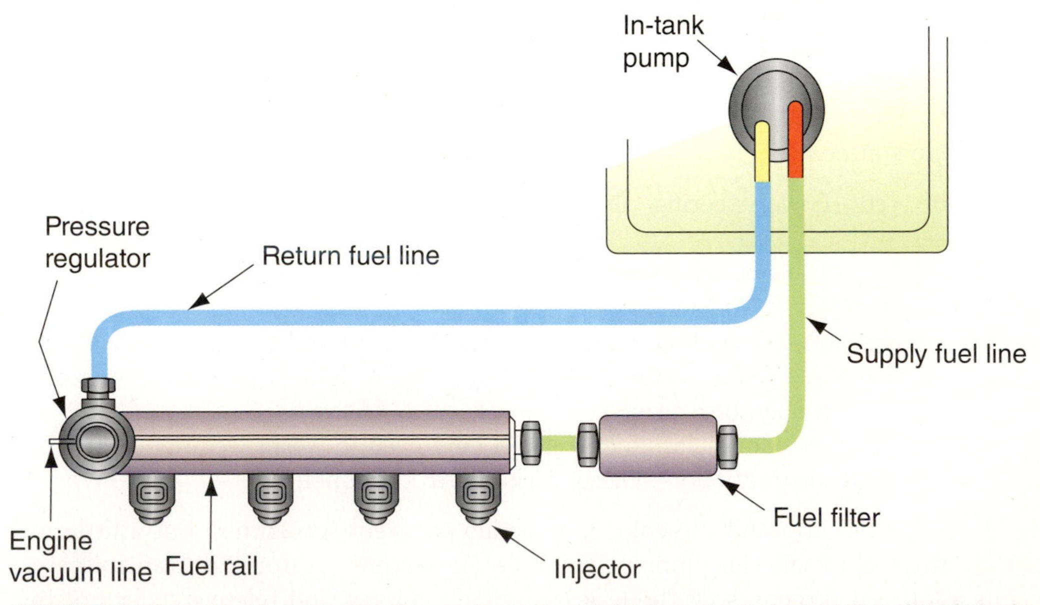

FIGURE 7-31 **There is a fuel pressure line, pump to throttle body, and a fuel return line, throttle body to tank.**

FIGURE 7-32 **A fuel rail connects each fuel injector to the pressurized supplied fuel from the pump.**

Air Intake

The air system is fairly easy to check. Air usually enters the intake system behind one of the headlights and travels through some flexible or rigid tubing to the air cleaner housing (Figure 7-33). Many times there is a **mass airflow** (MAF) **sensor** located in the air tubing (Figure 7-34). Any air entering the engine that does not flow through the sensor will result in poor engine performance. The sensor measures the amount of air entering the engine and the powertrain control module (PCM) selects the amount of fuel based on that measurement. Inspect air ducts for cracks or improper mounting and fastening.

The **MAF** measures the amount of air entering the engine's intake system.

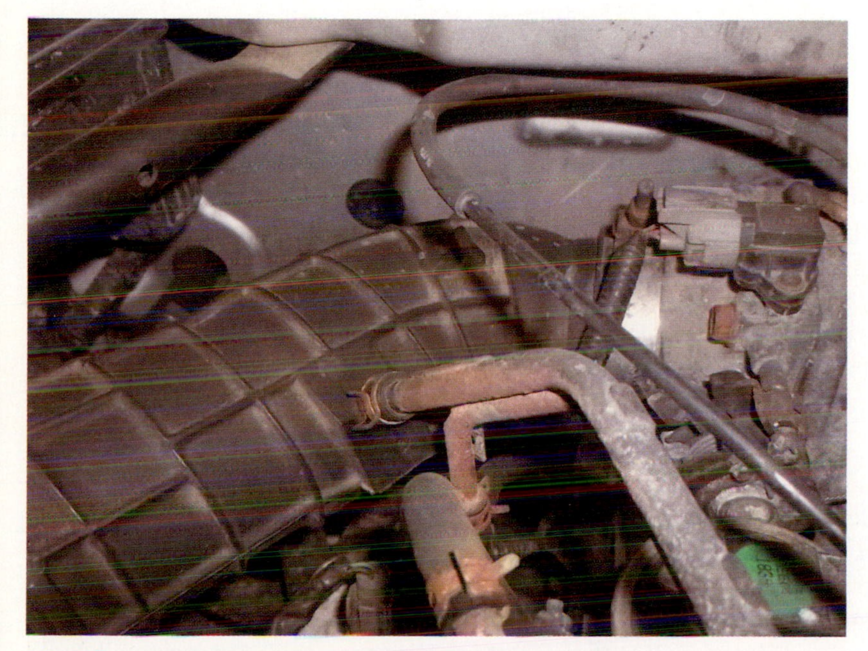

FIGURE 7-33 **Cracked or missing air ducts can allow unfiltered air into the engine and may bypass the airflow sensor.**

FIGURE 7-34 **Mass airflow sensors measure the volume of air and sometimes the temperature and density of the air entering the engine.**

The **PCV system** is used to vent the pressure created inside the engine by heat and pressure. Failure to vent the engine will cause oil leaks.

SERVICE TIP: Customers typically purchase the cheapest air filter available and later complain that the engine is operating poorly. This may be caused by the air filter. Some of the cheap air filters coat the filter fibers with oil. The oil works well in collecting dust and dirt entering the filter. However, this oil will also dissipate over time only to collect on the sensing grid of a MAF sensor. The oil coating now reduces the sensitivity of the grid and transmits incorrect airflow data to the PCM. If a customer complains of a poorly running engine, check the air filter condition and the sensing grid of the MAF. Should the feel or visual inspection of the grid indicate an oil film, then the condition may be resolved with a gentle cleaning of the grid and installing the correct air filter.

FIGURE 7-35 **The clogged air filter on the left restricts the volume of air entering the engine. This is very critical on fuel-injected engines, both gasoline and diesel.**

Vacuum leaks like the ones mentioned earlier also allow air to enter the system without going through the sensor. The end result is a lean mixture because of the extra air plus a wrong air volume measurement being sent to the PCM.

The air cleaner in the housing cleans the air before it enters the intake manifold (Figure 7-35). A dirty air filter restricts airflow and results in poor engine performance by causing a rich mixture. While checking the air filter, look for any oil or oil residue in the filter housing. The presence of oil there indicates a problem with the **positive crankcase ventilation (PCV) system**, which is covered in Chapter 8, “Engine Preventive Maintenance.”

A further note concerns industrial and equipment engines. Many of the old ones are still running strong. They use what is called an “oil bath” air cleaner. There are no paper fibers here, only a half-quart of engine oil and a metal mesh similar to steel wool. The incoming air passes through the mesh drawing some oil into the mesh. The oil captures the dust and debris in the air. If working on this type of engine, replace the oil after cleaning the bath and wash out the steel mesh. The best method I have found is to use carburetor cleaner, rinse in the hottest running water available, and then blow dry with compressed air.

Oil

Oil leaks are noticeable by black, oily dirt around the engine covers and on the sides of the engine. Each cover on the engine has some type of seal or sealing material between it and its mating components. Leaks result when the seal is broken or damaged. The PCV system may also cause an engine oil leak. The system is supposed to vent the pressure in the oil sump of a running engine (Figure 7-36). If the system is inoperative, pressure will build and oil will be forced through or around gaskets and seals or through the PCV valve.

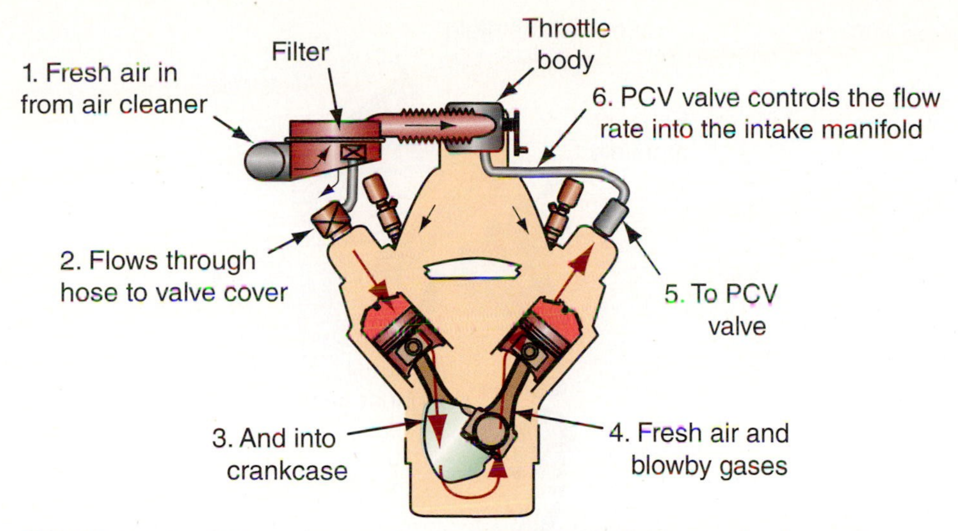

FIGURE 7-36 **The PCV system vents the crankcase vapors into the engine for burning. Clogged PCV systems can lead to serious engine oil leakage.**

Oil will drain to the lowest portion of a component before dropping on the ground. Leaks around seals on the front of the engine will cause the drive belts to be oily. The belts may also sling oil onto the underside of the hood and body panels in the area of the belts.

Coolant

Coolant leaks can be identified by the green or reddish color of the liquid and may occur because of loose clamps or water pump seals. The **radiator**, radiator cap, and hoses may also develop leaks. The radiator is mounted at the front of the vehicle and is connected to the engine by two large hoses, one at or near the top and one at the bottom (Figure 7-37). While checking for coolant leaks, inspect the two radiator hoses for dry rot or other damage. If a coolant leak is suspected but not visible, a coolant pressure tester can be used (Figure 7-38). The cooling system can be checked with a hand pump and adapter equipment. This system will be covered in Chapter 8, "Engine Preventive Maintenance."

A **radiator** is connected to a vehicle's engine by two hoses. Damage or rot to these hoses can cause coolant leaks.

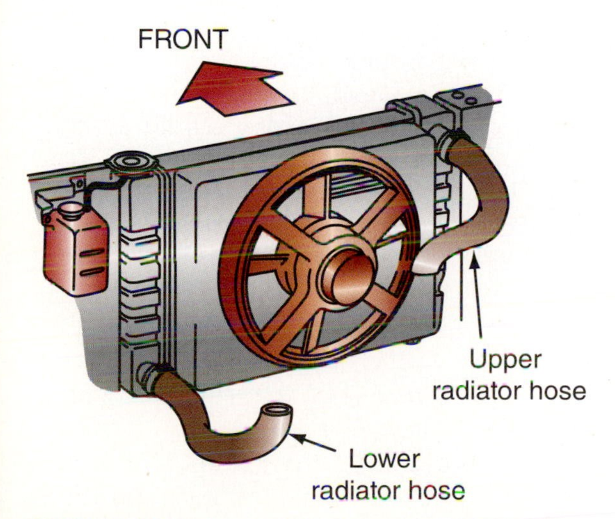

FIGURE 7-37 **The two radiator hoses transfer engine coolant from the engine to the radiator and back.**

FIGURE 7-38 **This is a common coolant system pressure tester.**

CAUTION:
The exhaust system gets hot very quickly after the engine is started. Extreme care must be taken when checking or working around the exhaust system. Serious burn injury could result.

TERMS TO KNOW

Baking soda
Carburetor
Chassis ear
Compressor
Control arm
Corrosion
Drive axle
Electronic fuel injection (EFI)
Fuel rail
Mass airflow sensor (MAF)
Positive crankcase ventilation (PCV)
Radiator
Ratio
Seepage
Subframe
Tachometer
Throttle body
Transaxle
Transverse
Water pump

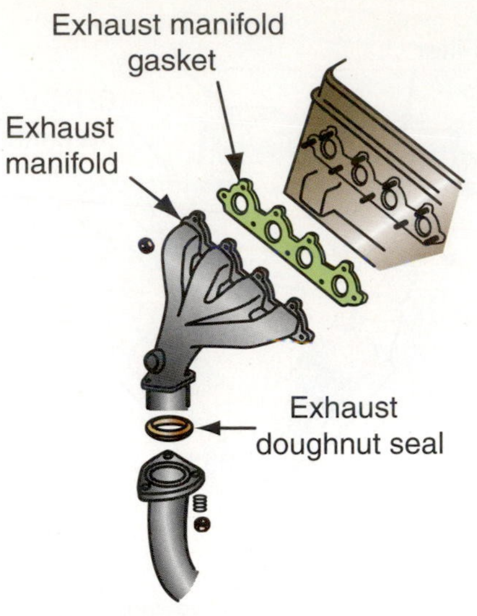

FIGURE 7-39 **A common exhaust leak point is between the exhaust manifold and the cylinder head. It is not uncommon for the exhaust manifold gasket to fail as vehicles age.**

Exhaust

Sound is the best and safest method to check the exhaust system for leaks. The sound of a leaking exhaust is different from almost any other sounds from the vehicle. If the sound is coming from the engine area, it usually means the leak is at the connection between the exhaust manifold and engine or between the manifold and exhaust pipe (Figure 7-39).

CASE STUDY

A customer complained of adding coolant every week or so. He stated that he could not find any coolant on the garage floor or inside the engine compartment. The technician performing the diagnosis noticed that the overflow reservoir was completely full, but the engine was only warm. Further testing with a pressure tester revealed no leaks, including the radiator cap. The technician located the end of the overflow reservoir drip hose. There was an indication of coolant in the end of the hose. It was decided that the cap was allowing coolant out of the system as designed, but not drawing it back into the system as it was supposed to do. As a result, the overflow would fill to capacity as the car was running and the excess would drain into the air stream flowing under the car. Since the driver checked the coolant regularly, the vehicle never overheated. But the driver did not check under the vehicle immediately after stopping. A replacement radiator cap solved the problem.

ASE-STYLE REVIEW QUESTIONS

1. *Technician A* says pressure and vacuum can be directly influenced by a leak.

   Technician B says pressure and vacuum are used to move the air and fuel into the engine.

   Who is correct?

   A. A only
   B. B only
   C. Both A and B
   D. Neither A nor B

2. Theories of operation are being discussed.

   Technician A says Newton's Laws affect the movement of air and fuel.

   Technician B says the laws of motion can affect thermodynamics.

   Who is correct?

   A. A only
   B. B only
   C. Both A and B
   D. Neither A nor B

3. *Technician A* says a dirty air filter can affect thermodynamic theory.

   Technician B says thermodynamics will be affected by fuel flow through the fuel filter.

   Who is correct?

   A. A only
   B. B only
   C. Both A and B
   D. Neither A nor B

4. Leaks are being discussed.

   Technician A says a leak could cause problems within the system if hydraulic theory cannot be maintained.

   Technician B says leaks may cause problems because the proper application of pressure is not maintained.

   Who is correct?

   A. A only
   B. B only
   C. Both A and B
   D. Neither A nor B

5. The air/fuel mixture is being discussed.

   Technician A says a leak in a vacuum line could cause the PCM to set the amount of fuel injected incorrectly.

   Technician B says a crack in the air intake ducting will result in a rich mixture.

   Who is correct?

   A. A only
   B. B only
   C. Both A and B
   D. Neither A nor B

6. All of the following statements concerning leaks are false EXCEPT:

   A. Leaks will not cause a system pressure drop.
   B. Seepage is not considered a leak.
   C. Leaks will not cause damage to flexible components of the vehicle.
   D. Leaks are never internal.

7. Application of electrical theory will be affected by each of the following EXCEPT:

   A. incorrect conductors.
   B. damaged controls.
   C. overcharged battery.
   D. undercharged battery.

8. When inspecting a leak under the vehicle, all of the following steps are important to keep in mind, EXCEPT:

   A. Oil leaks will look very dirty.
   B. Automatic transmission leaks will look very clean.
   C. Gravity may cause you to think that the oil pan is leaking when in reality it may be the valve cover leaking.
   D. Coolant is often similar in color to engine oil.

9. A test drive with the customer:

   A. allows a technician to always try driving different vehicles.
   B. can be very useful when diagnosing a problem.
   C. should always be short.
   D. is not important.

10. *Technician A* says fluid seepage on a rear wheel is acceptable.

    Technician B says power-steering fluid may be the same color as transaxle fluid.

    Who is correct?

    A. A only
    B. B only
    C. Both A and B
    D. Neither A nor B

JOB SHEET 16

Name ________________________ Date ____________

Locating the Source of a Leak

NATEF Correlation

This job sheet addresses the following NATEF tasks:

Inspect engine assembly for fuel, oil, coolant, and other leaks; determine necessary action.

Objective

Upon completion and review of this job sheet, you should be able to locate the source of a leak.

Tools and Materials Needed

Basic tool set
Flashlight
Mechanic mirror
Service manual

CAUTION: The engine, engine components, or other vehicle components may be hot. Do not touch any component until it has cooled. Injury can result from touching a hot component.

Procedures

Task Completed

1. Start a repair order on the vehicle being inspected. ☐
2. Identify the type of liquid and system leaking. ☐
   Color and smell of liquid ____________________
   Suspected system ____________________
3. Raise the hood and inspect any visible components of the suspected system. Use the flashlight and mirror as needed. ☐
   Remarks ____________________
   ____
4. Inspect the sides and covers of the engine and other components on or near the suspected system components. ☐
   Remarks ____________________
   ____
5. Is the leak source visible from this view? If not, prepare the vehicle for lifting. ☐
   Remarks ____________________
   ____
6. For the purpose of this job sheet, assume that the vehicle has to be lifted. Lift the vehicle to a good working height. ☐
7. Observe the point at which the liquid drops from the vehicle's components. ☐

Task Completed

☐ **8.** Use the light and mirror to trace the liquid up or over the components until a dry area is found. A dry area is defined as any component that does not show signs of the leaking liquid.

NOTE: It may take some effort and time to locate the highest point of the leak. Practice with the flashlight and mirror to become accustomed to the different angles of sight with the mirror.

☐ **9.** Identify the source of the leak. Students may want the assistance of the instructor or an experienced technician for the proper terminology at this point in their training.

Remarks ____

____

☐ **10.** Lower the vehicle and complete the repair order. A recommendation for repair or service can be made if desired. Use a service manual for information as needed.

Remarks ____

____

____

____

____

____

____

____

Instructor's Response ____

____

____

____

JOB SHEET 17

Name ______________________________ Date ______________

Underhood Inspection

NATEF Correlation

This job sheet addresses the following NATEF task:

Inspect engine assembly for fuel, oil, coolant, and other leaks; determine necessary action.

Objective

Upon completion and review of this job sheet, you should be able to visually inspect the engine compartment, determine the general condition of the engine, and make maintenance recommendations.

Tools and Materials Needed

Assigned vehicle

Flashlight or drop light

Mirror

Service manual

Procedures

CAUTION: The engine, engine components, or other vehicle components may be hot. Do not touch any component until it has cooled. Injury can result from touching a hot component.

CAUTION: Exercise caution when working around a hot engine. Serious burn injuries could result.

CAUTION: Keep the fingers and hands clear of electrical cooling fans. They could switch on even with the engine and ignition key set to off. Injuries to the fingers or hands could result when the fan is running.

CAUTION: Do not remove the radiator cap on a warm or hot engine. Hot coolant could be expelled rapidly and cause serious burn and/or eye injury.

Task Completed

1. Start a repair order on the vehicle. Also use a vehicle inspection form if one is available. ☐
2. Move the vehicle to a well-lit area and allow the engine to cool if necessary. ☐
3. Raise the hood and record your general impression of the engine, its mounted components, and the overall sense of the engine compartment. ☐

   ____

   ____

4. Locate the battery and record the visual condition of the battery and its cables. ☐

   ____

   ____

Task Completed

☐ 5. Use the light if necessary to inspect and record the condition of all drive belts. Check each belt for the proper tension.

☐ 6. Inspect the battery and its cables. Record your overall impression of the battery and its connections. Record the cold-cranking amps of the battery.

____

____

☐ 7. Inspect the parts of the spark plugs and other ignition components that are visual and record your findings.

____

____

☐ 8. Inspect the top and sides of the engine (and transaxle if equipped) that are visible. Are there signs of leakage of any type? If so, what liquid is leaking? Record your overall impression of the engine.

☐ 9. Check the following fluid levels: engine oil, power-steering and brake fluids, transmission and transaxle fluid level (if automatic), and coolant level. Are any fluids low and which one (s)? Do the respective fluids have the correct color and are not burned, discolored, or dirty?

____

____

☐ 10. Remove the radiator cap and inspect the coolant in the radiator. Does it appear to be the right color for the antifreeze used? Is it at the proper level?

____

____

☐ 11. Ensure the ignition key is off and the engine is cold. Spin the cooling fan by hand. Is it freewheeling without binding or dragging noises?

____

____

☐ 12. Inspect the front and rear of the radiator as much as possible. On air-conditioning equipped vehicles, the front side of the condenser core blocks a view of the radiator. Are there bent or missing radiator or condenser fins? Is there excessive trash or road debris on the front side of the radiator or condenser? Are there signs of leakage from either the radiator or condenser?

____

____

☐ 13. Based on your visual inspection, what repairs, if any, are recommended? Consult the service manual for guidance. If available, complete a vehicle inspection (21-point safety check) form. If a form is not available, record your recommendation on the repair order with reason(s) why the service is recommended.

____

____

Instructor's Response ____

____

____

____

JOB SHEET 18

Name ______________________________ Date ______________

PERFORMANCE CHECK OF THE ENGINE

CAUTION: The engine, engine components, or other vehicle components may be hot. Do not touch any component until it has cooled. Injury can result from touching a hot component.

NATEF Correlation

This job sheet addresses the following NATEF task:

Identify and interpret engine performance concern; determine necessary action.

Objective

Upon completion and review of this job sheet, you should be able to perform a performance check of the engine at idle speed, determine the general condition of the engine, and make maintenance recommendations.

CAUTION: Exercise caution when working around a hot engine. Serious burn injuries could result.

Tools and Materials Needed

Assigned vehicle

Service manual

INSTRUCTOR NOTE: An instructor, senior student, or an experienced technician should be on hand to observe and supervise a student's first attempt at performance checking an engine or vehicle.

Procedures

Task Completed

1. Start a repair order on the vehicle. ☐
2. Move the vehicle to a well-lit area and allow the engine to cool if necessary. Chock the wheels, place an automatic transmission in PARK (manual in NEUTRAL), set the parking brakes. ☐
3. Raise the hood and check the fluid levels in the engine and the cooling system. Are they correct? ☐

   ____

   ____

   NOTE: All tasks in this job sheet are done with the engine at idle speed, transmission in PARK (NEUTRAL).

4. Turn the ignition switch to RUN, ENGINE OFF. Record which warning or caution light is illuminated. Are all lights correct? If not, stop and consult the instructor and the service manual. ☐

   ____

   ____

5. Start the engine and allow it to idle. ☐

Task Completed

☐ **6.** Record your impression of the engine and starter when the ignition switch was in the start position. Was the cranking and starting noise "normal?" Did the starter respond correctly? Did the engine crank over easily and start up within 15 seconds? Did all of the warning or cautions lights go out when the engine started (within 15 seconds)? If answer to any question is NO, shut down the engine and consult the instructor and service manual. Record your observations.

____

____

☐ **7.** If the answer to all questions in step 4 was yes, proceed with this step. Does the engine operate quietly? If equipped with a tachometer, what is the registered engine speed? According to the service manual, what is the desired engine speed? If too high or too low, what do you recommend with regard to repairs?

____

____

☐ **8.** Is there any excessive vibration that can be felt when the engine is running?

____

____

☐ **9.** Check around the vehicle for clearance of objects and personnel. Ensure the parking brakes are on and the transmission is in PARK (NEUTRAL).

NOTE: Step 9 is only performed on vehicles equipped with an automatic transmission. Do not engage a drive gear on a manual transmission for this job sheet.

☐ **10.** Apply the service brakes and shift to REVERSE. Did the engine speed remain stable? Did you feel the transmission engage? Shift back to PARK. Record your observation and recommend any repairs you feel are needed.

____

____

☐ **11.** Turn the ignition key to OFF. Record your overall impression of the engine idle operation. Complete the repair order and recommend any maintenance that should be performed and why.

____

____

____

____

Instructor's Response ____

____

____

JOB SHEET 19

Name ______________________________ Date ______________

Undercarriage Inspection

NATEF Correlation

This job sheet addresses the following NATEF task:

There are no specific task(s) directly related to an undercarriage inspection, but there are related tasks in each NATEF task area that may be used as a guide.

Objective

Upon completion and review of this job sheet, you will be able to perform an undercarriage (under car) inspection and make maintenance recommendations.

Tools and Materials Needed

Assigned vehicle

Vehicle inspection sheet if available

Service manual

CAUTION: The engine, engine components, or other vehicle components may be hot. Do not touch any component until it has cooled. Injury can result from touching a hot component.

Procedures

Task Completed

1. Start a repair order on the vehicle. ☐
2. Move the vehicle to a lift. An above-ground, two-post lift is best suited for this job sheet. Set the parking brakes, chock the wheels, and place the transmission in PARK (NEUTRAL). ☐

WARNING: Consult the service manual for the proper lift points on the assigned vehicle.

3. Set the lift and raise the vehicle until wheels are clear of the floor. Shake the vehicle to ensure it is solidly supported. ☐
4. Reach (enter) into the vehicle and release the parking brakes. Shift the transmission to NEUTRAL if necessary. It may be necessary to switch the ignition to RUN and apply the service brakes. DO NOT start the engine. ☐
5. Lift the vehicle until easy access is gained to the undercarriage. ☐
6. Lower the lift until the manual safety locks engage. ☐

   NOTE: For the purpose of clarity, this job sheet will start at the front of the vehicle and proceed to the rear, performing checks on the left then the right as appropriate.

7. Observe the bottom of the engine and transaxle, if equipped, for leaks of any types. Observe the lower radiator and hoses for leaks and serviceability. Were any leaks or apparent damage to either component inspected? If yes, what do you recommend with regard to repairs? ☐

   ____

   ____

Task Completed

☐ **8.** Observe the components near or attached to the left front wheel including the tire tread and hydraulic lines and hoses. It may be necessary to pull or push accessible steering components to feel for any looseness or damage. Spin the wheel and listen for any noise or observation of wobble. If this is an FWD vehicle, inspect the drive axles and the CV joint boots. Note any damaged or apparently worn or broken parts.

____

____

☐ **9.** Repeat step 7 on the right front wheel.

____

____

☐ **10.** If this is a RWD vehicle, perform an inspection of the transmission and its external components for leaks or damage. Record your results.

____

____

☐ **11.** Beginning at the backside (rear) of the engine, inspect the exhaust system from front to rear. Is the exhaust suspended on serviceable hanger? Is there any black soot around the connections or anywhere on the pipes, converter, and muffler? Attempt to move the exhaust side to side and up and down. Were there any indications that any portion of the exhaust system was hitting the frame, body, or other vehicle component?

☐ **12.** Once at the rear of the vehicle, inspect the filler and breathing tubes from the filler cap to the fuel tank. Are there any signs of leakage or deterioration?

☐ **13.** Follow the brake lines and fuel lines back to the area of the engine. Are there any signs of leakage or deterioration?

☐ **14.** Return to the left rear wheel and repeat step 7.

☐ **15.** Repeat step 7 on the right rear wheel.

☐ **16.** If this is RWD vehicle, inspect the seals on the axle housing and around the wheels for signs of lubricant leaks.

☐ **17.** Lower the vehicle until the wheels are just clear of the floor. Shift the transmission to PARK. Lower the vehicle to the floor, ensuring the wheel chocks are properly placed.

☐ **18.** Based on your observations, what repairs are recommended and why? Complete the repair order.

____

____

____

____

Instructor's Response ____

____

____

____

JOB SHEET 20

Name ______________________________ Date ________________

Calculating the Pressures and Forces within a Hydraulic Circuit

NATEF Correlation

This job sheet addresses the following NATEF task:

Diagnose pressure concerns in the brake system using hydraulic principles (Pascal's Law).

Objective

Upon completion and review of this job sheet, you should be able to calculate the values of a hydraulic circuit and make suggestions on possible faults or repairs.

Tools and Material Needed

Calculator

Procedures

Task Completed

NOTE: The problem presented in this job sheet is theoretical and may/may not represent real-world hydraulic problems.

1. The following information is provided by the service manual and initial diagnosis on a nonworking circuit. ☐

   Four-wheel disc brakes with a diagonal split system; master cylinder piston (2) area 0.5 square inch each; caliper piston (8) area 1.5 inch each.

   For the student's information: A diagonal split system has each master cylinder piston supplying equal fluid/pressure to each side of the split. One split includes the LF and RR, the other split includes the RF and LR.

   Assume you have done performance and inspection checks of the brake system and have determined that the system is not supplying the necessary forces to safety slow the vehicle. There are no oblivious leaks.

   What are some of the possible faults for this circuit? What would be your next step to locate the fault(s) if any?

   ____

   ____

2. The following information was obtained from making hydraulic measurements and observations. ☐

   The brake pedal is applied with 50 ft.lb of force. A pressure gauge inserted into the brake system at each of the calipers indicate the pressures are different when the brakes are applied and held on.

   LF 20 psi; RF 25 psi; LR 25; RR 20

   Using Pascal Law, $F = PA$, calculate the theoretical pressures that should be present at each wheel. What are they?

   LF ____________ RF ____________ LR ____________ RR ____________

   $F = PA$, $P = F/A$, $A = F/P$

Task Completed

P = F (brake pedal input force) / A (area of the master cylinder piston) = hydraulic pressure (psi) throughout system.

☐ **3.** Comparing your calculation to the measured pressure, which component(s) is/are at fault in this system?

____

____

☐ **4.** What is your suggested repair?

____

____

Instructor's Response ____

____

____

____

Chapter 8

Engine Preventive Maintenance

Upon completion and review of this chapter, you should understand and be able to describe:

- How to perform an oil and filter change.
- How to reset the engine oil life indicator or reminder.
- How to inspect underhood fluids.
- How to change air and fuel filters.
- How to inspect and service the cooling system.
- How to inspect and replace a serpentine drive belt.
- How to inspect the timing chain or belt.
- How to remove and replace the timing chain or belt.
- How to adjust valves.
- How to inspect and service exhaust systems.
- How to inspect and service emission control devices.
- How to use a scan tool to check for diagnostic trouble codes (DTC).
- How to perform engine vacuum and pressure tests.
- How to interpret engine vacuum and pressure tests.
- The steps of an express service.
- How to recognize special procedures for servicing hybrid electric vehicles.

Introduction

The technician's first jobs will probably be lubrication service and tires. While performing a routine oil change, the technician should observe the general condition of the vehicle and the engine. The customer may not realize that there is a potential problem. Driving the vehicle to the lift gives the technician a chance to get a feel for the brakes, steering, and engine operating condition. When a defect is suspected, notify the customer or the service writer.

Most entry-level technicians with sufficient supervision can perform the services covered in this chapter and the remainder of this book. Keep in mind that shop managers or service writers will not assign more complicated jobs to a new technician until they feel comfortable with the technician's skill and knowledge.

SAFETY TIP: The services performed in this chapter may require the technician to work on or with hot components or liquids. **Heat-resistant gloves** that cover the hands *and* forearms are available at local parts stores. Remember, heat absorbed by the outside of a material can burn the skin if touched.

Changing the Engine Oil and Filter and Checking Fluid Levels

Almost every technician starts his or her first job as lube technician and will continue to perform this basic service throughout his or her career. If any task is considered to be a basic and very necessary vehicle service, it is the scheduled engine oil change.

The brand and type of oil are strictly a customer's choice. However, the technician may advise the customer on the type of engine oil recommended by the vehicle manufacturer.

Before starting the service, get the oil and filter. Usually the type of oil is determined by the engine specifications, but the customer may have a preference for a brand or type

SAFETY TIP:
(continued)
If the glove material is wet, it will have no resistance and the heat will travel straight through to the skin. Even when using protective clothing, the technician must be careful. Eye protection should always be worn. Always use the locks on a lift or use jack stands when working under the vehicle.

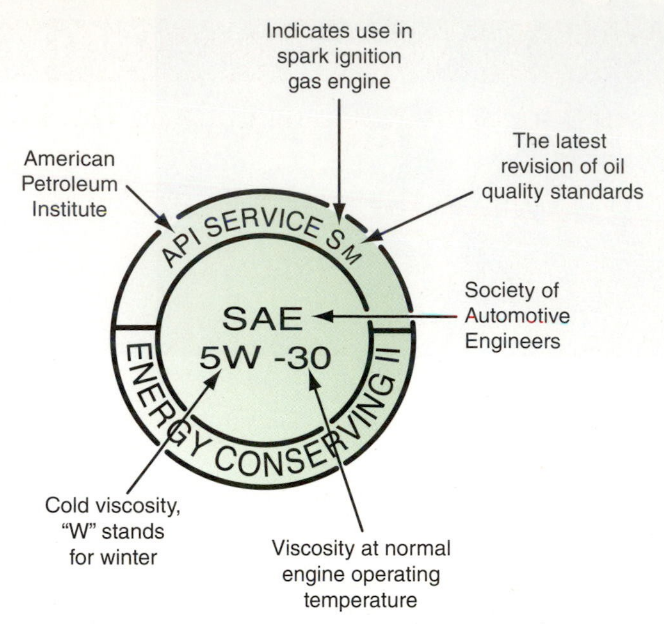

FIGURE 8-1 **Definition of the ratings shown on an oil container.**

Heat-resistant gloves and forearm sleeves are usually made of woven kelvar. If the kelvar becomes wet, the heat will be absorbed by the liquid and transferred to the skin.

(Figure 8-1). Check the repair order for customer preference or any additional tasks the customer may have requested. The customer may have heard a noise or other symptom that she or he would like to have checked.

Move the vehicle to the bay. As the vehicle is operated, note the performance of the engine, brakes, and other systems. This short drive may point to something that requires a closer inspection. Raise the hood and make a quick inspection for the oil filter. Every engine has the oil filter in a different location, so look carefully (Figure 8-2). Lift the vehicle to a comfortable working height. Refer to Photo Sequence 14 for the typical procedure for changing oil and filter.

WARNING: Do not remove the wrong drain plug. A few vehicles have a drain plug for the automatic transaxle. Accidentally draining the wrong fluid will cost the shop some money, but if the mistake is not caught before releasing the vehicle to the customer, damage to the engine and transaxle could occur.

Classroom Manual
page 184

SPECIAL TOOLS
Lift
Drain pan
Filter wrenches

FIGURE 8-2 **The oil filter can be easily removed.**

PHOTO SEQUENCE 14

Typical Procedure for Changing the Oil and Oil Filter

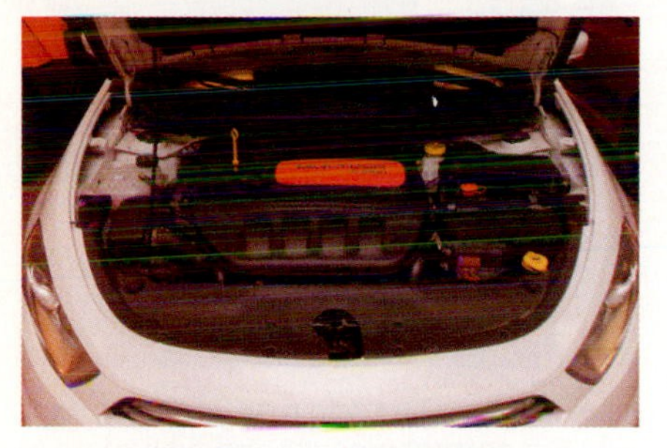

P14-1 Open the hood and pull the dipstick out slightly. Remove the oil cap.

P14-2 Raise the vehicle, locate the oil drain plug, and position the oil drain pan so that it will contain the oil. NOTE: The oil will come out with light pressure, and it will cause a stream to go to the side. Adjust for this by moving the oil drain pan to the side.

P14-3 Remove the oil drain plug, using a wrench. NOTE: Always use a properly sized wrench. You will find that over time the oil drain plug head will get rounded by the use of incorrect sized tools. It will need replacement if it is difficult to remove.

P14-4 Watch the oil coming out of the pan. The content, color, and clarity will give you clues to how often it is maintained and if there are any problems.

P14-5 Adjust the oil drain pan location as needed when the oil flow becomes less.

P14-6 After the oil is done draining, reinstall the oil drain plug with a torque wrench. Use the service manual to find the specification.

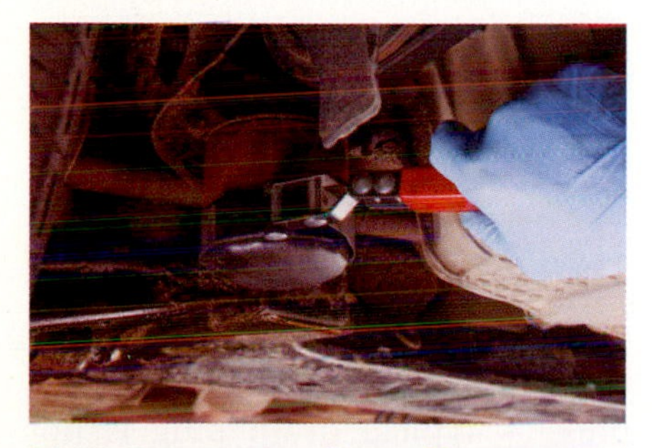

P14-7 Remove the oil filter. You may need to use an oil filter wrench if it is tight. There are several types to use. Oil filters are located at different places on different vehicles.

P14-8 Make sure that the oil filter gasket is removed with the old filter. Double check the engine side to make sure it is not there. CAUTION: If you do not remove the old oil filter gasket from the engine, when you install the new one, it will have two gaskets. This is called "double gasketing" an engine. This can cause oil to leak from the space in between the two gaskets at a very high rate and pressure.

P14-9 Lubricate the new oil filter gasket with new oil before installing it. Tighten the oil filter one-half turn past hand tight. Do not use any tools to tighten it.

P14-10 **Some vehicles have a cartridge style oil filter that requires a socket to remove.**

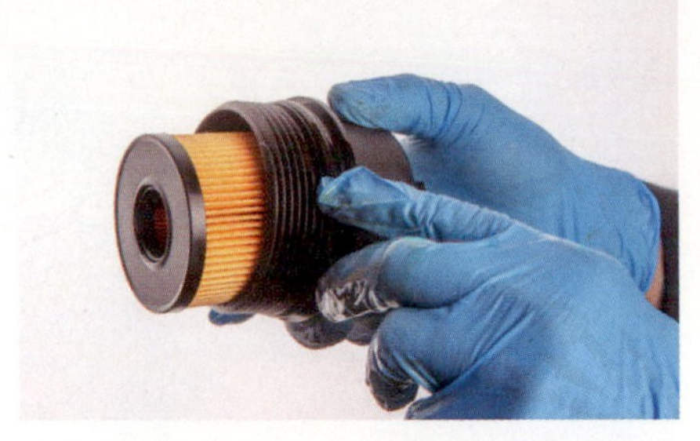

P14-11 **Similarly, lubricate the new oil filter O-ring if it is a cartridge style. Use a torque wrench to tighten the housing cover of a cartridge filter (if equipped).**

P14-12 **With the drain plug and filter tight, add the correct amount and type of oil to the engine.**

P14-13 **Check the level on the dipstick before starting the engine. If needed, add oil.**

P14-14 **Start the engine. Immediately watch the oil pressure light or gauge. The engine should have oil pressure within 5 seconds.**

P14-15 **If it does not have pressure, then turn the engine off and inspect. With the engine still running, walk around and look under the engine for any leaks.**

P14-16 **Turn the engine off, wait 30 seconds, and recheck the oil level. It should have gone down just a little bit, because oil is filling into the new filter. Adjust the oil level as needed.**

CAUTION:
Use hand and forearm protection when changing oil on a hot engine. Burns could result from the oil, filter, and engine components.

In many cases, the vehicle is hot and the customer is waiting, so speed is essential. Do not, however, forget basic safety procedures. Use heat-resistant gloves and eye protection. With the vehicle raised, move the drain container under the oil drain plug (Figure 8-3). Use the correct wrench to loosen the plug. A typical drain plug is not made of the strongest metal and can be rounded with a wrong wrench. A boxed-in wrench is suggested.

While the oil is draining, make a quick inspection of the undercarriage of the vehicle. Look for any leaks and bent or damaged components. On some vehicles, it is possible to see part of the brake disc and pads. Inspect the tire treads for wear. The tread should show even wear across the tire.

Check the drain plug for a gasket. Many plugs have a brass, paper, or plastic washer to help seal the drain hole (Figure 8-4). It is usually best to replace the gasket if one is used. Once the oil has completed draining, install and torque the drain plug. Do not overtighten the plug. The drain hole goes through thin metal and the threads can be stripped out.

FIGURE 8-3 An oil drain container.

CAUTION:
Always capture the used oil from the engine. Clean any spills immediately. Failure to do so may result in an injury to the technician or charges by the EPA for illegal waste disposal.

INTERNET TIP:
For proper disposal and control procedures for repair shop waste, information is available at the Web site for the Coordinating Committee for Automotive Repair. The Web site is http://www.greenlink.com.

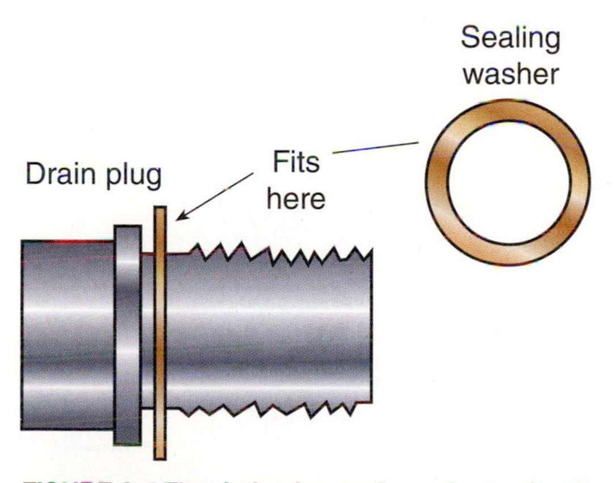

FIGURE 8-4 The drain plug seal may be made of brass, copper, or neoprene.

Locate the filter and determine the best type of **filter wrench** to be used. There are several types of wrenches available and more than one may work (Figure 8-5). The location of the filter and the components around it will determine which wrench will be the easiest to use. Position the drain pan directly under the filter and loosen it. At this point, the technician may have to work around a very hot exhaust and other hot components. Many filters are placed so the draining oil will flow down over engine and vehicle components. Be alert with the drain pan and watch for splashing oil. Work carefully and safely.

Filter wrenches are specifically designed to remove oil filters.

The **filter gasket** is usually an O-ring. A cartridge style oil filter requires a new O-ring as well (Figure 8-7). There is a specific amount of torque from the container to the filter housing.

While the filter is draining, double check that the old oil **filter gasket** came off with the old oil filter. Double check the engine side to make sure that it was removed. Leaving

FIGURE 8-5 **Examples of oil filter wrenches.**

it on will be catastrophic! If the old gasket is left on, you will end up with two gaskets (because the new oil filter has one). This is known as double gasketing the filter. It will cause a major leak and oil will spill everywhere when the engine is started, if you are lucky. In some cases the double gasket will not leak until after the vehicle has left the shop, causing a major oil leak and possibly losing oil pressure and causing major engine damage.

Next open the box for the new filter and a container of oil. Open the new oil container and use a finger to smear a light coat of oil on the filter's gasket (Figure 8-6). This will help remove it during the next service.

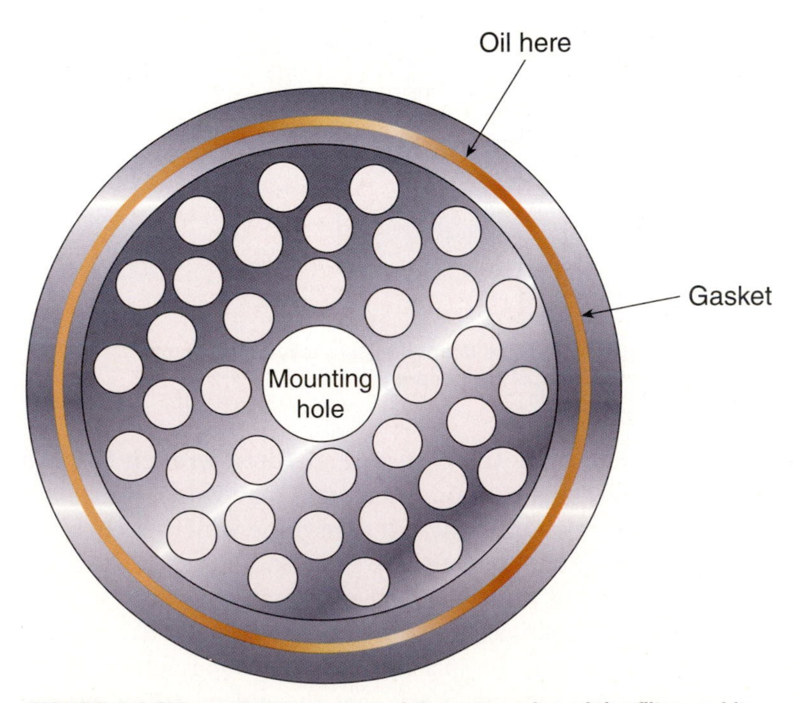

FIGURE 8-6 **This gasket goes around the outer edge of the filter and it should be lightly lubed before the filter is installed.**

FIGURE 8-7 **You will encounter different style oil filters. This is an example of the two most common types.**

Wipe the engine oil filter mounting surface with a clean rag and install the new filter. Hand tighten the filter to the specifications shown on the filter. In almost all cases, the filter is hand tightened about one-half of a turn after the seal contacts the mount. Do not use a wrench to tighten the filter or it will be very difficult to remove during the next service.

With the plug and filter in place, check the area for tools and remove the drain pan from under the vehicle. Lower the vehicle to the floor.

There are different methods used to install oil into an engine. Some shops have overhead drops that are plumbed to large oil containers in a storeroom (Figure 8-8). Follow the

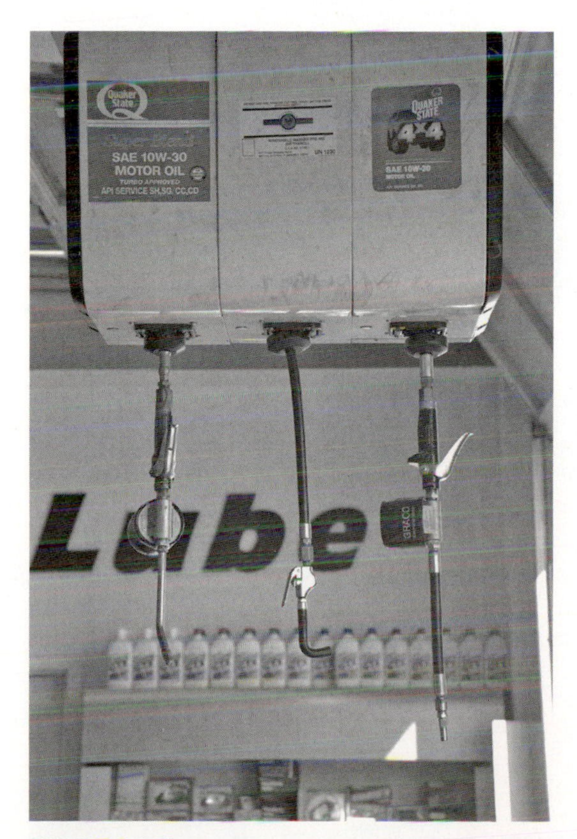

FIGURE 8-8 **Fleet repair and specialty shops have bulk containers and the oil or grease is pumped to each bay as needed.**

equipment instructions and shop policy to select the amount and type of oil. Many shops use quart containers, and the technician must pour each quart individually into a filler hole located in a valve cover.

Use a clean funnel to prevent oil from dripping on the engine. In the few situations when a funnel is not available, oil can be poured directly from the container. Most quart containers are shaped to reduce the amount of spillage. Notice where the pouring **spout** is attached to the container body (Figure 8-9). It is usually to one side at the top. Hold the container so the pouring spout is the highest point, placing the oil below the spout. The oil can be poured slowly and air can enter the container as the oil exits (Figure 8-10). Turning the container so the spout is down will cause a burping of the oil as air is drawn into the emptying container. Install the amount of oil specified and then replace the filler cap.

The offset position of the container's **spout** allows air to enter the container. It is suggested that the container not be squeezed to speed the process. This will block the airflow.

Many modern vehicles have a computer program that makes the driver aware of the need to change oil (Figure 8-11). If this is the case, you must reset the oil life indicator or reminder. Some vehicles require a special tool, connected like a scan tool. But most can be reset using the driver's controls on the instrument cluster or the information center. Even if the vehicle has a maintenance reminder, many shops still install the windshield sticker (Figure 8-12).

CAUTION:
Do not open a hot coolant system. Serious burn injuries could result.

Conduct a check of the fluid levels available under the hood. The most common are the brake master cylinder, windshield washer, coolant, and transaxle or transmission (Figure 8-13). The transaxle and transmission must be checked with the engine running, but the fluid can be checked for burning or age. This fluid should be reddish. If it is burned or dark in color, a service recommendation may be made to the customer.

FIGURE 8-9 **The spout of a quart container of engine oil is offset to facilitate pouring.**

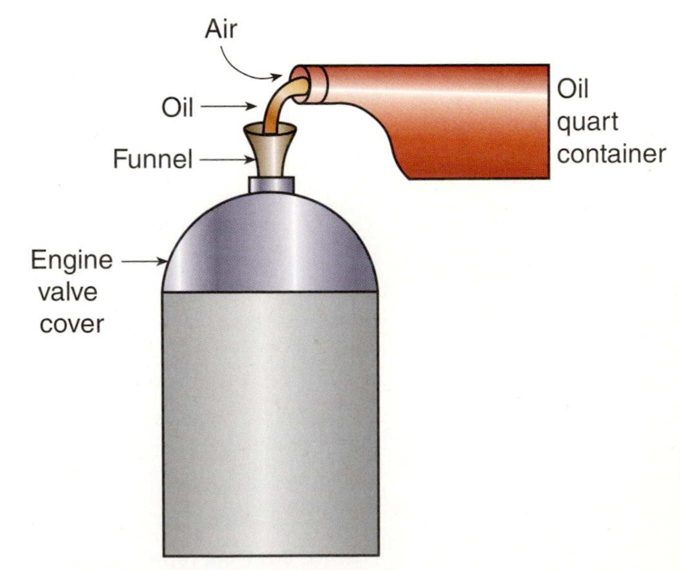

FIGURE 8-10 **The offset spout should be held as shown when adding oil to the engine.**

FIGURE 8-11 **Many newer vehicles have a computer program that makes the driver aware of the need to change oil.**

FIGURE 8-12 **Always put an oil sticker on the windshield or in the driver's door jam to remind the customer when to have the oil changed again.**

The coolant may be hot, so do not open the radiator. Check the coolant level in the overflow reservoir. It should be between the hot (high) and cold (low) marks (Figure 8-14). Removing a cap or observing the level through the translucent reservoirs provides a means to check the master cylinder and windshield washer fluids. A low brake fluid level may be correct based on the amount of brake wear. As the brakes wear, the fluid

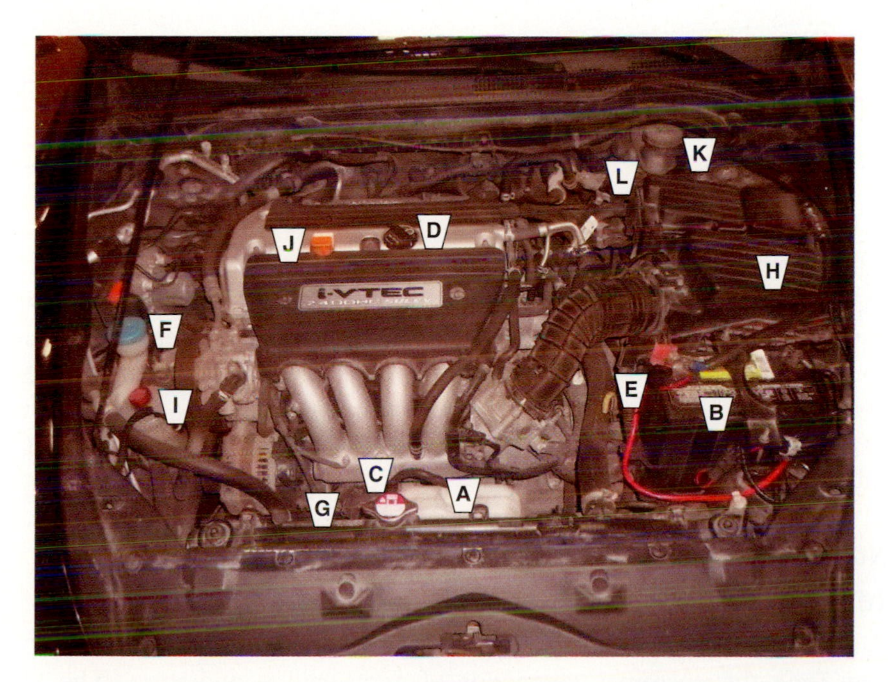

A. Engine coolant reservoir
B. Battery
C. Radiator fill cap
D. Engine oil fill cap
E. Automatic transmission dipstick (equipped)
F. Windshield washer reservoir
G. Engine cooling fans
H. Air cleaner
I. Power steering reservoir
J. Engine oil dipstick
K. Brake fluid reservoir
L. Clutch fluid reservoir (if equipped)

FIGURE 8-13 **In most shops, an oil change includes a check of the fluid level at each location shown here.**

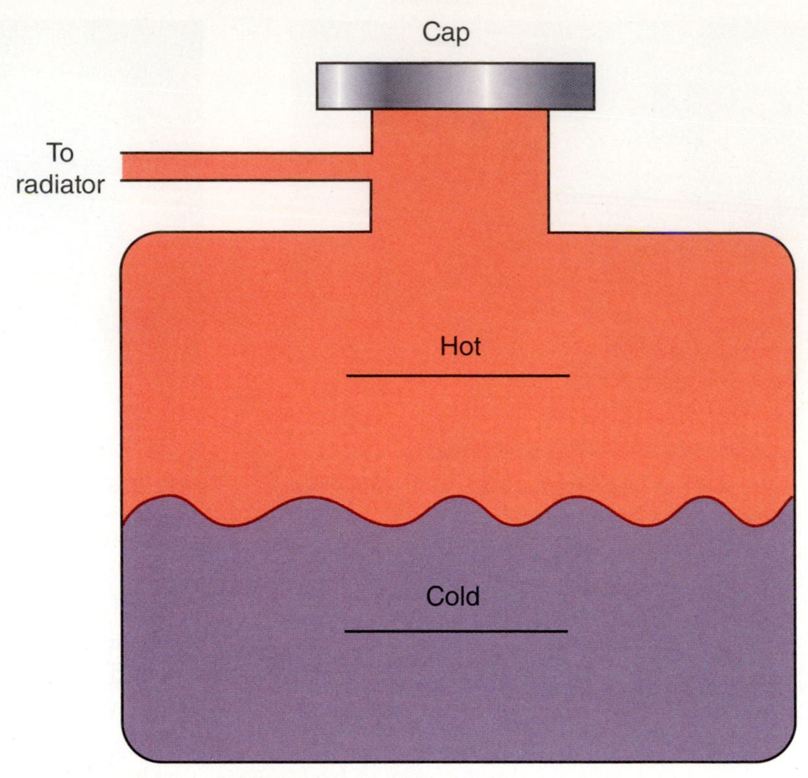

FIGURE 8-14 **The coolant should be visible through the reservoir and between the hot and cold mark on a warm engine.**

level will drop. The technician or service writer may consult with the customer if the brake fluid is low.

Top off is a term that means fluid is added to a system to bring it to the correct level.

Some shops will **top off** some fluids without charge to the customer. Generally, antifreeze and brake fluids are charged. With all of the fluids checked, start the engine and observe the oil warning light. It should go out within a few seconds after the engine is running. If it does not, shut the engine down immediately and correct the problem. When the light is out, shut off the engine and recheck the engine oil level. Some of the oil is used to fill the filter and the level may drop slightly. The level, with the filter full, should be at the full mark or slightly below the mark. Top off the oil as needed. Check under the vehicle for leaks. When the oil level is correct and no leaks are apparent, clean the vehicle of any oil drips. Complete the repair order and return the vehicle to the customer. Properly dispose of waste oil and the old filter.

Classroom Manual
page 204

Changing the Air and Fuel Filters

Air Filter

The air filter is usually checked during an oil change service. However, it should be checked any time the customer complains of poor engine operation. The filter will be located in a housing on top of the engine or in an area near a fender behind a headlight. The housing cover may be bolted down or retained with spring clips (Figure 8-15).

Select the correct replacement filter and open the housing. There may be electrical or vacuum connections that must be disconnected before the cover can be removed. Be sure that neither is damaged during removal. Remove the filter and inspect the housing for dirt, leaves, and oil. Dirt and leaves on the inlet side of the housing is normal. Clean the

FIGURE 8-15 **Snap clamps typically hold the air filter housing cover in place. Some use threaded fasteners for the same purpose.**

SERVICE TIP: Many of the vacuum hoses are made from hard plastic and break easily. Use both hands, one to pull the hose and one to hold the component, to pull the hose straight from the component. Do not bend or flex the hose. A small, pocket-size, flat-tipped screwdriver may be used to unlock electrical connections.

housing. If oil is present, this may indicate a problem with the PCV system. Further inspection of the PCV system should be accomplished.

In some vehicles, a separate air filter for the PCV is located in the engine air filter housing. This filter should also be replaced at this time. Install the new engine air filter and cover. If bolts are used, do not overtighten them. On the round units mounted directly to the top of the engine, the tightening of the bolts can be critical. In many cases, the cover on this type of housing applies force to the air filter. Overtightening can depress the filter and restrict airflow to the engine.

SPECIAL TOOLS

Line wrenches

Quick-disconnect tools

Catch pan for fuel

Hose-clamping tools

Vehicles that have a PCV valve should have it inspected every time the air filter is inspected. Remove the PCV valve (Figure 8-16) and shake it. If it rattles, it is still good and not stuck. If no noise is heard, replace it.

In other vehicles, no PCV valve is used. These vehicles use an orifice tube that goes from the valve cover to the intake manifold. You must inspect this hose by squeezing it to see if it is softened from being oil soaked. Often these hoses will collapse from oil saturation over time. Removing this hose is important to inspect the inside of it for damage or blockage.

Fuel Filters

CAUTION: Do not allow any hot components or flame-producing devices near the vehicle when the fuel system is being serviced. Dangerous fires could result.

Changing the fuel filter presents some safety and mechanical problems. Opening the line to remove the filter will allow some fuel to escape. A suitable catch container should be used, and the waste fuel must be stored in an appropriate container. The oil drain pan should *not* be used. Consult with a supervisor for instructions. The vehicle should be allowed to sit until the exhaust system is cool. It is preferable that the entire vehicle has cooled down.

The technical end involves relieving the fuel pressure on electronic fuel injection (EFI) systems and using tools to disconnect and close the lines. Consult the service manual to determine how the fuel pressure is relieved. It usually involves disabling the fuel pump and starting the engine. When the engine stalls, the fuel pressure has been removed but there will still be some fuel in the lines.

FIGURE 8-16 **Some PCV systems have an additional air filter within the engine's air filter housing. Change it at the same time the engine filter is changed.**

Some fuel filters are connected to the fuel lines with compression fittings requiring line wrenches for loosening (Figure 8-17). Others use quick-release fittings that can be unlocked with special tools or a small, flat-tipped screwdriver (Figure 8-18). Almost all filters have a small mounting bracket that is removed or loosened with a socket or screwdriver.

Before starting work, select the correct filter and check it for the bracket and replacement quick-release locks if used (Figure 8-19). Some types of quick-release locks are

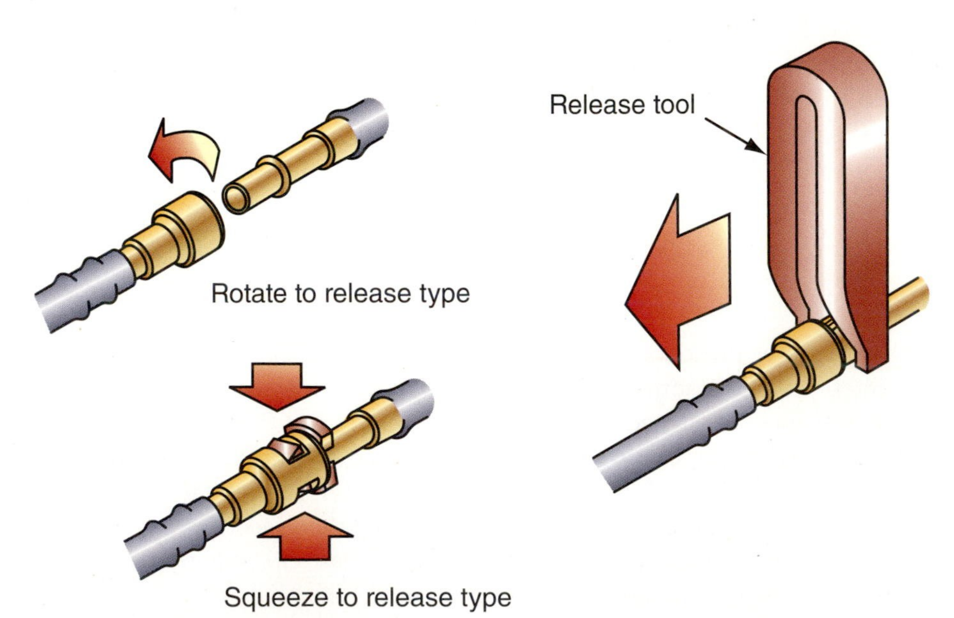

FIGURE 8-17 **Twist the fuel lines, if possible, at the quick-disconnect. This will loosen the connection for easier release.**

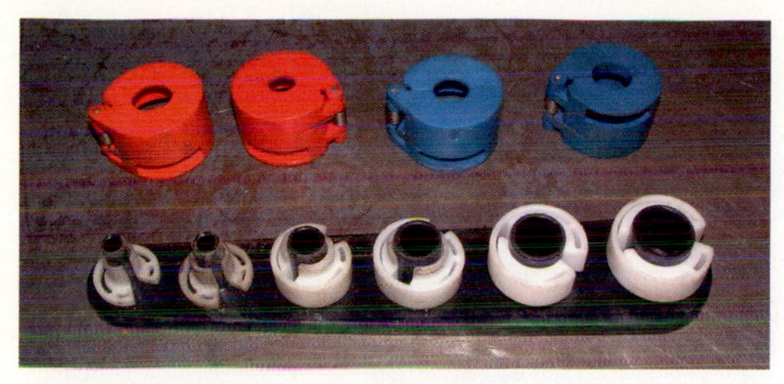

FIGURE 8-18 **There are many types of quick-disconnect tools available.**

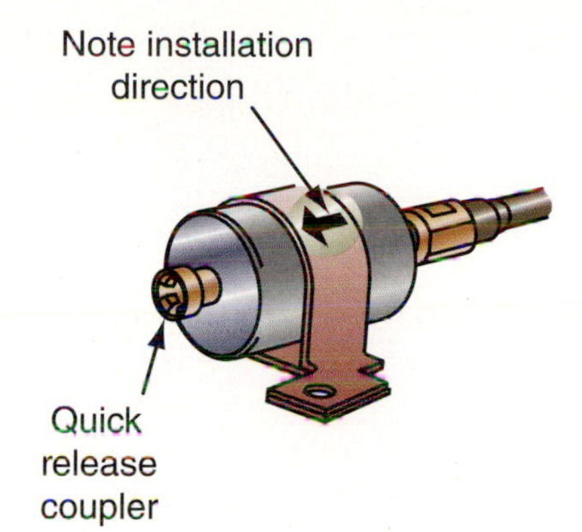

FIGURE 8-19 **Most fuel filters for EFI are mounted to the frame near the fuel tank. Some are mounted in the engine compartment.**

damaged during removal. Most EFI systems have fuel filters mounted on their frame rails and use the quick-release connections used for this discussion. Other types follow the same procedures except for filter location and type of connection.

Follow the instructions to relieve the fuel pressure and locate the fuel filter. Do not enable the fuel pump until the filter is replaced. If filter is under the vehicle, raise the vehicle to a good working height. Locate the filter and determine the tools needed. If the filter is connected to flexible hoses, same may be clamped shut with special tools. Check the service manual.

CAUTION:
If rags are used to catch some of the fuel, ensure that they are stored separately from other rags.

Clamp flexible hoses if possible and position the pan securely as two hands will be required to remove the filter (Figure 8-20). Remove or loosen the bracket. Loosen the fittings by prying the locks open with the tool. Twist the filter once or twice to loosen it on the lines. Fuel may be allowed to drain at this point. Allow as much as possible to drain. Grasp a fuel line to hold it in place as the filter is pulled from the line. The locks will have to be held open while separating the filter and hose. Allow the fuel to drain before separating the filter from the other line. There will still be fuel in the filter, so handle accordingly.

Ensure that the filter is mounted in the correct direction. The filter should be marked "inlet" and "outlet" or may only have one end labeled (Figure 8-21). The label may be just an arrow showing flow. The filter is installed with the outlet end toward the engine.

FIGURE 8-20 **This type of quick-disconnect does not require a tool. It can be done by hand.**

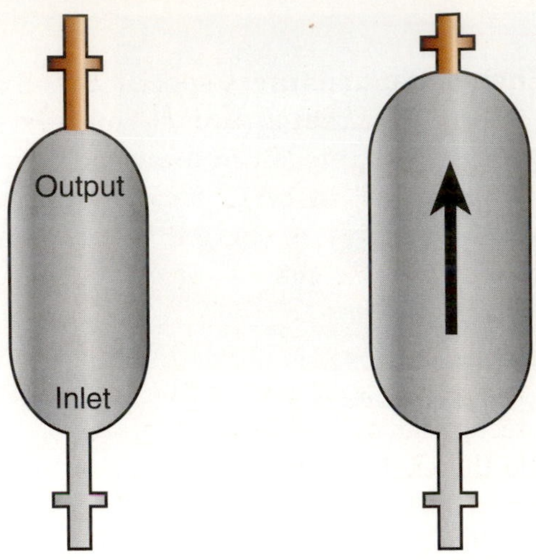

FIGURE 8-21 **The fuel filter may be labeled in several ways. Two examples are shown.**

Classroom Manual
page 190

CAUTION:
Do not open, service, or repair a hot coolant system. Serious injury can result.

Antifreeze is normally slick and oily. Metal dissolving in the coolant, however, causes it to feel like **slime**. You can also tell a lot about coolant by its color. Green is the color of traditional coolant. This type of coolant will last about 3 years. Extended life coolants are very common on modern vehicles. They will last anywhere from 3 to 6 years. Extended life coolants are available in many colors: red, pink, orange, yellow, and so on. But they are not green.

Install one end of the filter into the fuel line and be sure the lock slips in place. The short pipe at each end of the filter has a flange or collar. The lock must lock between the collar and the filter body. The connection should be checked by attempting to pull the filter from the line. The filter may move a short distance. Repeat the procedures with the other end. Install and tighten the bracket. Remove any hose clamps installed.

Lower the vehicle just enough to gain access to the door. Do not start the engine at this time. Enable the fuel pump by turning the ignition key from OFF to RUN two or three times. This will pressurize the system. Check the fuel filter connections for leaks. If dry, start the engine and allow it to run long enough to settle down and operate smoothly. Shut off the engine and check the filter for leaks. This procedure is easier if a second person is available to check for leaks or operate the running engine while the filter area is checked. With the service complete, check for tools and lower the vehicle.

Drain all of the fuel from the old filter. Properly dispose of the filter and waste fuel. Complete the repair order and return the vehicle to the customer.

The newest light vehicle EFI system is a *returnless fuel arrangement*. Instead of a pressure regulator situated near the injectors, pressure is controlled by the speed and time of the fuel pump. This is done by varying the voltage supply to the pump. For instance, during idling conditions, the pump's supply voltage is reduced, thereby lowering the fuel pressure and volume. As the engine load increases, more voltage is supplied and fuel flow is increased to meet demand. Systems of this type usually do not have external fuel filters that need periodic replacement. The filter is part of the pump and sending unit assembly inside the tank. Most newer Chrysler vehicles have this type of EFT.

Cooling System Inspection and Service

Inspections

Inspection of the cooling system consists of checking for external coolant leaks, hoses, belts if applicable, and the visual condition of the coolant. The coolant is usually a 50/50 mixture of water and antifreeze, which can be green, orange, or red in color. Check the inside of the cap for rust, corrosion, contamination, or **slime**. If any of these three conditions is present, the cooling should be flushed and new coolant installed.

WARNING: Engine manufacturers specify exactly which antifreeze to use in their engines. Many antifreezes are chemically designed for a specific brand or model of vehicle engine. Even many of the so-called one-size fits all antifreezes will not work well in many modern engines. They prevent freezing; however, they can corrode the coolant system or cause excessive electrolysis to build. Use the antifreeze that is recommended by the vehicle or engine manufacturer.

External cooling leaks are easy to spot and can be easily identified by the color of the antifreeze. An experienced technician should locate and repair internal cooling leaks. Check under the vehicle and inside the engine compartment for coolant. Do not forget to check the heater hoses and heater. A leaking heater will deposit coolant in the passenger compartment on the floorboard. The owner will usually complain of a coolant smell.

Hoses and Coolant Changes

WARNING: Some new engines require special procedures for refilling the coolant systems. Follow the manufacturer's instructions. Damage to an engine may occur if the system has air or is not completely full.

Most radiator hoses are easily replaced. The coolant must be drained, captured, and recycled or disposed of properly. Before starting the replacement process, read the service manual to determine the correct method for draining the system, refilling the system, and choosing the amount and type of antifreeze required (Figure 8-22). Many engines, including diesels, require special antifreeze.

Special equipment is available to make draining and refilling a coolant system easy or easier than the following procedures. One very simple system uses shop air pressure to create a vacuum within the coolant system and withdraw the coolant. The same system is used to force coolant back into the system. Other than the radiator cap, the system is not opened. This speeds up the work, eliminates replacing clamps, and sometimes time-consuming gaskets.

Drain the coolant into a suitable container using the **petcock** located on the lower side of the radiator (Figure 8-23). Do not loosen the cap at this time. Some vehicles have petcocks on the engine block. If this is to be a complete coolant change, those petcocks will have to be opened to drain all of the coolant. When the amount of draining coolant slows,

CAUTION:
Collect and dispose of the waste coolant according to shop policy. Wrongful disposal of waste coolant could damage the environment, injure or kill wild or domestic animals, and bring legal charges filed by the EPA against the shop owner.

A **petcock** is a drain plug for the coolant system.

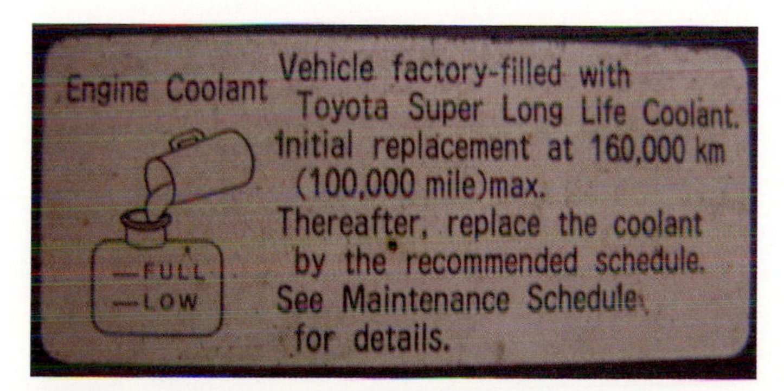

FIGURE 8-22 **This underhood label indicates that the vehicle requires an extended-life coolant. Always use the proper type of coolant.**

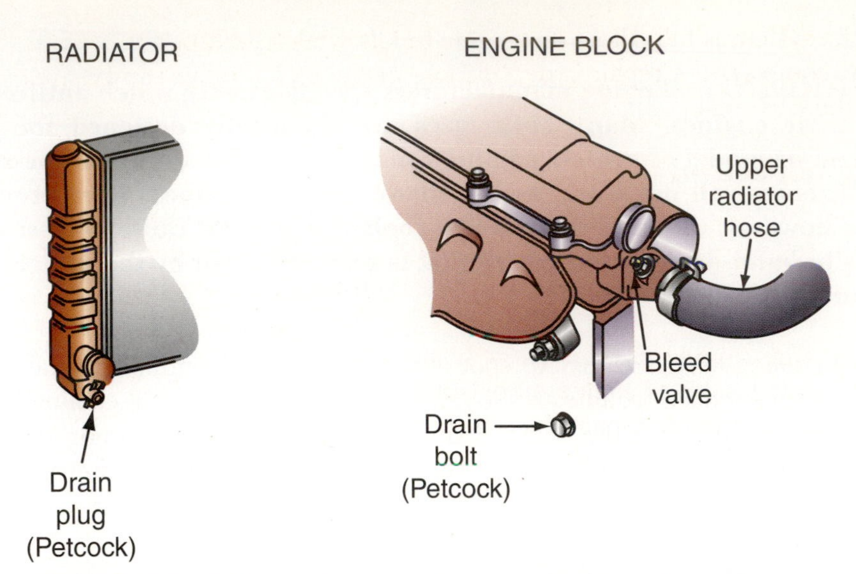

FIGURE 8-23 **Most light vehicles do not have a coolant drain plug on the engine block, but many heavy trucks and equipment do have this plug.**

remove the radiator cap. Initially leaving the cap on allows the coolant recovery tank to drain through the radiator.

Release the hose clamp using pliers, wrenches, or a screwdriver. The hose will probably be stuck to the mounts and will have to be twisted and pulled off. A possible solution is to split the hose where it fits over the mounts and slide a flat-tipped screwdriver between the hose and mount (Figure 8-24). Use a small piece of sandpaper or scraper to remove any corrosion from the mounts. Slip a new clamp over the new hose and slide the hose on the mount. Place the clamp about halfway between the end of the mount (feel it through the hose) and the end of the hose. Tighten the clamp until bits of the hose extend into the clamp's screw slots. Repeat the procedures for the other end and any other hoses. The same basic procedures also can be used to replace the heater hoses. Do not use excessive force to remove a hose from the heater. Heater hose connections are thin and can be damaged while trying to remove the hoses. Use a knife or another blade to slice the hose as mentioned above.

Close the petcocks and pour in the correct amount of antifreeze. Completely fill the cooling system with water. Follow the manufacturer's instructions to bleed air from the system. Some vehicles have a small bleeder screw at the upper radiator hose/engine connection for this purpose. Others require increasing the engine speed to 1,200–1,500 rpm, adding additional coolant, and capping the radiator before reducing the engine speed to idle. The heater temperature control should be placed in the HOT position during refilling. This will flush air from the heater and completely fill the system with coolant.

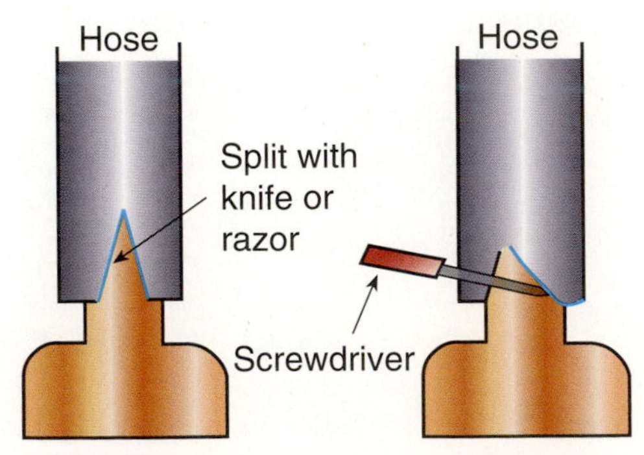

FIGURE 8-24 **Split the hose and pry it open for removal. Do not do this if the hose is to be reused.**

Once the system is full, allow the engine to idle while checking for leaks. If the repair is satisfactory, complete the repair order, dispose of the waste, clean the area, and return the vehicle to the customer.

SERVICE TIP: The condition of the antifreeze can be checked using a digital voltmeter. The red voltmeter lead is inserted into the coolant, while the black lead is attached to a good engine ground. If the voltmeter reads 0.2 volt or less, the antifreeze is good. A reading between 0.2 and 0.6 volt indicates the antifreeze is deteriorating and is borderline. A voltage value above 0.7 volt indicates the antifreeze must be changed. High voltage values can also indicate a blown cylinder head gasket. The voltage reading is the result of the chemical reaction between the engine block materials and the coolant. As the additives in the antifreeze deplete, the antifreeze becomes a weak acid. The result is the cooling system becoming a weak battery.

Coolant Inspection

The antifreeze content in the coolant may be tested with a coolant hydrometer, refractometer (Figure 8-25), or coolant test strips. The vehicle manufacturer's specified antifreeze content must be maintained in the cooling system because the antifreeze contains a rust inhibitor to protect the cooling system. Always check the vehicle or service manual for the type of coolant to add to the system. Inspect the coolant for rust, scale, corrosion, and other contaminants such as engine oil or automatic transmission fluid. Aside from a visual check of coolant condition, a paper test strip can be used to test the coolant pH level. This will give you a visual indication of whether the coolant has become too acidic and requires flushing. If the coolant is contaminated, the cooling system must be drained and flushed (Figure 8-26). If oil is visible floating on top of the coolant, the automatic transmission cooler may be leaking. This condition also contaminates the transmission fluid with

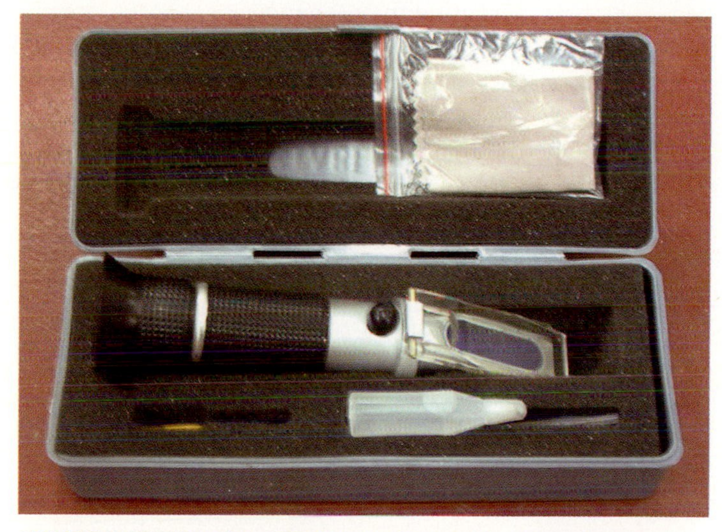

FIGURE 8-25 **This refractometer can be used to determine the antifreeze protection level of any type of coolant.**

FIGURE 8-26 **This coolant reservoir is covered in sludge; the system should be flushed and refilled with clean coolant.**

coolant. If the vehicle has an external engine oil cooler, it may also be a source of oil contamination in the coolant. These coolers may be pressure tested to determine if they are leaking. Testing or repairing radiators and internal or external oil coolers is usually done by a radiator specialty shop. Inspect the air passages through the radiator core and external engine oil cooler for contamination from debris or bugs. These may be removed from the heater core with an air gun and shop air, or water pressure from a water hose.

Cooling System Component Testing

Testing the Thermostat. If a customer brings in a vehicle with an overheating problem, it is possible the thermostat is not opening. In addition, an engine that fails to reach normal operating temperature can be caused by a faulty thermostat. A thermostat that is not opening will cause the coolant temperature gauge to rise steadily into the red zone after several minutes of operation. Sometimes the thermostat will stick closed intermittently. The customer may report that the coolant gauge spiked into the red and then came back down to a normal reading one or more times. The thermostat is inexpensive; problems caused by engine overheating are not. It is wise to replace the thermostat if you suspect that it is sticking. Thermostats have a high failure rate and are often replaced simply based on the symptoms described or as part of routine maintenance.

To attempt to verify a sticking thermostat, start the engine while it is relatively cool. Hold the upper radiator hose toward the radiator as the engine warms up. Periodically check the coolant temperature gauge to make sure the engine does not overheat. If the thermostat is working properly, the upper radiator hose should get quite hot and feel pressurized as the temperature gauge rises into the normal zone. If the gauge starts to read hot and you have not felt a distinct temperature increase in the hose temperature, the thermostat is sticking closed. Replace a faulty thermostat. You can also test for a stuck-open thermostat using a similar method. Start the engine after it has cooled. Hold the upper radiator hose toward the radiator end. Let the vehicle run for a few minutes. If the hose starts to get warm within the first few minutes of operation or before the engine reaches a normal temperature, it is sticking open (Figure 8-27).

Sometimes you may have to remove the thermostat to try to verify that it is faulty and causing the system overheating. Check the rating of the thermostat, and confirm it is the proper one for the engine application. Also confirm it was installed in the right direction. It

FIGURE 8-27 **The thermostat sits on the top front of this engine.**

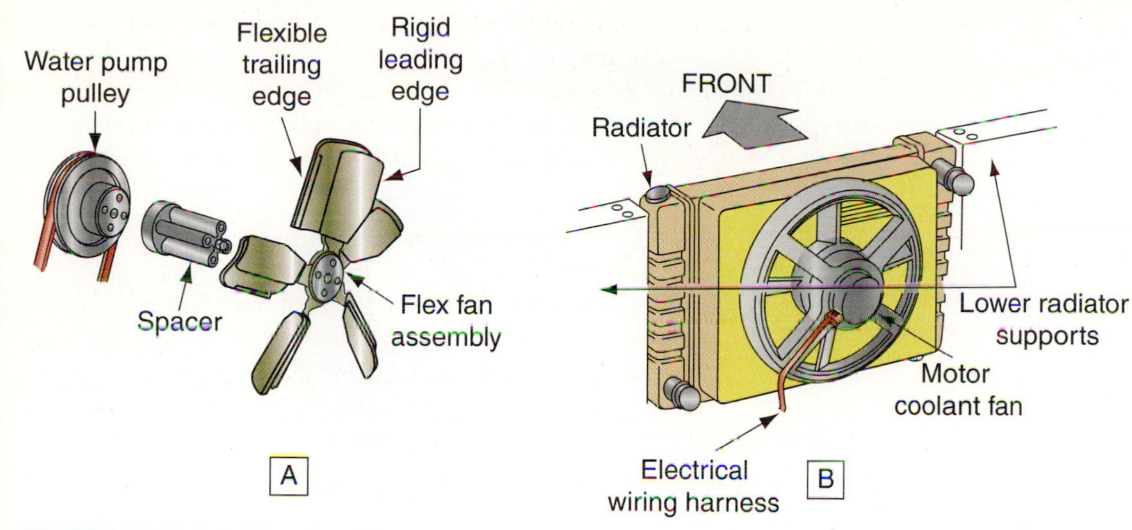

FIGURE 8-28 **(A) Belt driven fan. (B) Electric fan.**

is possible to test thermostat operation by submerging it in a container of water and heating the water while observing the thermostat. Use a thermometer so the temperature when the thermostat opens can be determined. At the rated temperature of the thermostat, it should begin to open.

Cooling Fan Inspection. The fan can be driven by a drive belt off of the crankshaft, or it can be operated electrically (Figure 8-28). Regardless of the design used, inspect the fan blades for stress cracks. The fan blades are balanced to prevent damage to the water pump bearings and seals. If any of the blades are damaged, replace the fan.

Cooling fans force airflow draw the radiator to help in the transfer of heat from the coolant to the air.

WARNING: An electric fan may start at any time, even after the engine has stopped. Either let the engine cool down or disconnect the fan electrically before touching the fan.

Electric fans are also inspected for damage and looseness. If the fan fails to turn on at the proper temperature, the problem could be the temperature sensor, the fan motor, the fan control relay, the circuit wires, or the powertrain control module (PCM). To isolate the cause of the malfunction, attempt to operate the fan by bypassing its control. On many modern vehicles, this can be done by using a scan tool to activate the fan. If the fan operates, the problem is probably in the coolant sensor circuit. Individual electrical component testing will be covered in later chapters of the book.

SPECIAL TOOLS

Serpentine belt tensioner tool

To direct airflow more efficiently, many manufacturers use a shroud (Figure 8-29). Proper location of the fan within the shroud is also required for proper operation. Generally, the fan should be at least 50 percent inside the shroud. If the fan is outside the shroud, the engine may experience overheating due to hot underhood air being drawn by the fan instead of the cooler outside air. If the shroud is broken, it should be repaired or replaced. Do not drive a vehicle without the shroud installed.

A **micro-vee** belt is a belt that is composed of multiple small vee grooves in it.

Inspecting and Replacing the Serpentine Drive Belt and Pulley System

Modern vehicles use a **micro-vee** belt called a **serpentine belt**. A micro-vee belt is like a vee belt that is used on older cars, but the vee grooves are much smaller and there are multiple grooves. A serpentine belt gets its name from the configuration it is in. Micro-vee belts are not always found in serpentine configuration.

The **serpentine belt** is so named because of its winding route among the various pulleys at the front of the engine.

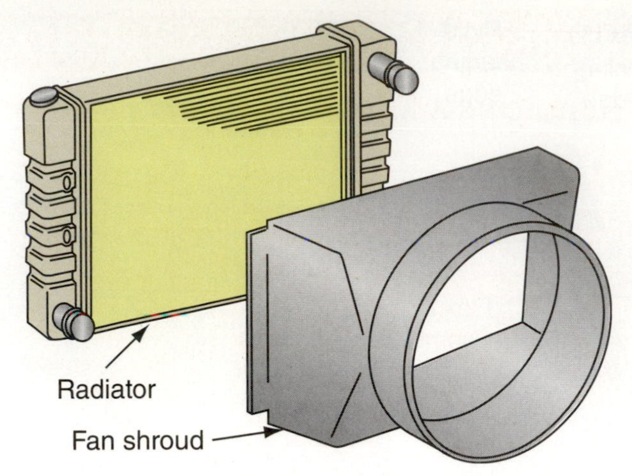

FIGURE 8-29 **The fan shroud is used to control airflow around the engine.**

Belts will stretch over time and start to crack with age. A stretched-out belt does not have the tension of a new belt, and it will cause a squeaking noise because it is not tight. It is the slipping on the pulleys that causes the squeaking noise. Some micro-vee belts are installed so that they are a stretch fit. These belts contain a chemical that shrinks the belt after installation. No tensioner or tightening is required. These belts wear out too, and you must cut them off to remove them.

Replacing a serpentine belt can be both complicated and simple. The method of removing the belt tension or even locating the tensioner is usually the most difficult. Before beginning the repair, locate a drawing of the belt's routing (Figure 8-30). This is sometimes located on an underhood label, but can be found in the owner's manual or the service manual. If a drawing is not available, it may be advisable to draw a diagram of the old belt's routing. It would appear to be a simple operation, but a single-belt system has many bends as the belt goes around each pulley. Without a drawing, errors in routing the new belt may cost the technician valuable time. Photo Sequence 15 highlights the typical sequence of tasks needed to remove and replace a serpentine drive belt.

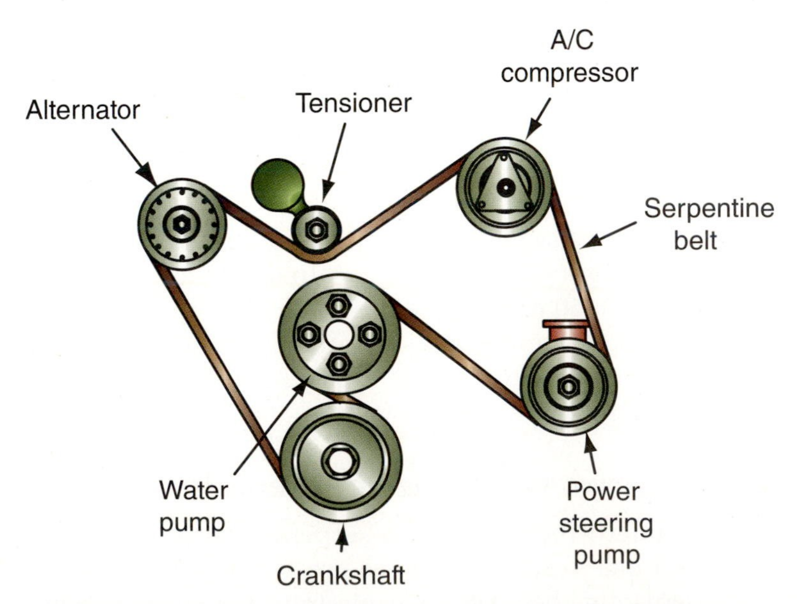

FIGURE 8-30 **Note that on this engine the alternator pulley would be the best place to make the initial and final step in replacing this belt.**

PHOTO SEQUENCE 15

Removing and Replacing a Serpentine Belt

P15-1 This photo shows the limited amount of workspace available for replacing a typical serpentine belt.

P15-2 The tensioner tool is used to move the automatic tensioner to put slack into the belt. The belt is removed from the easiest pulley to access, usually the alternator.

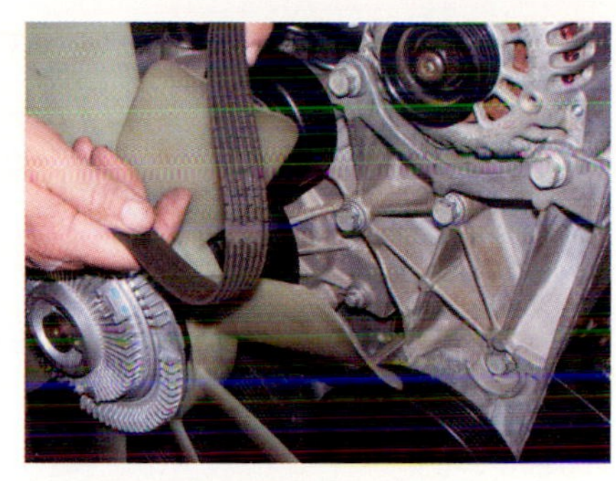

P15-3 (Engine is out-of-vehicle for clarity.) The belt is removed and bent backwards. This will reveal any cracks or wear and tear on the belt.

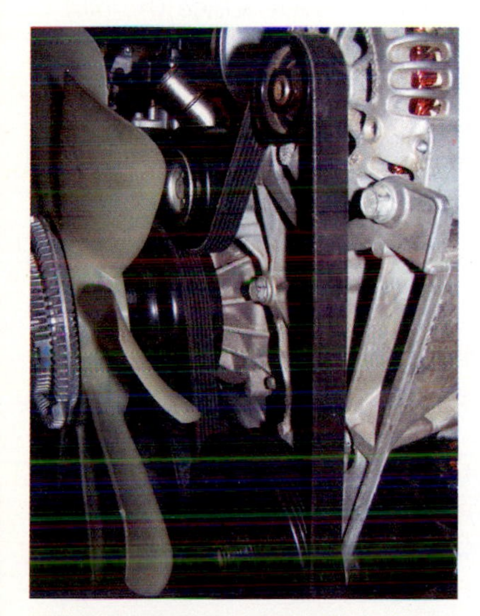

P15-4 (Engine is out-of-vehicle for clarity.) The belt is installed starting with the crankshaft pulley and around, under, or over each of the other pulleys except for the last one (same as the first from which removed).

P15-5 With the tensioner moved to provide sufficient slack in the belt, the belt can be looped over the last pulley. Before starting the engine, recheck the belt for proper placement around each pulley.

Classroom Manual
pages 170, 173

CAUTION:
Ensure that the tensioner and tool are securely mated to each other before attempting to relieve tension from the belt. The tensioner is under a great deal of force. A tool that is not placed properly could slip and cause injury or damage.

SERVICE TIP:
Before attempting to move the tensioner, note how the belt is threaded over it. If the belt contacts the top of the tensioner pulley, the tensioner will probably have to be swung down and away from the belt. Just the opposite is true if the belt touches the pulley on the lower side.

CAUTION:
Do not stand in the rotating plane of a belt. It can come off and cause injury.

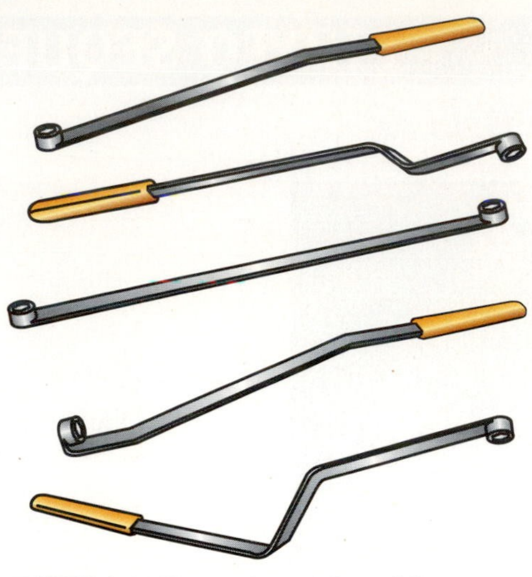

FIGURE 8-31 **Pictured are a few of the different belt tensioner tools available on the market.**

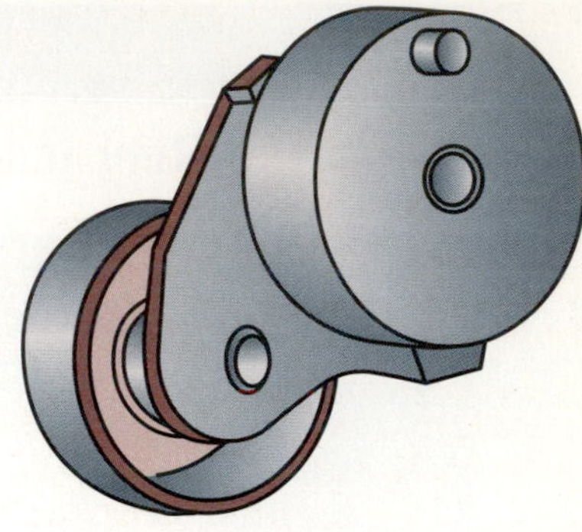

FIGURE 8-32 **Typical automatic belt tensioner.**

Locate the tensioner and determine the type of tool or wrench needed to remove tension from the belt. In some cases, a box-end wrench or a 3/8″ ratchet is all that is needed. A special tool is available for this task and will work in most cases (Figure 8-31).

Use a tool to force the tensioner back from the belt. A great deal of force is usually required. With the tension released, remove the belt from the pulley that is the easiest to access. It is also easiest to choose to remove the belt from a pulley without ribs on it. It will slide off instead of being lifted off. The alternator is usually the best option. This pulley is fairly small and is usually best placed for removing and installing the belt. Once the belt is loose, slowly allow the tensioner to move back into place. Sometimes the tool will hang on an engine component and need to be worked loose. Do not allow the tool to hold the tensioner as the belt is removed. It could come loose and cause an injury.

Remove the belt from the rest of the pulleys and from the vehicle. When the belt is removed you should inspect the tensioner and pulleys. Spin the pulleys by hand and listen for a grinding or squeaking bearing. Check each pulley for looseness and bearing movement. If the tensioner is a spring-loaded tensioner (Figure 8-32), use the removal tool to move it back and forth to check for movement and stickiness. Do not get in the habit of overlooking these items as they frequently require replacement.

With the belt removed, inspect the belt for cracks. The rule of thumb for replacement is three cracks in 3 inches. However, many newer belts are made of EPDM, a chemical that is used in the belt's construction. This type of belt will stretch and require replacement, but it will not crack. You must use a special tool (Figure 8-33) to check the depth of the belt.

Install the new belt starting at the crankshaft pulley. Maintain hand tension on the belt as it is routed and installed over all but the last pulley. The last pulley should be the easiest to access. Secure the belt while the tool is replaced on the tensioner.

This step is easier and safer to do with two people, but one person can do it. Move the tensioner until it is at the furthest end of travel. Pull the belt over the last pulley. Before allowing the tensioner to move, take a quick check to ensure that the belt is properly positioned on each pulley. Release the tensioner and recheck the belt. If the belt is not positioned properly on any pulley, remove the tension and place the belt properly. Don't start the engine until you are sure that the belt is on properly. Even if the belt is slightly off, it will not correct itself; instead it will get cut on the edge of the pulleys.

Remove the tensioner tool. Have a second technician crank the engine once or twice while observing the belt from the side. If the belt routing is correct, start the engine and

FIGURE 8-33 **Special belt depth tool is used to check a belt's condition.**

observe the belt operation. If the repair is correct, remove and store the tools, complete the repair order, and return the vehicle to the customer.

Inspecting the Timing Belt or Chain

Symptoms of a Worn Timing Mechanism

We mentioned that a worn timing chain may slap against the cover or against itself on a quick deceleration and make a rattling noise. Use a stethoscope in the area of the chain to confirm your preliminary diagnosis, and then proceed with the checks described in the next section to confirm the extent of the damage (Figure 8-34).

SERVICE TIP: A noisy timing chain that is slapping against the timing cover should be replaced right away to avoid failure and the need for more costly repairs.

FIGURE 8-34 **Listen for a slapping or rattling noise on deceleration.**

A customer may also report that the engine seems to have poor acceleration from starts but runs well, perhaps even better, at higher rpm. When there is slack in the chain, the camshaft timing is behind the crank. This is called retarded valve timing. This improves high-end performance while sacrificing low-end responsiveness.

Timing gears may clatter on acceleration and deceleration. The engine does not have to be under a load; just snap the throttle open and closed while listening under the cover for gear clatter. The customer is unlikely to notice the reduced low-end performance before the gears break from too much **backlash**.

Backlash is the small amount of clearance between the gear teeth that allows room for expansion as the gears heat up.

Customers with worn or cracked timing belts will notice nothing until their engines stop running or begin running very poorly. What you and they should be paying close attention to is the mileage on the engine and the recommended belt service interval. Skipping a recommended replacement is at the very least an invitation to meet a tow truck driver.

Symptoms of a Jumped or Broken Timing Mechanism

When a timing chain, belt, or gear breaks and no longer drives the camshaft(s), you can often tell by the sound of how the engine turns over. It cranks over very quickly, and no sounds even resembling firing occur (Figure 8-35). The engine has no compression. Hopefully, it has been towed to your shop; that's the only way it'll run again.

When a timing chain, belt, or gear skips one or more teeth, or jumps time, it sets the valve timing off significantly. Some engines will run when the timing is off one or even two teeth, but not well. Usually you can hear the problem as the engine cranks over. It turns over unevenly; the engine speeds up and then slows down as you crank. Idle quality, acceleration, and emissions will be affected. If the timing is off significantly, it will pop out of the intake or exhaust or backfire under acceleration. Often the engine will not even start.

Timing Belt Inspection

To inspect a timing belt, you really need to pull the top timing cover off of the engine. This is usually a plastic cover with an upper section and lower section. For inspection purposes, it only will be necessary to remove the upper cover (Figure 8-36). You will be looking for cracks similar to those in serpentine belts. Rotate the engine around to be able to view the whole belt. Look on the underside of the belt at the cogs or teeth that hold the belt in the sprocket. Any broken or missing teeth dictate immediate replacement. Figure 8-37 shows some typical timing belt failures. If the belt shows signs of oil or coolant contamination,

FIGURE 8-35 **This snapped timing belt left the customer stranded on the highway.**

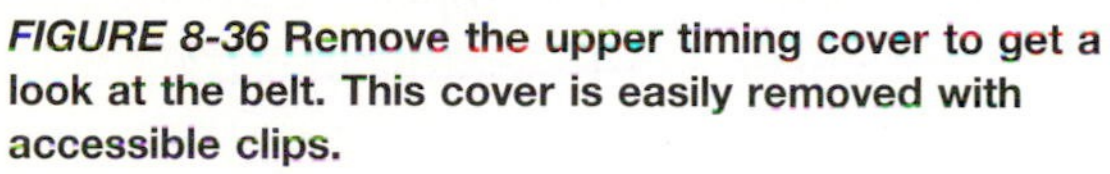

FIGURE 8-36 **Remove the upper timing cover to get a look at the belt. This cover is easily removed with accessible clips.**

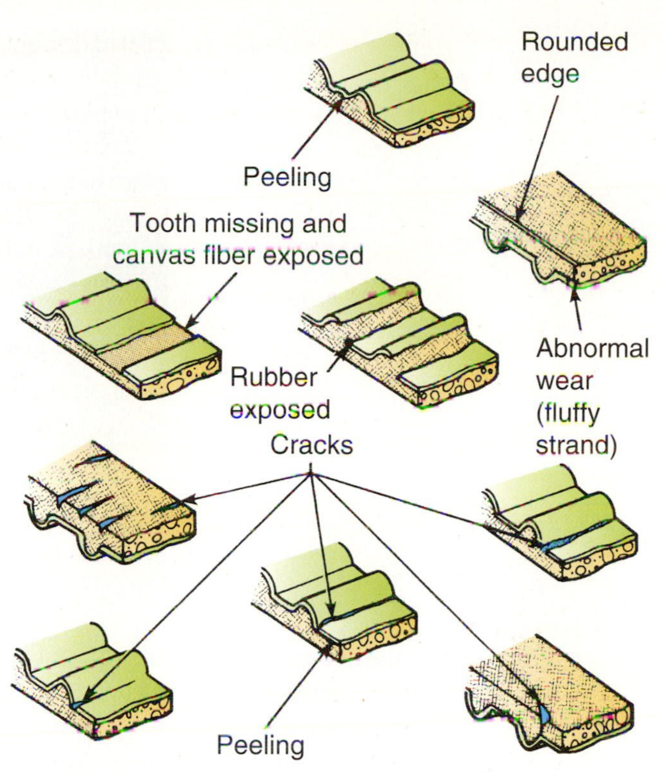

FIGURE 8-37 **Any fault with the belt warrants replacement.**

the belt should be replaced and the leak must also be fixed as part of the repair. Be sure to diagnose the cause of the leak and include that procedure and part(s) in your estimate to the customer. As we've discussed, there is no reason to try to get a few more months out of a damaged belt; it could result in a very expensive repair.

CUSTOMER CARE: Many maintenance manuals provided with new vehicles have a long list of components and systems that are supposed to be inspected at regular intervals. It may be, for example, that at every 60,000 miles when the spark plugs are replaced, the timing belt is also supposed to be inspected. A careless technician may ignore this and other portions of the 60,000-mile service and just replace the new components. A true professional will perform all the tasks recommended by the manufacturer; they are listed for a reason. By performing full service, you will be providing your customers with the excellent quality repair work they are paying for.

Tensioner Inspection

The pulley type tensioner used with belts is also commonly replaced when the service interval is long, every 90,000 miles, for example. If the timing belt is being replaced for the second time and the tensioner was not replaced the first time, you should definitely recommend replacement (Figure 8-38). An engine that requires belt replacement every 60,000 miles and is being serviced at 120,000 miles deserves a new tensioner to prevent

FIGURE 8-38 **Replace the belt tensioner pulley if it has excessive mileage or if you can feel any roughness or wobbling in the bearing.**

premature failure. To inspect the tensioner for faults, check for scoring on the pulley surface. Scratches, sharp edges, or grooves can damage a new belt rapidly. Also check the pulley's bearing. Rotate it by hand, and feel for any roughness. If you can hear it scratching or feel coarseness, replace it. Also check the pulley for side play; grasp it on its ends and check for looseness. If it wobbles at all, replace it.

A timing belt automatic tensioner is also typically replaced during belt replacement. If there are any signs of oil leakage, replacement is mandatory. If the belt replacement interval is relatively frequent, it may be possible to use the automatic tensioner through the life of two timing belts. The best advice is to follow the recommendation of the manufacturer and carry out any inspections or measurements indicated.

Timing Mechanism Replacement

The actual repair procedures for service of timing chains, belts, and gears vary immensely. Some overhead cam (OHC) timing chains can be replaced in a few hours with the engine in place; others require the engine to be removed from the vehicle. It is common to have to remove the cylinder head or the timing cover. Timing belts can generally be replaced with the engine in the vehicle. They can be simple and straightforward or time-consuming and a little tricky. A timing belt on a transversely mounted V6 engine in a minivan, for example, may leave you little room to access the front end of the engine. Replacing the timing mechanism on cam-in-the-block engines using either a chain or a gear set is usually uncomplicated. In every case, it is critical that you set all the shafts into perfect time with the crankshaft. As we've discussed, setting the valve timing off by one tooth will impact engine drivability and emissions. Similarly, an engine with balance shafts not timed properly will vibrate noticeably. Often it is too easy to miss the timing of a shaft by one tooth if you aren't paying close attention and double-checking your work. This is important and time-consuming service work; you want to do it only once! Many technicians will replace the camshaft and crankshaft seals while they have access to them during the job. If the water pump is driven by the timing belt, you will usually replace that as well. Always check idle quality and engine performance after a replacement job to be sure the engine is operating as designed. This text cannot offer a replacement for the manufacturer's service procedures. Instead it offers a couple of specific examples of the procedures for replacing different timing mechanisms. You'll still need manufacturers' specific information to achieve proper timing mark alignment and to perform the job in the most efficient manner. You may also need special tools to hold the camshaft sprockets in position while replacing the belt. Photo Sequence 16 shows how a timing belt is replaced.

PHOTO SEQUENCE 16

Typical Procedure for Replacing a Timing Belt on an OHC Engine

P16-1 Disconnect the negative cable from the battery prior to removing and replacing the timing belt.

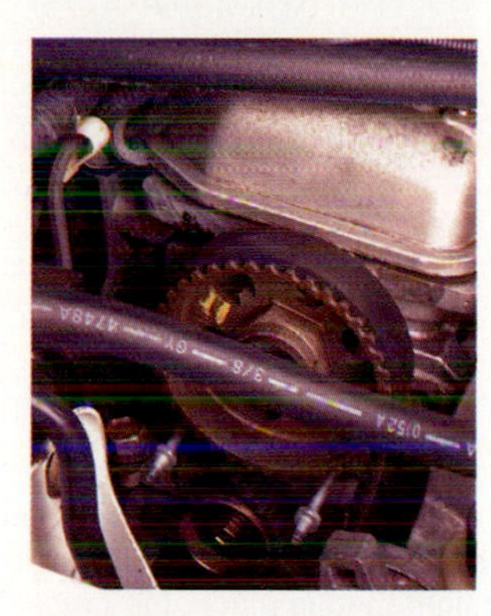

P16-2 Carefully remove the timing cover. Be careful not to distort or damage it while pulling it up. With the cover removed, check the immediate area around the belt for wires and other obstacles. If some are found, move them out of the way.

P16-3 Align the timing marks on the camshaft's sprocket with the mark on the cylinder head. If the marks are not obvious, use a paint stick or chalk to mark them clearly.

P16-4 Carefully remove the crankshaft timing sensor and probe holder.

P16-5 Loosen the adjustment bolt on the belt tensioner pulley. It is normally not necessary to remove the tensioner assembly.

P16-6 Slide the belt off the camshaft sprocket. Do not allow the camshaft pulley to rotate while doing this.

P16-7 To remove the belt from the engine, the crankshaft pulley must be removed. Then the belt can be slipped off the crankshaft sprocket.

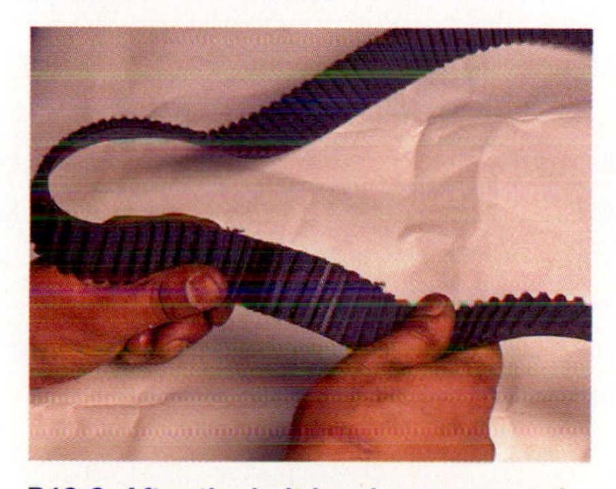

P16-8 After the belt has been removed, inspect it for cracks and other damage. Cracks will become more obvious if the belt is twisted slightly. Any defects in the belt indicate it must be replaced. Never twist the belt more than 90 degrees.

P16-9 To begin reassembly, place the belt around the crankshaft sprocket. Then reinstall the crankshaft pulley.

PHOTO SEQUENCE 16 (CONTINUED)

P16-10 Make sure the timing marks on the crankshaft pulley are lined up with the marks on the engine block. If they are not, carefully rock the crankshaft until the marks are lined up.

P16-11 With the timing belt fitted onto the crankshaft sprocket and the crankshaft pulley tightened in place, the crankshaft timing sensor and probe can be reinstalled.

P16-12 Align the camshaft sprocket with the timing marks on the cylinder head. Then wrap the timing belt around the camshaft sprocket, and allow the belt tensioner to put a slight amount of pressure on the belt.

P16-13 Adjust the tension as described in the service manual. Then rotate the engine through two complete turns. Recheck the tension.

P16-14 Rotate the engine through two complete turns again; then check the alignment marks on the camshaft and the crankshaft. Any deviation needs to be corrected before the timing cover is reinstalled.

CUSTOMER CARE: When repairing an engine with a broken timing mechanism, it is important to know whether the engine is an interference engine or not. If it is, advise the customer of the potentially very high added cost of removing the cylinder head and replacing valves. Also make sure the customer understands that even more serious damage, such as a valve stuck into a piston, could have occurred. Even on freewheeling engines, it's wise to inform the customer that valve damage is possible. Discuss these issues with the customer or service advisor before offering an estimate for the work.

ADJUSTING VALVES

Some engines have a valvetrain system that must be periodically adjusted. To perform a periodic adjustment, you will remove the valve cover(s) and follow the same procedures. The adjustment must be correct to provide long valve life and excellent engine performance. If there is too much clearance (lash) in the valvetrain, it will clatter and the valves won't open as much as they should. Too little valve lash can cause valve burning and poor engine performance. Some hydraulic lifters are adjustable and are set after replacement or reinstallation. Other hydraulic lifters are not adjustable. Mechanical lifters require periodic adjustment. One method of adjustment involves loosening or tightening an adjustment screw on the rocker arm (Figure 8-39). Another process involves measuring the lash and adjusting the clearance by changing the size of a shim installed between the camshaft and the follower. We will look into the adjustment method using adjustable rocker arms. For more detailed explanations of valve adjustment, refer to *Today's Technician: Engine Repair and Rebuilding*, 5th Edition.

Adjustment Intervals

Adjustable hydraulic lifters should be adjusted during installation of new or used lifters or whenever the valvetrain has been disassembled. An example would be after an on-the-car valve seal replacement. After reinstalling the rockers, the lifters must be adjusted. This initial adjustment should last as long as the valvetrain components or until they are disassembled again. If during normal service the valvetrain becomes noisy, check the lifter adjustment before condemning the lifters if no visible damage is seen.

Mechanical or solid lifters require periodic adjustment. They should be adjusted at the specified interval. This may be as little as every 15,000 miles or as long as every 90,000 miles. Adjustment every 30,000 or 60,000 miles is a common recommendation. If the valves clatter, they should be adjusted regardless of the specified mileage.

Adjusting Valves with Adjustable Rocker Arms

Another common method of adjusting mechanical valvetrains is through the rocker arm. One side of the rocker may ride on the camshaft or a solid lifter. The other side can work on the valve. One end of the rocker arm has a lock nut and an adjusting screw. To adjust the valve, you can loosen the locking nut and turn the screw in or out to tighten or loosen the valve adjustment, respectively (Figure 8-40). The clearance specifications will be similar

FIGURE 8-39 **These rocker arms have adjusting screws held in place with a locking nut. Use a feeler gauge to measure the clearance between the rocker arm and the valve tip. Adjust the screw to the correct clearance and tighten the lock nut.**

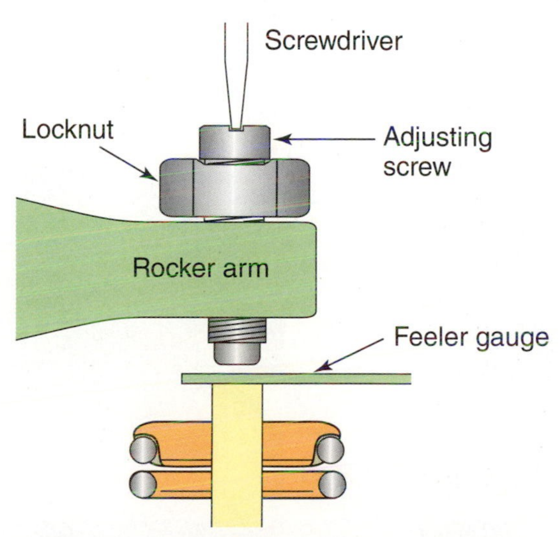

FIGURE 8-40 **Loosen the lock nut on the adjusting screw, fit the appropriate feeler gauge between the screw and the valve tip, and adjust the screw until some drag is felt on the feeler gauge.**

to those with shim-adjusted valve lash. Photo Sequence 17 outlines the typical procedure for adjusting valves on an OHC engine. To adjust the valves:

1. Obtain the lash specifications and be sure the engine is at the desired temperature.
2. Rotate the engine to top dead center (TDC) number 1 and note which valves can be adjusted using the service information.
3. Using feeler gauges, determine the clearance on the valve you are adjusting. The feeler gauge should fit in the clearance under the adjustment screw with light drag (Figure 8-41).
4. If the valve needs adjustment, loosen the lock nut just enough to turn the screw. Insert the desired thickness (clearance) feeler gauge under the lash adjuster, and *lightly* tighten the adjuster screw while alternately feeling the clearance. When there is just the right amount of drag, tighten the lock nut (Figure 8-42).
5. Recheck your adjustment. This is essential as the adjustment can change as the lock nut pulls the adjuster screw up. It may be necessary to adjust the valve a little snugly and then have it loosen up as you tighten the lock nut. It is essential to recheck the adjustment after tightening the lock nut.

PHOTO SEQUENCE 17

Typical Procedure for Adjusting Valves

P17-1 **Prepare the vehicle and assess the task.**

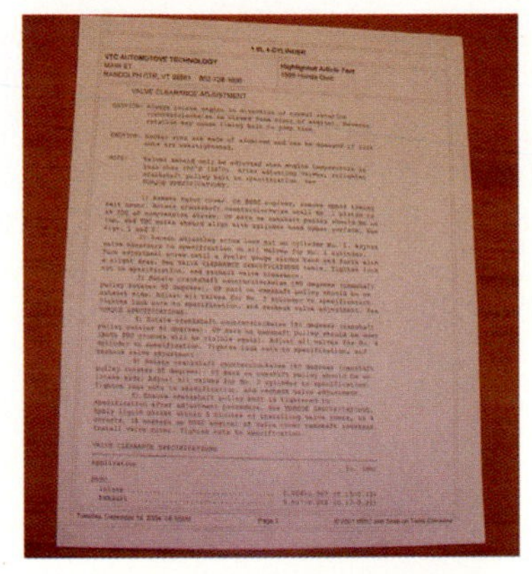

P17-2 **Read through and follow the service information procedures.**

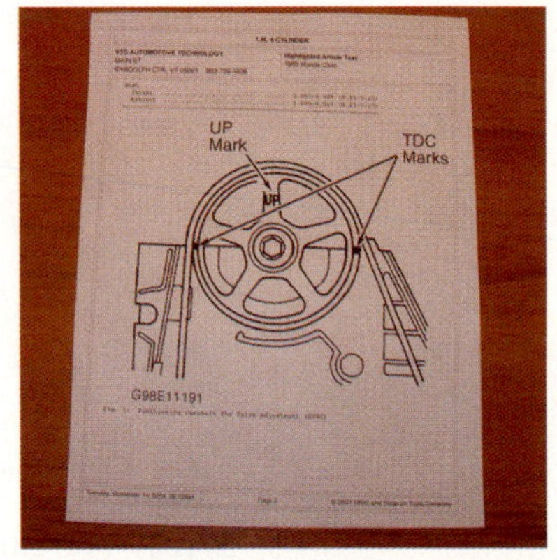

P17-3 **This service information provides a diagram of how to line up the camshaft to adjust the valves.**

P17-4 **Remove the necessary wiring and linkage to remove the valve cover.**

P17-5 **Remove the valve cover following any sequence specified.**

P17-6 **Remove the upper timing belt cover to access the timing marks on the camshaft sprocket.**

P17-7 With the valve cover and timing cover off you have plenty of room to work.

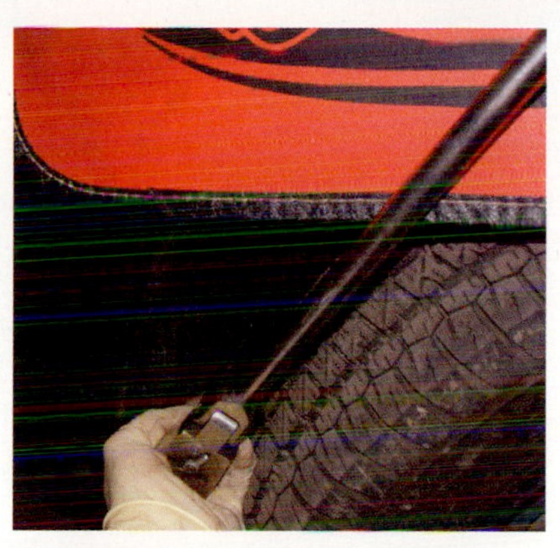

P17-8 Rotate the engine through the access port in the wheel well to rotate the engine to line up the marks on the camshaft. Do not rotate the engine over with the camshaft sprocket bolt.

P17-9 Check the adjustment of the valves using feeler gauges and compare to specifications.

P17-10 Loosen the lock nut and turn the adjusting screw to achieve the proper adjustment. Double check your work after you retighten the lock nut.

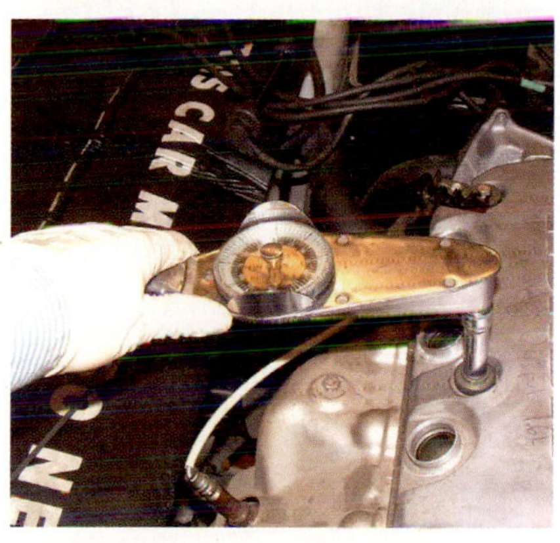

P17-11 Properly torque the new valve cover gasket and valve cover into place.

P17-12 Be sure to reinstall all hoses, connectors and linkages you disconnected to remove the valve cover.

P17-13 Make sure your work is clean and that the engine runs smoothly and quietly before you call the job complete.

FIGURE 8-41 **Check the clearance first to be sure adjustment is needed.**

FIGURE 8-42 **Three hands to hold the wrench, the screwdriver, and the feeler gauges would make this job easier.**

SPECIAL TOOLS

Special Tools
Torque wrench
Alignment dowels

The **front pipe**, **muffler**, and **tailpipe** are the three main components used to channel exhaust gases from the engine to the rear bumper and reduce exhaust noises.

A **diesel particulate filter** is used to capture soot and incinerate it during regeneration cycles.

The engine computer uses a strategy to spray a fine mist of **diesel exhaust fluid (DEF)** in the exhaust during certain periods to help clean up the exhaust emissions and burn the soot in the filter.

6. Repeat for all the other valves adjustable at TDC number 1. Rotate the crank 360 degrees, and complete the adjustment of the valves.
7. Properly install the valve cover and start the engine. It should turn over smoothly.
8. Allow the engine to warm up and listen for any excessive valvetrain clatter and feel for rough running.

Exhaust System Inspection

Most parts of the exhaust system, particularly the **front pipe**, **muffler**, and **tailpipe**, are subject to rust, corrosion, and cracking. Broken or loose clamps and hangers can allow parts to separate or hit the road as the car moves.

Some newer diesel engines use a different type of exhaust system that is essential to its emission system. This is known as an after-treatment system. This type of exhaust uses a **diesel particulate filter** that captures soot to incinerate it during regeneration cycles. The engine computer uses a strategy to spray a fine mist of **diesel exhaust fluid (DEF)** in the exhaust during certain periods. This helps clean up the exhaust emissions and burn the soot in the filter. This reduces the black smoke typically found in the exhaust of a diesel engine.

WARNING: If the engine has been running, exhaust system components may be extremely hot. Wear protective gloves when working on these components and keep flammable objects away from hot areas.

WARNING: Exhaust gas contains poisonous carbon monoxide (CO) gas. This gas can cause illness and death by asphyxiation. Exhaust system leaks are dangerous for customers and technicians.

Any exhaust system inspection should include listening for hissing or rumbling that would result from a leak in the system (Figure 8-43). An on-lift inspection should pinpoint any of the following types of damage:

1. Holes, road damage, separated connections, and bulging muffler seams.
2. Kinks and dents.

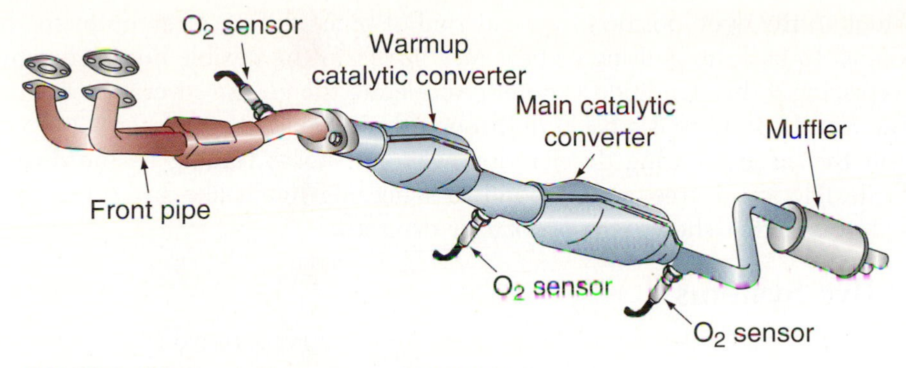

FIGURE 8-43 **Exhaust system for an OBD II vehicle.**

3. Discoloration, rust, soft corroded metal, and so forth.
4. Torn, broken, or missing hangers and clamps.
5. Loose tailpipes or other components.
6. Bluish or brownish catalytic converter shell, which indicates overheating.

Inspecting the Presence and Condition of Emission Control Devices

Each of the emission systems outlined in the Classroom Manual needs periodic servicing and replacement. In state and local areas requiring emissions testing, a vehicle will not pass inspection if emissions components are missing or damaged, even if the exhaust emissions are within standards. In this section, basic inspection procedures on some emission systems are discussed. Systems not listed usually require an experienced technician and electronic testers. All systems will require an experienced technician for detailed testing.

Classroom Manual
page 197

Secondary Air Injection

The drive belt for the secondary air injection is inspected as other belts. As the engine is operated, there should be no exhaust noises in or near the secondary air **diverter valve**. If there is exhaust noise, check the flexible hose running from the valve to the metal line going to the exhaust manifold or catalytic converter (Figure 8-44). There is a check valve inside the metal line that prevents exhaust from entering the flexible hose and valve. It may

CAUTION:
Do not attempt to touch or repair hot engine components. Serious burn injury could result. It is best to allow the vehicle to cool before beginning the tasks in this section.

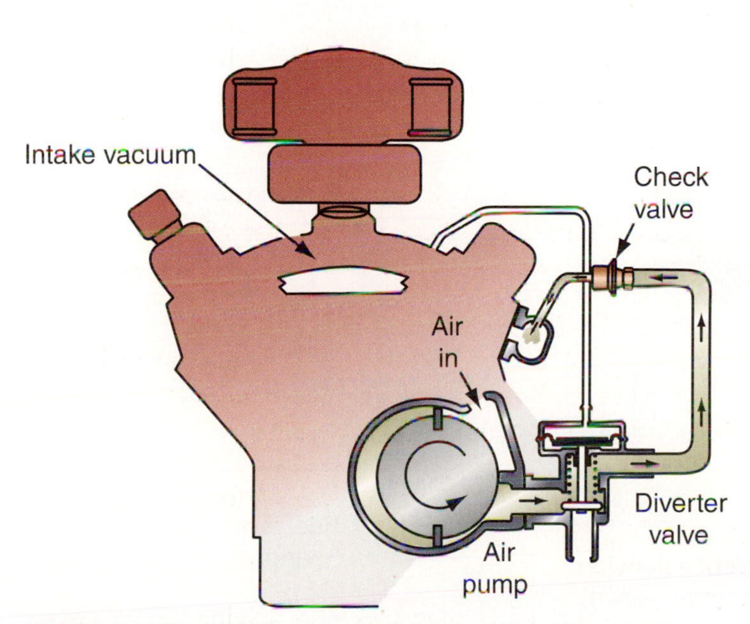

FIGURE 8-44 **On the secondary air injection system, if the check valve fails it often leads to a complete failure of the system.**

become stuck in the open position over a period of time and gases can enter the hose. If it allows exhaust to back up, pulsing or heat may be felt in the flexible hose. The check valve requires replacing if this condition is noted. Accelerate the engine several times and quickly release the throttle. If there is a backfire through the exhaust, the diverter valve may not be shutting off the airflow during deceleration. Vacuum hoses leading to the diverter valve should be flexible, crack-free, and respond to squeezing much like any other rubber-type hoses. Replace any that show signs of wear or dry rotting.

Evaporative Systems

Removing the fuel tank filler cap is the first check of the evaporative (EVAP) system. If the sound of moving air occurs as it is loosened, the cap is probably working. The other end of the system is located in the engine compartment and has vacuum hoses that need to be checked (Figure 8-45). The remainder of the system can only be inspected for leaks. Further tests to diagnose the EVAP system require electronic test equipment. Always replace the fuel filler cap with one of identical design and operation.

CAUTION:
Use care when working around an operating converter. The catalytic converter may reach temperatures of 1,400°F. Second- and third-degree burns on hands and arms could result.

Catalytic Converters

The catalytic converter (Figure 8-46) cannot be checked completely, but some simple checks can be made. Start the engine and allow it to warm. While the engine is operating, check the end of the tailpipe. If there is soot in the pipe or the exhaust gas is black, an overly rich air/fuel mixture is indicated. Too much fuel in the exhaust will clog a converter and restrict the exhaust flow. After the engine has warmed to operating temperature, attempt to observe the exhaust pipe between the converter and exhaust manifold. A pipe that is beginning to turn or has turned red also indicates a clogged exhaust system. Do not touch the pipe or any other component of the exhaust. A noise from near or inside the converter may indicate a loose heat shield or the material in the converter may be coming apart or loose.

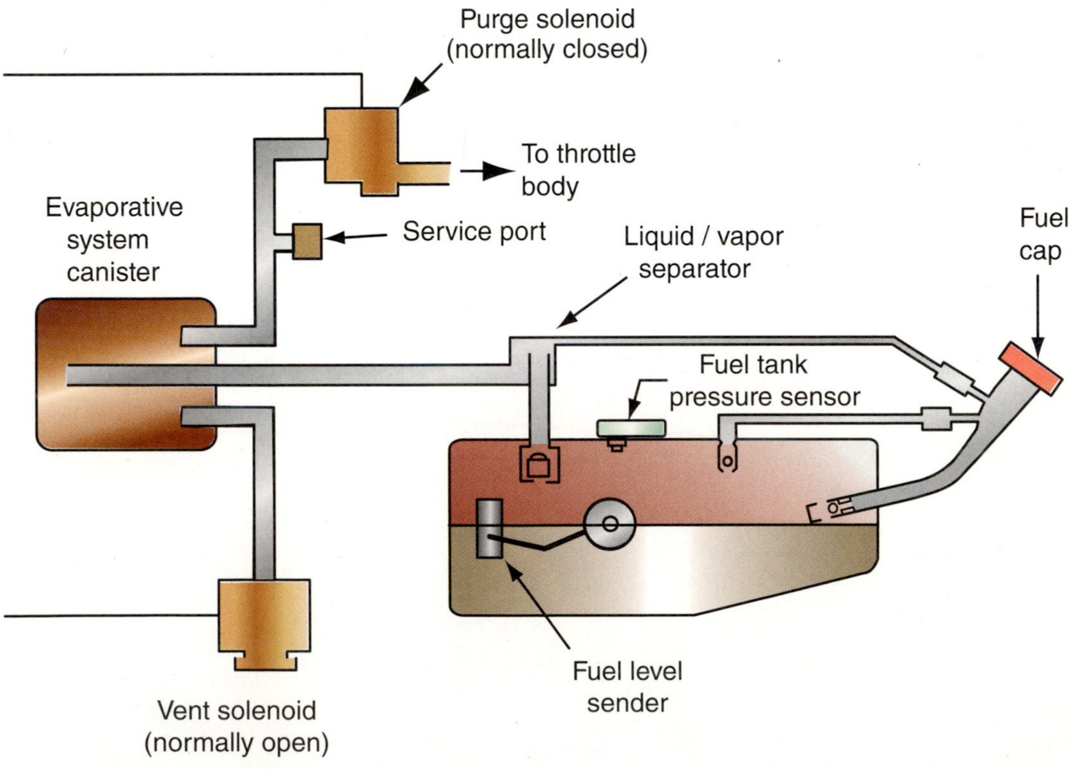

FIGURE 8-45 **An older but typical EVAP system. The biggest difference between the old system and the newer ones is that the various valves are now computer controlled.**

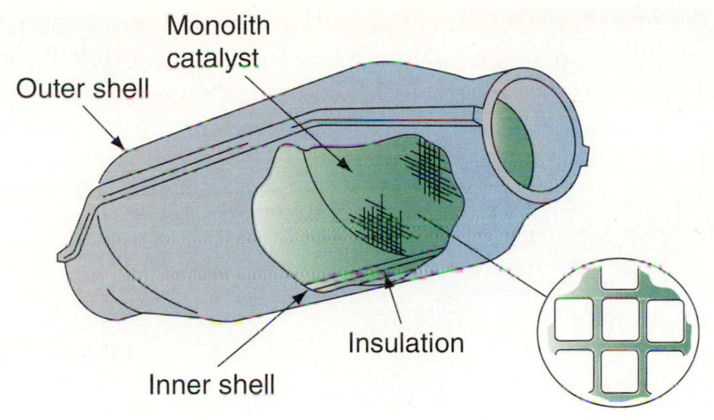

FIGURE 8-46 **Monolythic-type catalytic converter.**

The remainder of the emission systems are checked at this point for their presence. Some of the components may not be visible and may require specialized testing equipment and training. As you grow into an experienced technician and student, you will learn how to diagnose these components.

Scanning for Diagnostic Trouble Codes (DTC) and Monitors

Classroom Manual page 207

The use of a scan tool like the ones discussed in Chapter 3 can aid the technician during the diagnostics process. Since the late 1980s, vehicle computers have been designed to provide technicians with some insight into the electronic world on the vehicle. The first system was known as On-Board Diagnosis I (OBD I), and it grew into the present system known as OBD II. OBD II first arrived on model year 1996 vehicles. Not all vehicles were OBD II compliant in 1996, most notably Honda. Certain characteristics of an OBD II vehicle are the underhood certification label, the location and shape of the 16 pin Diagnostic Link Connector (DLC) (Figures 8-47 and 8-48), having a post-catalytic converter oxygen sensor, having heated oxygen sensors, monitoring systems, saving freeze frame data, and having one of a few chosen computer languages and programs to convey the information to a scan tool. Another common characteristic with OBD II vehicles is the **MIL**. This stands for **malfunction indicator lamp**, also known as the "check engine" or "service engine soon" light on the dash (Figure 8-49).

MIL (malfunction indicator lamp) is a dash-mounted warning light commonly referred to as the check engine light.

FIGURE 8-47 **The DLC on an OBD II vehicle is located under the driver's side dash.**

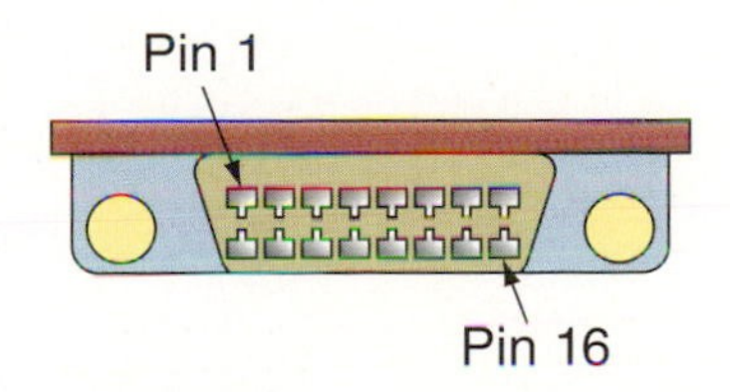

FIGURE 8-48 **An OBD II DLC.**

FIGURE 8-49 **The malfunction indicator light (MIL) will come on when the PCM has detected a problem that could allow emissions to increase; this certainly includes misfire.**

SERVICE TIP:
Some vehicles, such as those manufactured by Ford, will go through an OBD self-test. This self-test will turn on and actuate certain components under the hood. Don't be alarmed when you hear this happening, and always watch the scan tool screen for what is happening.

Diagnostic trouble codes (DTCs) are alphanumeric fault codes that represent a circuit failure in a monitored system.

OBD II provides more accurate data and has the ability to **monitor** certain electronic components and systems for wear or damage. To retrieve this information a technician must use a scan tool.

A scan tool is capable of retrieving **diagnostic trouble codes (DTCs)**, parameter identification data (PID), freeze frame data, and inspecting the monitor status. Scan tools come in two different forms. A generic scan tool will work on all OBD II vehicles, but it will only scan the generic side of OBD II. This will get you the minimum amount of data needed to diagnose and repair the vehicle (Figure 8-50). A manufacture scan tool (Figure 8-51) can only connect and work with the manufacture's vehicles that it was designed for. However, a manufacture scan tool can get you more data than a generic scan tool (also called Global OBD II scan tool) and can perform some tasks that no other scan tool can.

With the scan tool connected to the DLC, power it on. Some scan tools use the vehicle's battery and some use their own battery. If you will be using the scan tool for a while, you may want to connect a battery charger to the vehicle or disable the daytime running

FIGURE 8-50 **Scan tool communicates with the PCM and displays DTCs and engine data.**

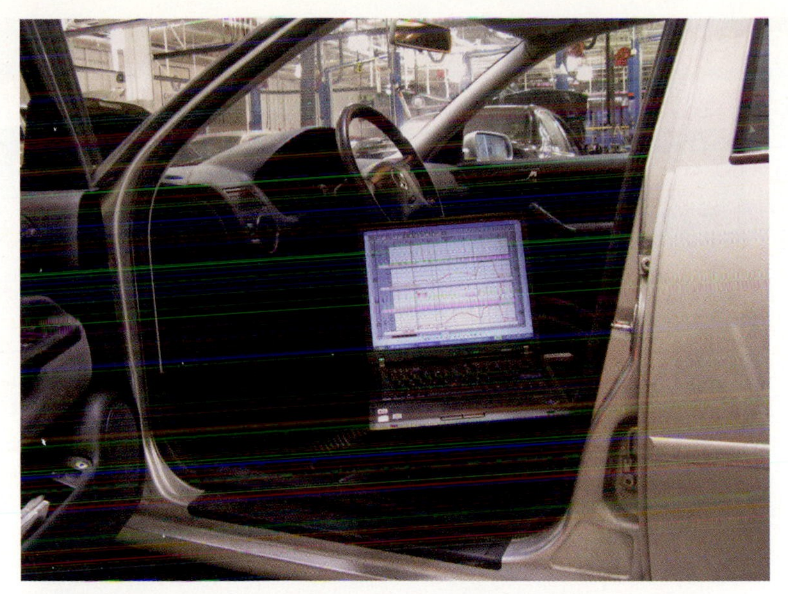

FIGURE 8-51 **Laptop computers are now often used as a scan tool.**

lights. The two computers will communicate and any stored and pending DTCs can be displayed on the scan tool's display screen. You can now record the codes shown on the screen. OBD II will have five digits. There will be a letter followed by four numbers. All OBD II codes are the same for all manufactures. To decipher the codes you will want to consult the service manual.

An example deciphering a code goes like this. If the code is P0304, the "P" stands for powertrain (meaning engine and transmission). The "0" stands for a generic OBD II code, meaning a generic OBD II scan tool can see this code and it is the reason the MIL is illuminated. The "3" is the subsystem. In this case the "3" means it is an ignition or misfire code. The "04" is the variable code. In this case the "04" means cylinder #4. So this code is telling the technician that there is a misfire on cylinder #4. Photo sequence 18 shows the typical procedure for retrieving a DTC using a generic (Global) OBD II scan tool.

SERVICE TIP:
If a NO COMMUNICATION message is displayed, check the vehicle setup and entry on the scan tool and tool/vehicle connection and the vehicle's battery voltage.

PHOTO SEQUENCE 18

Typical Procedure for Using a Scan Tool to Retrieve DTCs

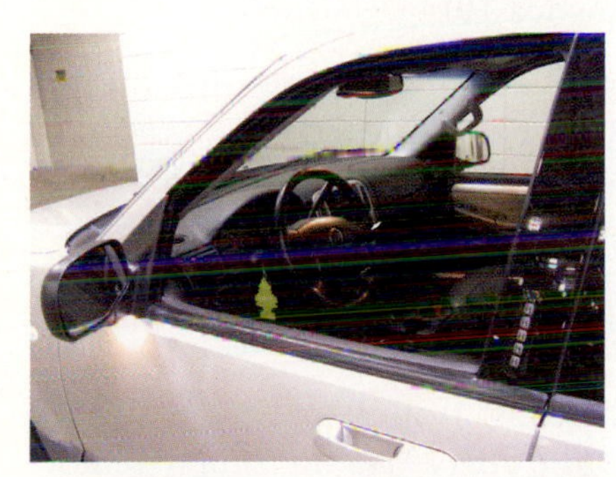

P18-1 **Roll the driver's side window down, turn the ignition to the off position, and block the wheels (or set the parking brake).**

P18-2 **Select the proper scan tool data cable for the vehicle you are working on.**

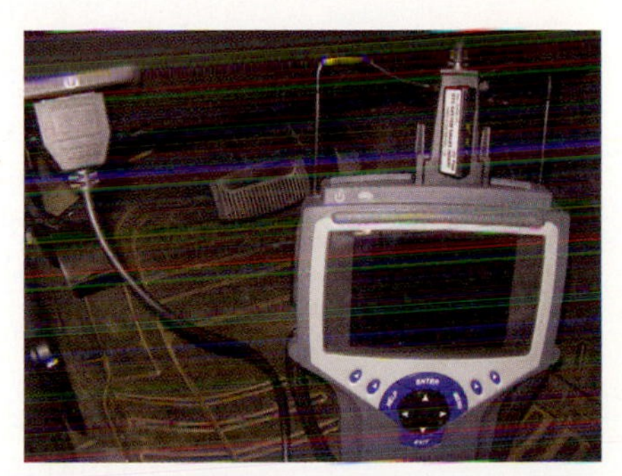

P18-3 **Connect the scan tool data cable to the DLC on the vehicle.**

PHOTO SEQUENCE 18 (CONTINUED)

P18-4 **Turn the ignition key to the KOEO (key on, engine off) position. Some scan tools will automatically turn on at this point; others may need to be turned on.**

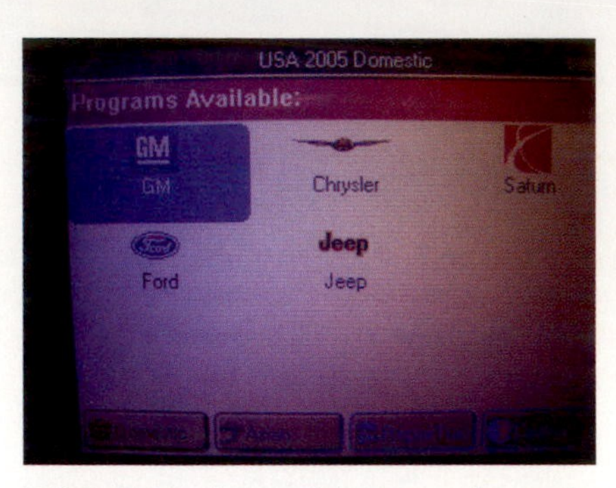

P18-5 **Enter the model year, make, and selected VIN components into the scan tool.**

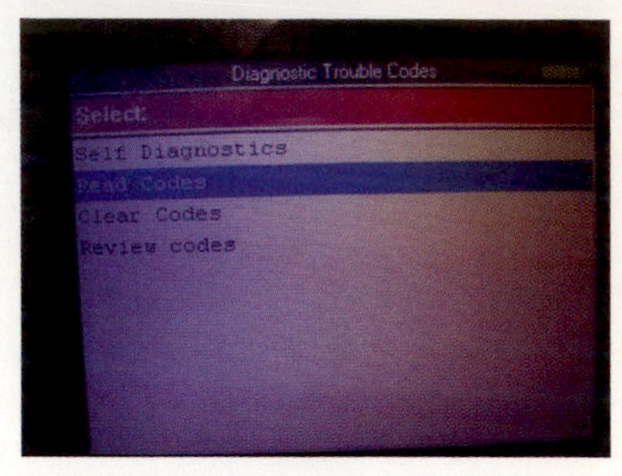

P18-6 **Maneuver through the scan tool's menu, and select "read/retrieve DTC."**

P18-7 **Record the DTCs on the repair order.**

Engine Pressure Testing

Cranking Compression Test

A cranking compression test is the most commonly used type of engine pressure testing. It will test the pressure output of each individual cylinder. It can clearly determine if the valves, rings, and head gasket(s) are sealing the combustion chamber properly (Figure 8-52). When any of those components are faulty, the engine may not be able to develop adequate compression. The manufacturer will provide a specification for the appropriate cranking compression pressure. Low compression will cause less productive combustion. You and the customer will notice that the engine lacks power or misfires on a cylinder(s). An OBD II code that indicates a misfire may lead to this test series.

To perform a cranking compression test (Photo Sequence 19):

1. Let the engine warm up to normal operating temperature. (This ensures that the pistons have expanded to properly fit and seal the cylinder.)
2. Let the engine cool for 10 minutes so you don't strip any threads; then remove the spark plugs. Remember to keep them in order so you can read them.
3. Block the throttle at least part way open so the engine can breathe.
4. Disable the fuel system by removing the fuel pump fuse or relay to prevent contamination of the engine oil.
5. Disable the ignition system so you do not damage the coils. You can either remove the ignition fuse or remove the low-voltage connector to the coil(s). Do not simply remove the plug or coil wires and let them dangle. The coil can be damaged as it puts out maximum voltage trying to find a ground path for the wires.
6. Install a remote starter or have an assistant help you.

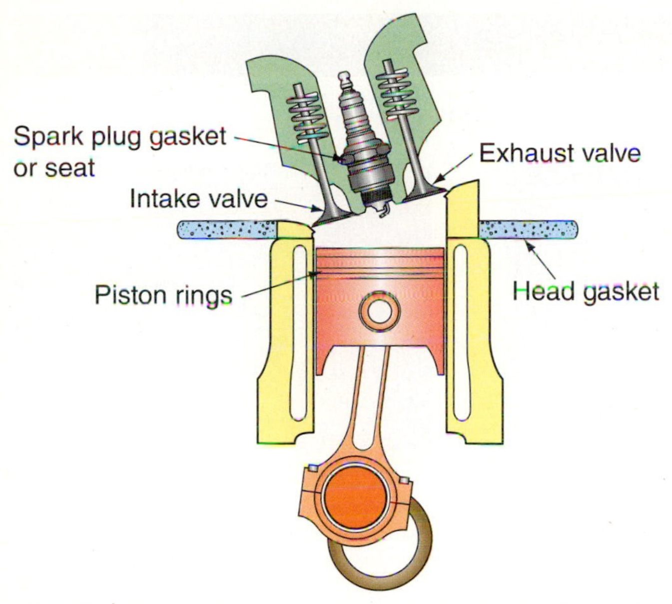

FIGURE 8-52 **A cranking compression test will detect leaks at the valves, rings, pistons, and head gasket.**

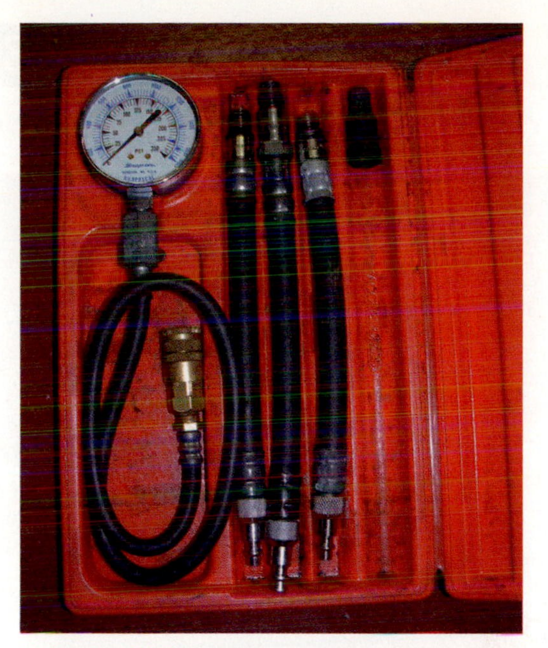

FIGURE 8-53 **This style compression tester has adaptors that thread into the spark plug hole.**

7. Carefully thread the compression tester hose into the spark plug hole and attach the pressure gauge to the adaptor hose (Figure 8-53).
8. Crank the engine over through five full compression cycles. Each cycle will produce a puffing sound as engine compression escapes through the spark plug ports.
9. Record the final pressure.
10. Repeat the test on each of the other cylinders.

Analyzing the Results

With all the pressure readings in front of you, compare them to the manufacturer's specifications. Typical compression pressures on a gas engine range between 125 psi and 200 psi. The readings should be close to the specification and within 15 to 20 percent of each other. Some manufacturers specify only that the compression be over 100 psi and that each reading is within 20 percent of the others. To calculate 20 percent easily, take the normal reading (say, 160 psi) and drop the last digit (16 psi). Multiply by 2 (32 psi), and that is 20 percent. If the readings were 160, 150, 100, and 155, the third cylinder is more than 32 psi different than the other three. This indicates a significant problem with cylinder number three that must be identified and corrected.

If one or more cylinders are below specification, it is likely caused by worn rings, burned valves, or valves sticking open from carbon on the seats, a faulty head gasket, or a worn or broken piston. Perform a wet compression test to help find the probable cause. When all the cylinders are slightly low, it usually indicates a high mileage engine with worn rings. Two adjacent cylinders with low compression readings may indicate a blown head gasket leaking compression between two cylinders. If all the cylinders are near zero or very low, suspect improper valve timing. A timing belt, chain, or gears that have "jumped" time by slipping a tooth will allow the valves to be open at the wrong time, causing low compression readings.

SERVICE TIP: A "blown" head gasket that is allowing coolant into the combustion chamber may be doing an adequate job of sealing compression. If an engine passes a compression test, it does not prove that the head gasket is sealing properly.

Wet Compression Test

To help determine the cause of a weak cylinder, perform a wet compression test (Figure 8-54). Use an oil can and put two squirts of oil into the weak cylinder. Reinstall the compression gauge, crank the engine over five times, and record the reading. If the low reading cylinder increases to almost normal compression pressure, it is likely that

PHOTO SEQUENCE 19

Typical Procedure for Performing a Cranking Compression Test

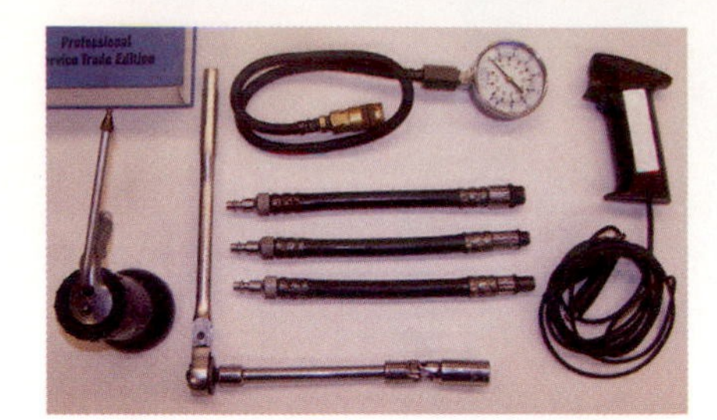

P19-1 **Tools required to perform this task: compression tester and adapters, spark plug socket, ratchet, extensions, universal joint, remote starter button, battery charger, service manual, and squirt can of oil.**

P19-2 **Follow the service manual procedures for disabling the ignition and fuel systems.**

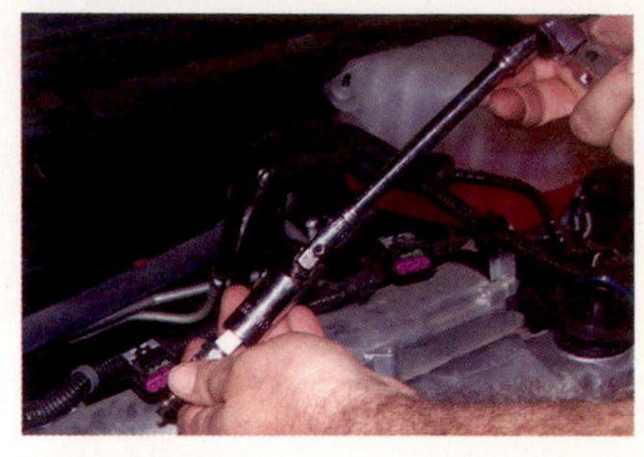

P19-3 **Clean around the spark plugs with a low-pressure air gun, and remove all of the spark plugs from the engine.**

P19-4 **Install the remote starter button.**

P19-5 **Carefully install the compression tester adapter into the first cylinder's spark plug hole.**

P19-6 **Connect the battery charger across the battery, and adjust to maintain 13 to 14.5 volts. This is needed to allow the engine to crank at constant speeds throughout the test.**

P19-7 **With the throttle plate held or locked into the wide-open throttle position, crank the engine while observing the compression reading.**

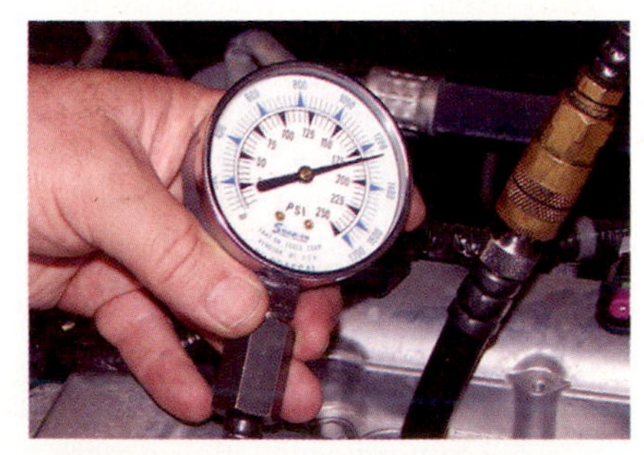

P19-8 **Continue to rotate the engine through four compression strokes as indicated by the jumps of the needle.**

P19-9 **Record the reading after the fourth compression stroke. Continue until all cylinders have been tested, and compare the test results. Compare test results with the manufacturer's specifications.**

worn rings are causing the cylinder leakage. The extra oil in the cylinder helps the rings to seal better during the test. If a valve is burned or being held open by carbon or if there is a hole in the piston, a film of oil will not dramatically improve the compression pressure. Note that a wet compression test may not be as effective on horizontally opposed engines, because the oil will seal only the lower half of the cylinder wall.

FIGURE 8-54 **This cylinder showed low compression of only 100 psi while cranking dry and wet. A leakdown test will verify a leaking valve.**

Look at the examples that follow:

Compression test results					
Cyl. #1	Cyl. #2	Cyl. #3	Cyl. #4	Cyl. #5	Cyl. #6
175 psi	165 psi	170 psi	80 psi	170 psi	170 psi
Wet compression test					
Cyl. #4					
160 psi					

This is a clear indication that the rings are severely worn on cylinder number four. If the wet compression test had shown a rise to only 105 psi, you would suspect a burned valve or leaking valve. To definitively determine the cause of low cranking compression, perform a cylinder leakage test (Figure 8-55).

FIGURE 8-55 **Listen for air in the coolant reservoir for indications of a leaking head gasket.**

Running Compression Test

It is possible that the cranking compression can show good results but the engine still is not mechanically sound. A cranking compression test does not do a good job of testing the valvetrain action. There is so much time for the cylinder to fill with air when the engine is spinning only at cranking speeds of about 150–250 rpm. A more accurate way to test the loaded operation of the valvetrain is to perform a running compression test. Running compression readings will be quite low compared to the cranking compression pressures because there is much less time to fill the cylinders with air.

If you have identified a weak cylinder through power balance testing, check that cylinder and a few others to gain comparative information. In most cases, it is worth the time to check all the cylinders. To perform the running compression test:

1. Remove the spark plug only from the test cylinder.
2. Disable the spark to that cylinder by grounding the plug wire or removing the primary wiring connector to the individual coil.
3. Disable the fuel to the test cylinder by disconnecting the fuel injector connector.
4. Install the compression gauge into the spark plug port.
5. Start the engine and let it idle.
6. Release the pressure from the gauge whose reading reflects cranking compression. Record the new pressure that develops at idle.
7. Rev the engine to 2,500 rpm. Release the idle pressure, and record the new pressure that develops at 2,500 rpm.
8. Remove the gauge and reinstall the spark plug. Remember to reattach the plug wire or coil and the fuel injector connector.
9. Repeat for several or all cylinders.

Analyzing the Results

Generally, the running compression will be between 60 and 90 psi at idle and between 30 and 60 psi at 2,500 rpm. The most important indication of problems, however, is how the readings compare to each other.

Look at the following readings:

	Cyl. #1	Cyl. #2	Cyl. #3	Cyl. #4
Idle	70 psi	75 psi	45 psi	70 psi
2,500 rpm	40 psi	40 psi	20 psi	45 psi

Cylinder number three is barely below the general specifications, but it is clearly weaker than the other cylinders. These results warrant investigation of the problem in cylinder number three. The most likely causes of low running compression readings are:

- Worn camshaft lobes
- Faulty valve lifters
- Excessive carbon buildup on the back of intake valves (restricting airflow) (Figure 8-56)
- Broken valve springs
- Worn valve guides
- Bent pushrods

Use a stethoscope, vacuum testing, and visual inspection under the valve covers to help locate the reason for the low reading.

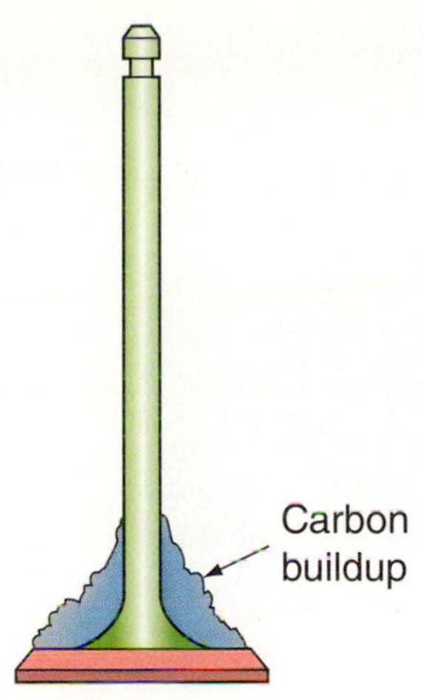

FIGURE 8-56 **Excessive carbon buildup on the back of intake valves caused by a bad valve seal or guide leakage prevents the engine from pulling in enough air to make adequate power. The added weight can also cause the valve to float at higher rpm.**

A **cylinder leakage test** places air pressure into a combustion chamber so you can listen for leaks from possible faulty components.

Cylinder Leakage Testing

Once you have identified a cylinder(s) with low cranking compression, you can use a **cylinder leakage test** to pinpoint the cause of the problem. To perform a cylinder leakage test, you put pressurized air in the weak cylinder through the spark plug hole when the valves are closed. You can then listen for leakage of air from the combustion chamber at various engine ports to determine the source of the leak. This is an excellent way to gather more information before disassembling the engine or when deciding whether a replacement engine should be installed. A cylinder leakage test is also a very thorough way of evaluating an engine when someone is considering purchasing a vehicle. You can even perform a cylinder leakage test on a junkyard engine to assess its value before purchasing it.

SERVICE TIP: If the piston and connecting rod are not exactly at TDC, the engine may actually rotate as you apply air pressure. Inline engines need to be almost exactly at TDC, while V-type engines can be a little off. If the engine starts to turn while you apply air pressure, try to move the cylinder closer to TDC. If you still cannot get the cylinder to stop moving, then try other things like holding the ratchet (watch your hand), using a piston position locater, or checking the timing marks on the crankshaft pulley. Each vehicle is very different, but you may be able to use a bar in a location where it stops a pulley from moving.

CUSTOMER CARE: A customer was anxious to buy a good-looking, performance-modified Honda. He was finally convinced to allow us to perform a cylinder leakage test after we noticed a bit of blue smoke coming from the tailpipe. A leakdown test showed very serious ring wear on two cylinders. While he still bought the car, he paid $1,500 less for it than he had planned.

Cylinder Leakage Test

The cylinder leakage test can help you evaluate the severity of an engine's mechanical problem and pinpoint the cause. You can test one low cylinder to find a problem, or you can test each cylinder to gauge the condition of the engine. Like a compression test, the cylinder leakage test will detect leaks in the combustion chamber from the valves, rings, or pistons and the head gasket. From this test, however, you can clearly identify which is the cause of the leak.

To perform a cylinder leakage test (Photo Sequence 20):

1. Let the engine warm up to normal operating temperature.
2. Let the engine cool for 10 minutes, and then remove the spark plug from the cylinder to be tested.

PHOTO SEQUENCE 20

Typical Procedure for Performing a Cylinder Leakage Test

P20-1 **Loosen the ignition cassette to access the spark plugs.**

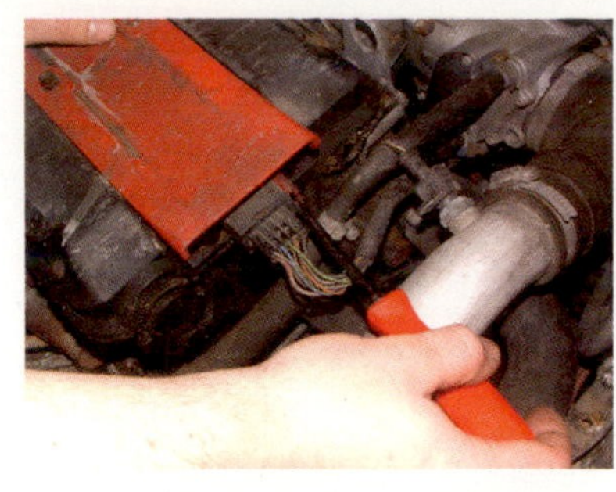

P20-2 **Disconnect power to the ignition cassette to avoid damaging the coils.**

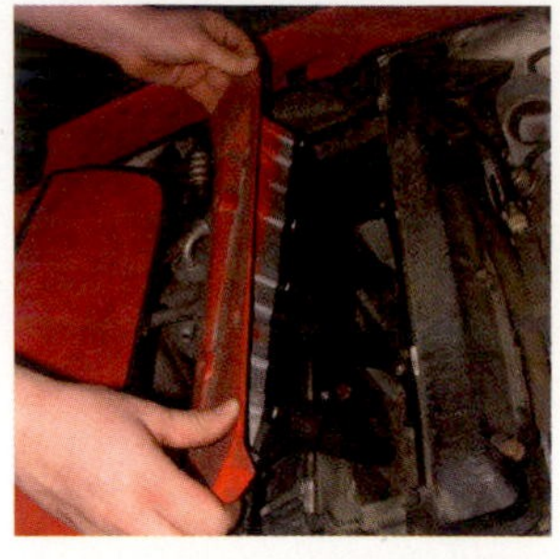

P20-3 **Remove the ignition cassette.**

P20-4 **Remove the spark plugs.**

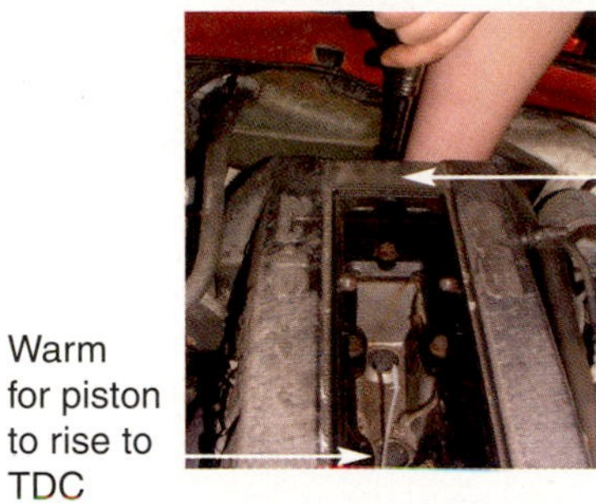

P20-5 **Rotate the engine to TDC compression for the test cylinder so that the piston is at TDC compression.**

P20-6 **Hook up shop pressure to the leakage tester, and set to 0 percent leakage.**

P20-7 **Connect adaptor to test cylinder and read leakage. This cylinder is leaking slightly over 30 percent.**

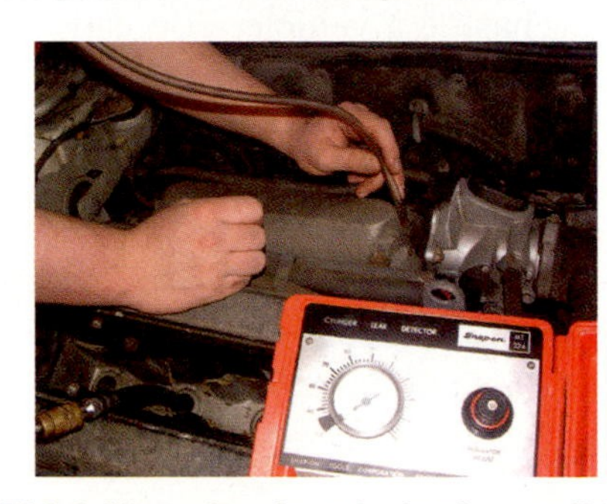

P20-8 **Listen for air at the intake manifold for leakage past an intake valve.**

P20-9 **Listen for air at the oil fill to detect leakage past the rings. In this case you could feel the air blowing out the fill pipe.**

P20-10 **Listen for air in the coolant reservoir for indications of a leaking head gasket.**

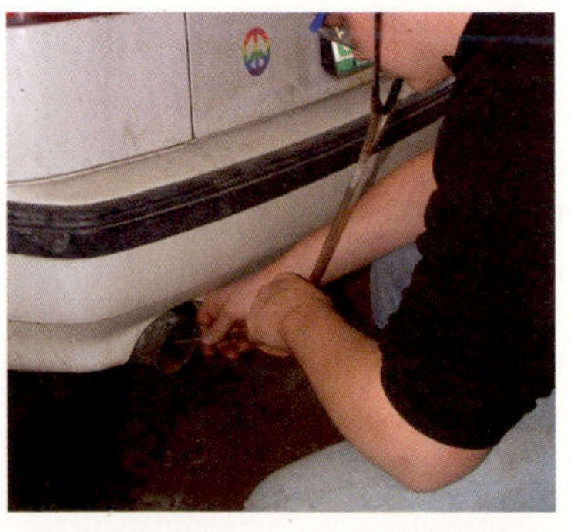

P20-11 **Listen for air at the tailpipe to find leakage past an exhaust valve.**

P20-12 **The moderate blue smoke from the tailpipe substantiates the diagnosis of worn rings on this high-mileage vehicle.**

3. Remove the radiator cap and the PCV valve. This prevents damage to the radiator or engine seals if excessive pressure leaks into the radiator or the engine's crankcase.
4. Turn the engine over by hand until the cylinder is at TDC on the compression stroke. This ensures that the valves will be closed and the rings will be at their highest point of travel. The cylinders wear the most at the top because of the heat and pressure of combustion, so if the rings are leaking, they will show the greatest evidence at the top. There are a few different methods of getting a cylinder to TDC compression, depending on the engine.
   a. Have an assistant turn the engine over slowly while watching the piston come up to TDC. You should be able to feel and hear blowing out of the spark plug hole if it is coming up on the compression stroke. If you begin the leakage test and show 100 percent leakage, you are on TDC exhaust. Rotate the engine 360 degrees and retest.
   b. Remove the valve cover, and watch the rockers or camshaft as you turn the engine over. On the intake stroke, the intake valve will open and then close. Then both the valves will remain closed while the piston comes up to TDC. If the engine has rocker arms, move them up and down to be sure they both have lash (clearance) between the arm and the pushrod. If the engine has an overhead camshaft(s), the followers will be on the base circle, not the lobes of the camshaft.
   c. If the vehicle has a distributor, remove the cap and rotate the engine until the rotor points at the test cylinder's firing point.
5. With the cylinder at TDC compression, thread the cylinder leakage tester adaptor hose into the spark plug hole.
6. Apply air pressure to the leakage tester, and calibrate it according to the equipment instructions. On the tester shown, you turn the adjusting knob until the gauge reads 0 percent leakage without it attached to the cylinder.
7. Connect the leakage tester to the adaptor hose in the cylinder while watching the crankshaft pulley. If at all the engine rotates at all, reset it to TDC. When the engine is right at TDC, it will not rotate with air pressure applied.
8. Read the percentage of leakage on the tester and gauge as follows:
   - Up to 10 percent leakage is excellent; the engine is in fine condition.
   - Up to 20 percent leakage is acceptable; the engine is showing some wear but should still provide reliable service.
   - Up to 30 percent leakage is borderline; the engine has distinct wear but may perform reasonably well.
   - Over 30 percent leakage shows a significant concern that warrants repair.
9. With the air pressure still applied, use a stethoscope and listen at the following ports:
   - At the tailpipe: Leakage here indicates a burned or leaking exhaust valve.
   - At the throttle or a vacuum port on the intake manifold: Leakage here indicates a burned or leaking intake valve.
   - At the radiator or reservoir cap: Leakage here proves that the head gasket is leaking.
   - At the oil fill cap or dipstick tube: Leakage here indicates worn rings.
   - If the leakage at the oil fill is near 100 percent, suspect damage to a piston (Figure 8-57).

SERVICE TIP: There will always be some leakage past the rings; a light noise is to be expected. Excessive ring wear will cause distinct blowing out of the cap or dipstick hole. You will be able to hear the noise and feel the pressure.

FIGURE 8-57 **This destroyed piston showed 100 percent leakage.**

10. Release the pressure, remove the adaptor, and rotate the engine over to the next test cylinder. Repeat as indicated. Refer to Photo Sequence 20 for detailed pictures and directions on performing a cylinder leakage test.

CUSTOMER CARE: While a cylinder leakage test can identify the primary cause of low compression readings, remember to keep your mind and eyes open to other problems when repairing the engine. Do not automatically inform a customer that an exhaust valve is leaking and a valve job will cure the problem. Consider the engine mileage, oil condition, pressure and usage, and degree of leakage past the rings. If the engine has high miles, it is very likely that the rings are significantly worn. By replacing the faulty valve(s) and restoring the others to like-new condition, you will increase the compression and combustion pressures. The change may be significant enough that the old rings will not seal as well as they had been before the valve job. In this case, the customer would be back complaining about blue smoke from oil consumption or about the engine still lacking adequate power. This may very well be an example of when a replacement engine would be the most cost-effective repair for the customer. Evaluate the whole situation, and make your recommendation to the customer. If you offer the customer repair options, be very clear about explaining the possible consequences.

TERMS TO KNOW

Backlash
Cylinder leakage test
Diagnostic trouble code (DTC)
Diesel exhaust fluid (DEF)
Diesel particulate filter
Diverter valve
Filter gasket
Filter wrench
Front pipe
Heat-resistant gloves
Load test
Malfunction indicator lamp (MIL)
Micro-vee
Muffler
Petcock
Serpentine belt
Slime
Spout
Tailpipe
Top off

ASE-STYLE REVIEW QUESTIONS

1. *Technician A* says a plugged PCV may be indicated by oil in the air filter housing.
   Technician B says that some vehicles do not have a PCV valve.
   Who is correct?
   A. A only
   B. B only
   C. Both A and B
   D. Neither A nor B

2. OBD II scan tools are being discussed.
   Technician A says a generic (Global) OBD II scan tool will connect to all OBD II vehicles and be able to read the stored DTCs.
   Technician B says that a manufacture scan tool can perform more tasks than a generic (Global) OBD II scan tool can.
   Who is correct?
   A. A only
   B. B only
   C. Both A and B
   D. Neither A nor B

3. Belts are being discussed.
   Technician A says that a serpentine belt is really just a micro-vee belt that is in a certain configuration.
   Technician B says that some belts do not crack as they age and wear out.
   Who is correct?
   A. A only
   B. B only
   C. Both A and B
   D. Neither A nor B

4. The fuel system is being discussed.
   Technician A says the pressure must be released on all vehicles before removing the fuel filter.
   Technician B says poor engine operation may result if the inlet end of the fuel filter is toward the fuel tank (filter is installed backwards).
   Who is correct?
   A. A only
   B. B only
   C. Both A and B
   D. Neither A nor B

5. Coolant that is green in color is:
   A. Considered a traditional coolant.
   B. Commonly found on new cars.
   C. A long-life coolant.
   D. None of the above

6. The lubrication system is being discussed.
   Technician A says an oil filter gasket should have a light coat of RTV applied.
   Technician B says the old oil filter gasket may stick to the filter mounting on the engine.
   Who is correct?
   A. A only
   B. B only
   C. Both A and B
   D. Neither A nor B

7. The cooling system is being discussed.
   Technician A says coolant must be mixed with antifreeze and water.
   Technician B says a slimy feel to the coolant indicates the coolant should be changed and the system flushed.
   Who is correct?
   A. A only
   B. B only
   C. Both A and B
   D. Neither A nor B

8. *Technician A* says the oil drain plug must be torqued to specifications.
   Technician B says the oil pan plug may require a new gasket.
   Who is correct?
   A. A only
   B. B only
   C. Both A and B
   D. Neither A nor B

9. An OBD II vehicle is being discussed.
   Technician A says all OBD II vehicles have a 10 pin DLC.
   Technician B says the OBD II DLC is located under the dash.
   Who is correct?
   A. A only
   B. B only
   C. Both A and B
   D. Neither A nor B

10. An engine test that can tell you the pressure of each individual cylinder is called a:
    A. Compression test.
    B. Vacuum test.
    C. Power balance test.
    D. Cylinder leakage test.

JOB SHEET 

Name ________________________ **Date** ____________

Performing an Engine Lubrication Service

NATEF Correlation

This job sheet addresses the following NATEF tasks:

Inspect engine assembly for fuel, oil, coolant, and other leaks; determine necessary action.
Perform oil and filter change.

Objective

Upon completion and review of this job sheet, you should be able to perform a routine engine lubrication service.

Tools and Materials Needed

Drain pan
Filter wrenches
Box-end wrenches
Service manual

Procedures

Task Completed

1. Move the vehicle to the lift. Record any noise or obvious defects found during the movement. ____ ☐
   ____
   ____
   ____
2. Engine Identification: ☐
   Displacement ____________ Block ____________ Valvetrain ____________
3. Determine the type and amount of oil and the filter: ☐
   Oil ____________________ Amount ____________________
   Filter/customer preferred brand ________________________
4. Raise the vehicle and place the drain pan under the oil pan plug. ☐
5. Remove the plug and drain the oil. Note the condition of the oil. ☐
   ____
6. Check the plug gasket and replace if necessary. ☐
7. Install the plug after the oil has drained. Torque to specifications. ☐
8. Move the drain pan if necessary. Remove and drain the oil filter. Did the gasket come off with the old filter?
   ____
9. Clean the oil filter mount, lube the new gasket, and install the new filter. ☐

Task Completed

☐ **10.** Lower the vehicle.

☐ **11.** Install the proper amount of oil into the engine.

☐ **12.** Start the engine. Watch the light or gauge. If no oil pressure is indicated within a few seconds, shut down the engine and locate the problem.

☐ **13.** If the oil pressure is correct, shut down the engine.

☐ **14.** Check for leaks at the oil filter and oil pan plug.

☐ **15.** Check the oil level and top off as needed.
Was oil required? ____________________ If yes, how much? ____________________

Instructor's Response ____

____

____

____

JOB SHEET 22

Name ______________________________ Date ______________

Checking the Battery and Generator

NATEF Correlation

This job sheet addresses the following NATEF task:

Perform charging system output test; determine necessary action.

Objective

Upon completion and review of this job sheet, you will be able to test the battery and generator using a digital multimeter (DMM) and make suggestions on possible faults or repairs.

CAUTION: Use extreme care when working around a hot engine. Serious burn injuries could result if contact is made with hot components.

Tools and Materials Needed

Assigned vehicle

DMM

Service manual

Procedures

Task Completed

1. Use the service manual to collect the following information. ☐

   Recommended battery size by cold-cranking amps ______________________

   Generator size based on maximum ampere output ______________________

WARNING: Ensure the DMM leads are clear of the cooling fan and drive belts. Keep hands clear of moving engine parts and the cooling fan. Damage to the DMM or personal injury could result.

2. Place the vehicle on level ground or floor and set the parking brake. Shift the transmission to PARK (NEUTRAL on a manual). Ensure the ignition key is in the OFF position. Chock the wheels as soon as you exit the vehicle. ☐

3. Raise the hood and perform an inspection of the battery and generator drive belts. Record your observations and note any other conditions that may point to problems other than those typically associated with the starting and charging systems. ☐

   ____

   ____

4. Connect the DMM to the battery for an open circuit voltage test (red on positive post, black on negative post). Record the measurement. Is it within the normal range for a fully charged battery? If not, what tests or repairs do you suggest be performed at this point? ☐

   ____

   ____

Task Completed

☐ 5. Connect the meter to the battery for a voltage drop test on the positive side of the battery (red on terminal, black on post). Record the measurement and then do the same test on the negative side. Is there a voltage drop on either terminal? If so, what possible fault would cause a voltage drop at this point?

____

____

☐ 6. Use the DMM to perform a voltage leakage test on the battery. Is there a reading of voltage and what is the average of the readings? What do you recommend if there is a measurement of voltage leakage?

____

____

☐ 7. An assistant can help perform the generator output test, but is not required.

☐ 8. Connect the DMM to the battery for an open circuit test.

☐ 9. Start the engine and allow it to idle. Measure and record the voltage present at the battery. Turn on a heavy load, either the air conditioning or bright headlights. Measure and record the battery voltage. Does there seem to be a problem with the generator at this point? If so, what do you recommend for the next test or a possible repair?

____

____

☐ 10. Shut down the engine and close the hood. Based on the measurements just completed, what are your recommended actions for the starting and charging systems?

____

____

Instructor's Response ____

____

____

JOB SHEET 23

Name ______________________________ Date ______________

Perform an Express Service

Objective

Upon completion and review of this job sheet, you should be able to perform an express service.

NOTE: The NATEF tasks involved here are numerous as are the various tasks to be performed. This job sheet will list the general tasks typically performed by express lube technicians along with preparation of a repair order. Many of the individual tasks will include a reference (Job Sheet 1, for example) to a job sheet in this Shop Manual for the student to use as necessary. NATEF tasks are listed on the individual job sheets for reference. It is expected that the instructor will have locally produced repair orders and inspection sheets pertinent to local businesses. A below-ground lube pit will be used in this job sheet as the service area. Additional instructions and procedures will be required if an above-ground lift is used.

NOTE: Procedures below are general and may require adjustment based on shop policy and equipment available.

Tools and Materials Needed

Service information
Repair order
Inspection sheet
Battery tester
Tire depth gauge
Tire pressure gauge
Oil, oil filter, oil filter wrenches
Grease gun

Procedures

Task Completed

1. Greet the customer and determine type of service requested. ☐
2. Initialize the repair order (Job Sheet 1). Secure an inspection sheet and record the service tag number, mileage, and make of vehicle. ☐
3. Attach the service tags on rear-view mirror and ignition key. Perform the first phase of the inspection in the service lane and note results on the inspection sheet. ☐

   Interior lights; horn; parking brakes; lit warning lights with engine running; exterior lighting; tire pressure, tread condition, and tread depth; and exterior condition of the vehicle.

   NOTE: Some shops require that certain inspection findings also be recorded on the repair order.

4. Inform instructor of any defects found. ☐

Task Completed

☐ **5.** Move vehicle to express service bay and drive onto express lube pit. Note and record operating condition of engine and drivetrain as vehicle is moved. If an above-ground lift is used, refer to Job Sheet 2.

☐ **6.** Determine type and amount of oil and oil filter designation. (Job Sheet 6).

NOTE: The following is based on two-person team: top and pit.

☐ **7.** Pit person: Drain the engine oil and replace the oil filter (Job Sheet 21).

☐ **8.** Top person: Upon notification that the drain plug and oil filter are in place, install correct amount and type of engine oil (Job Sheet 21).

☐ **9.** Pit person: Inspect the complete undercarriage for leaks, damage, and other defects (Job Sheet 16, Job Sheet 19). Record findings on inspection sheet.

☐ **10.** If possible, measure brake pad wear. This may have to be done from top and with wheel assemblies removed. Record findings on inspection sheet if measurement is made.

☐ **11.** Lubricate steering and suspension components (Job Sheet 29).

☐ **12.** Top person: Inspect each of the following and record findings on inspection sheet (Job Sheet 17).

Belt(s) condition and tension; battery performance check (Job Sheet 22); fluid levels (master cylinder, coolant, power steering, automatic transmission, windshield washer); leaks and type of leak (Job Sheet 16); and overall visual impression of engine compartment.

NOTE: Some shops offer free top off of some fluids.

☐ **13.** Top person: If necessary, lift vehicle and remove wheel assemblies to measure brake pad or shoe wear. Record measurement and make estimate, in percentage, of wear remaining on inspection sheet.

☐ **14.** Either person: If vehicle passes inspection or customer refuses additional services, complete technician portion of repair order, inspection sheet, and return vehicle to service lane.

☐ **15.** Either person: Remove the disposable mat and seat covers, complete repair order and turn in repair order, inspection sheet, and ignition key to instructor.

Instructor's Response ______________________________

____

____

____

Chapter 9

Inspecting and Servicing the Electrical System

Upon completion and review of this chapter, you should understand and be able to describe:

- How to read and use a wiring diagram.
- How to use a test light in an automotive circuit.
- How to use a DMM to test an automotive circuit.
- How to use a fused jumper wire to test an automotive circuit.
- How to inspect circuit protection devices.
- How to inspect and test switches and relays.
- How to diagnose an open circuit.
- How to diagnose a short circuit.
- How to diagnose a high resistance circuit.
- How to inspect, service, and test the battery.
- How to inspect, service, and test the starter and circuit.
- How to inspect, service, and test the alternator and circuit.
- How to inspect, service, and test the lighting components and circuit.
- How to inspect, service, and replace secondary ignition circuit components.
- How to perform a power balance test.
- The special procedure for working safely around hybrid electric vehicles.

Introduction

Similar to what was emphasized in the Classroom Manual, understanding the electrical system is just as important as knowing how to test, inspect, repair, and replace problems associated with it. As an entry-level technician you will be expected to perform some basic electrical checks, inspections, and repairs. This chapter outlines those basic procedures and emphasizes safety.

When diagnosing electrical systems, always try to remember Ohm's law and apply it. The voltage across a circuit is equal to the current through a circuit times the resistance of a circuit. These verbs are underlined because they will help you understand how to measure electrical circuits.

Understanding a Wiring Diagram

Probably the most overlooked and misunderstood part of any electrical diagnosis or test by an entry-level technician is the wiring diagram. The wiring diagram can be complicated and difficult to read at first. One of the best strategies for reading a wiring diagram is trying to remember how an electrical component works. Once you remember how it works, trying to trace the wires will seem easier.

The first step is to find the correct wiring diagram. This usually means that you have to look up the right vehicle. This is done by accessing the vehicle identification number (VIN) and entering it into the service manual software or program correctly. Entering it incorrectly or slightly off will result in slightly incorrect wiring diagrams. But even slightly incorrect diagrams will cause headaches. The best example is trying to follow an alternator's wiring harness for a vehicle. If the vehicle came with two different four-cylinder engine options, then incorrectly entering the VIN will result in having the incorrect wire colors, circuit numbers, or harness locations. Here's a good example: A Saturn SL-2 has two different 1.9 L engines. One had a single overhead camshaft and the other had a dual overhead camshaft. The alternators are different, and the wiring harness for each is different. The only difference was in the VIN, so incorrectly entering in the information for the eighth digit will cost you a lot of time and grief.

The next step is to find the information in the service manual that will help you read the diagram. Each manufacturer uses slightly different color codes and symbols for the wiring diagram. Let's look at Figure 9-1 as an example. In this wiring diagram, the number 181 (by the fuse) stands for the circuit number. The "Br/O" is shorthand for a wire that is brown with an orange strip. Never take for granted that each manufacturer will use the same notations. The letter "B" for one manufacturer could mean black and for another it could mean the color blue. The two arrows by the letters "C001" are symbols for female

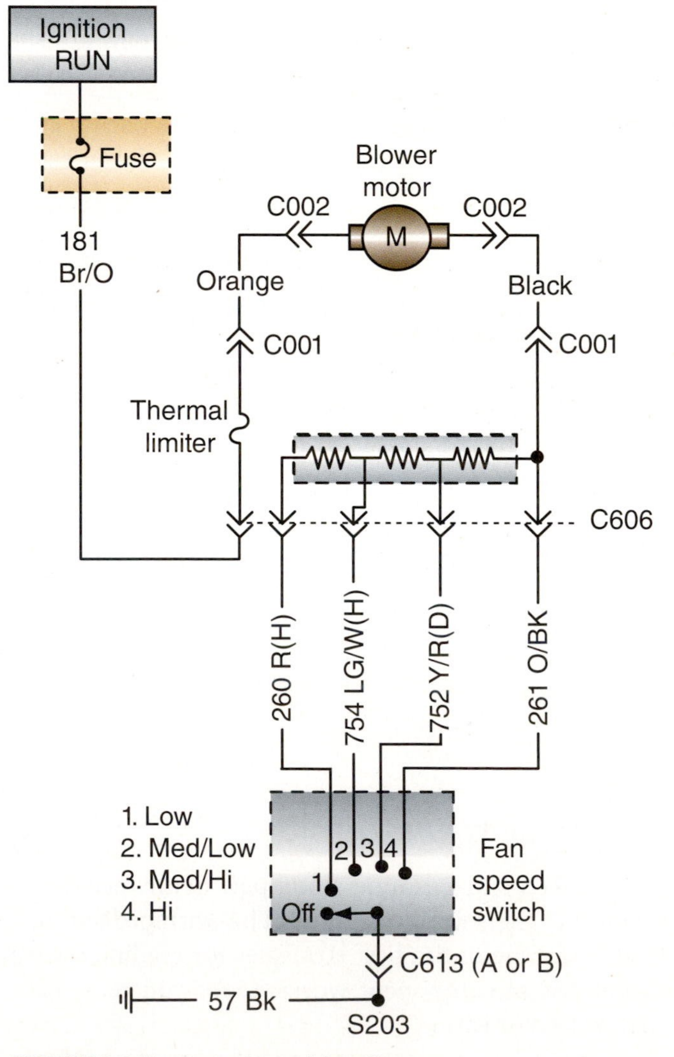

FIGURE 9-1 **Heater system wiring diagram.**

and male spade connectors, while other connectors are represented by other symbols. The "C001" stands for the number of the connector. When you look at the fuse, the dots that touch the fuse are connectors, but do you notice how they are inside the dashed lined box? That means that the wires go into the fuse box, and then the connector is on the inside. The dashed line is also interesting. It means there is more than what is drawn. Look at Figure 9-1: have you ever opened up a fuse box in a car and found only one fuse? No. That's what the dashed line means. There's more to it. Did you notice that the power side (also known as the B+ or hot side) is on the top of the diagram? This is common. It is just as common as to have the ground on the bottom. With practice, you will be able to read wiring diagrams easily. Table 9-1 outlines the different electrical symbols that are commonly used in the automotive industry.

USING A TEST LIGHT

There are two types of test lights commonly used in automotive diagnostics and repair: non-powered and self-powered. A **test light** (Figure 9-2) is a tool the technician uses when he or she needs to "look" for electrical power in a circuit. The key here is the word *look*. A non-powered test light will light up when there isn't full voltage, but it will do so with less brightness because there is less voltage. Sometimes this is hard to see. And, it can be misleading. Some computerized circuits and components will only work with 12 volts. So if a circuit is measuring 10 volts because of a defect, the test light will illuminate, but the circuit will not have enough power to turn on the component. With some experience you will be able to tell when the test light is not bright enough, but that's not what a test light is used for. It is there to test for the presence of voltage, the amount.

A **test light** is connected between a positive conductor and ground. It will only show voltage presence not amount.

The **self-powered** test light (also known as a self-powered continuity tester) has an internal battery that powers the light bulb (Figure 9-3). This device will work with a different color light when the test light is in the opposite direction. This type of tester can work on a vehicle with a dead battery.

A **self-powered** test light is connected across a portion of the circuit. If the lamp lights then there is continuity between the test points.

To use a test light you have to connect the clip to ground and then move the tester tip around the circuit; wherever it lights up, voltage is present (Figure 9-4). It is important to limit the use of a test light to non-computerized circuits. A test light will draw current from the system to light up the bulb. This may draw more current from a computer circuit than was originally designed, thus burning the circuit. It is good practice to limit the use of a test light to non-computerized circuits.

USING A DIGITAL MULTIMETER

The digital multimeter (DMM) is the most effective and most used tool by an automotive technician. It can measure voltage, resistance, amperage, and sometimes more than that. Most voltmeters are sold in digital format (Figure 9-5). Analog format DMMs are not used in modern vehicles with computerized equipment. DMMs that are compatible with computerized circuits have internal impedance rating of 10 mega ohms or greater. Always check the tool for this specification before working with it on a modern computer controlled vehicle.

To measure voltage, you need to connect the red and black test leads to the appropriate sockets in the meters, making sure to match the colors and symbols. You can use a voltmeter to measure the voltage across or **voltage drop**. Voltage is always measured "across" a circuit (remember Ohm's law). Figure 9-6 shows how the voltage drop of a resistor is measured across the circuit. Note that only a limited amount of current will flow through the meter and that the majority of current will flow through the bulb (because the meter has high resistance). Figure 9-7 shows how to make several voltage checks in the system. Note that the negative test lead never leaves the ground connection of the

Voltage drop is a measurement of the voltage across a portion of the circuit.

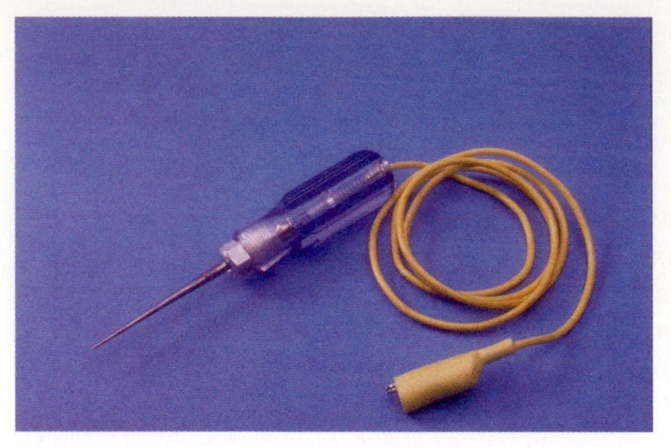

FIGURE 9-2 Typical test light used to probe voltage in a circuit.

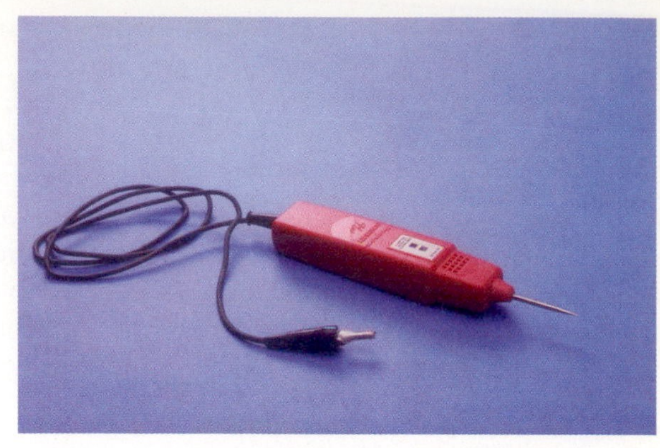

FIGURE 9-3 Typical self-powered continuity tester.

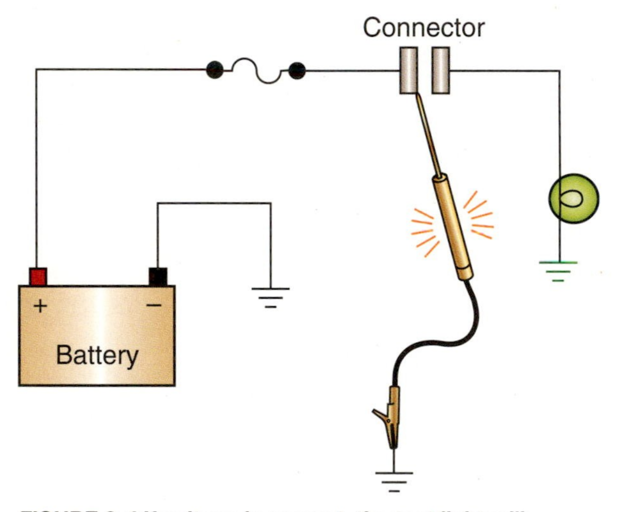

FIGURE 9-4 If voltage is present, the test light will illuminate.

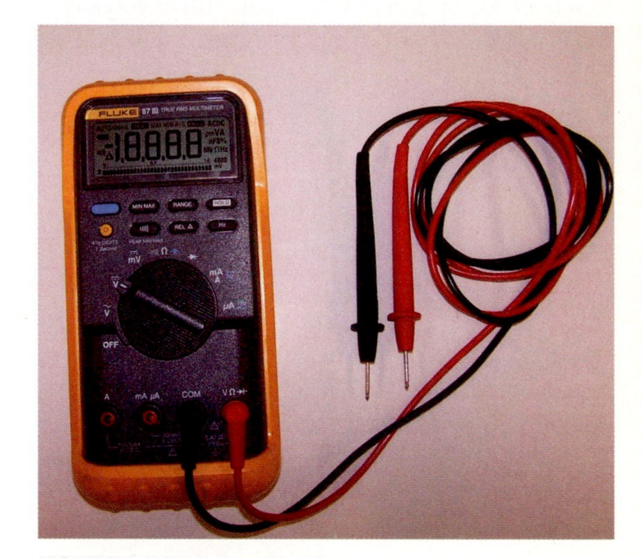

FIGURE 9-5 Digital multimeter.

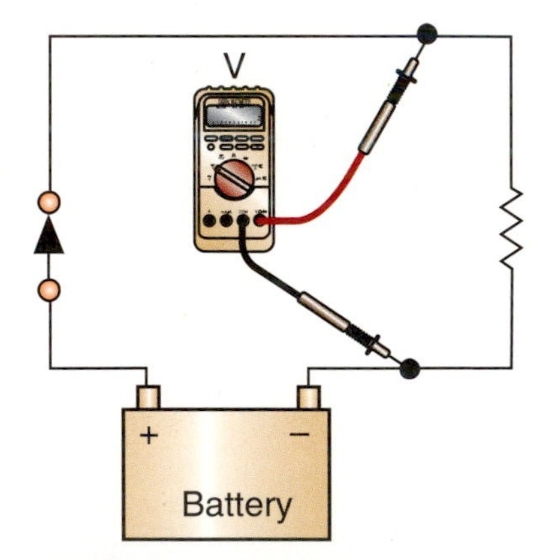

FIGURE 9-6 Connecting a voltmeter in parallel to the circuit. This will register the voltage drop across the resistor.

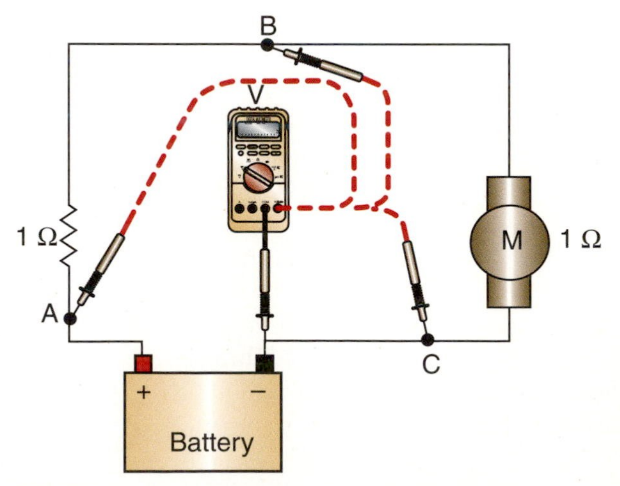

FIGURE 9-7 Checking voltage in a closed circuit.

TABLE 9-1 **COMMON ELECTRICAL AND ELECTRONIC SYMBOLS USED IN WIRING DIAGRAMS**

COMPONENT	SYMBOL	ALTERNATE
Ammeter	 A	
AND Gate	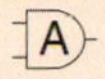 A	
Antenna		
Attenuator, Fixed		
Attenuator, Variable		
Battery	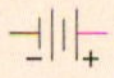	
Capacitor, Feedthrough		
Capacitor, Fixed, Nonpolarized	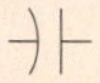	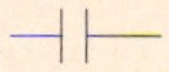
Capacitor, Fixed, Polarized	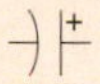 +	
Capacitor, Ganged, Variable		
Capacitor, General	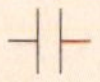	
Capacitor, Variable, Single		
Capacitor, Variable, Split-Stator		
Cathode, Cold		
Cathode, Directly Heated		
Cathode, Indirectly Heated	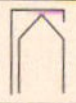	
Cavity Resonator	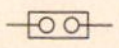	
Cell	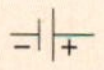	
Choice Bracket	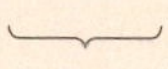	
Circuit Breaker or PTC device	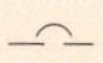	
Clockspring		
Coaxial Cable		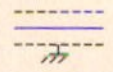
Coil		
Crystal, Piezoelectric	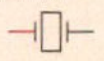	
Delay Line		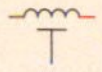
Diode		
Diode, Gunn		
Diode, Light-Emitting		
Diode, Photosensitive	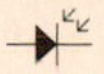	
Diode, Photovoltaic		

TABLE 9-1 (Continued)

Diode, Pin		
Diode, Varactor		
Diode, Zener		
Directional Coupler		
Dual Filament Lamp		
Exclusive-OR Gate		
Female Contact		
Ferrite Bead		
Fuse		
Fusible link		
Gauge		
Ground, Chassis		
Ground, Earth		
Handset		
Headphone, Double		
Headphone, Single		
Heating element		
Hot Bar	BATT A0	
Inductor, Air-Core		
Inductor, Bifilar		
Inductor, Iron-Core		
Inductor, Tapped		
Inductor, Variable		
In-Line Connectors	2 C123 2 C123	
Integrated Circuit		
Inverter		
Jack, Coaxial		
Jack, Phone, 2-Conductor		
Jack, Phone, 2-Conductor Interrupting		
Jack, Phone, 3-Conductor		
Jack, Phono		

TABLE 9-1 (Continued)

Key, Telegraph		
Lamp, Neon		
Male Contact		
Microphone		
Motor, One speed	M	
Motor, Reversible	M	
Motor, two Speed	M	
Multiple connectors	8 5 2 C123	
NAND Gate		
Negative Voltage Connection	–o–	
NOR Gate		
Operational Amplifier		
OR Gate		
Outlet, Utility, 117-V		
Outlet, Utility, 234-V		
Oxygen Sensor		
Page Reference	(BW-30-10)	
Piezoelectric Cell		
Photocell, Tube		
Plug, Phone, 2-Conductor		
Plug, Phone, 3-Conductor		
Plug, Phono		
Plug, Utility, 117-V		
Plug, Utility, 234-V		
Positive Voltage Connection	–o+	
Potentiometer		
Probe, Radio-Frequency		
Rectifier, Semiconductor		
Rectifier, Silicon-Controlled		
Rectifier, Tube-Type		
Relay, DPDT		

TABLE 9-1 (Continued)

Relay, DPST	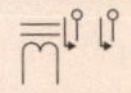	
Relay, SPDT	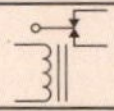	
Relay, SPST	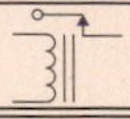	
Resistor		
Resonator	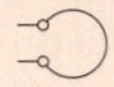	
Rheostat, Variable Resistor, Thermistor	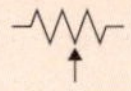	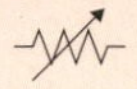
Saturable Reactor	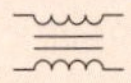	
Shielding		
Signal Generator	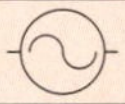	
Single Filament Lamp		
Sliding Door Contact	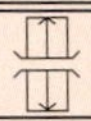	
Solenoid	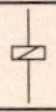	
Solenoid Valve	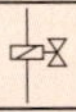	
Speaker	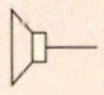	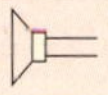
Splice, External	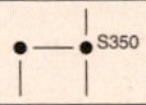 S350	
Splice, Internal	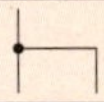	
Splice, Internal (Incompleted)	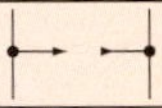	
Switch, Closed	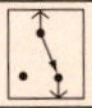	
Switch, DPDT	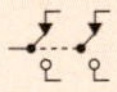	
Switch, DPST		
Switch, Ganged	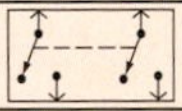	
Switch, Momentary-Contact	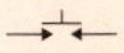	
Switch, Open	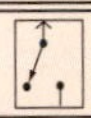	
Switch, Resistive Multiplex	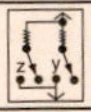	
Switch, Rotary	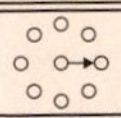	
Switch, SPDT		
Switch, SPST		
Terminals		
Test Point		
Thermocouple		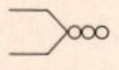

TABLE 9-1 (Continued)

Thyristor		
Tone Generator	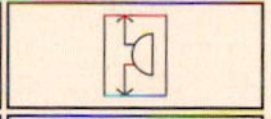	
Transformer, Air-Core		
Transformer, Iron-Core		
Transformer, Tapped Primary		
Transformer, Tapped Secondary	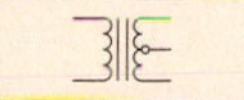	
Transistor, Bipolar, npn		
Transistor, Bipolar, pnp		
Transistor, Field-Effect, N-Channel	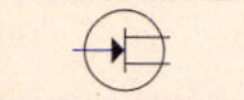	
Transistor, Field-Effect, P-Channel		
Transistor, Metal-Oxide, Dual-Gate		
Transistor, Metal-Oxide, Single-Gate		
Transistor, Photosensitive		
Transistor, Unijunction	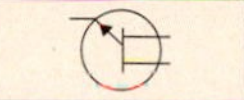	
Tube, Diode		
Tube, Pentode		
Tube, Photomultiplier	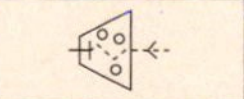	
Tube, Tetrode	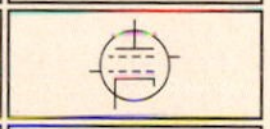	
Tube, Triode		
Unspecified Component	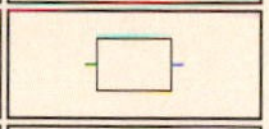	
Voltmeter	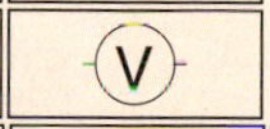 V	
Wattmeter	 W	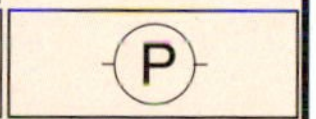 P
Wire Destination In Another Cell		
Wire Origin & Destination Within Cell	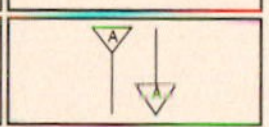	
Wires	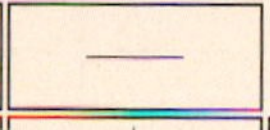	
Wires, Connected, Crossing		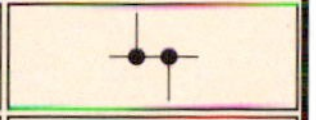
Wires, Not Connected, Crossing	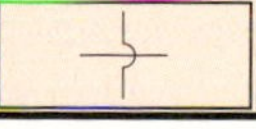	

Ground reference voltage testing is used to determine the voltage within a circuit.

battery. This is called **ground reference voltage testing**. The loss of voltage due to resistance in the wires, connectors, and loads is called voltage drop. Voltage drop is the amount of electrical energy that is converted to another form of energy.

For example, to make a lamp light, electrical energy is converted to heat energy; it is the heat that makes the lamp light up. To make voltage checks across the system, you need to determine what is positive and what is negative (or ground). The load in the system is always the splitting device.

Using Figure 9-7 as the example, the motor is the load in the circuit. Everything before the load is the positive side (including the resistor). The meter at point A will measure only slightly less than the battery voltage because there is only a small section of wire between it and the positive terminal and there are no loads in it. At point B, the voltage reading will measure less because voltage has dropped across the resistor. And at point C, the voltage will be very small because almost all of the voltage was used by the circuit and the load. If you add the voltage drops at points A, B, and C, you should get the voltage of the battery.

Using the DMM to measure resistance is very similar to measuring voltage. The internal battery of the DMM is used for this purpose. To start, you must set the test leads in the correct places (usually the same place as voltage, but each meter is different). Then set the dial or buttons to measure resistance. Touch the two test leads together to calibrate the meter. Next, remove the component from the circuit or de-energize the circuit (Figure 9-8). This has to be done because otherwise you will be measuring the resistance of the entire circuit and the wiring. Most service manual specifications for electrical component testing are for the component only, not for the component and circuit together.

Infinite means forever or too large. In automobile circuit testing it means the value being measured is too great for the test device being used.

Connect the meter test leads in parallel with the component or circuit to read the resistance. You can measure the resistance of almost anything that doesn't have voltage applied to it. A reading of O.L. means that the reading is "outside of limits." Another term used for this is **infinite**. This means that the meter cannot read the resistance between the test leads. If you are measuring a circuit, this is an open circuit. If you are measuring a component, there is an open circuit in the component (Figure 9-9). If you are measuring a switch, it may just be in the off position. Turning the switch to the on position will show a resistance reading. A common misconception arises when measuring the resistance of a component and the DMM shows a reading of 0.00 (zero) ohms. This does not mean that there is no resistance; it simply means that the reading is so low that it is rounded down. The reading is likely 0.001 ohms or less, and the meter will not display a reading

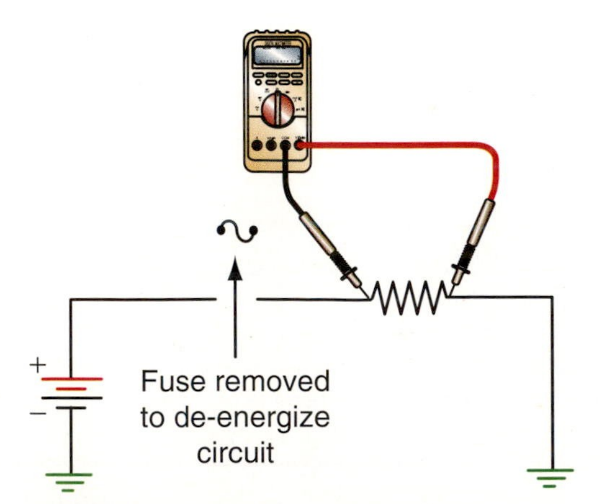

FIGURE 9-8 **Measuring resistance with an ohmmeter. The meter is connected in parallel with the component being tested after power is removed to the circuit.**

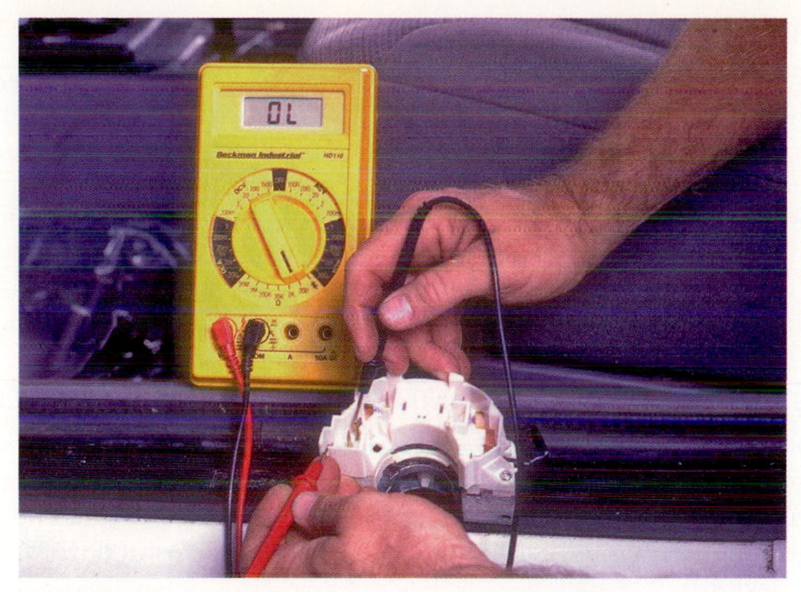

FIGURE 9-9 **A digital ohmmeter showing an infinite reading.**

that low. A reading of 0.00 ohms still means the component or circuit has resistance, but a very low amount (which in many cases is good).

There are two ways to measure current. Remember Ohm's law? Current must be measured "through" the circuit. One way of making the current go through the circuit is by interrupting the circuit and making all current go through the DMM (Figure 9-10). Unlike measuring voltage or resistance (where the DMM is placed in parallel with the circuit or component), you must have the current going through the meter to measure current (amperage).

In order to keep you and the tool safe, you must use caution and read the tool's amperage fuse rating. Then look at the circuit's fuse. Never place a DMM in a circuit where the fuse rating is greater than the rating of the DMM.

The second way to measure current (and maybe the safer way) is to use an inductive ammeter (also known as an amp clamp). This tool connects with the DMM and clamps around a wire. There is no need to open up the circuit. This comes in handy when a large harness cannot be easily removed for testing or has multiple other wires that need to be connected for testing (Figure 9-11).

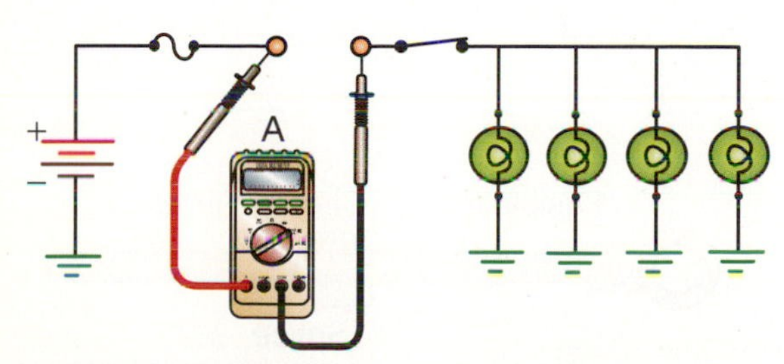

FIGURE 9-10 **Measuring current flow with an ammeter. The meter must be connected in series with the circuit.**

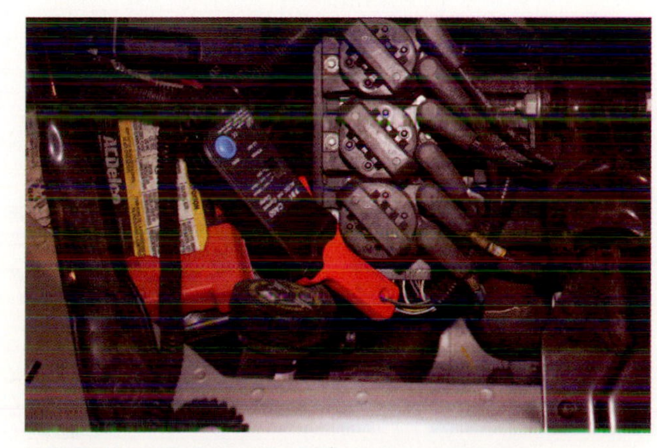

FIGURE 9-11 **An inductive pickup for an ammeter.**

Using a Fused Jumper Wire

A very common test tool, and one easily made by yourself, is the fused jumper wire. It can be used as an extension for testing using a DMM or as a wire replacement. If you suspect that a wire is broken or a switch is not working properly, you can bypass it and verify that the remainder of the circuit is working properly (Figures 9-12 and 9-13).

Jumper wires can be used anywhere in the circuit to bypass what would normally be a wire or anything of low resistance. They cannot be placed in the circuit in place of a load. When using the fused jumper wire, always check to see the size of the fuse the circuit you are testing normally uses. Then install that fuse into the jumper wire and test.

Inspecting Power Distribution and Circuit Protection Devices

There are many kinds of protection devices in automotive systems. Each one is designed to "turn off" the circuit if too much current starts to flow through it. Some of the protection devices will turn back on after they cool down (like a circuit breaker), and others will need to be replaced (fuses and fusible links). In this section, you will learn how to test each type of protection device.

To test a fuse, the best method is to visually inspect it for a blown or open link (Figure 9-14). This isn't always the easiest thing to see, so testing with a DMM is required. Remove the fuse and test the resistance of it. O.L. means that the fuse is open and requires replacement (Figure 9-15). A small resistance reading (usually 0.1 ohms) means it is good. Anything higher than that usually means there is some corrosion. Fuses don't wear out. (This is a very common misconception.) If they are blown, then there is a problem in the circuit.

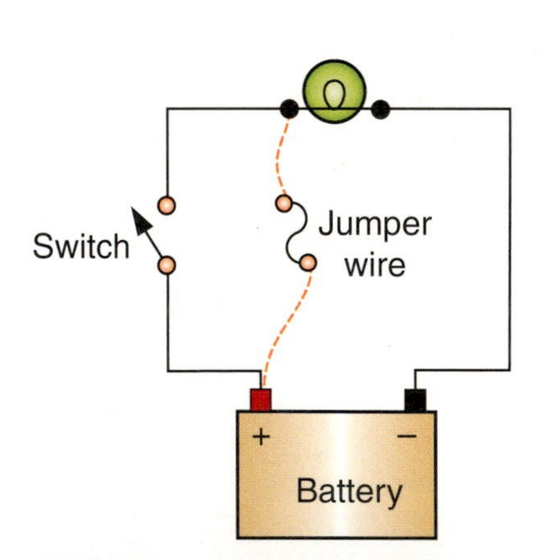

FIGURE 9-12 **Using a fused jumper wire to bypass the switch.**

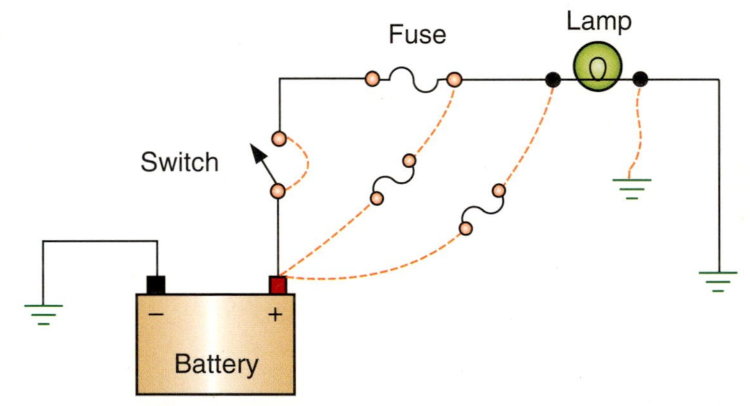

FIGURE 9-13 **Examples of locations a jumper wire can be used to bypass a portion of the circuit, and to test the ground circuit. Remember, if you bypass the circuit fuse, the jumper wire should be fitted with a fuse. Never bypass the load component.**

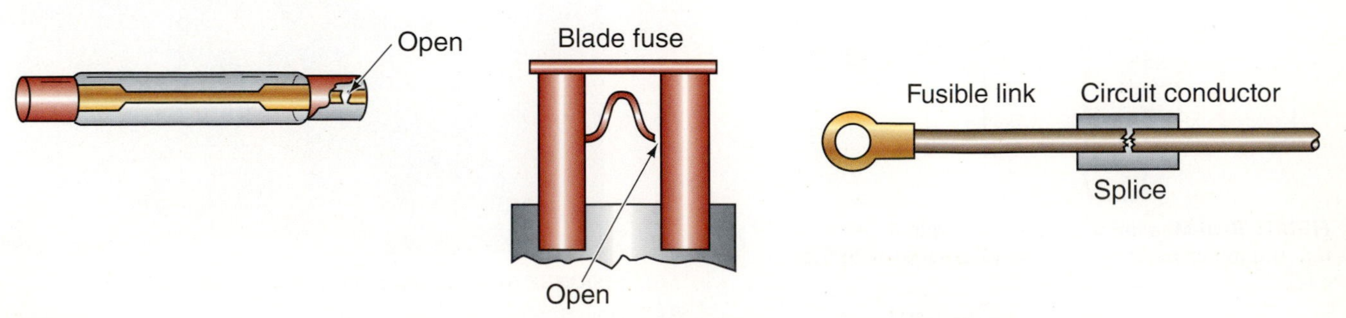

FIGURE 9-14 **A fuse or fusible link can have a hidden fault that cannot be seen by the technician.**

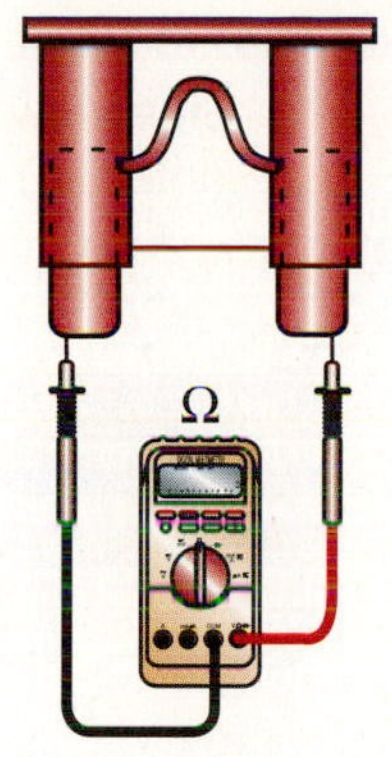

FIGURE 9-15 **A good fuse will have zero resistance when tested with an ohmmeter.**

You can also test the fuse while installed by checking the voltage drop across it (Figure 9-16). If there is 12 volts on one side and 0 volts on the other, it is open. If there is 12 volts on both sides, then it is good and the difference in voltage (voltage drop) is too small to be displayed on the meter. Fusible links are not as common, but to test them you would use an ohmmeter just like you would with a regular fuse (Figure 9-17). The only difference is that you have to find the metal connector end of the wire to test it.

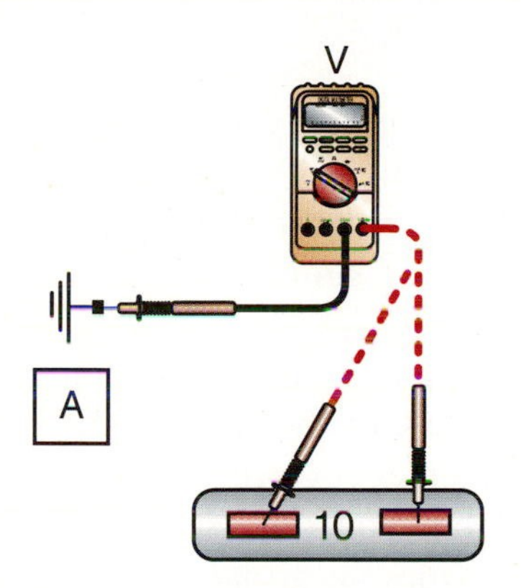

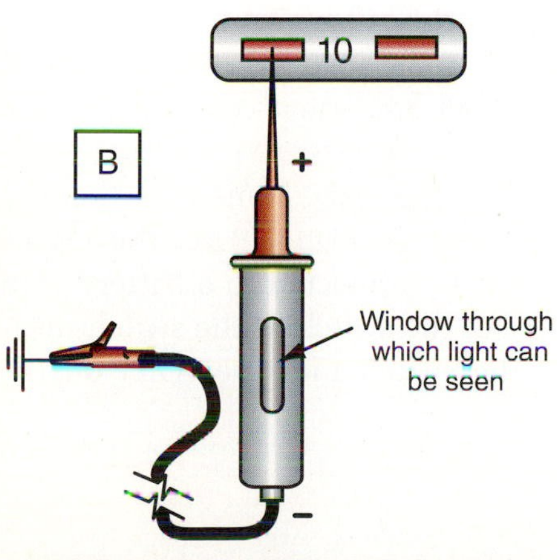

FIGURE 9-16 **(A) Voltmeter test of a circuit protection device. Battery voltage should be present on both sides. (B) The test light will illuminate on both terminals if the fuse is good.**

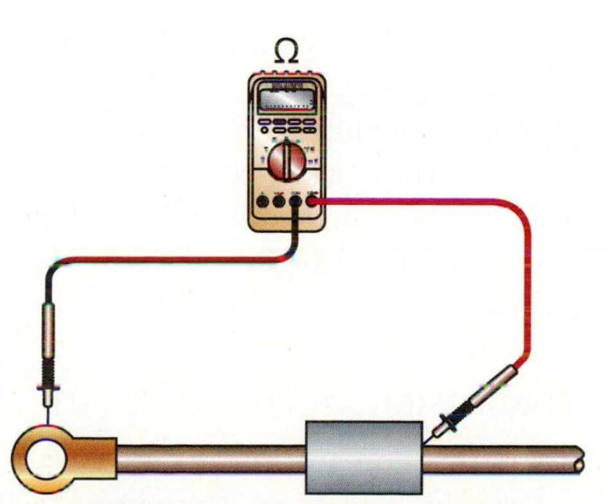

FIGURE 9-17 **A fusible link can be tested with an ohmmeter, once it is disconnected from power.**

Non-cycling circuit breaker that is reset by removing from power

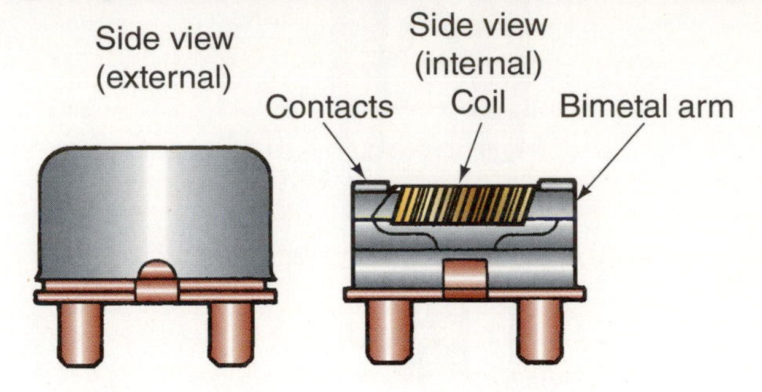

Manual resetting circuit breaker.

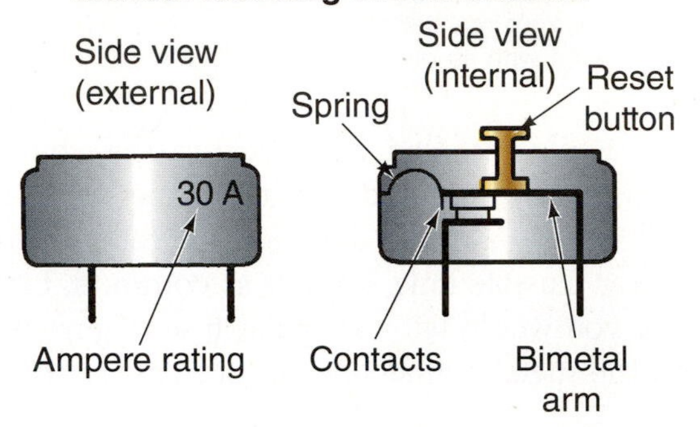

FIGURE 9-18 **Operation of a self-resetting circuit breaker.**

To test a breaker you need to remove it from the vehicle and use the DMM set to ohms. Test the resistance of the breaker. The readings should be similar to a fuse. Some circuit breakers have manual resets on them (Figure 9-18).

Testing a relay requires you to use all of the strategies mentioned up to this point. Start by testing the relay without a DMM. Place your hand on the top of the relay and attempt to turn the circuit on. You should be able to feel and hear the contact closing when the coil is energized, assuming it is working properly. If you hear this, you can assume that the relay coil and the control side circuit (low current side) are working properly (power and ground) and that there is a problem with the high current side. If you don't hear or feel anything, remove the relay and test the relay connectors (not the relay itself) for power, ground, and resistance. In Figure 9-19, the DMM measures the voltage of the relay when it is installed. Remember that the coil (between C and D) is the load in a circuit, and it controls the switch (between A and B). The switch and load are tested according to the procedures outlined earlier in this chapter.

If everything checks out in the relay connectors, then test the relay. Remember that the relay is a moving part, eventually moving parts will wear out. In Figure 9-20, the DMM measures the resistance of the coil; in Figure 9-21, the relay is connected to a battery with fused jumper wires and the DMM measures the resistance of the switch. If the switch measures 0 ohms (like in Figure 9-21), then it is good. If the switch measures O.L. on the DMM, it is bad.

Diagnosing an Open Circuit

It is possible to test for *opens* using a DMM, test light, or fused jumper wire. An open circuit can be defined in many ways. A broken wire and broken component like a blown light bulb are both examples of an open circuit. When a switch is in the open (or off) position, the circuit is open because current is not flowing through it.

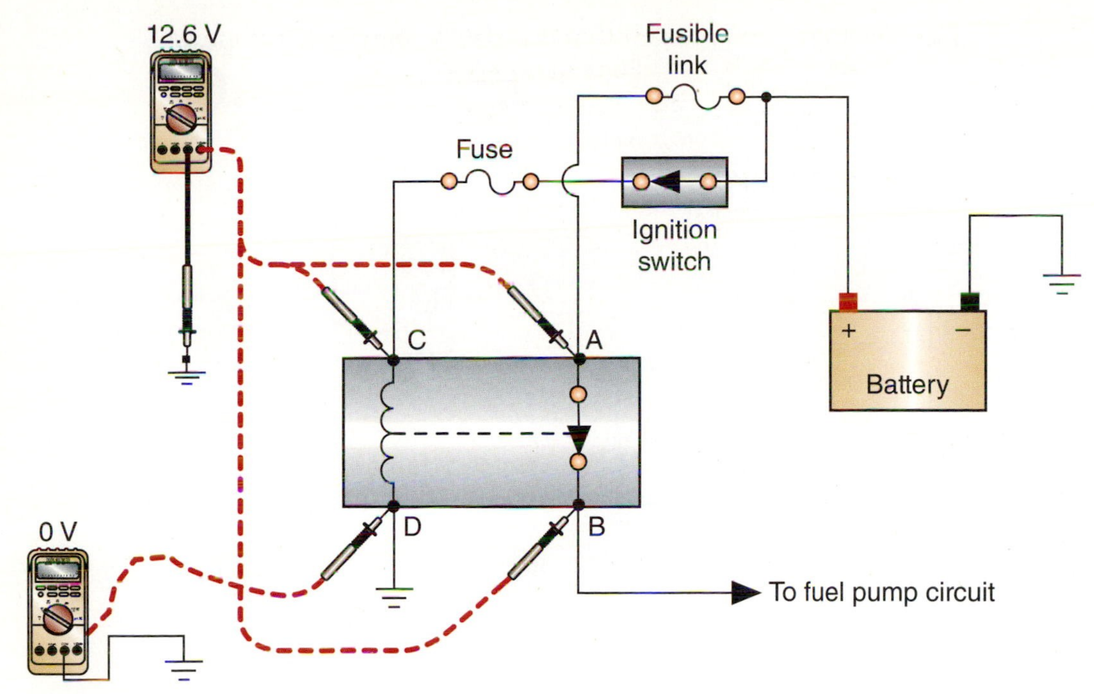

FIGURE 9-19 **Testing relay operation with a voltmeter.**

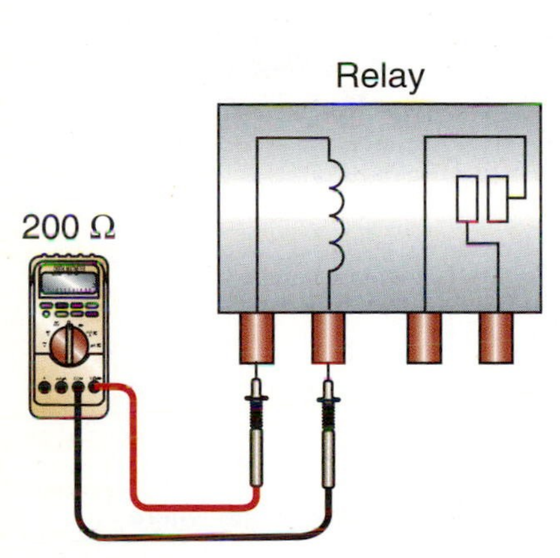

FIGURE 9-20 **Testing the resistance of the relay coil.**

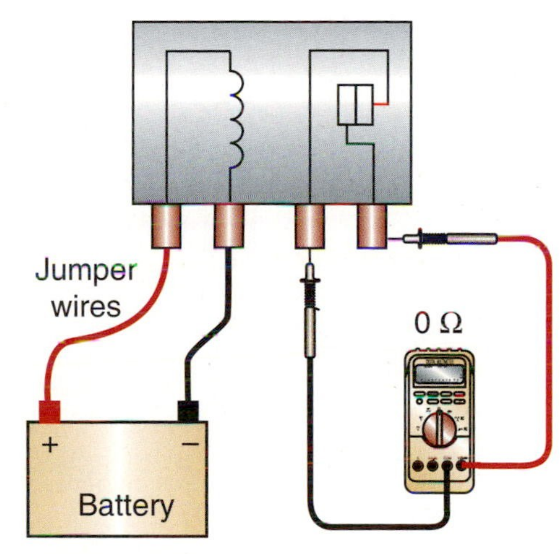

FIGURE 9-21 **Bench testing a relay.**

When something doesn't work or doesn't operate as intended, it can be an open circuit. To test for this, it is best to do a visual inspection by checking the fuses, circuit protection, and inspecting the components and wiggling the connectors. If everything checks out, then proceed to turn on the DMM and measure voltage starting at the battery. If the battery has enough voltage continue testing; if not, connect a jumper box.

Start by testing at the load that does not work. Test the load itself for resistance. A bad light bulb is an open circuit. Figure 9-22 shows the DMM checking for positive voltage at the load in the circuit. This is the best technique to use. If voltage is found here (in the case of the picture, it is), then move to test the negative (ground) side of the circuit. This will help you determine which side to start the testing on.

Let's use the same circuit. Look at Figure 9-23. It is drawn with an open circuit (broken wire). When the DMM is connected to the same place, it reads 0 volts. This helps us understand that the problem is on the power side. *Note:* Look at both circuits and notice

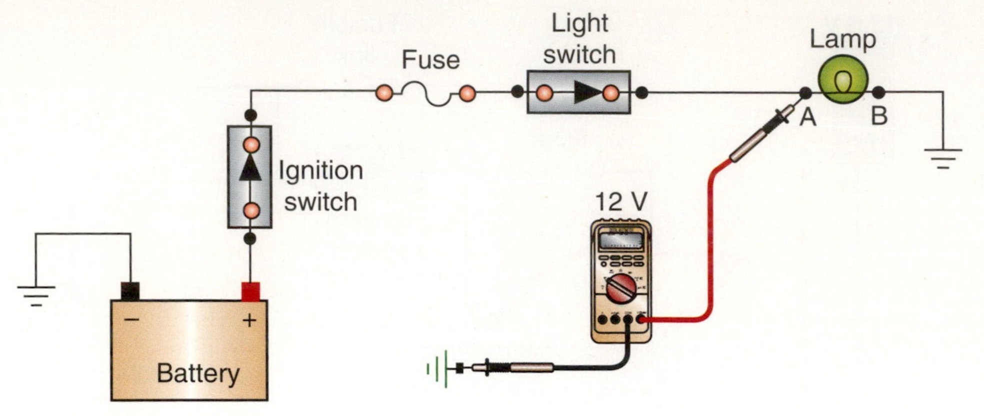

FIGURE 9-22 **Locating an open by testing for voltage.**

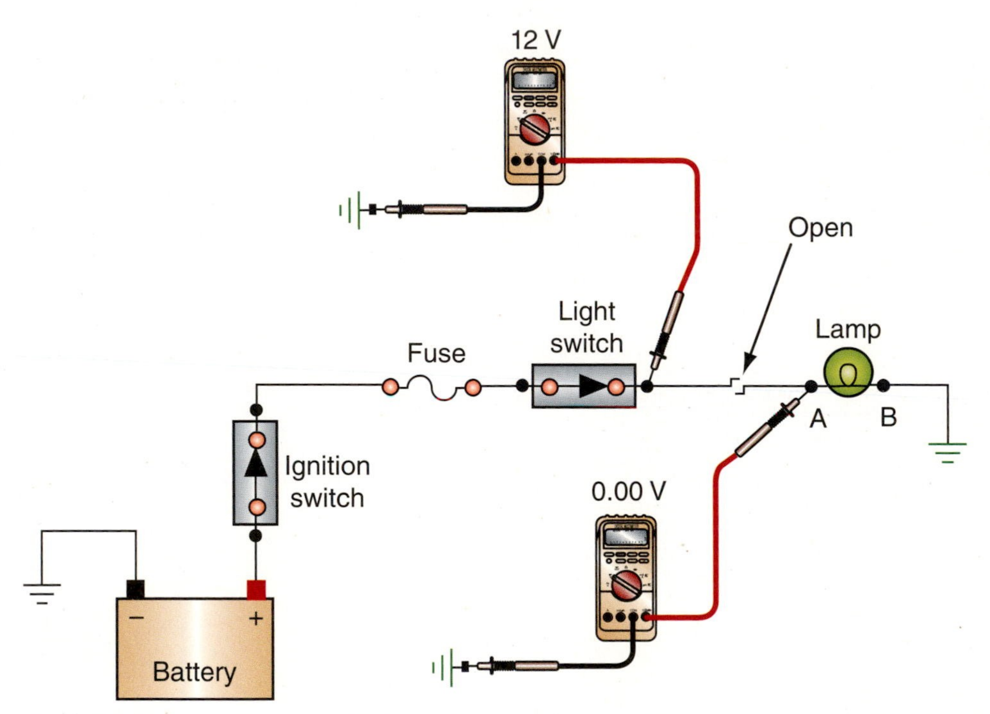

FIGURE 9-23 **An open is present between the point where voltage was measured and where it was not.**

that both of them have the switch closed. In this case, the switch must be turned on in order to properly test the circuit. Figures 9-24 and 9-25 show the normal and abnormal (open) operation of the same circuit.

Diagnosing a Short Circuit

A short circuit is when the current has found a lower resistance pathway (a shorter pathway). It can result when two power-side copper wires touch each other (short to power) or when two copper wires (one power and one ground) touch each other (short to ground). It can also result when a copper wire touches a metal ground (short to ground). A short to ground is when the current bypasses the component (because the ground circuit has less resistance). This lower resistance increases the current flow in the circuit and usually blows the fuse or circuit breaker. Diagnosing this can be tricky because it is an open circuit (because of the blown fuse). Remember that fuses don't wear out. If they are blown, there is a problem.

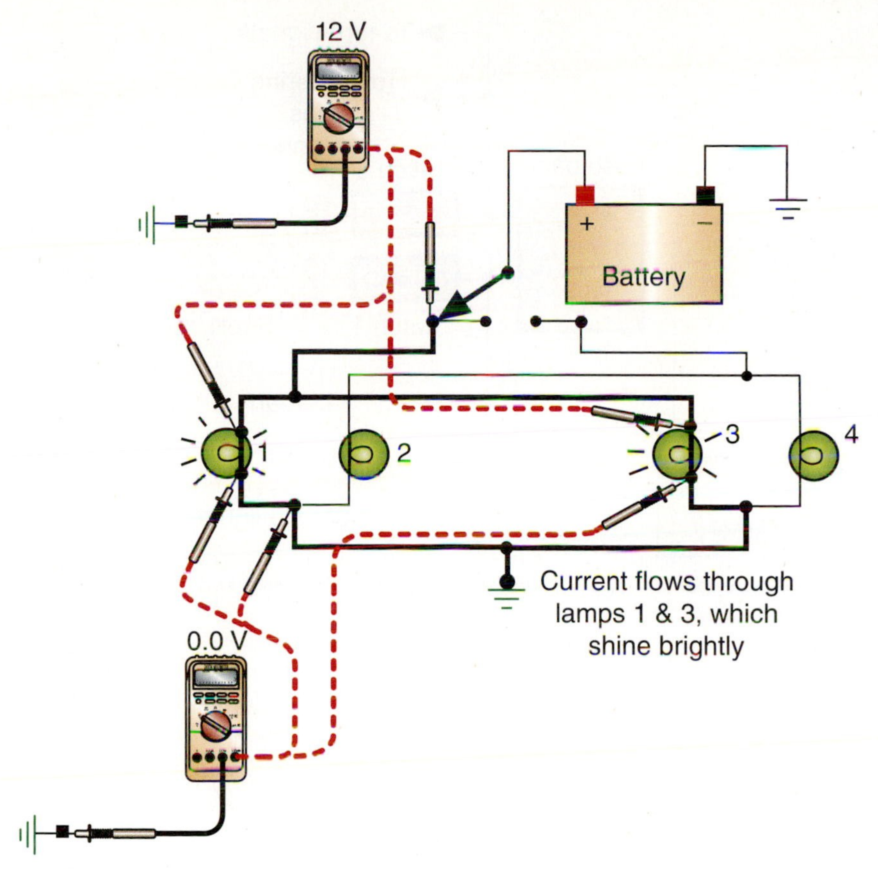

FIGURE 9-24 **Properly operating complex parallel circuit.**

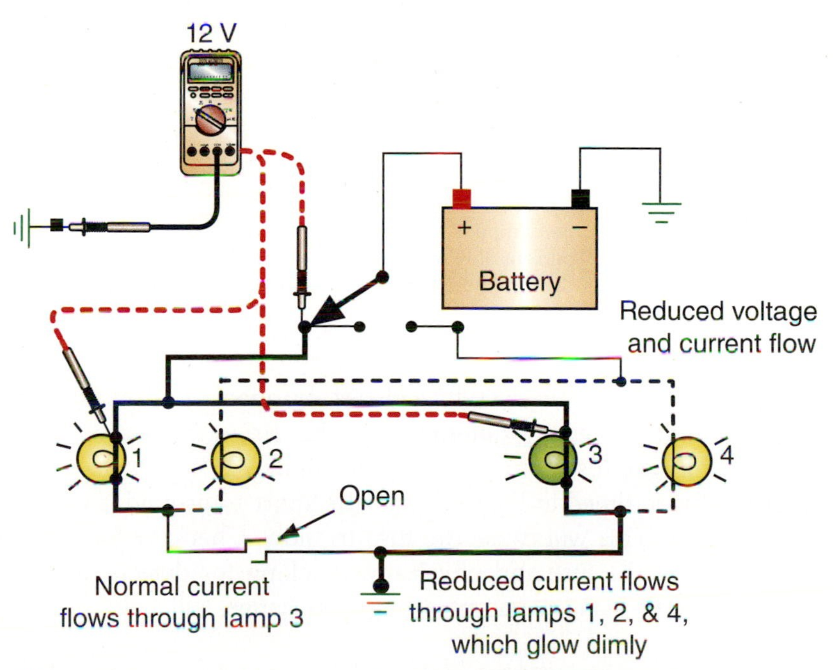

FIGURE 9-25 **An open in the ground circuit can convert the circuit to a series circuit. The dashed black line represents the resulting path to ground.**

A short to power will cause two power wires to bypass a switch in the circuit. This usually makes a component not turn off or not work correctly in different switch positions.

In Figure 9-26 a copper to copper short is shown. The short is circled in the drawing. This type of short is a short to power and will cause both lamps to light when switch 1 is

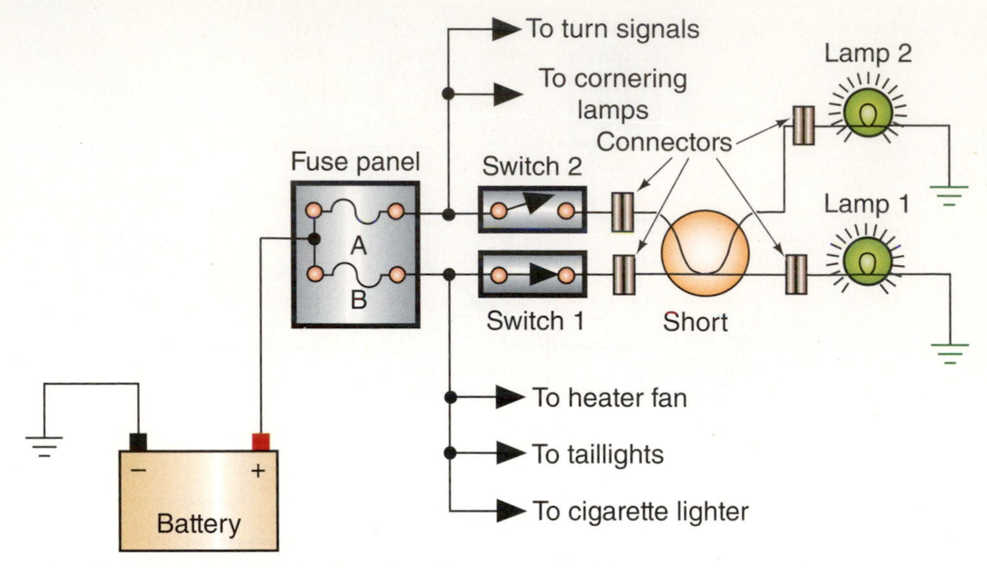

FIGURE 9-26 **A copper-to-copper short (short-to-power) between two circuits.**

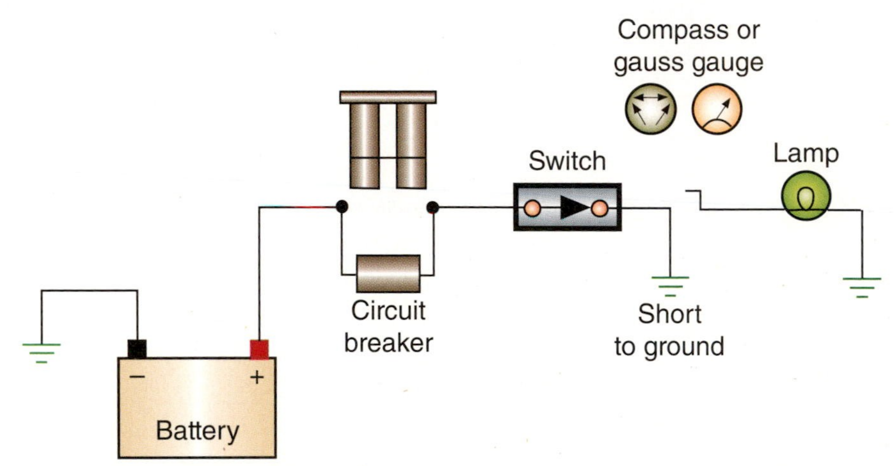

FIGURE 9-27 **The needle of a compass or gauss gauge will fluctuate over the portion of the circuit that has current flowing through it. Once the short to ground has been passed, the needle will stop fluctuating.**

closed and nothing to occur when switch 2 closes or opens. In fact, switch 2 could be removed and it wouldn't make any difference to the circuit's operation. Depending on how much current is drawn, fuse B may or may not blow.

A short to ground is outlined in Figure 9-27. The short to ground is before the component (load) in the circuit. This will cause the fuse to blow. Checking for a short to ground is best done by removing the fuse and physically checking for defects. Replacing the fuse with a circuit breaker and using a compass or gauss gauge will help you find the short (Figure 9-28).

Diagnosing a High Resistance Circuit

If a circuit performs lesser or not as designed, it could have high resistance. A bulb that is dimmer than usual is an example of a circuit with high resistance. You can use a DMM to check for high resistance, but the most effective method is to measure for voltage drop. This method will provide you with a quick and reliable test result.

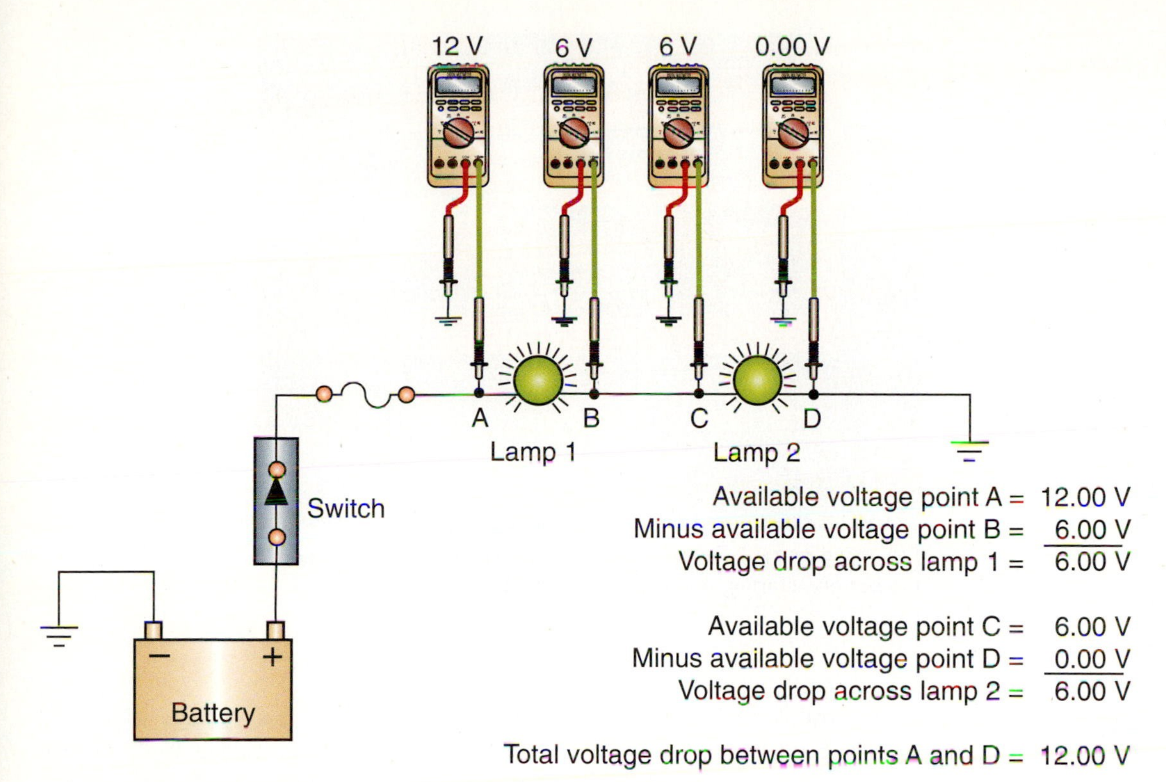

FIGURE 9-28 **Using available voltage to calculate voltage drop over a component. This method is used if the wires of the circuit are too long to test with standard tests leads.**

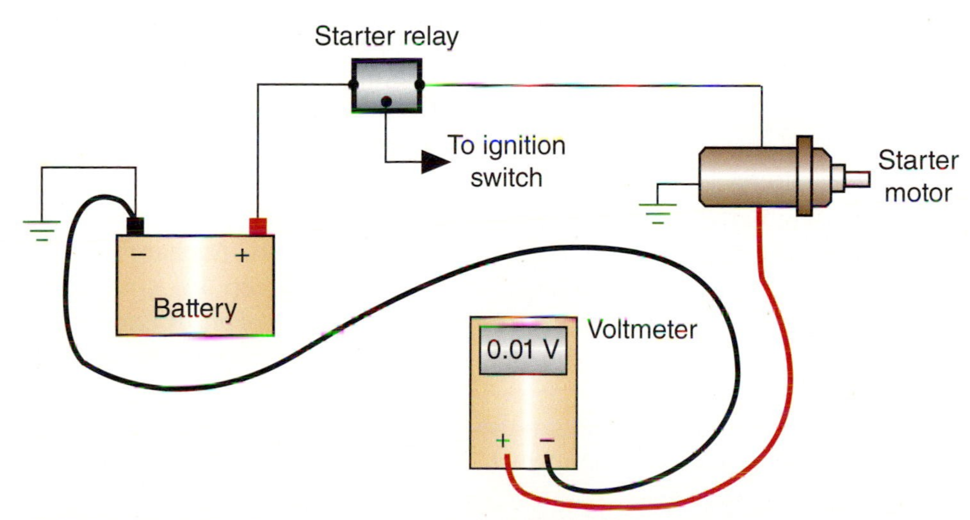

FIGURE 9-29 **Testing the ground side of the starter motor circuit for high resistance by measuring voltage drop. Notice the voltmeter connections.**

The best way to think of voltage drop is to look at Figure 9-29. The DMM is placed across the circuit and the values add up to the same voltage as the battery. When testing for high resistance you are usually testing for corrosion and other issues, although a voltage drop test can tell you if you have an open or short circuit.

The most common method of voltage drop testing is depicted in Figure 9-30. This is a test for the ground circuit of the starter. When testing for voltage drop, the circuit must be energized and working. This is why it is so effective of a test. So in the case of

FIGURE 9-30 **This volt-amp tester (VAT) can test the capacity of the battery, starting and charging system.**

Figure 9-30, the starter must be cranking. The steps for performing a voltage drop test can be found in Photo Sequence 21 (using a voltage drop test to locate high circuit resistance).

Inspecting, Servicing, and Testing the Battery

Battery Safety

Understanding batteries is important, but you must first understand how to handle them safely. There are many types of batteries that can be used in vehicles, but the most common type of battery used is the 12-volt lead-acid type. These batteries have metal plates inside that are made of lead and lead dioxide. In order to create the chemical reaction needed to produce electricity, an electrolyte acid is used.

The acid inside a battery can be unsafe if not handled properly. You must take care not to spill the battery, turn it upside down, or overcharge it. A charging battery will emit hydrogen gas from its vents. This gas is explosive, so keep any sparks away from it. You should also not charge or jump start a battery that is frozen. A battery will freeze when it becomes discharged and cold. Charging a frozen battery could cause it to explode.

When handling a battery you should also take care not to touch the acid with your skin. Wear the proper safety equipment, such as gloves. Acid will irritate your skin, and it can cause holes in your clothing.

You also need to be aware of the high-voltage batteries in hybrid and electric vehicles. These batteries have a charge level on them that could be lethal. More information about hybrid and electric battery safety is covered in Chapter 14.

Battery Testing

Because of its importance, the battery should be checked whenever the vehicle is brought into the shop for service. A faulty battery can cause engine performance problems, engine starting problems, or leave a customer stranded. Checking the battery regularly can save diagnostic time and prevent vehicle breakdowns. By performing a battery test series, we can determine from the state of charge and output capacity of the battery if it is good, in need of recharging, or must be replaced.

PHOTO SEQUENCE 21

Using a Voltage Drop Test to Locate High Circuit Resistance

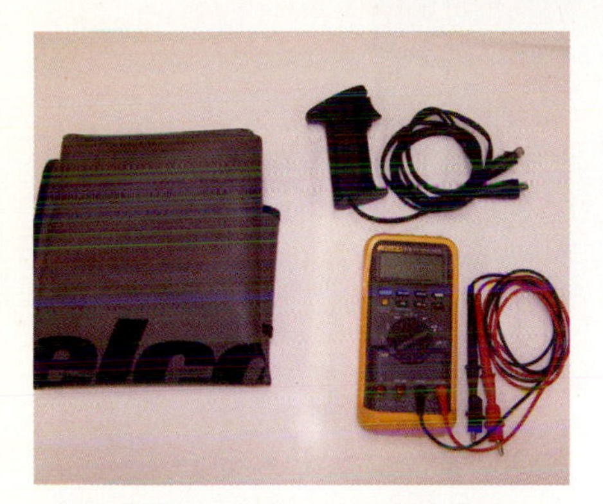

P21-1 Tools required to test for excessive resistance in a starting circuit are fender covers, a digital volt ohm meter (DVOM), and a remote starter switch.

P21-2 Connect the positive lead of the meter to the positive battery post. If possible, do not connect the lead to the cable clamp (terminal).

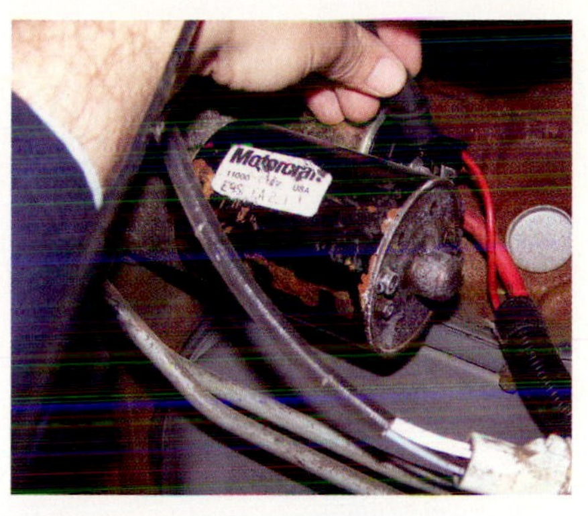

P21-3 Connect the negative lead of the meter to the main battery terminal on the starter motor.

P21-4 To conduct a voltage drop test, current must flow through the circuit. In this test, the ignition system is disabled and the engine is cranked using a remote starter switch.

P21-5 With the engine cranking, read the voltmeter. The reading is the amount of voltage drop.

P21-6 If the reading is out of specifications, test at the next connection toward the battery. In this instance, the next test point is the starter side of the relay.

P21-7 Crank the engine and touch the negative test lead to the starter side of the relay. Observe the voltmeter while the engine is cranking.

P21-8 Test in the same manner on the battery side of the relay. This is the voltage drop across the positive circuit from the battery to the relay.

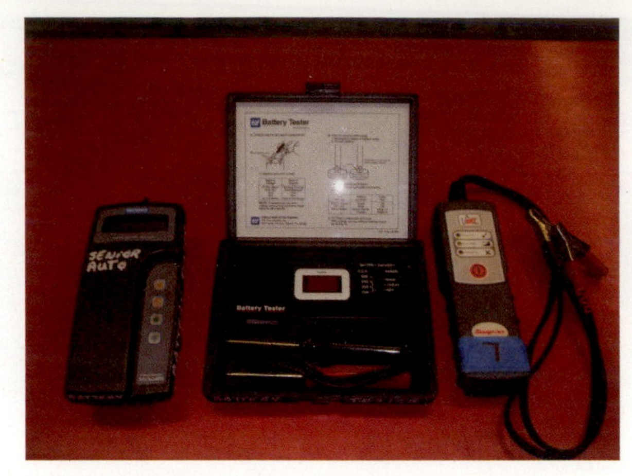

FIGURE 9-31 **An assortment of battery testers.**

There are many different manufacturers of test equipment designed for testing the battery (Figure 9-31). Always follow the procedures given by the manufacturer of the equipment you are using.

SPECIAL TOOLS

Fender covers

Safety glasses

Battery Inspection

Before performing any electrical tests, the battery should be inspected, along with the cables and terminals. The complete visual inspection of the battery will include the following items:

1. *Battery date code.* This provides information as to the age of the battery.
2. *Condition of battery case.* Check for dirt, grease, and electrolyte condensation. Any of these contaminants can create an electrical path between the terminals and cause the battery to drain. Also check for damaged or missing vent caps and cracks in the case. A cracked or buckled case could be caused by excessive tightening of the hold-down fixture, excessive underhood temperatures, buckled plates from extended overcharged conditions, freezing, or excessive charge rate.
3. *Electrolyte level, color, and odor.* If necessary, add distilled water to fill to 1/2-inch level above the top of the plates. After adding water, charge the battery before any tests are performed. Discoloration of electrolyte and the presence of a rotten egg odor indicate an excessive charge rate, excessive deep cycling, impurities in the electrolyte solution, or an old battery.

SPECIAL TOOLS

Safety glasses

Voltmeter

Terminal pliers

Terminal puller

Terminal and clamp cleaner

Fender covers

WARNING: Do not attempt to pry the caps off of a maintenance-free battery; this can ruin the case. You cannot check the electrolyte level or condition on maintenance-free batteries.

WARNING: Always wear safety glasses when working around a battery. Electrolyte can cause severe burns and permanent damage to the eye.

4. *Condition of battery cables and terminals.* Check for corrosion, broken clamps, frayed cables, and loose terminals (Figure 9-32).
5. *Battery abuse.* This includes the use of bungee cords and 2 × 4s for hold-down fixtures, too small a battery rating for the application, and obvious neglect to periodic maintenance. In addition, inspect the terminals for indications that they have been hit upon by a hammer and for improper cable removal procedures; also check for proper cable length.

FIGURE 9-32 **These battery connections could easily cause a no-start condition.**

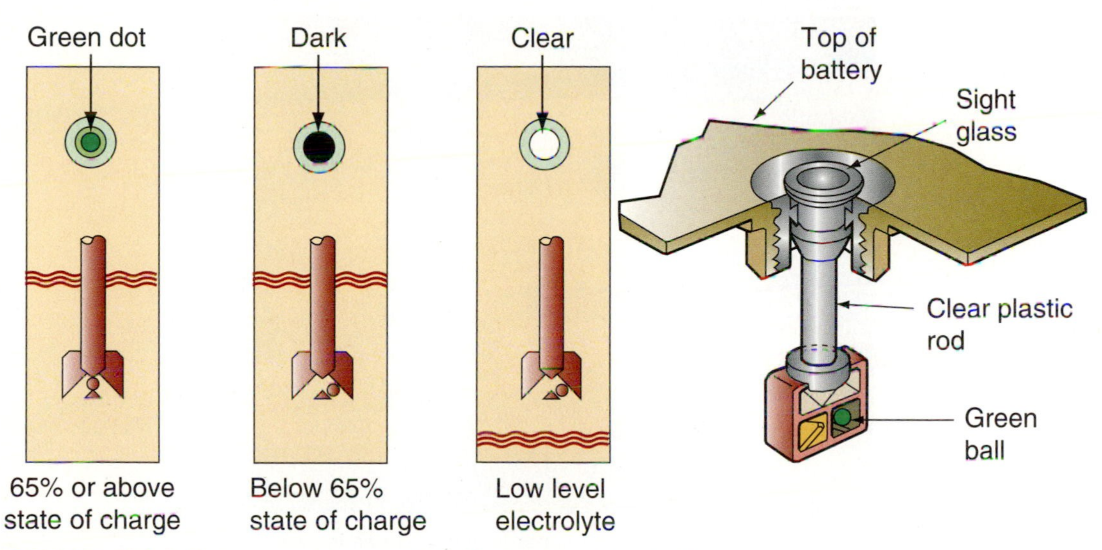

FIGURE 9-33 **A built-in hydrometer indicates the battery state of charge.**

6. *Battery tray and hold-down fixture.* Check for proper tightness. Also check for signs of acid corrosion of the tray and hold-down unit. Replace as needed.
7. If the battery has a built-in hydrometer, check its color indicator (Figure 9-33).

Green indicates the battery has a sufficient charge, black indicates the battery requires charging before testing, clear indicates low electrolyte level, and yellow indicates the battery should be replaced.

When the battery and cables have been completely inspected and any problems corrected, the battery is ready to be tested further. For many of the tests to be accurate, the battery must be fully charged.

A **conductance test** sends a low-current AC signal into the battery. The return signal is then analyzed.

SERVICE TIP: Grid growth can cause the battery plate to short out the cell. If there is normal electrolyte level in all cells but one, that cell is probably shorted.

Battery Testing Series

Battery Conductance Test

A newer method of testing a battery is called a **conductance test**. A conductance tester will test the battery's ability to conduct electricity (Figure 9-34). This is a good indicator of a battery's health and whether it will be able to provide adequate service when installed

FIGURE 9-34 **This type of tester will test the conductance of the battery. It is quick and can usually be done in less than a minute.**

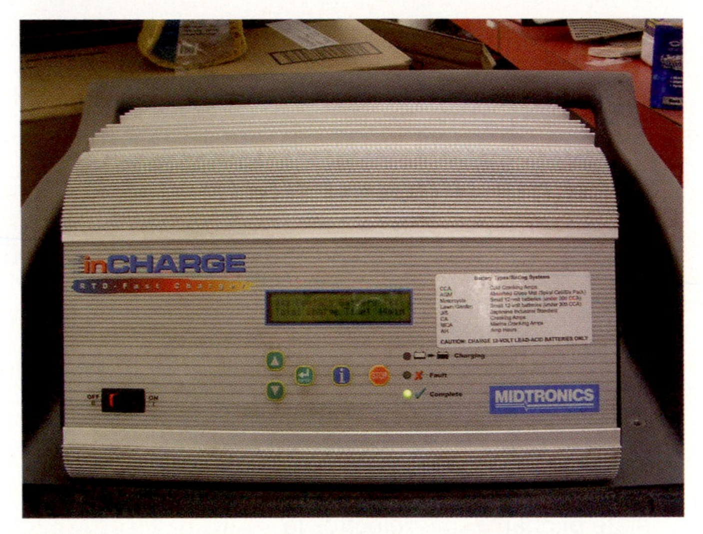

FIGURE 9-35 **This Midtronics tester is becoming an industry standard.**

in the vehicle. A conductance test is considered to be more accurate than the previous method of testing batteries. During a conductance test, the tester will send a low-current AC signal into the battery. The return signal that is generated is then analyzed. Some conductance battery testers also include a battery charger. These types of testers can recharge a battery if it is not ready to be tested accurately (Figure 9-35). During recharging, the tester can measure key items that will indicate if the battery will be strong enough to hold a good charge when it is done.

Battery Capacity Test

A **capacity (load) test** uses a carbon pile to draw energy out of the battery. The voltage level of the battery should not drop below 9.6 volts during the test.

A **capacity test** (also known as a load test) is still a valid battery test, but it has been proven to be less accurate in determining if a battery will provide good service. A carbon pile tester is used to perform this test (Figure 9-36). There are many types and brands of testers available, but all of them have a voltmeter and an ammeter. A load test machine works by connecting the high current clamps of the tester to the battery terminals and the inductive pickup lead (amp clamp) to the negative tester cable. The battery must be fully charged before performing the test. For best results, the battery electrolyte should be

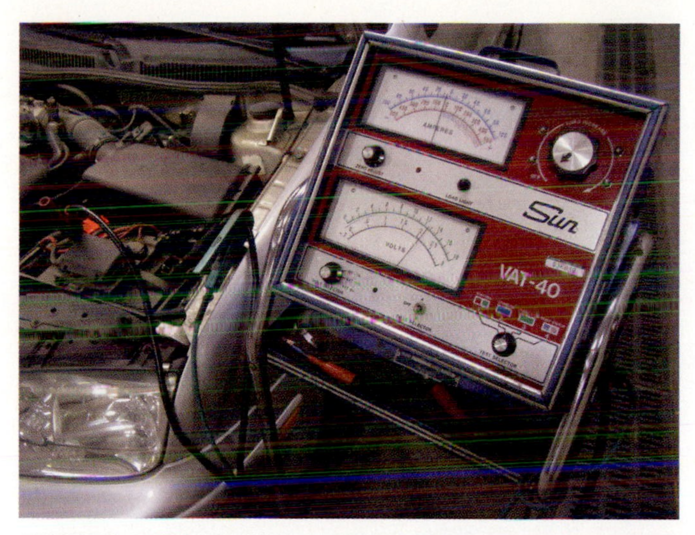

FIGURE 9-36 **This is a VAT machine. It contains a carbon pile in the backside and can perform a capacity (load) test.**

close to 80°F (26.7°C). The tester then draws half the battery's cold-cranking ampere (CCA) rating from the battery for 15 seconds during engine cranking. On some testers this is done automatically and on others there is a control knob for the amperage. If the voltmeter does not drop below 9.6 volts during the 15-second test, then the battery passes the load test.

The **battery terminal voltage drop test** checks for poor electrical connections between the battery cables and terminals.

Battery Terminal Voltage Drop Test

This simple test will establish the condition of the terminal connection. It is a good practice to perform this test any time the battery cables are cleaned or disconnected and reconnected to the terminals. By performing this test, comebacks due to loose or faulty connections can be reduced.

Set your DMM to DC volts. Connect the negative test lead to the cable clamp, and connect the positive meter lead to the battery terminal. Disable the ignition system to prevent the vehicle from starting. This may be done by removing the ignition coil primary wires (Figure 9-37). (You can also remove the ignition system fuse, powertrain control module [PCM] fuse, or fuel pump fuse.) On vehicles with **distributor**-less ignition, disconnect the coil primary wires (these are the small gauge wires connected to the coils).

SPECIAL TOOLS

Digital multimeter

Fender covers

Safety glasses

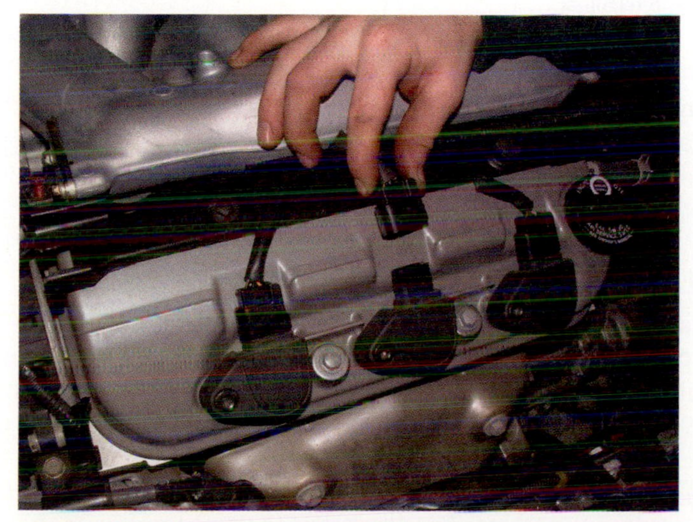

FIGURE 9-37 **Disconnecting the coil(s) primary connections to prevent the engine from starting and to protect the coil.**

FIGURE 9-38 **Use battery terminal pullers to remove the clamp from the battery post. Do not pry the clamp.**

FIGURE 9-39 **The terminal cleaning tool is used to clean the clamp and terminal.**

The **open circuit voltage test** is used to determine the battery's state of charge.

CAUTION:
Always refer to the manufacturer's service manual for the correct procedure for disabling the ignition system. Using the wrong procedure may damage ignition components.

Crank the engine and observe the voltmeter reading. If the voltmeter shows over 0.5 volt, there is a high resistance at the cable connection. Remove the battery cable using the terminal puller (Figure 9-38). Clean the cable ends and battery terminals or *replace* as needed (Figure 9-39).

Open Circuit Voltage Test

To obtain accurate test results, the battery must be stabilized. If the battery has just been recharged, perform the capacity test, then wait at least 10 minutes to allow battery voltage to stabilize. Connect a voltmeter across the battery terminals, observing polarity (Figure 9-40). Measure the open circuit voltage, taking the reading to the one-tenth volt.

To analyze the open circuit voltage test results, consider that a battery at a temperature of 80°F, in good condition, should show at least 12.45 volts (Figure 9-41). If the state of charge is 75 percent or more, the battery is considered charged (Figure 9-42).

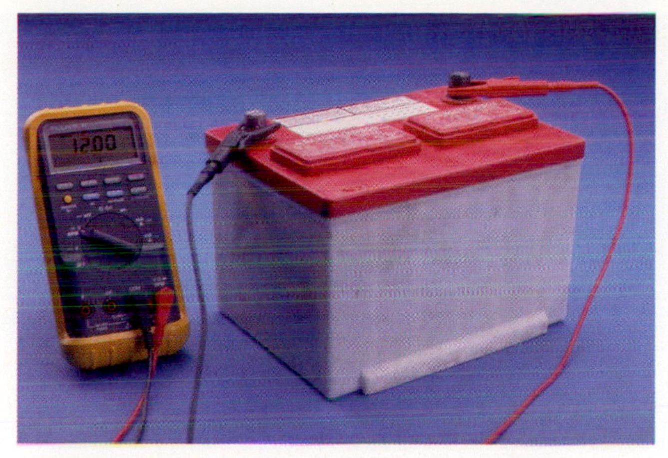

FIGURE 9-40 **Open circuit voltage test using a DMM.**

Open Circuit Voltage	State of Charge
12.6 or greater	100%
12.4–12.6	70–100%
12.2–12.4	50–70%
12.0–12.2	25–50%
0.0–12.0	0–25%

FIGURE 9-41 **The results of the open circuit voltage test indicate the state of charge.**

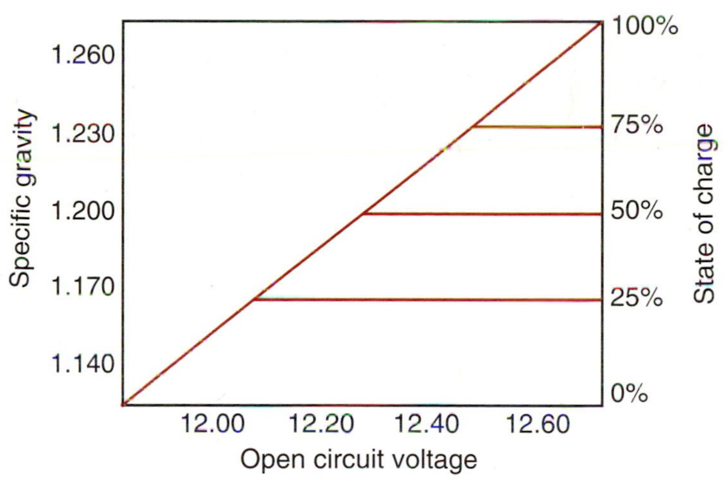

FIGURE 9-42 **Open circuit voltage test results relate to the specific gravity of the battery cells.**

Battery Case Test

A dirty battery can allow voltage to leak across the top of the battery and cause discharging. To test a battery for this condition, place the positive voltmeter lead on a battery post. Place the negative lead at spots across the top of the battery. Repeat the test on the other battery post. Any reading above 0.5 volt indicates that the battery is dirty and allows discharging. Clean the top of the battery with a mixture of baking soda and water or use a battery cleaning solvent. Retest to verify your repair.

SPECIAL TOOLS

Volt-ampere tester (VAT)
Electrolyte thermometer
Safety glasses
Fender covers

Absorbed Glass Mat Battery Precautions

An absorbed glass mat (AGM) battery is a lead-acid battery that is sealed using special pressure valves. This is truly a maintenance-free battery that cannot and should never be opened. Overcharging is especially harmful to AGM batteries and can shorten their service life. You should use a good battery charger that does not allow more than 16 volts, usually the manufacture of the battery, or the dealership, will have a special charger. A conductance tester is the only type of tester that will work on AGM batteries. You need to select that you are testing an AGM battery when using the tester (Figure 9-43).

FIGURE 9-43 **AGM batteries have to be tested using a conductance tester.**

SPECIAL TOOLS

Battery charger

Voltmeter

Safety glasses

Fender covers

CAUTION:
Some engines turn counterclockwise as viewed from the front crankshaft pulley. These engines are rare, but damage may occur if they are turned the other way. Check the service manual for warnings or ask your instructor.

Hydrostatic lock is the result of attempting to compress a liquid in the cylinder. Since liquid is not compressible, the piston is not able to travel in the cylinder.

Inspecting, Servicing, and Testing the Starter and Circuit

The starter motor must be capable of rotating the engine fast enough so it can start and run under its own power. The starting system is a combination of mechanical and electrical parts working together to start the engine. The starting system includes the following components:

1. Battery
2. Cable and wires
3. Ignition switch
4. Starter solenoid or relay
5. Starter motor
6. Starter drive and flywheel ring gear
7. Starting safety switch

Starting System Troubleshooting

Customer complaints concerning the starting system generally fall into four categories: no-crank, slow cranking, starter spins but does not turn the engine, and excessive noise. As with any electrical system complaint, a systematic approach to diagnosing the starting system will make the task easier. First, the battery must be in good condition and fully charged. Perform a complete battery test series to confirm the battery's condition. Many starting system complaints are actually attributable to battery problems. If the starting system tests are performed with a weak battery, the results can be misleading, and the conclusions reached may be erroneous and costly.

Before performing any test on the starting system, begin with a visual inspection of the circuit. Repair or replace any corroded or loose connections, frayed wires, or any other trouble sources. The battery terminals must be clean, and the starter motor must be properly grounded.

The diagnostic chart shows a logical sequence to follow whenever a starting system complaint is made (Figure 9-44). The tests to be performed are determined by whether or not the starter will crank the engine.

If the customer complains of a no-crank situation, attempt to rotate the engine by the crankshaft pulley nut. Rotate the crankshaft in a clockwise direction through two full rotations, using a large socket wrench. If the engine does not rotate, it may be seized due to operating with lack of oil, **hydrostatic lock**, or broken engine components.

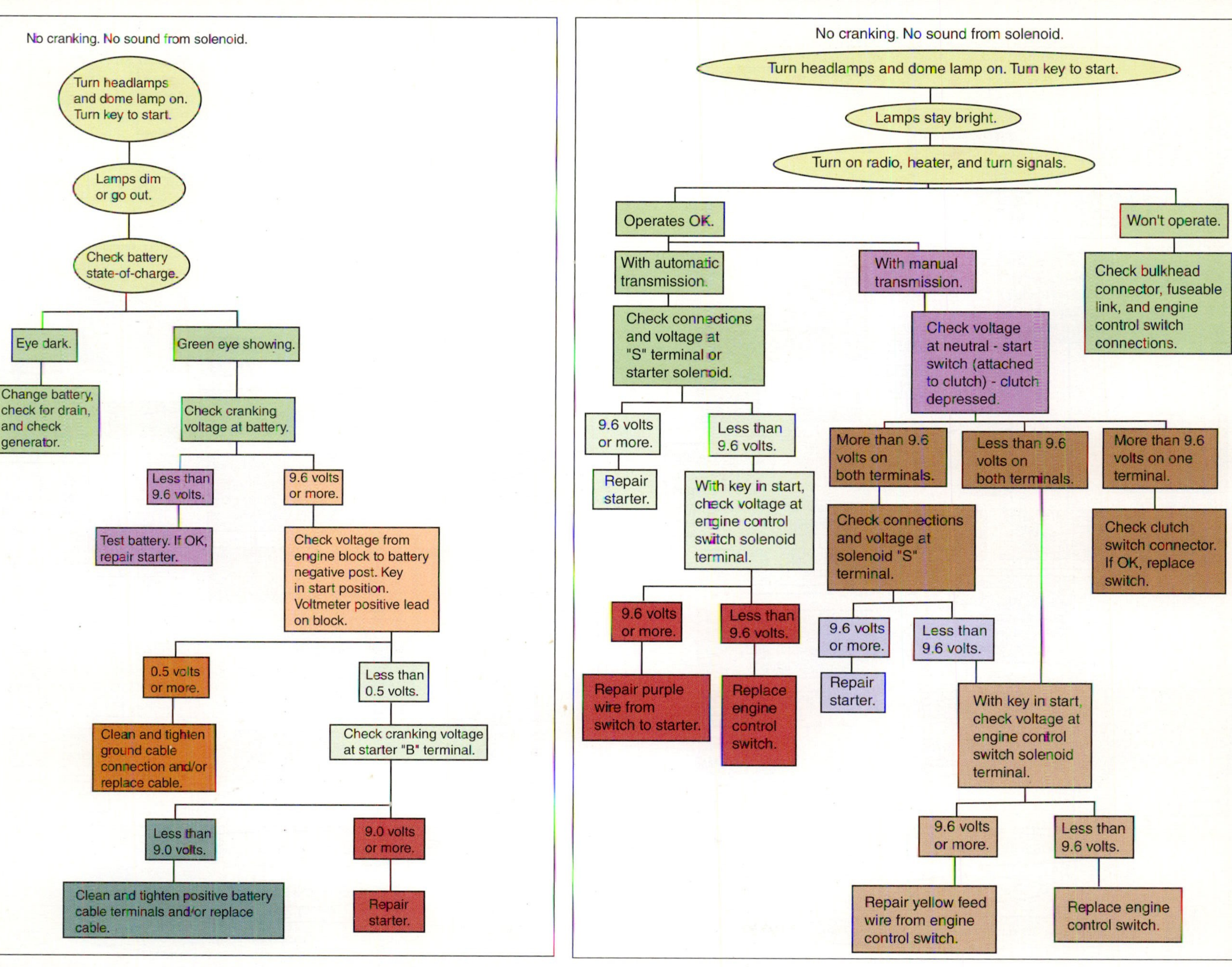

FIGURE 9-44 **Diagnostic chart used to determine starting system problems.**

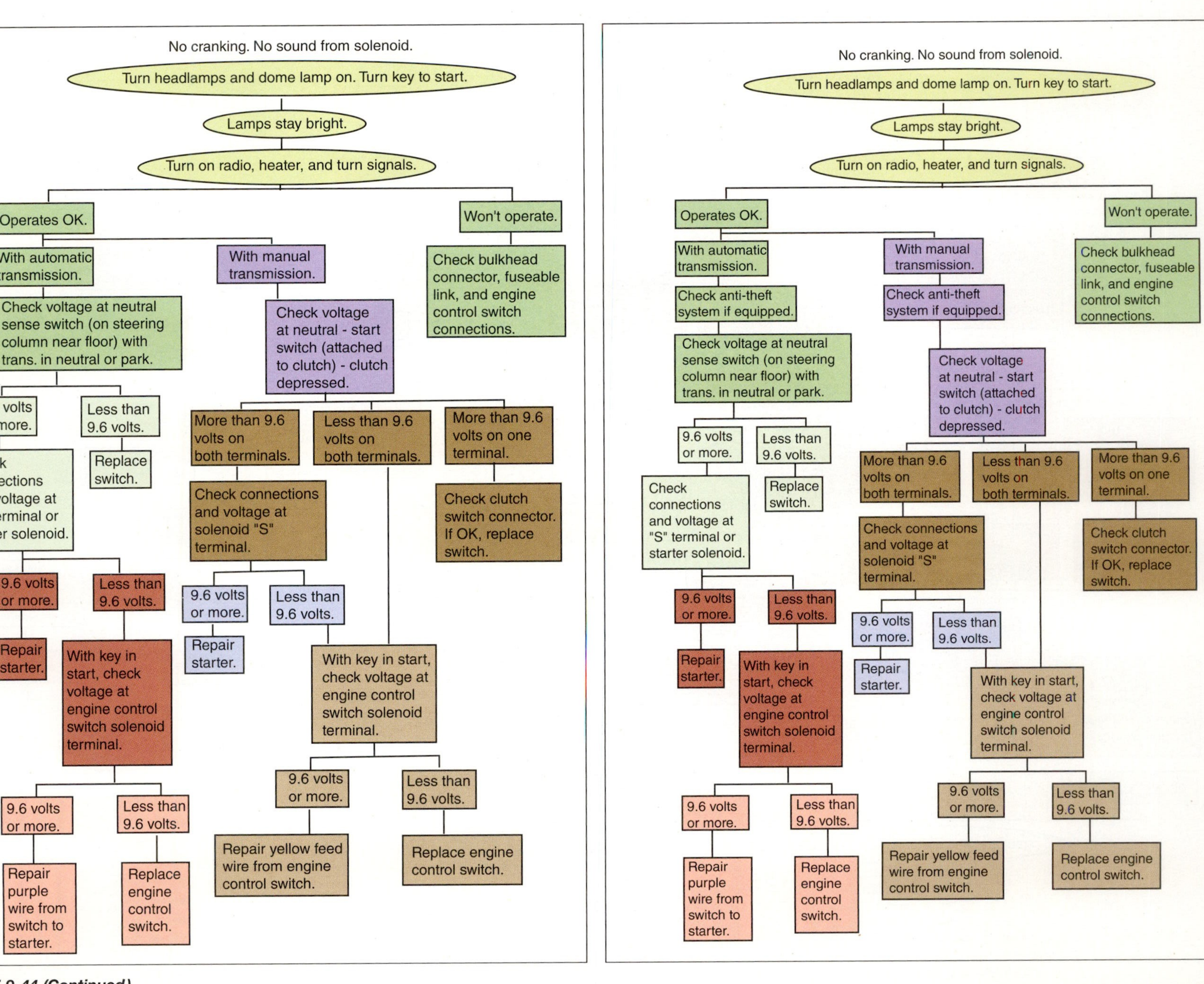

FIGURE 9-44 (Continued)

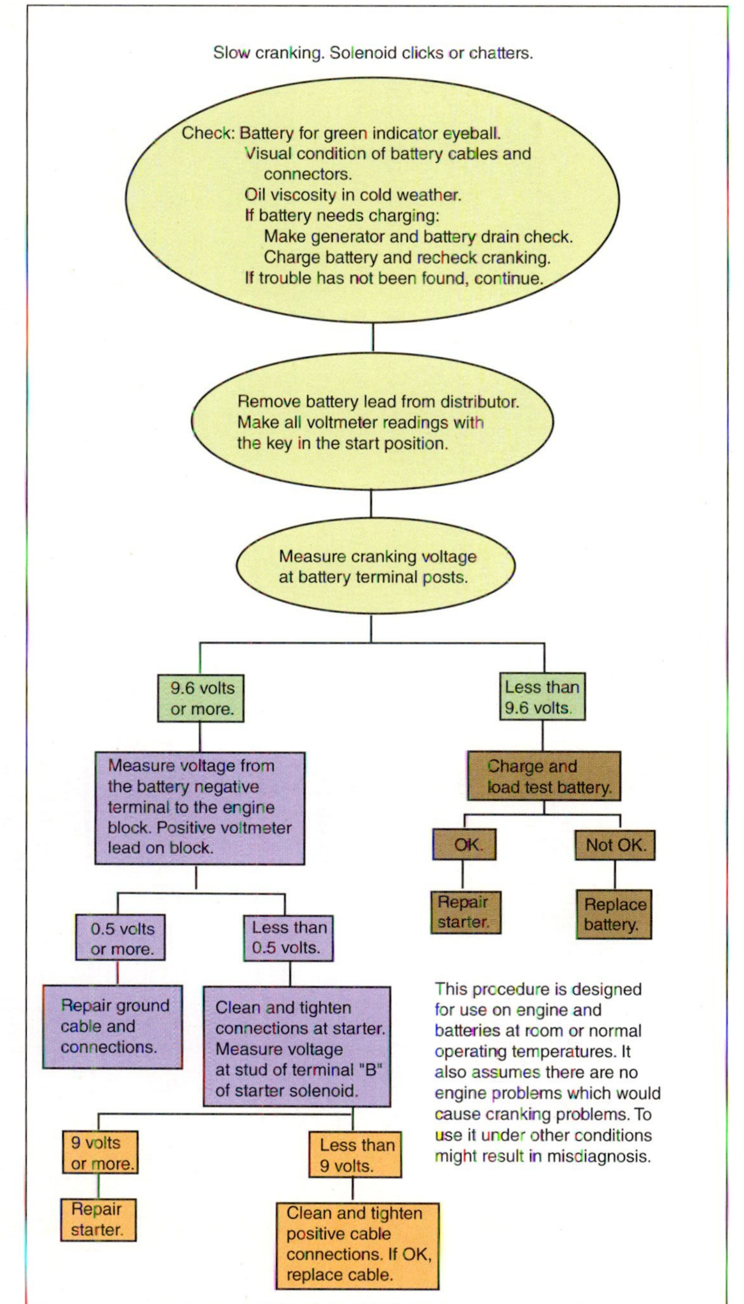

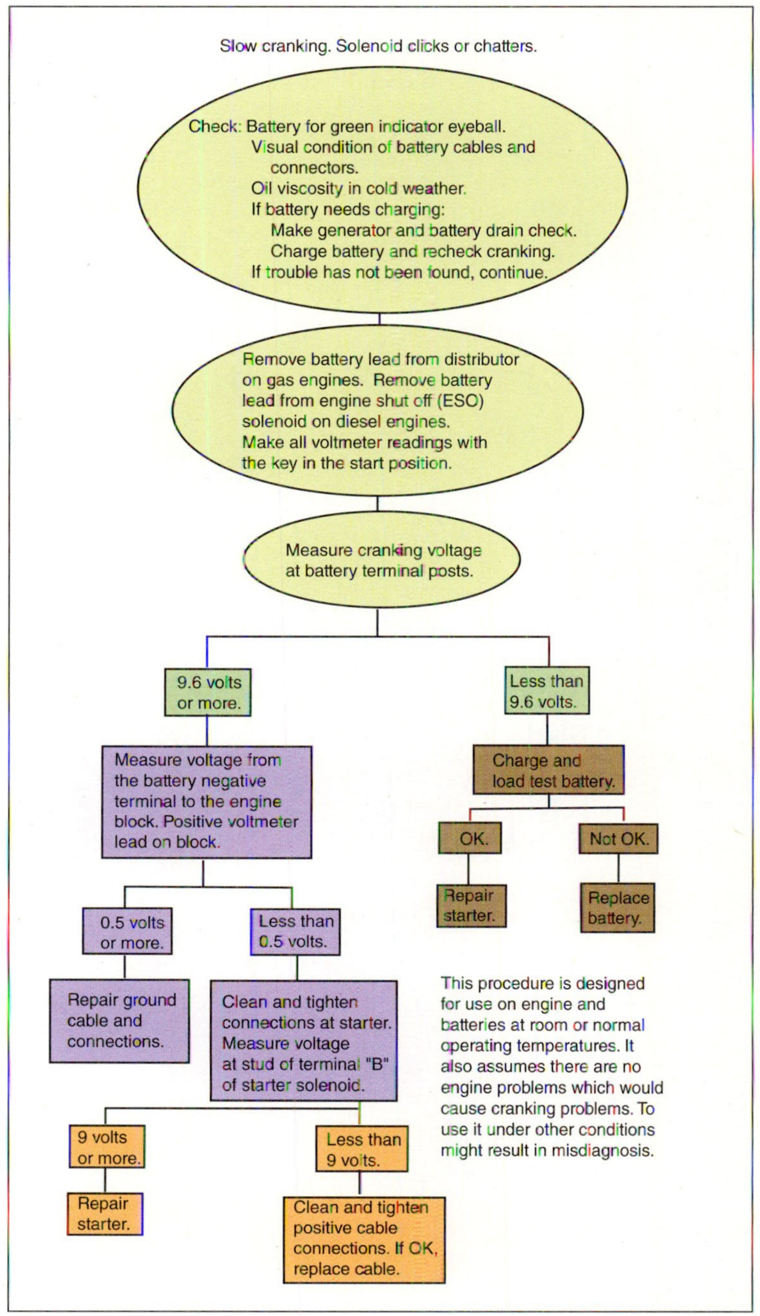

FIGURE 9-44 (Continued)

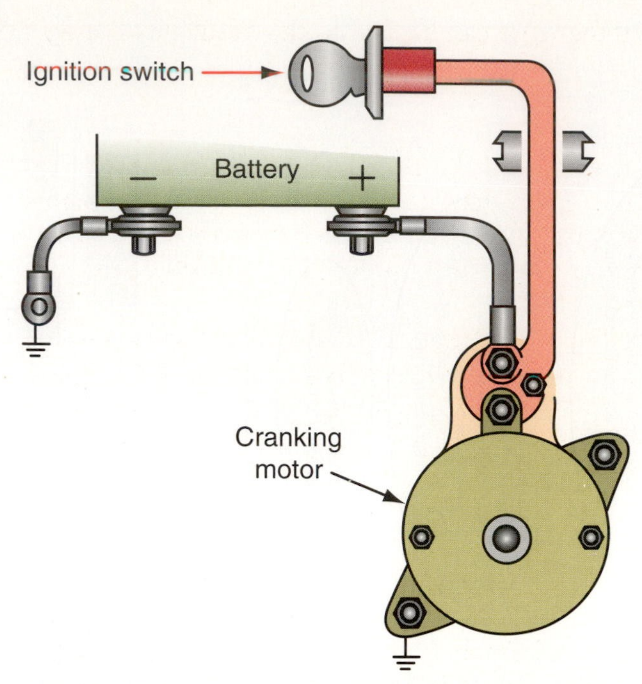

FIGURE 9-45 **Excessive wear, loose electrical connections, or excessive voltage drop in any of these areas can cause a slow crank or no-crank condition.**

A slow crank or no-crank complaint can be caused by several potential trouble spots in the circuit (Figure 9-45). Excessive voltage drops in these areas will cause the starter motor to operate slower than required to start the engine. The speed with which the starter motor rotates the engine is important to starting the engine. If the speed is too slow, compression is lost and the air/fuel mixture draw is impeded. Most manufacturers require a speed of approximately 250 rpm during engine cranking.

If the starter spins but the engine does not rotate, the most likely cause is a faulty starter drive. If the starter drive is at fault, the starter motor will have to be removed to install a new starter or drive mechanism. Before faulting the starter drive, also check the starter ring gear teeth for wear or breakage, and for incorrect gear mesh of the ring gear and starter motor pinion gear.

Most noises can be traced to the starter drive mechanism. The starter drive can be replaced as a separate component of the starter.

SPECIAL TOOLS

Fender covers

Battery cable puller

The **starter quick test** will isolate the problem area; that is, if the starter motor, solenoid, or control circuit is at fault.

Testing the Starting System

As with the battery testing series, the tests for the starting system are performed on the volt-ampere tester (VAT). Since the starter performance and battery performance are so closely related, it is important that a full battery test series be done before trying to test the starter system. If the battery fails the load test and is fully charged, it must be replaced before doing any other tests.

Starter Quick Testing

If the starter does not turn the engine at all and the engine is in good mechanical condition, the **starter quick test** can be performed to locate the problem area. To perform this test, make sure the transmission is in neutral, and set the parking brake. Turn on the headlights. Next, turn the ignition switch to the START position while observing the headlights.

There are three things that can happen to the headlights during this test:

1. They will go out.
2. They will dim.
3. They will remain at the same brightness level.

If the lights go out completely, the most likely cause is a poor connection at one of the battery terminals. Check the battery cables for tight and clean connections. It will be necessary to remove the cable from the terminal and clean the cable clamp and battery terminals of all corrosion.

If the headlights dim when the ignition switch is turned to the START position, the battery may be discharged. Check the battery condition. If it is good, then there may be a mechanical condition in the engine preventing it from rotating. If the engine rotates when turning it with a socket wrench on the pulley nut, the starter motor may have internal damage. A bent starter armature, worn bearings, thrown armature windings, loose pole shoe screws, or any other worn component in the starter motor that will allow the armature to drag can cause a high current demand.

If the lights stay brightly lit and the starter makes no sound (listen for a deep clicking noise), there is an open in the circuit. The fault is in either the solenoid or the control circuit. To test the solenoid, bypass the solenoid by bridging the BAT and S terminals on the back of the solenoid. The B1 terminal is a large terminal on the solenoid with a heavy gauge wire coming from the battery. Also on the solenoid, the S terminal has a smaller gauge wire coming to it from the ignition switch or starter relay. Use heavy jumper cables to connect the two terminals. If the starter spins, the ignition switch, starter relay, or control circuit wiring is faulty. If the starter does not spin, the starter motor or the solenoid is faulty.

SERVICE TIP: If the engine does not crank and the headlights do not come on, check the fusible link.

The **current draw test** measures the amount of current the starter draws when actuated. It determines the electrical and mechanical condition of the starting system.

WARNING: The starter will draw up to 400 amperes. The tool used to jump the terminals must be able to carry this high current and it must have an insulated handle. A jumper cable that is too light may become extremely hot, resulting in burns to the technician's hand.

SPECIAL TOOLS

Fender covers

VAT-40 Jumper wires

Current Draw Test

If the starter motor cranks the engine, the technician should perform the **current draw test**. The following procedure is used for volts or amperes tester:

1. Connect the large red and black test leads on the battery posts, observing polarity.
2. Zero the ammeter.
3. Connect the ampere inductive probe around the battery ground cable. If more than one ground cable is used, clamp the probe around all of them (Figure 9-46).
4. Make sure all loads are turned off (lights, radio, etc.).
5. Disable the ignition system to prevent the vehicle from starting. This may be done by removing the ignition coil secondary wire from the distributor cap and putting it to ground, by removing the ignition system fuse, fuel pump fuse, PCM fuse, or by disconnecting the primary wires from the coils of an electronic ignition (EI) system.
6. Crank the engine and note the voltmeter reading.
7. Read the ammeter scale to determine the amount of current draw.

After recording the readings from the current draw test, compare them with the manufacturer's specifications. If specifications are not available, as a rule, correctly functioning systems will crank at 9.6 volts or higher. Amperage draw is dependent upon engine size. Most V8 engines have an ampere draw of about 200 amperes, six-cylinder engines about 150 amperes, and four-cylinder engines about 125 amperes.

SERVICE TIP: If the starter windings are thrown, this is an indication of several different problems. The most common is that the driver is keeping the ignition switch in the START position too long after the engine has started. Other causes include the driver opening the throttle plates too wide while starting the engine, resulting in excessive armature speeds

SERVICE TIP:
(*continued*)
when the engine does start. Also the windings can be thrown because of excessive heat buildup in the motor. The motor is designed to operate for very short periods of time. If it is operated for longer than 15 seconds, heat begins to build up at a very fast rate. If the engine does not start after a 15-second crank, the starter motor needs to cool for about 2 minutes before attempting to start the engine again.

Ignition timing refers to the timing of spark plug firing in relation to position of the piston.

CAUTION:
Always refer to the manufacturer's service manual for the correct procedure for disabling the ignition system. Using an improper procedure may damage ignition components.

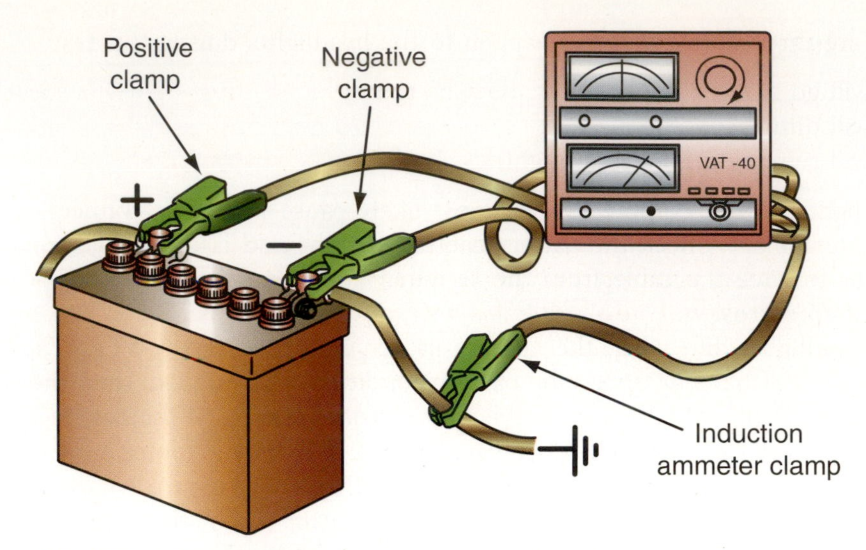

FIGURE 9-46 **Test lead connections to perform the starter current draw test.**

Higher-than-specified current draw test results indicate an impedance to rotation of the starter motor. This includes worn bushings, a mechanical blockage, internal starter motor damage, and excessively advanced **ignition timing**. Lower-than-normal current draw test results indicate excessive voltage drop in the circuit, a faulty solenoid, or worn brushes.

If the current test results are high and the engine does not turn, the starter may not need replacement. Hydrostatic lock or engine seizure may have occurred. In order to check for this, you should put a socket and ratchet on the crankshaft pulley bolt and attempt to turn the engine over manually by hand. If the engine turns easily, it is most likely a problem with the starter. If the engine does not turn, it is an internal engine problem.

To determine the type of internal problem, start by removing the spark plugs from the engine and attempt to turn the crankshaft again. If it starts to turn easier, look for coolant, fuel, or oil that might be coming out of the spark plug holes (Figure 9-47).

These fluids may have gotten into the combustion chamber by a blown head gasket, leaking fuel injector, or a cracked cylinder head or block. If the engine does not turn and no fluids come out of the spark plug hole, the engine may be mechanically seized.

Since the readings obtained from the current draw test were taken at the battery, these readings may not be an exact representation of the actual voltage at the starter motor. Voltage losses due to bad cables, connections, or relays (or solenoids) may diminish the amount of voltage to the starter. Before removing the starter from the vehicle, these should be tested.

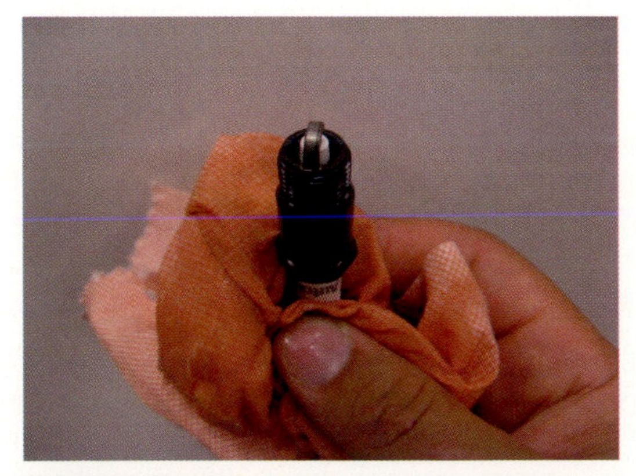

FIGURE 9-47 **This spark plug is fouled with coolant.**

Starter Motor Removal

If the tests indicate the starter motor must be removed, the first step is to disconnect the battery from the system.

WARNING: Remove the negative battery cable. It is a good practice to wrap the cable clamp with tape or enclose it in a rubber hose to prevent accidental contact with the battery terminal. If the positive cable is removed first, the wrench may slip and make contact between the terminal and ground. This action may overheat the wrench and burn the technician's hand.

SPECIAL TOOLS

Fender covers

Jumper cables

The specification for current draw is the maximum allowable, and the specification for cranking voltage is the minimum allowable.

It may be necessary to place the vehicle on a lift to gain access to the starter motor. Before lifting the vehicle, disconnect all wires, fasteners, and so on that can be reached from the top of the engine compartment. Disconnect the wires leading to the solenoid terminals. To prevent confusion, it is a good practice to use pieces of tape to identify the different wires.

CAUTION: Do not operate the starter motor for longer than 15 seconds. Allow the motor to cool between cranking attempts. Operating the starter for more than 15 seconds may overheat and damage starter motor components.

WARNING: Check for proper lift pad-to-frame contact after the vehicle is a few inches above the ground. Shake the vehicle. If there are any unusual noises or movement of the vehicle, lower it and reset the pads. If the lift pads are not properly positioned on the specified vehicle lift points, the vehicle may slip off the lifts, resulting in technician injury and vehicle damage.

On some vehicles, it may be necessary to disconnect the exhaust system to be able to remove the starter motor. Spray the exhaust system fasteners with a penetrating oil to assist in removal. Loosen the starter mounting bolts and remove all but one. Support the starter motor, remove the remaining bolt, and then remove the starter motor.

WARNING: The starter motor is heavy. Make sure that it is secured before removing the last bolt. If the starting motor is not properly secured, it may drop suddenly, resulting in leg or foot injury.

The shims are used to provide proper pinion-to-ring gear clearance.

To install the starter motor, reverse the procedure. Be sure all electrical connections are tight. If you are installing a new or remanufactured starter, remove any paint that may prevent a good ground connection. Be careful not to drop the starter. Make sure that it is supported properly. Refer to Photo Sequence 4 for the typical starter removal and reinstallation.

Some General Motors' starters use shims between the starter motor and the mounting pad (Figure 9-48). To check this clearance, insert a flat blade screwdriver into the access slot on the side of the drive housing. Pry the drive pinion gear into the engaged position. Use a piece of wire that is 0.020 inch in diameter to check the clearance between the gears (Figure 9-49).

If the clearance between the two gears is excessive, the starter will produce a high pitched whine while the engine is being cranked. If the clearance is too small, the starter will make a high pitched whine after the engine starts and the ignition switch is returned to the RUN position.

SERVICE TIP: The major cause of drive housing breakage is too small a clearance between the pinion and ring gears. It is always better to have a little more clearance than too small a clearance.

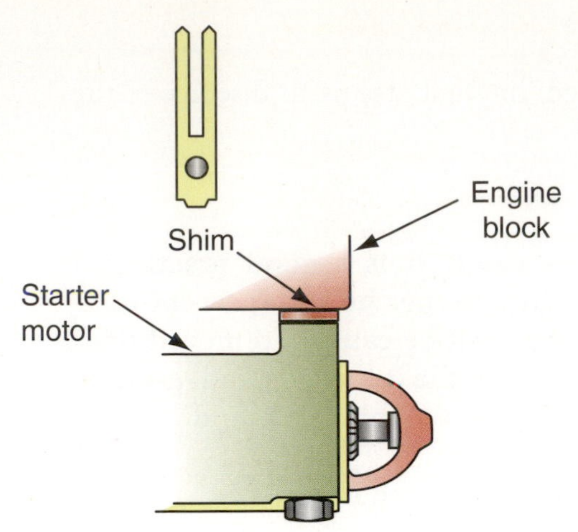

FIGURE 9-48 **Shimming the starter to obtain the correct pinion-to-ring gear clearance.**

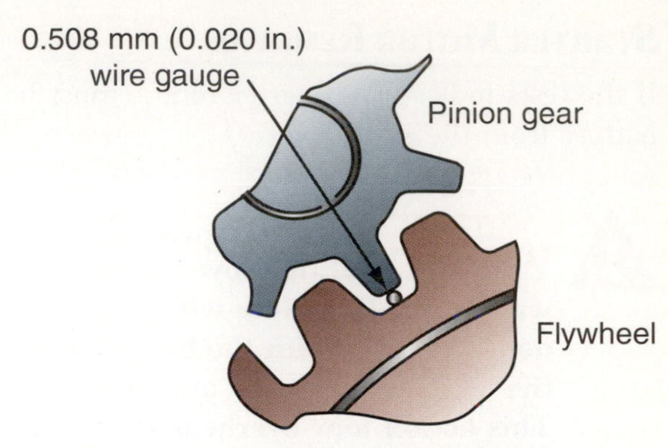

FIGURE 9-49 **Checking the clearance between the pinion gear and ring gear.**

Starter Circuit Tests Using a DMM

At several points in the circuit, you will need to test the individual components and wires. To perform a voltage drop test, follow the steps in Figure 9-50. Remember to treat the starter circuit just like any other circuit. It can have problems with open circuits, short circuits, and high resistance. Two other common designs to be made aware of when testing the circuit are outlined in Figure 9-51.

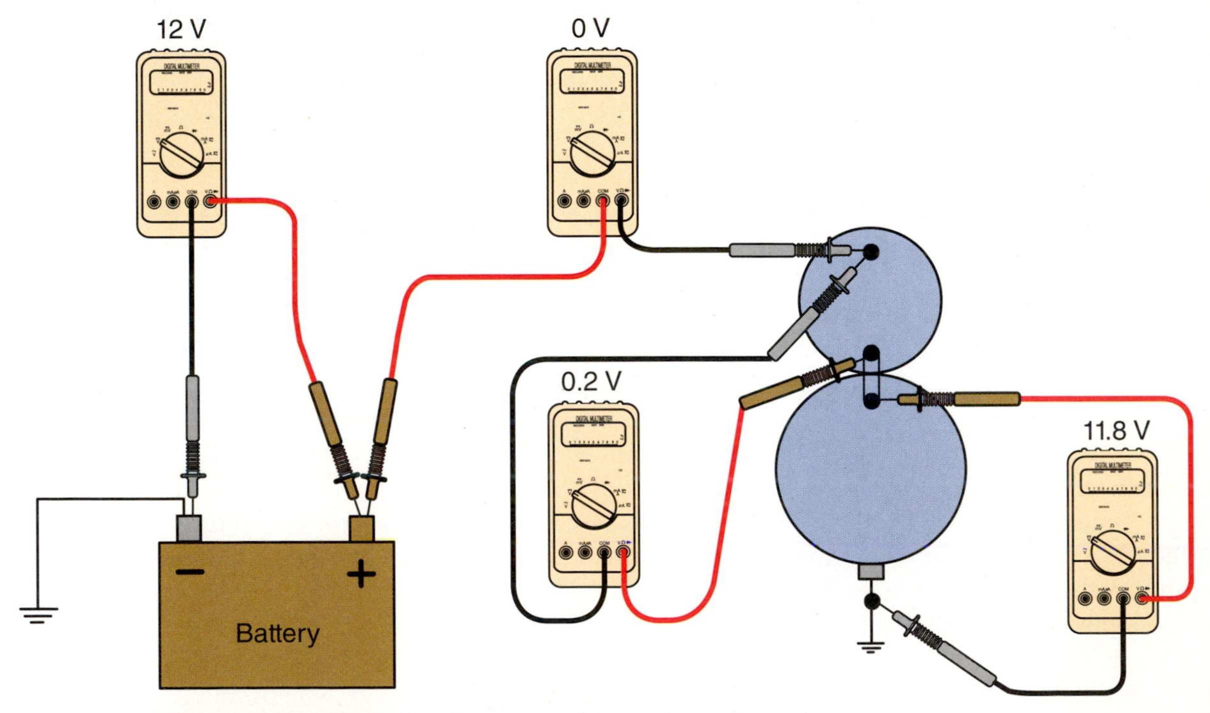

FIGURE 9-50 **Voltage drop testing to identify sources of excessive resistance.**

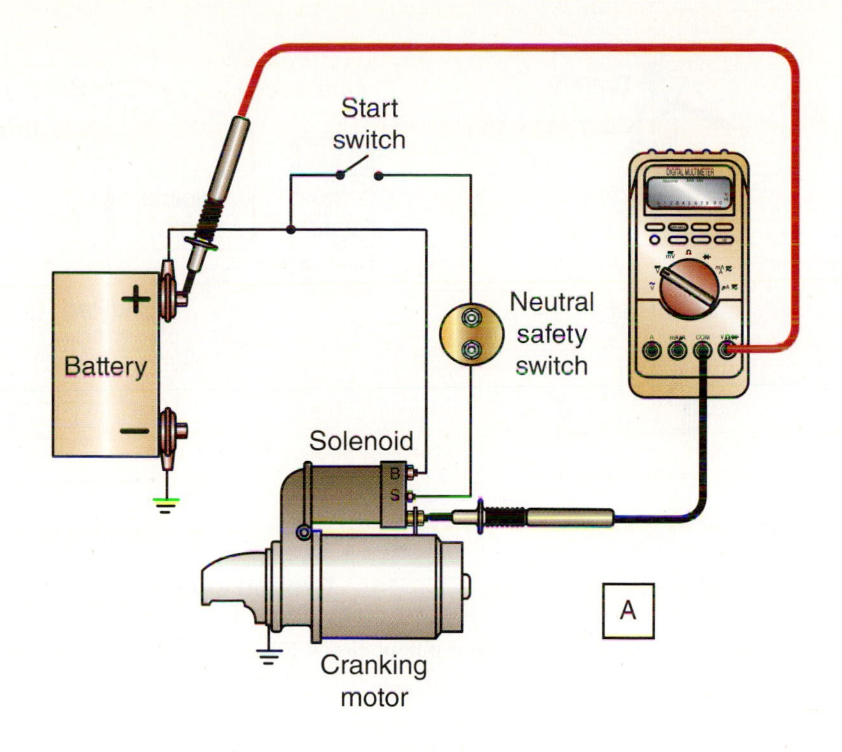

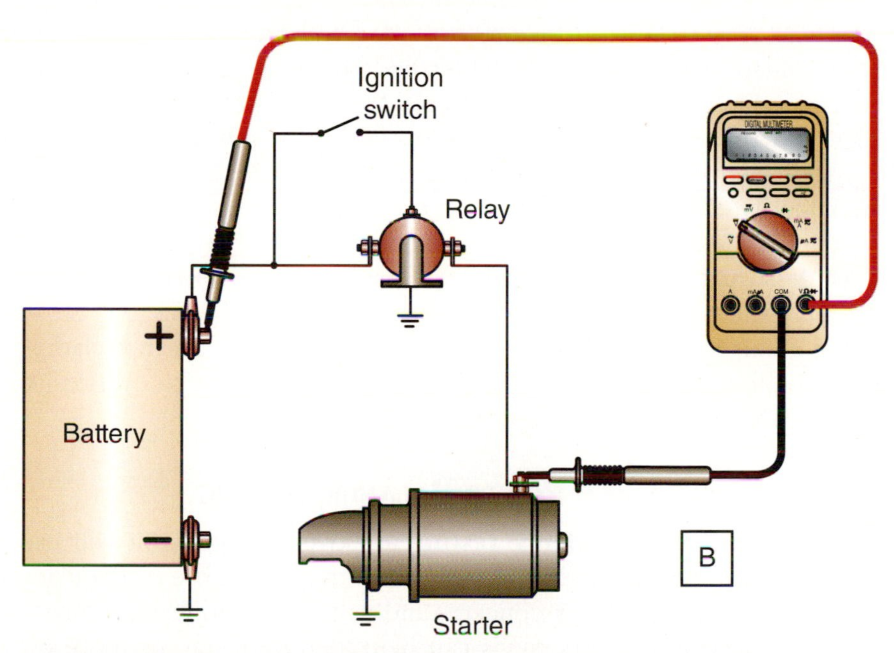

FIGURE 9-51 **(A) Test lead connections for a starter mounted solenoid; (B) test lead connections for relay-controlled systems.**

Inspecting, Servicing, and Testing the Alternator and Charging Circuit

Testing the charging system requires knowledge of the circuit and some of the basic electrical testing tools and procedures previously outlined in this chapter.

The most effective test using a DMM is done by performing both a B+ voltage drop test and a ground side voltage drop test (Figure 9-52). Place the DMM at the appropriate

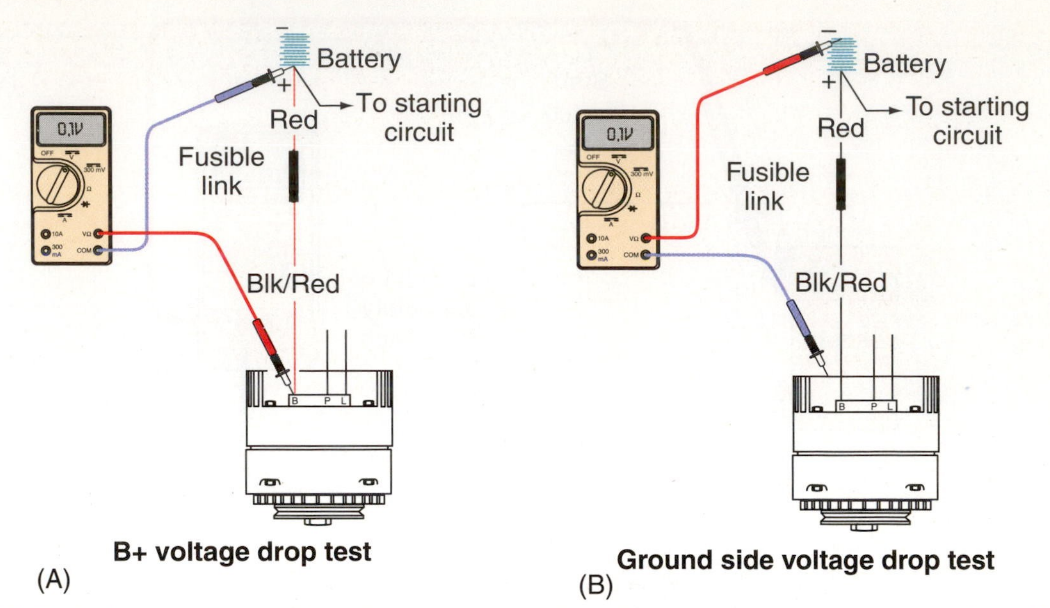

FIGURE 9-52 **Voltage drop testing of the generator output circuit (A) and the generator ground circuit (B).**

places on the system. Then turn the engine on and perform these tests. You should perform these voltage drop tests with the system loaded and unloaded. A loaded test is when you request as much current from the battery as possible, placing the alternator under stress. This means that you should turn on all electrical loads like the headlights, defroster, radio, and fan and run the engine at 2,500 rpm. *Unloaded* means just what you think it means: the system with almost nothing turned on and at idle. You have to do both tests to check the integrity of the wiring.

Always refer to the manufacture's specification, but generally you can expect a good system to have less than a 0.7 volt drop on the insulated (power) side and less than a 0.2 volt drop on the ground side.

To perform a current draw output test, you need to connect an amp clamp and a load to the vehicle's battery. Be careful when performing this test not to overload the system. Refer to Photo Sequence 22 for the steps to perform this test.

Spark Plug Evaluation and Power Balance Testing

A **power balance test** evaluates approximately how much each cylinder is contributing to the overall engine power. It is used to help identify a weak cylinder.

Spark plug reading and **power balance testing** are two methods you will use to gather information about how the engine is operating. You may use these tests during a routine tune up, when there is a driveability concern, and as part of gathering information about an engine mechanical failure. Spark plugs can help identify if one or more cylinders are acting up or if all cylinders are affected. They can also help guide your diagnosis toward a fuel, ignition, or mechanical issue, for example. Similarly, a power balance test can determine how much each cylinder is contributing and pick out the malfunctioning cylinder(s).

Spark Plugs

The spark plug must be in good condition to allow the spark to jump the gap. If the electrodes are worn rounded or covered with oil or gas, or if the gap is too large, the spark will not be strong enough to ignite a flame front, if it sparks at all. Spark plugs are generally one of two basic types: the older type plug with nickel alloy steel electrodes and the newer, more common, platinum tip electrode plugs. Most new vehicles use the platinum

Performing the Charging System Output Test

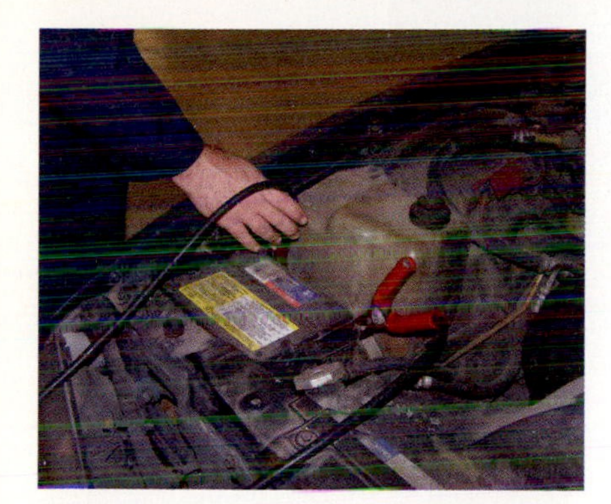

P22-1 Connect the large red and black cables across the battery, observing polarity.

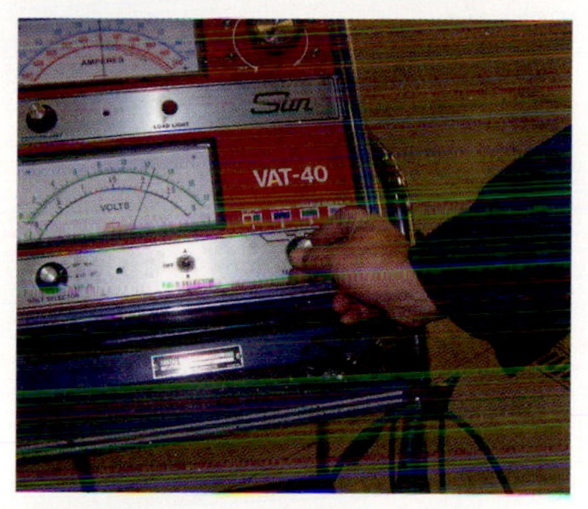

P22-2 Select CHARGING.

P22-3 Zero the ammeter.

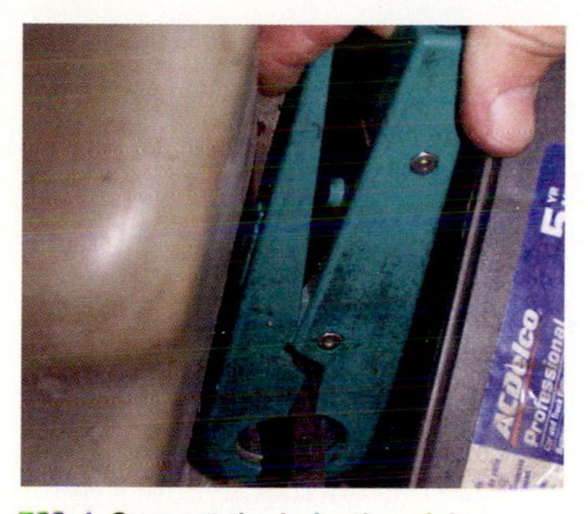

P22-4 Connect the inductive pickup around all battery ground cables.

P22-5 With the ignition switch in the RUN position, engine not running, observe the ammeter reading. This reading indicates how much current is required to operate any full-time accessories.

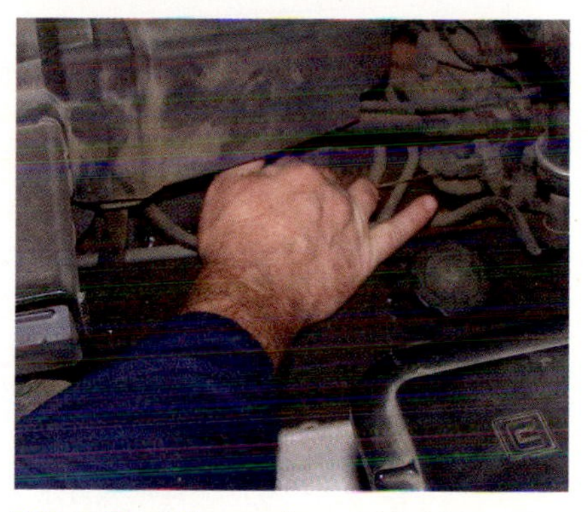

P22-6 Start the engine and hold between 1,500 and 2,000 rpm.

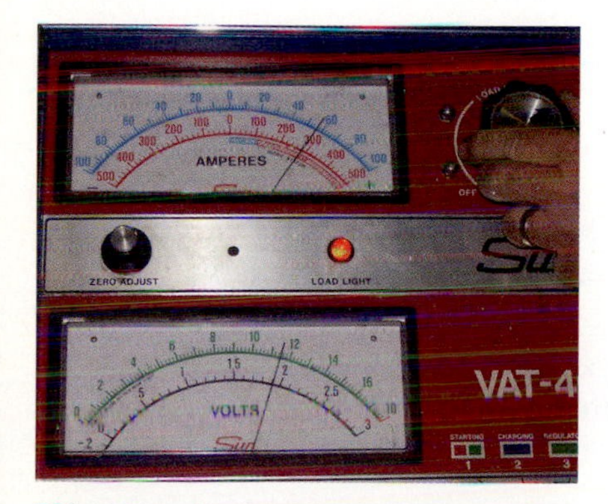

P22-7 Turn the load knob for the carbon pile slowly until the highest possible ammeter reading is obtained. Do not reduce battery voltage below 12 volts.

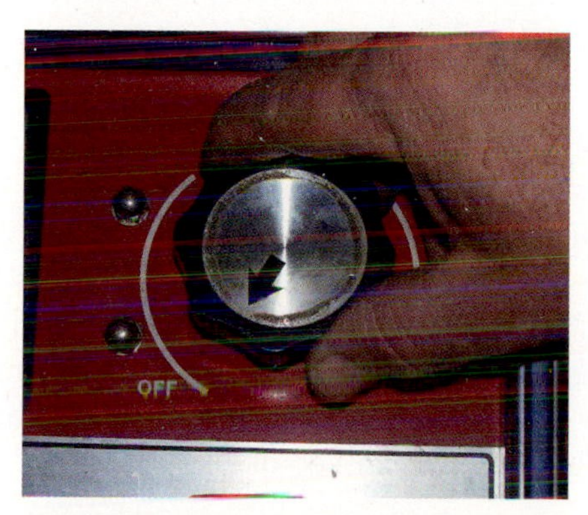

P22-8 Return the load control knob to the OFF position.

P22-9 The highest reading indicates maximum current output.

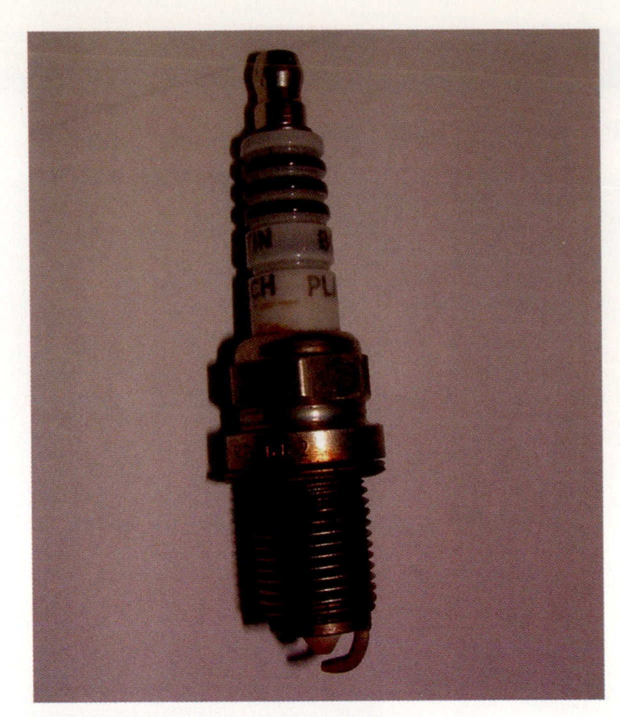

FIGURE 9-53 **A platinum tip spark plug.**

plugs as original equipment (Figure 9-53). The platinum tip plugs have a service interval of 60,000 miles or longer. The older style plugs must generally be replaced every 30,000 miles.

Many times customers do not bring their cars in for service until there is a driveability concern. Spark plugs have a high failure rate; check them early on in your diagnosis. As a spark plug wears, the gap between the electrodes becomes so wide that it may not allow the spark to jump the gap. This can cause partial or total misfiring on that cylinder. Ignition misfire is usually noticeable at low rpm under heavy acceleration. The engine may buck and shudder when the cylinder(s) misfires. It is often difficult to see wear on a platinum plug; the electrodes are small and do not show much visible degradation. Many technicians will replace rather than inspect and reinstall platinum plugs with over 30,000 miles on them. Given today's labor rates, this is often more cost effective for the consumer. If the engine is running poorly and you have the spark plugs out, it may be wise to perform a compression test, which is described later in this chapter.

Spark Plug Removal and Installation

CAUTION:
Do not remove spark plugs from a hot aluminum cylinder head; this can easily damage the threads in the head. Allow the engine to cool down adequately before removal of spark plugs.

While spark plug removal is a very common task in the automotive repair shop, it is not always a simple one. Many times you will have to disassemble parts of the engine to access the spark plugs. There are vehicles that require you to remove an engine mount or remove the turbocharger just to access the plugs. Every time you replace a set of spark plugs you want to be sure you do not damage the plugs or the threads while installing them. Always be sure to replace the whole set of plugs. If one has failed, the others are likely not far behind.

If access to the plugs is not simple, read through the service information to pick up any steps that will make the job easier. This task often requires a universal joint and an assortment of extensions (Figure 9-54).

Blow away any sand and grit from around the spark plug holes before removing the plugs, to prevent this from falling into the cylinder head. Remove all plugs and keep them in order, so you can evaluate the condition of each plug and correlate it to

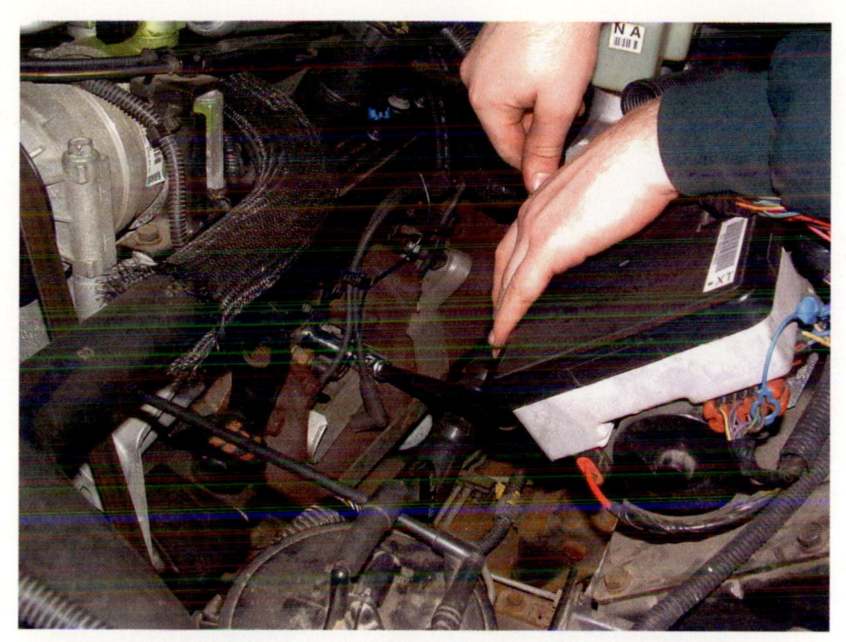

FIGURE 9-54 **This spark plug is readily accessible; not every engine makes it this easy.**

the correct cylinder. Keep the spark plug wires (if applicable) organized, so you return them to their correct positions. Carefully inspect the ends of the plug wires. They can easily be damaged during removal and may require replacement. Also look closely at the plug wires for signs of electrical leakage or arcing to a ground path. If the wire's insulation is damaged, you can often see a light gray or white residue on the wire where the high voltage arcs to a ground—on the valve cover, for example, rather than across the spark plug gap.

Check the spark plug gap before installing the new set of plugs. Sometimes the gap gets closed if the plugs have been dropped. Note that some platinum spark plugs do not have adjustable gaps. Clean the area around the spark plug hole. If you get grease on the electrodes during installation, you will likely wind up with a misfiring spark plug. Install the new plugs by hand; *do not* use air tools. If you can't reach the plug itself, turn it in a few turns using only the extension.

Finish tightening the plugs to the correct torque specification using a torque wrench. After your work is complete, always give the vehicle a thorough road test to be sure the engine runs perfectly.

SERVICE TIP: In one semester, three students had come to me with a misfiring problem. In each case, they had recently replaced spark plugs using a popular, low-cost brand of spark plugs. Installing higher quality or original equipment manufacturer (OEM) spark plugs repaired each of the three late-model vehicles. Use OEM or high-quality spark plugs on your customers' cars to prevent comebacks.

CAUTION: *Never* use air tools to remove or install spark plugs. It takes a lot longer to repair a spark plug hole than it does to loosen plugs with a ratchet.

Spark Plug Reading

Each spark plug can tell you a story about what is occurring in the cylinder. During a routine service, you can use this information to help determine if the engine is mechanically sound. If one spark plug comes out caked with oil deposits from oil leaking past the rings or valve seals, you will be able to let the customer know that the engine is in need of more serious work. When performing driveability diagnosis, the spark plug can help guide your diagnosis. If one spark plug were wet with fuel, you would first confirm that the ignition system is delivering adequate spark to the plug. And whenever you are trying to narrow down the cause of an engine failure, analyze the spark plugs as part of your pre-teardown investigation. The more information you have going into the job, the more confident you can be that you will find the real cause of the problem.

Spark plugs that are wearing normally should show a light tan to almost white color on the electrodes, with no deposits (Figure 9-55). The electrodes should be square and the gap

SERVICE TIP: When the plugs are very difficult to reach, it is helpful to put a piece of vacuum line on the end of the spark plug. This flexible connection allows you to reach the spark plug hole.

SERVICE TIP:
(continued)
Turn the plug in a few turns using the end of the vacuum line. If you start to cross thread the plug, the vacuum line will spin on the spark plug and prevent thread damage.

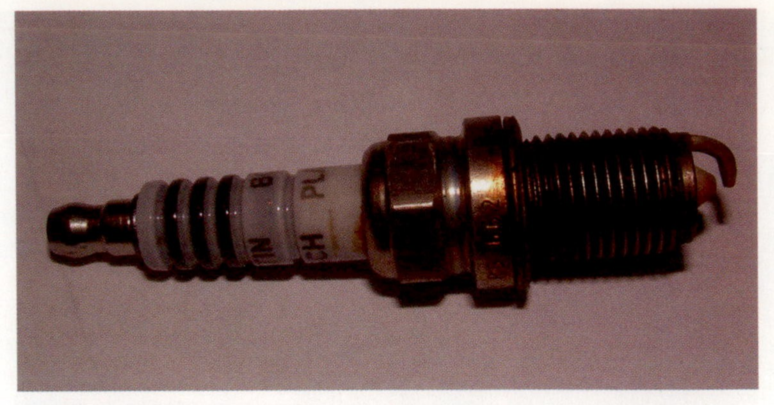

FIGURE 9-55 **This spark plug is wearing nicely; it has no deposits and is a light tan color.**

CAUTION:
No technician has a torque wrench built into his elbow, no matter how adamantly he insists he can torque something properly. Spark plug torque specifications are very light on many new engines, and a stiff elbow can easily strip the spark plug threads on an aluminum cylinder head.

FIGURE 9-56 **Notice the wide gap and rounded electrodes on these worn spark plugs.**

SERVICE TIP:
The classic symptoms of an ignition misfire are bucking and hesitation on hard acceleration. A fuel-related problem, such as a restricted fuel filter, is more likely to cause a steady lack of power as engine speed slowly increases.

should be at or near at the specified measurement. You will see plugs that have a red or orange tint on them. This is the result of fuel additives found in certain fuels or in fuel system cleaning additives.

A spark plug that has many miles on it but is wearing normally will still be light tan in color. There should be only very light deposits, if any, on the plug (Figure 9-56). On non-platinum plugs, you will be able to see that the corners of the electrodes are rounded. The gap will usually be wider than specified. On platinum plugs, the tip may look brand new even when the plug is not firing well. Often you will have to judge wear by the miles since the last change. If the plug is not firing at all, it will be gas-soaked and smell like fuel.

Fuel-fouled plugs will be wet with gas when you remove them from the engine. They will smell like fuel and may have a varnish on the ceramic insulator from heated fuel. The most common cause of this is inadequate spark, though a leaking fuel injector will also flood the plug. If the valves are burned or the compression rings are badly worn, the engine may lack adequate compression to make the air/fuel mixture combustible. A compression check will either verify or rule out that problem.

When a vehicle is burning oil, the spark plugs will develop tan to dark tan deposits on and around the electrode. As the oil is heated during combustion, it bakes onto the plug, leaving a residue. You will have to rule out any other possible causes of oil consumption, such as a plugged PCV system, badly worn valve seals, or a faulty turbocharger, but this is usually a telltale sign that the engine is mechanically worn. If you find this problem during a routine service, it is important to warn the customer about the failing condition of his engine. If the oil is wet and thick on the plug, look for a cause of heavy oil leakage into the cylinder. On rare occasions a head gasket will crack between an oil passage and the cylinder, allowing oil into just one cylinder. This can cause an oil-soaked plug. Similarly, a hole blown in a piston can cover a plug with oil, but you will have other indicators of a serious mechanical failure.

Refer to the spark plug chart in Figure 9-57 to see what the spark plug inspection can tell you about the engine's condition.

SERVICE TIP: Some spark plugs are advertised to last for over 100,000 miles. This doesn't guarantee they'll last that long; these platinum plugs have a high failure rate over 50,000 miles and often become welded into the cylinder head by 75,000 miles. A 60,000-mile service interval is a reasonable compromise that is likely to serve the customer better.

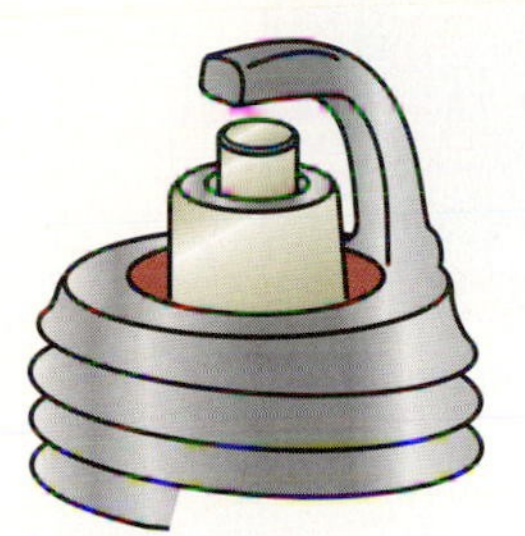

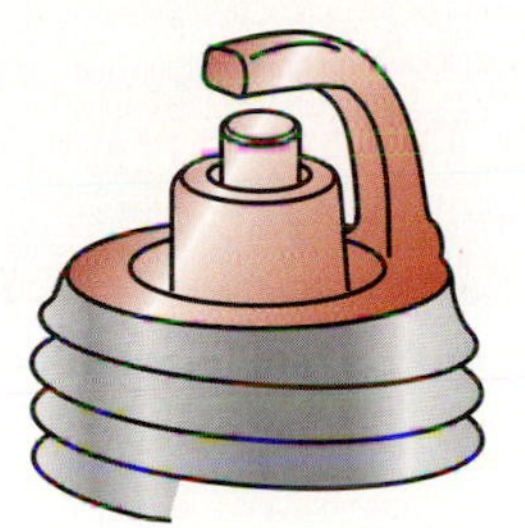

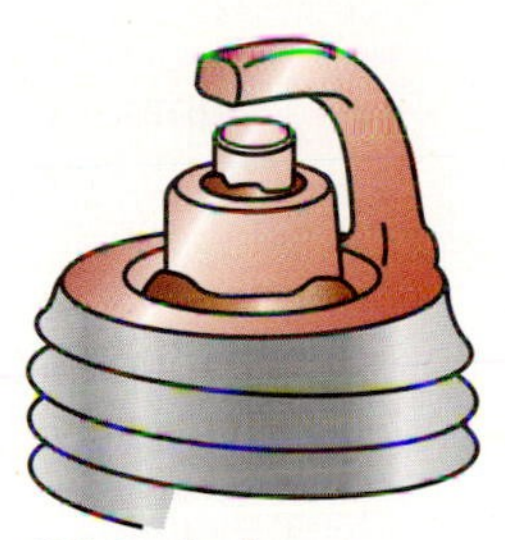

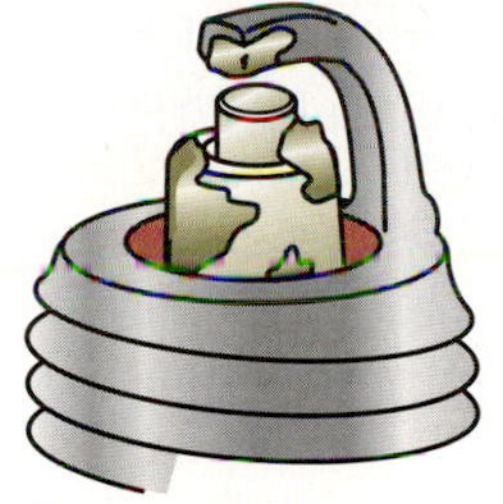

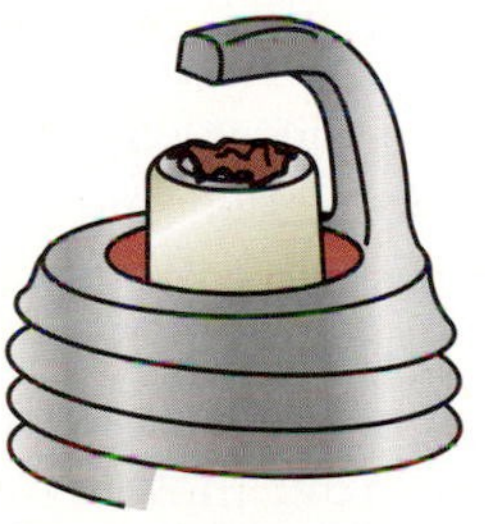

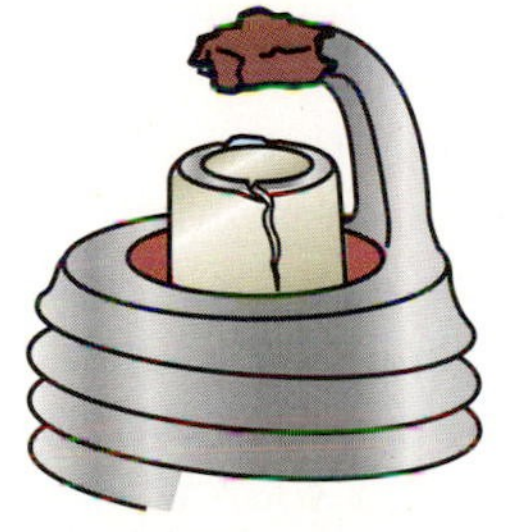

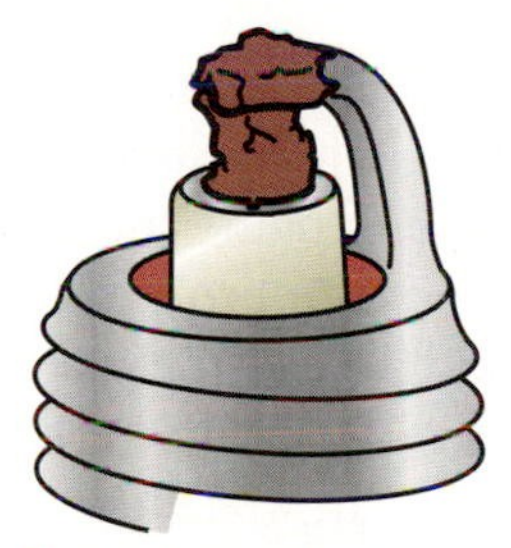

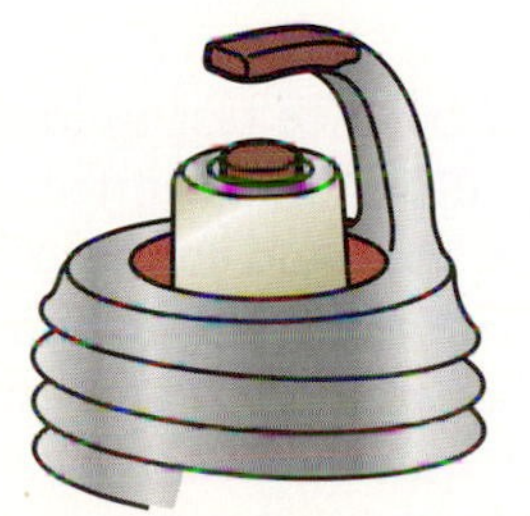

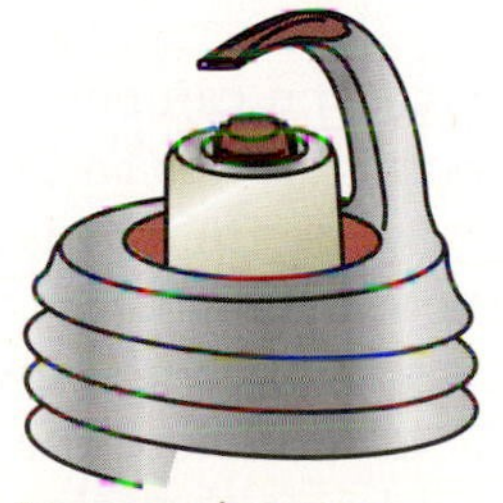

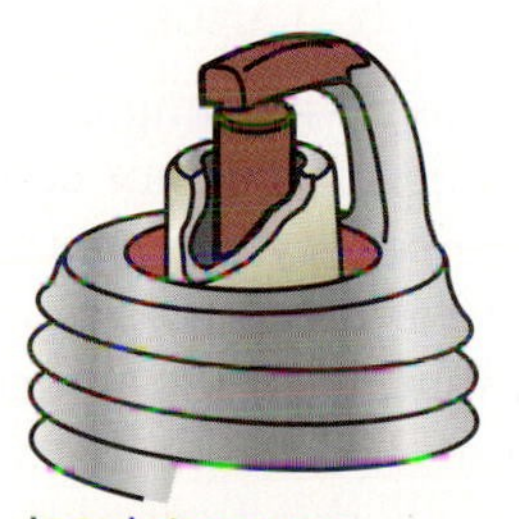

FIGURE 9-57 **A good inspection of used spark plugs can reveal much information about the internal operation of the engine and its subsystems.**

FIGURE 9-58 **A faulty engine coolant temperature sensor can cause the engine to run rich.**

Black-colored, soot-covered spark plugs are found when the air–fuel mixture is too rich. This means there is too much fuel and not enough air. This can be caused by something as simple as a plugged air filter or as involved in the powertrain control system as an inaccurate engine coolant temperature (ECT) sensor (Figure 9-58). An engine that does not have good compression can also create carbon because combustion will not be complete. The key is to notice the condition. Locate the cause before considering a maintenance service to be complete. Write down your observations if you are evaluating an engine for serious mechanical repairs.

Blistered or overheated spark plugs are a sign of potentially serious trouble in the combustion chamber. The porcelain insulator around the center electrode will actually have blisters in it, or there may be pieces of metal welded onto the electrode. It is possible that the wrong spark plugs were installed in the engine. Otherwise, it is likely that pinging or detonation is occurring. If there is no engine damage yet, it is critical to find the cause of the problem before returning the vehicle to the customer. If you find this as you are diagnosing an engine mechanical failure, it is essential to record this information and find the source of the problem during your repair work. A cooling system passage with deposits in it could be causing one corner of a combustion chamber to be getting so hot that detonation occurs. If you were to replace the damaged piston and rings but not repair the cause of the problem, the customer would be back soon with a repeat failure. Your thorough evaluation of the engine and analysis of the causes of the symptoms will ensure customer satisfaction.

Power Balance Testing

On a rough-running or misfiring engine, you can perform a power balance test to identify a cylinder that is not equally contributing power to the engine. To perform the test, you disable one cylinder at a time while noting the change in engine rpm and idling condition. The more the rpm drops or the rougher the engine runs, the more the cylinder is contributing. If a cylinder is producing little or no power, the change in rpm and engine running will barely be noticeable. You must note the rpm drop quickly as the cylinder is initially disabled; today's PCMs will boost the idle almost immediately to compensate for the drop. You will also be able to feel and see the roughness of the engine as a functional cylinder is disabled. We'll discuss a few different methods of performing a power balance test.

Power Balance Testing Using an Engine Analyzer or Scan Tool

Many engine analyzers can perform a manual or automated power balance or cylinder efficiency test. With the analyzer leads connected to the ignition system as instructed in the user's manual, select the power balance test from the menu (Figure 9-59). The tester will automatically disable the ignition to one cylinder at a time for just a few seconds and display the rpm drop. The analyzer disables spark in the firing order; an engine with the firing order 1-3-4-2 would be tested on cylinder number 1 first, then number 3, number 4, and number 2. On some analyzers, the power balance results will be displayed in bar-graph format for easy comparison of cylinder contribution.

On some vehicles, the PCM is programmed to be able to perform a power balance test. The manufacturers and some aftermarket scan tools will be able to access this test. The only hookup required is to the diagnostic link connector (DLC). Then select the power balance test from the menu system. At the prompt, the scan tool will communicate with the PCM to initiate its disabling of cylinders one by one. The rpm drop for each cylinder will be displayed after the test.

CAUTION:
You must be careful not to disable spark for long periods of time and must allow the engine to run on all cylinders for one minute between tests. Failure to follow these guidelines can overload the catalytic converter with raw fuel and damage it from overheating.

Manual Power Balance Testing

Even without an engine analyzer or a capable scan tool, you can perform a power balance test. The procedure will require that you manually disable either fuel or spark to each cylinder. It is also important that you check the vehicle for diagnostic trouble codes (DTCs) after performing this test and clear the codes if any are present. You can access DTCs using a scan tool; the code can help you locate the source of the problem. If the PCM detects a misfire, it is likely to set a DTC. When the malfunction indicator light comes on, a DTC has been set. Using the menu on a scan tool, you can erase any DTCs when your testing is complete.

On vehicles that use spark plug wires to deliver the spark to the plug, you can short the spark to ground on each cylinder in turn to test the power contribution of cylinders. One way to do this is to use an insulated pair of pliers (Figure 9-60). Carefully pull the wire off the plug and hold the wire to the block or head to allow it to find a ground path. It is important that the spark can find an easy path to ground; do not hold the wire more than a half an inch away from a good metal conductor. The coil can be

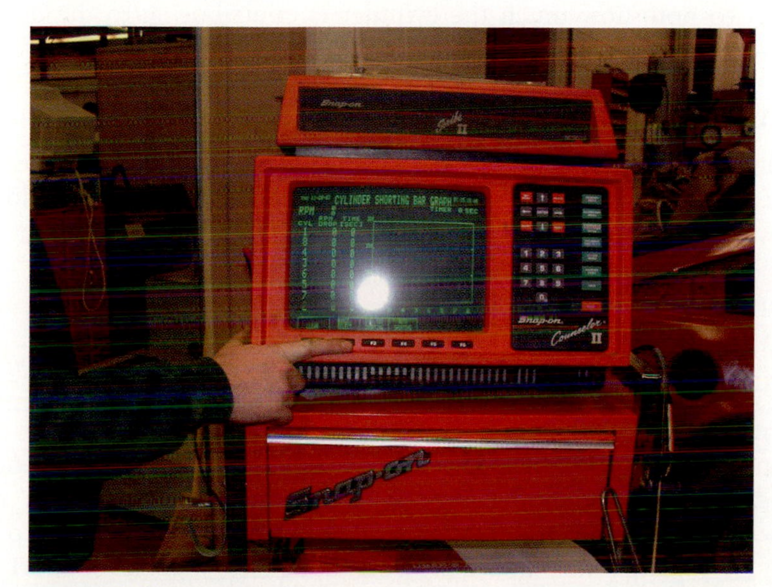

FIGURE 9-59 **This engine analyzer shorts one cylinder at a time and displays the engine rpm drop.**

FIGURE 9-60 **Use insulated pliers to remove a live plug wire. Ground the wire to a metal component immediately to prevent harming the coil.**

FIGURE 9-61 **During a power balance test, disconnect the low voltage primary wiring to the individual coils and record the rpm drop. Be sure to clear any DTCs after your testing.**

damaged if it uses all its possible power to find a path to ground. Sometimes the spark will track through the insulation of the coil and destroy it. As you remove the spark plug wire, you should notice a clear change in rpm and idle quality if that cylinder is contributing adequate power.

Many newer engines use coil-on-plug ignition systems. Each spark plug has its own coil and you cannot safely remove the coil and ground it. Instead, locate the connector going into the coil (Figure 9-61). While the engine is running, you can briefly remove the connector to prevent the coil from firing. As you unplug the coil, monitor the rpm drop on a tachometer. Repeat this for each cylinder and compare your results. This testing is very likely to trigger the malfunction indicator lamp (MIL) and set a DTC, so be sure to clear the DTCs after your testing.

Another way to perform a manual power balance test is to unplug the connectors to the fuel injectors one by one while listening to and watching the engine rpm. By preventing fuel from entering the cylinder, you can momentarily disable the cylinder. This allows you to note how much the cylinder is contributing to the overall power of the engine.

Analyzing the Power Balance Test Results

When an engine is running properly, each cylinder should cause very close to the same rpm drop when disabled. Cylinders that are lower by just 50 rpm should be analyzed. Look at the results below:

Cyl. #1	Cyl. #3	Cyl. #4	Cyl. #2
125 rpm	150 rpm	50 rpm	125 rpm

Cylinder number 4 is definitely not contributing equally to the power of the engine. The rpm changed very little when it was disabled, meaning that the engine speed is barely affected by that cylinder. There could be many causes of the problem, but you now know which cylinder to investigate. The problems could relate to fuel delivery; perhaps the fuel injector was not delivering adequate fuel to support combustion.

An ignition system problem, something as simple as a spark plug with a very wide gap from worn electrodes, could also cause this. A cylinder with low compression will also produce less power and be suspect during a power balance test. Compression testing and cylinder leakage testing will help you to determine whether the weak cylinder has an engine mechanical problem. They can also provide information about the specific cause.

SPECIAL TOOL

Lineman's gloves

Special Procedures for Working Safely Around Hybrid Electric Vehicles

The service of hybrid electric vehicles (HEVs) is limited to technicians who have received factory training from the manufacturer or are master technicians who have received training and education from a certified program. This text cannot train you to work on a HEV. Working on HEVs is not an entry-level technician's job; however, you must be aware of the safety precautions. As a technician, you should be aware of the safety precautions associated with these vehicles. If you perform towing, you must be fully aware of the safety precautions before approaching a damaged HEV. Your education may well place you at the front of the line for training on these advanced technology vehicles.

Safety must be your first consideration when approaching an HEV (Figure 9-62). The voltages on some HEVs may reach 500 volts and regularly store 360 volts. Voltage varies with make, but over 30 volts is potentially fatal to humans; so we are dealing with a potentially lethal mechanism. All high-voltage cables are identifiable by a bright orange coating. Never touch these wires while the electrical system is live. All high-voltage components are labeled as such (Figure 9-63). The battery packs are well insulated and protected in an accident. Nickel metal hydride (NiMH) batteries, as used in many of today's HEVs, are sealed in packs designed not to leak even in the event of a crash. If they do leak, the electrolyte can be neutralized with a dilute boric acid solution or vinegar.

SERVICE TIP: Something as simple as a blown high-voltage fuse could cause a vehicle not to drive. You'll still need to follow special safety precautions for testing them, though. They are rated between 400 and 600 amps.

Understanding proper operation of the vehicle you are working on is essential. Though you should have received factory training, much of this information can be found in the owner's manual. The vehicle may be "ready" and able to move with a tap of the accelerator, even when neither the electric motor nor the gasoline engine is running. The owner's manual will provide you with special operating characteristics of the vehicle and basic safety warnings.

SERVICE TIP: It is not uncommon for a customer with a new technology vehicle to come in with service questions. Most consumers do not take the time to read the owner's manual or view the operational CD provided. When you are working on a make of vehicle, be as familiar with it as possible, so you can competently answer questions a customer might have about his new vehicle.

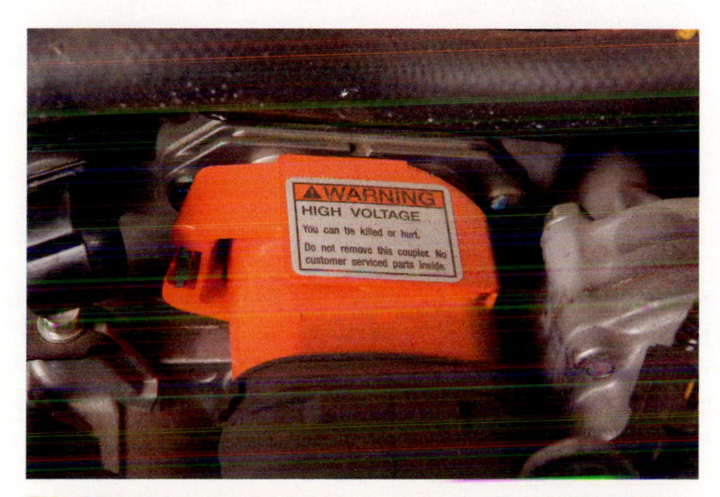

FIGURE 9-62 **Use caution! Look carefully at this warning label; it says "you will be killed" if you do not use appropriate safety precautions.**

FIGURE 9-63 **The brightly colored big cables hold high voltage; components that use high voltage have warning labels.**

Servicing the Toyota Prius

First, familiarize yourself with the layout of the vehicle components (Figure 9-64). The figure identifies the location of the following components:

Component	Location	Description
1. 12-volt battery	LR of trunk	Typical lead-acid battery used to operate accessories
2. HV battery pack	Rear seat back support, in trunk	274 NiMH battery pack
3. HV power cables	Undercarriage and engine compartment	Thick, orange-colored cables carrying high voltage (HV) DC and AC current
4. Inverter	Engine compartment	Converts DC to AC for the electric motor and AC to DC to recharge the battery pack
5. Gasoline engine	Engine compartment	Powers vehicle and powers generator to recharge battery
6. Electric motor	Engine compartment	AC permanent magnet (PM) motor in transaxle to power vehicle
7. Electric generator	Engine compartment	AC generator in transaxle used to recharge battery pack

Only a certified technician should disable the high voltage system by disconnecting the negative terminal of the 12-volt auxiliary battery. This is located in the left rear of the trunk. Disconnecting the auxiliary battery shuts down the high voltage circuit. For additional protection, wait 5 minutes, and then remove the service plug located through an access port in the trunk (Figure 9-65). Wear insulated rubber **lineman's gloves** when removing the service plug. Always carefully check the gloves for pin holes. Blow into them like a balloon and see that they don't leak any air. Even a very small hole could allow a deadly spark to penetrate to you. After removing the service plug, wait at least 5 minutes to allow full discharge of the high voltage condensers inside the inverter.

Lineman's gloves are highly insulated rubber gloves that protect the technician from lethal high voltages.

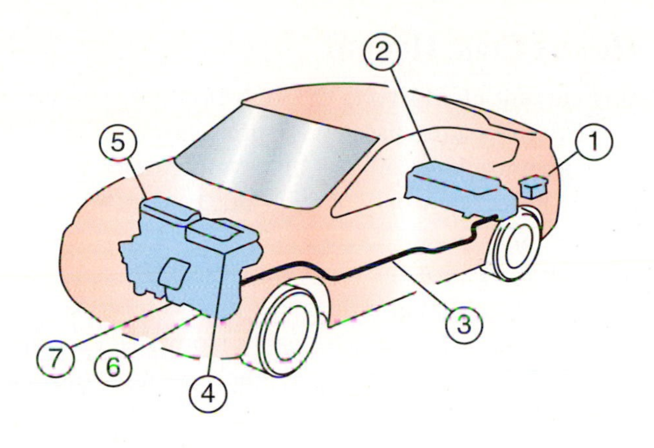

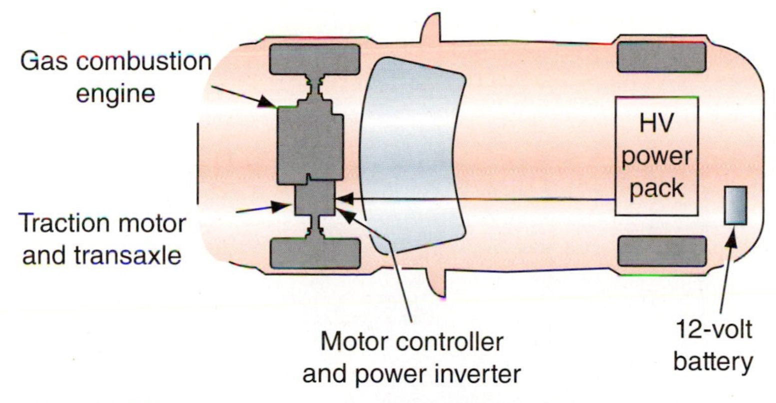

FIGURE 9-64 **Locations of components on the Toyota Prius.**

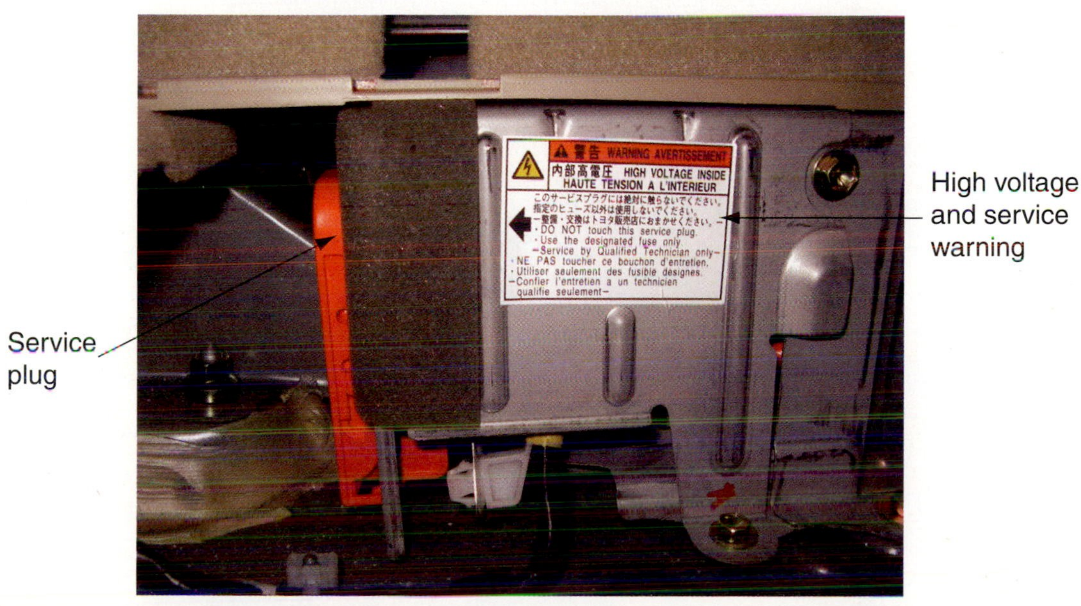

FIGURE 9-65 **The high-voltage service plug is located in the trunk. Use lineman's gloves to switch power off.**

Once the high voltage system has fully discharged, you may diagnose and repair the Prius gas engine as you would any other engine. This is a newly designed engine; be sure to use proper procedures and service specifications.

Servicing The Honda Civic Hybrid

The Civic hybrid runs current at 144 volts through brightly colored orange cables. Its Integrated Motor Assist (IMA) System uses a 1.3-liter gasoline engine and a DC brushless motor (Figure 9-66). You must shut off the high voltage electrical circuits and isolate the IMA system before servicing its components. Do not touch the high voltage cables without lineman's gloves (Figure 9-67). To isolate the high voltage circuit, turn the high voltage battery module switch to OFF. Secure the switch in the off position with the locking cover before servicing the IMA system. Wait at least 5 minutes after turning off the battery module switch, and then disconnect the negative cable from the 12-volt battery. Test the voltage between the high-voltage battery cable terminals with a voltmeter. Do not disconnect them until the voltage is below 30 volts.

HEV Warranties

In order to reassure the public about the new technology and encourage consumer confidence, most HEV manufacturers offer extensive warranties on the battery packs

FIGURE 9-66 **Honda's IMA system "engine" compartment.**

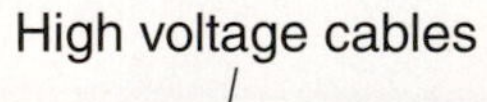

FIGURE 9-67 **The high-voltage cables are routed through the engine compartment, but you can't miss them due to their bright orange color. Disable the high voltage system before touching them.**

and the hybrid components. You should always consult the manufacturer's service information before recommending expensive repairs to consumers of these vehicles, even as the vehicles age and you begin to see them in independent repair facilities. A few examples are:

Ford Escape	8 years/80,000 miles on battery pack*
Honda HEV	8 years/80,000 miles on battery pack*
Toyota Prius	8 years/100,000 miles on battery pack* and on the hybrid power plant, including the engine

*Battery packs must be warranted for 10 years/150,000 miles in "green states" (CA, MA, VT, NY, ME)

CASE STUDY

A friend of mine asked me to jump start her vehicle. While doing so she explained the battery had been replaced several times along with a new starter but still there was a cranking problem. Later that day I used a DVOM to check the starting system. A voltage drop test on the battery to starter cable revealed a 6-volt drop. An inspection of the cable found heavy corrosion under the cable covering near the battery connection. Replacement of the cable solved the problem. Always check the cables and connections when dealing with the starting and charging system, particularly on older vehicles.

TERMS TO KNOW

Battery terminal voltage drop test
Capacity test
Conductance test
Current draw test
Distributor
Ground reference voltage testing
Hydrostatic lock
Ignition timing
Infinite
Lineman's gloves
Open circuit
Open circuit voltage test
Power balance testing
Self-powered test light
Short circuit
Starter quick test
Test light
Voltage drop

ASE-STYLE REVIEW QUESTIONS

1. Which of the following describes the function of the starter solenoid?
   A. An electromagnetic device that transfers high current to the starter motor.
   B. An electromagnetic device that moves and engages the starter.
   C. Neither A or B.
   D. Both A and B.
2. The rotor in an alternator:
   A. Is a stationary component.
   B. Controls the speed of the alternator.
   C. Is touching the rectifier.
   D. Is a rotating component.
3. A Hybrid Electric Vehicle uses a DC to DC converter to:
   A. Charge the 12-volt battery.
   B. Turn the headlights on.
   C. Power the alternator.
   D. Start the engine.
4. A flywheel is mounted to the:
   A. Crankshaft.
   B. Camshaft.
   C. Cylinder block.
   D. Transmission housing.
5. A fully charged conventional automotive battery has a specific gravity of:
   A. 1.460
   B. 1.265
   C. 1.225
   D. 1.190
6. The battery stores energy in which form?
   A. Chemical.
   B. Electrical.
   C. Mechanical.
   D. Kinetic.
7. Two technicians are discussing relay function. *Technician A* says that a relay can be controlled by a switch. *Technician B* says that a relay can be controlled by a computer. Who is correct?
   A. A only
   B. B only
   C. Both A and B
   D. Neither A nor B

8. A recombination battery:
   A. Uses a gel instead of acid.
   B. Contains flat cells and plates.
   C. Can operate only in the upright position.
   D. None of the above.

9. The source of electrical power for everything in the car is the:
   A. Alternator.
   B. Starter.
   C. Computer.
   D. Battery.

10. A computer circuit consists of:
    A. Inputs.
    B. Outputs.
    C. Processing device.
    D. All of the above.

JOB SHEET

Name ______________________________ Date ______________

TESTING ELECTRICAL CIRCUITS USING TEST LIGHTS AND A DIGITAL MULTIMETER (DMM)

NATEF Correlation

This job sheet addresses the following NATEF tasks:

Demonstrate proper use of a digital multimeter (DMM) when measuring source voltage, voltage drop (including grounds), current flow, and resistance.

Check operation of electrical circuits with a test light.

Objective

Upon completion and review of this job sheet, you should be able to test and diagnose circuits using a test light or DMM.

Tools and Materials Needed

Wiring Diagram

Vehicle with a designated test circuits

Test Light

Self-powered test light

Digital multimeter (DMM)

Fender cover

Service manual

Safety glasses

Procedures

Task Completed

1. Determine the following information from the service manual. ☐

   Vehicle ____________ Make ____________ Model ____________ Year ____________

   Headlight and tail light circuits' lay out in vehicle

   Are the conductors in these circuits in the same electrical conduits as the air bag circuit? If yes, consult your instructor ______________________________

   Will any of the tests here involve working around air bag components? If yes, consult your instructor. ______________________________

   Is either circuit an electronic circuit? ______________ If yes, disregard the use of any test light. Use only the digital multimeter.

WARNING: Do not work on or near air bag circuits without clearing your work with the instructor.

WARNING: Before working on a circuit with a test light, ensure that it is not an electronic circuit or part of one. A test light can damage electronic components.

CAUTION:
Disable the air bag system by disconnecting the battery's negative cable. Wait the for the air bag capacitor to discharge following the manufacturer's specific instructions . Serious injury could result from a deploying air bag.

CAUTION:
Electrical power must be provided for this circuit test. If the air bag system was completely disarmed, disconnect all of the air bag system's bright yellow connections before reconnecting the battery. After the test(s) are completed, disconnect the battery negative cable, re-connect all the yellow connections and then reconnect the battery. Consult your instructor before disconnecting or connecting any air bag circuits.

Task Completed

Headlight Circuit

☐ 2. Connect the test light across the battery terminals. Did the test lamp light and is it bright?

____

☐ 3. Set the DMM to read DC voltage.

☐ 4. Connect the DMM across the battery terminals. What voltage reading did you get?

____

☐ 5. What is your next action if the battery voltage is less than 11-volts?

____

☐ 6. Use the wiring diagram to locate the easiest point to access a connection in the circuit to be tested. Disconnect the circuit.

☐ 7. Turn the circuit on (power on)

☐ 8. Connect the test clip to a good ground.

☐ 9. Touch the test light's probe to the electrical terminal on the vehicle harness. What happened to the test light and what does it indicate?

____

☐ 10. Remove the test light. Leave the power on.

☐ 11. Connect the DMM's black probe to a good ground. Ensure the DMM is set to read DC voltage.

WARNING: Do not connect an ohmmeter or a DMM set to the ohm's scale to a powered circuit. The test could be damaged.

☐ 12. Probe the vehicle harness terminal with the DMM's red lead. What is the result and what does it indicate?

____

Task Completed

13. The next several steps require that the power is off within the circuit portion being tested. Follow the WARNING instructions. ☐

WARNING: Do not power up the circuit being tested for continuity. Using a continuity tester to test a powered circuit will damage the tester and possibly damage the circuit.

14. Ensure the power to the circuit is off. Locate the next circuit connection towards the load and disconnect it. ☐
15. Use the self-power test light for this test. Connect the test light's clip to a terminal at one end of the isolated portion of the circuit. Use the test light's probe to touch the terminal at the other end of the isolated circuit. ☐
16. Observe the test light lamp and record your results and what do the results mean? ☐

____

17. Set the DMM to read resistance. ☐
18. Connect one lead of the DMM to one end of the isolated circuit and the other lead to the opposite end. Record the results and what do the results mean? ☐

____

19. Reconnect the all the circuit connections you unplugged. ☐
20. Turn on the circuit's power to ensure the circuit is working properly. Turn off the circuit power and replace any items moved to gain access to the circuit. ☐

CAUTION:
If the air bag was disabled consult your instructor and follow the manufacturer's procedures to enable the air bag system. Once the system is enable start the engine and check to see if the air bag warning lamp is lit. If so consult your instructor.

Stop (Brake) Light

WARNING: Before working on the stop light circuit determine if the under dash connection is near the air bag circuit. If so consult your instructor.

21. Disconnect the left side brake light inside the trunk. It may be necessary to move the trunk carpet or paneling for access. Ensure that any fasteners are not damaged. ☐
22. Connect the test light clip to a good ground. ☐
23. Have an assistant apply the service brakes and hold the pedal in place. ☐
24. Probe each of the stop light terminals on the vehicle harness. Which terminal is the power conductor for the stop light? Did the test lamp light up? If not, what is the most probable cause? ☐

____

Task Completed

☐ **25.** Once the power terminal is determined, set the DMM to read DC voltage.

☐ **26.** Connect the DMM's black lead to a good ground and touch the red lead to the power terminal. Ensure the brake pedal is applied. What is your voltage reading?

____

☐ **27.** Release the brake pedal and re-install any carpeting or paneling removed for access to the stop light connection.

☐ **28.** Check the operation of the stop light and head light circuit. Ensure any other items removed to gain access to either circuit are installed properly.

Instructor's Response ____

____

____

____

JOB SHEET 25

Name ______________________ Date ______________

INSPECT, SERVICE, AND TEST THE STARTER AND STARTER CIRCUITS

NATEF Correlation

This job sheet addresses the following NATEF tasks:

Demonstrate proper use of a digital multimeter (DMM) when measuring source voltage, voltage drop (including grounds), current flow, and resistance.

Confirm proper battery capacity for vehicle application; perform battery capacity test; determine necessary action.

Perform starter circuit voltage drop tests; determine necessary action.

Inspect and test starter relays and solenoids; determine necessary action.

Remove and install starter in a vehicle.

Inspect and test switches, connectors, and wires of starter control circuits; determine necessary action.

Objective

Upon completion and review of this job sheet, you should be able to test and diagnose starter circuits using a DMM, determine any faults, and replace a starter motor.

CAUTION: Disable the fuel and ignition system before checking the starter circuits. If the engine should start, serious injury to persons or damage to the vehicle or equipment could occur.

Tools and Materials Needed

Hand tools
Floor jack with safety stands or vehicle lift
Wiring Diagram
Vehicle with a standard starter control circuits
Digital multimeter (DMM)
Fender cover
Service manual
Safety glasses

Procedures

Task Completed

1. Determine the following information from the service manual. ☐
   Vehicle Make______________ Model______________ Year______________
   Correct battery for test vehicle
   Location of the starter motor______________
   Location of starter relay if any______________
   Location of safety devices within the starter control circuit______________
   Manual or automatic transmission?______________
2. Ensure the wheels are chocked and the fuel and ignition systems are disabled. ☐
3. Inspect the alternator drive belt(s) for tightness and serviceability. ☐

Task Completed

☐ 4. Is the battery the correct one for this vehicle? If not, can tests be conducted?

____

☐ 5. Set the DMM to read DC voltage.

☐ 6. Connect the test leads across the battery posts. Does the battery have at least 12-volts? If not, what is your next step?____

☐ 7. Place the DMM's black lead to the battery's positive and the red lead to the battery's positive cable terminal. Record the results.

____

☐ 8. Repeat step 7 on the negative post and negative cable terminal. Record the results.

____

☐ 9. Lift the vehicle to gain access to the input cable on the starter motor.

☐ 10. Connect the DMM's black lead to a good ground.

WARNING: If the starter is working at all, ensure that the hands and tools are clear of any moving parts like a crankshaft-driven belt.

☐ 11. Have an assistant turn the ignition to start and probe the starter positive terminal with the red lead. Record the result. What is your next action if the voltage does not equal battery voltage?

____

____

____

☐ 12. Assuming there is not battery voltage at the starter, fashion a jumper wire long enough to reach from the starter relay output to under the vehicle. Connect one end of the jumper wire at the relay's output and the other end to the black lead of the DMM.

☐ 13. Have an assistant hold the ignition switch in the start position while probing the starter's positive terminal with the red lead. What is the result? Is the DMM reading zero or near zero voltage? If the reading exceeds more than 0.2-volts what is your repair recommendation?

____

____

☐ 14. If the reading in step 13 is below 0.2-volts, lower the vehicle to gain access to the interior and engine compartment.

☐ 15. Locate the starter relay and the conductor from the ignition switch.

☐ 16. Disconnect the ignition switch conductor. Set the DMM to read DC voltage. Connect the black lead to a ground and probe the conductor terminal with the red lead.

☐ 17. Hold the ignition switch in the start position. Did you get a positive voltage reading? If not, set the DMM to ohms. Did you get a resistance reading of zero or near zero? Typically the starter relay is grounded by the ignition switch. Checking for voltage first will prevent damage to a DMM that is set to the ohm scale and connected to an energized conductor.

____

____

Task Completed

18. Based on the tests results you have found so far and after consulting the instructor and service information, what do you recommend as the next test or the repair needed? ☐

___

___

Starter Motor Replacement

WARNING: Before beginning the replace of electrical components ensure the battery's negative cable is disconnected. Damage to the vehicle or persons could occur.

19. Disconnect the battery's negative cable and positon it where it will not accidently come in contact with the negative terminal. ☐
20. Lift the vehicle to gain access to the starter motor. Remove any heat shields or other items blocking access. ☐
21. Disconnect the starter's positive input cable and any other electrical connections. ☐
22. Use the correct wrenches to remove the upper bolts holding the starter motor to the engine block/transmission housing. Leave the lowest bolt in place while removing the others. ☐
23. Loosen the last bolt and hold the starter motor in place as the bolt is completely removed. On heavier units it is best to have an assistant support the starter motor as the last bolt is removed. ☐
24. Remove the starter motor. Compare the old motor to the new one. If they are the same, positon the new motor into place. ☐
25. While holding the starter motor in place start and finger tighten the lowest bolt sufficient to hold the motor in place. ☐
26. Start each of the other bolts in place before tightening any of them. Once all are in place tighten each bolt to torque specifications. ☐
27. Connect all electrical connections to the starter motor. Replace any items removed for access. ☐
28. Lower the vehicle and connect the battery negative cable. ☐
29. Check the starter system for operation. ☐

Instructor's Response ___

___

___

___

JOB SHEET 26

Name ________________________ Date ______________

Inspect, Service, and Test the Alternator and Alternator Circuits

NATEF Correlation

This job sheet addresses the following NATEF tasks:

Perform charging system output test; determine necessary action.

Inspect, adjust, or replace generator (alternator) drive belts; check pulleys and tensioners for wear; check pulley and belt alignment.

Remove, inspect, and re-install generator (alternator).

Perform charging circuit voltage drop tests; determine necessary action.

CAUTION: Do not attempt to work on the alternator when the engine is hot. Serious burns could result from touching the hot surfaces of the engine or exhaust. Allow the engine to cool before beginning charging systems diagnostics.

Objective

Upon completion and review of this job sheet, you should be able to test and diagnose alternator circuits using a DMM, determine any faults, and replace an alternator.

Tools and Materials Needed

Hand tools

Floor jack with safety stands or vehicle lift

Wiring Diagram

Vehicle with a standard alternator control circuits

Digital multimeter (DMM)

Fender cover

Service manual

Safety glasses

Procedures

Task Completed

1. Determine the following information from the service manual. ☐

   Vehicle ____________ Make ____________ Model ____________ Year ____________

   Correct battery for test vehicle

   Location of the alternator________________________________

   Location of alternator protection devices if any________________________________

   Location of safety devices within the alternator's control circuit________________________________

2. Ensure the wheels are chocked, transmission in parked (neutral for manual). ☐

3. Inspect the alternator drive belt(s) for tightness and serviceability. ☐

4. Is the battery the correct one for this vehicle? If not, can tests be conducted? ☐

   ____

5. Inspect the battery and cable for cleanliness and serviceability. ☐

6. Inspect the alternator electrical connections for looseness and damage. ☐

Task Completed

☐ **7.** Set the DMM to read DC voltage.

☐ **8.** Set the DMM to read DC voltage. Connect the black lead to the battery's positive post and the red to the positive cable terminal.

☐ **9.** What is the voltage drop between the two leads? The reading should be zero or near zero ohms. If not there is excessive resistance between the post and the terminal.

____

☐ **10.** Repeat steps 8 and 9 on the negative post and terminal. What is the reading?

____

☐ **11.** Connect the DMM across the battery terminals, red lead to positive, black lead to negative.

☐ **12.** With the DMM connected across the terminals, have an assistant start the engine and allow it to idle. What is the voltage reading? If the reading is 12-volts or less, what you recommend as the next test or repair? Consult your instructor and the service information for guidance as needed.

____

☐ **13.** If the initial voltage is below 12-volts or it dropped when the electrical loads were added prepare for the next test by shutting down the engine and access the wiring diagram.

☐ **14.** On the wiring diagram find the voltage output for the alternator and then find it physically on the back of the alternator.

☐ **15.** Clear area around the alternator output terminal so the read lead of the DMM can be touched safety to the terminal of the output wire with the engine running.

☐ **16.** Ensure the DMM is set to DC voltage and connect the black lead to a good ground.

☐ **17.** Have the assistant start the engine and allow it to idle.

☐ **18.** Touch the red lead of the DMM to the alternator's output and read the voltage. What is the voltage reading? If the alternator's output reading is more than battery voltage, what is the most probable cause of the dropped voltage? What would be your recommendation in that case?

____

☐ **19.** If the alternator's output voltage and battery voltage does not show a charge without or with electrical loads, replace the alternator.

____

____

Instructor's Response ______________________________

____

____

____

JOB SHEET 

Name ______________________________ Date ______________

USE WIRING DIAGRAMS TO LOCATE, INSPECT, TEST, AND SERVICE CIRCUIT PROTECTIONS DEVICES, SWITCHES AND CONDUCTORS

NATEF Correlation

This job sheet addresses the following NATEF tasks:

Use wiring diagrams to trace electrical/electronic circuits.

Inspect and test fusible links, circuit breakers, and fuses; determine necessary action.

Inspect and test switches, connectors, and wires of starter control circuits; determine necessary action.

Objective

Upon completion and review of this job sheet, you should be able to use the wiring diagrams to locate, test, and service circuit protection devices, relays, switches, and conductors for serviceability.

Tools and Materials Needed

Wiring Diagram
Digital multimeter (DMM)
Test light
Service manual
Safety glasses
Assortment of good and bad circuit protection devices, switches, relays, and conductors (to be used before testing is done on a vehicle.)

NOTE TO INSTRUCTOR: The protection devices with a visible means of detecting the devices' serviceability should have that portion taped or painted over to prevent the student from doing a visible inspection. The short piece of conductor can be wrapped to cover any damage and should have a terminal at each end.

Procedures

Task Completed

1. Determine the following information from the service manual. ☐

   Vehicle Make______________ Model______________ Year______________

   Location of the protection devices for the following circuits: headlight, stop light, turn signals, air conditioning, blower motor, cooling fan motors, fuel pump, horn. Record the location and amperage for each device.

   ____
   ____
   ____
   ____
   ____

Task Completed

Location of the brake light switch, clutch switch (if equipped), power steering pressure switch, air conditioning low pressure switch. Record their location.

Begin the tests off-vehicle using the instructor's supplied devices.

☐ **2.** Set the DMM to ohms scale and test each device for serviceability. Record your results for each device measured. Determined if the device is usable.

Fuse 1 ________ Fuse 2 ________

Fuse 3 ________ Fuse 4 ________

Circuit breaker 1 ________ Circuit breaker 2 ________

Circuit breaker 3 ________ Circuit breaker 4 ________

Relay 1 ________ Relay 2 ________

Switch 1 ________ Switch 2 ________

Fusible 1 ________

Conductor ________

☐ **3.** Before testing the on-vehicle devices, ensure that none are attached directly to an electronic circuit. If one is in an electronic circuit, which tester should be used?

☐ **4.** Locate the fuse panel housing the fuses and circuit breakers.

☐ **5.** Remove the fuse housing panel.

☐ **6.** Locate and test the following protection devices in-place on the vehicle using a test light.

Headlight, Air conditioning, Stop light, Blower motor, Horn, Cooling fan motors, Turn signals, Fuel pump

☐ **7.** What is the specified amperage for each?

Headlight ________ Air conditioning ________

Stop light ________ Blower motor ________

Horn ________ Cooling fan motors ________

Turn signals ________ Fuel pump ________

☐ **8.** Are the correct protection devices in the correct slot? If not, replace with the correct device.

☐ **9.** Connect the test light clip to a good ground. Touch to each test point for each device. Record the results as good or bad. Good is indicated when the lamp lights at both test points on any single device. If a device is bad, which electrical defect is indicated?

Task Completed

Headlight ________________ Air conditioning ____________

Stop light ________________ Blower motor ____________

Horn ________________ Cooling fan motors ____________

Turn signals ________________ Fuel pump ____________

10. Use the wiring diagram to locate each of the following components. ☐

Air conditioning and horn: relays, switches, easy to access conductor terminal and/or splices

NOTE: The following tests requires the use of a DMM unless the service manual states otherwise.

11. Locate the relay and low pressure switch for the air conditioning system. ☐

12. Remove the relay from its mounting and determine which terminals are for the interior coil. Many times the interior schematic is printed on the outside of the relay case. ☐

13. Set the DMM to read ohms. ☐

14. Connect one lead to each coil terminal. What is the resistance reading? ☐

____

15. Move the DMM leads to the relay's other terminals. One will be battery (B+) power in and the other battery out. Measure the resistance between the two. Is the relay serviceable? ☐

____

16. Use the relay schematic and/or the wiring diagram to locate the battery (B+) cavity in the fuse housing. Double check to make sure the right cavity is selected. ☐

17. Set the DMM to read DC voltage. Connect the DMM black lead to a good ground. ☐

18. Probe the B+ cavity and record the reading. Record the reading. Is it within one volt of battery voltage? It may be necessary to do an open voltage measurement of the battery to make the comparison. ☐

____

19. Locate the low pressure switch for the air conditioning and unplug the electrical connection. If this AC system has a combined high/low pressure switch, do not test. Instead use the stop light switch under the dash. The same technique applies. ☐

20. Set the DMM to read ohms. ☐

21. The terminals on the switch are usually small pins and difficult to get an accurate connection with the DMM leads. It may be necessary to make a short pigtail to plug into the switch with bare conductor ends so a good connection between the DMM and AC switch can be achieved. If using the stop light switch, a pigtail is probably not required. ☐

22. Connect the DMM leads, one to each switch terminal. What is the reading? Is the switch open or closed? According to the service information what should it be now? ☐

____

23. Locate and disconnect the electrical connection at the AC compressor. ☐

Task Completed

☐ **24.** With the DMM set to read resistance, connect the leads, one to each of the compressor terminals. It may be necessary to use a pigtail again. Record the reading. Is the circuit open or closed?

___

WARNING: Ensure that the ignition key is OFF and the blower motor and AC switches are OFF. The DMM can be damaged if a powered circuit is connected to the leads with the DMM in ohms mode.

☐ **25.** Move one of the DMM leads to a good ground. Probe the other lead to each terminal of the vehicle's AC compressor harness connection. Which terminal shows a resistance? What this mean?

___

☐ **26.** Reconnect the AC electrical connection.

___

Instructor's Response ___

___

___

___

JOB SHEET 28

Name ______________________ Date ____________

Inspect, Service, and Test the Battery

NATEF Correlation

This job sheet addresses the following NATEF tasks:

Perform battery state-of-charge test; determine necessary action.

Confirm proper battery capacity for vehicle application; perform battery capacity test; determine necessary action.

Maintain or restore electronic memory functions.

Inspect and clean battery; fill battery cells; check battery cables, connectors, clamps, and hold-downs.

Objective

Upon completion and review of this job sheet, you should be able to test and diagnose problems related to the battery.

Tools and Materials Needed

Hand tools
VAT 40/60 or other battery load tester
Digital multimeter (DMM)
Service information
Fender cover
Service manual
Safety glasses

Procedures

Task Completed

1. Determine the following information from the service manual. ☐

   Vehicle __________ Make __________ Model __________ Year __________

   What is the correct battery for test vehicle ______________

   Location of the battery ______________

2. Ensure the wheels are chocked, transmission in parked (neutral for manual). ☐

3. Is the battery the correct one for this vehicle? If not, can tests be conducted? ☐

   ____

4. Begin the inspection of the battery and its related components. Record your results. ☐

   Before proceeding are there any odors of possible gases resulting from a discharged or overly charged battery. If yes, stop and consult your instructor.

   ____

   Battery cleanliness ______________

   Battery cable terminals status ______________

   Visible portions of the battery cables ______________

   Alternator drive belts ______________

   General condition of the engine/engine compartment ______________

Task Completed

If not a maintenance-free battery, remove caps and check fluid levels ______________

____

☐ **5.** Make a quick check of the battery's voltage by connecting the DMM across the battery terminal, red to positive, black to negative. Who is the voltage? Is it sufficient to properly test the battery?

____

☐ **6.** If the battery voltage is sufficient to continue the testing. If not, stop and consult your instructor.

☐ **7.** Measure voltage between the battery posts and their corresponding cable terminals. Is there excessive voltage drop between the posts and the terminals? If so, repair as needed.

____

☐ **8.** Review the operating instructions for the battery tester. Connect the battery tester to the battery as specified.

WARNING: Do not put a high current load on the battery longer than that specified by the tester's instructors. Serious damage to the battery or tester could occur.

WARNING: If this is a very new model vehicle with many electronic systems, consult the service manual for the proper test procedures. It may be necessary to disconnect the battery from the vehicle wiring before conducting a capacity or conductive test.

☐ **9.** Follow the battery's tester instruction, apply the test load to the battery and observe the results.

____

☐ **10.** What does this capacity test indicate of this battery?

____

☐ **11.** Prepare the conductive tester (may require a different tester) to perform a battery conductive test.

☐ **12.** Connect the conductive tester to the battery as specified by the tester instructions.

☐ **13.** Perform the conductive test on the battery following the tester's instructions. Record the results.

____

☐ **14.** Based on the test results and your inspection, is the battery serviceable enough for vehicle operation. What factors would cause you to recommend a new battery?

____

☐ **15.** Disconnect and store the tester(s). Prepare the vehicle for the customer.

Instructor's Response ____

____

____

____

Chapter 10

Servicing Brake Systems

Upon completion and review of this chapter, you should understand and be able to describe:

- The different types of automotive brake fluids and methods to test them.
- How to diagnose brake systems and perform a basic inspection.
- The safety precautions associated with ABS and air bag systems.
- How to test, inspect, and replace brake power boosters.
- How to test, inspect, and replace master cylinders.
- How to inspect, clean, and replace disc brake pads.
- How to inspect, clean, adjust, and replace drum brake shoes.
- How to machine brake drums and rotors.
- How to inspect, clean, and adjust parking brake cables.
- The general replacement procedures for brake components.

Introduction

The brake system provides safe, controllable slowing and stopping of the vehicle. Considering that most vehicles weigh between one and two tons and may be traveling at interstate highway speeds, major failure of a component can lead to a life-threatening or at least an unnerving situation. Though brakes are fairly simple to repair, brake system service requires the technician to accurately install and test components to prevent a possible accident.

This chapter will cover basic brake diagnosis and repairs. Since antilock brake system (ABS) and traction control systems generally require the use of a scan tool and special procedures for bleeding and installing some components, repairs to those systems will not be discussed. For specific information on ABS and traction control systems, the technician may refer to *Today's Technician: Automotive Brake Systems*, 5th Edition, Delmar Cengage Learning, for detailed instruction on the two systems' operation, testing, and repair.

Classroom Manual
pages 251, 252

Brake Fluid Inspection and Servicing

Before beginning the diagnosis of a brake system, it would be wise to consult the service manual for the type of brake fluid that is to be used. There have been instances where adding a few ounces of fluid to top off the system have led to expensive repairs done at the shop's expense.

There are three basic compounds used for brake fluids: **glycol-based**, silicone-based, and petroleum-based. Installation of the wrong fluid into a system results in quick and complete deterioration of all the brake components that have a rubber seal. European vehicle manufacturers used a petroleum-based fluid in some of their older vehicles. Petroleum-based brake fluid has been discontinued for many years, but a classic vehicle refurbisher may encounter it.

Glycol is a member of a group of alcohols. It should not be confused with prefix "glyco," which is a combined form of glycerol, a sugar.

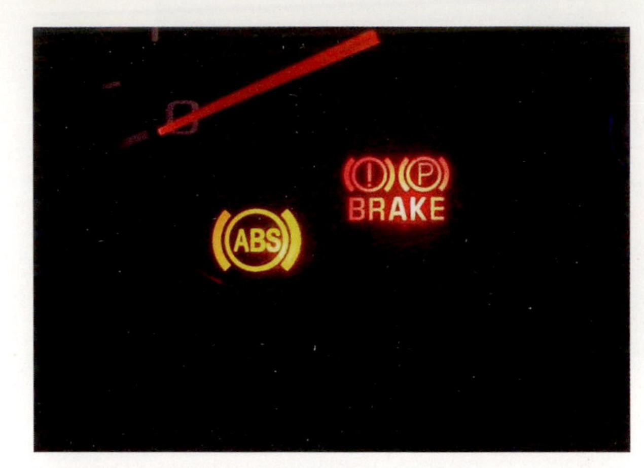

FIGURE 10-1 **Silicone-based brake fluid cannot be mixed successfully with glycol-based fluids and never used in current ABS.**

Silicone-based brake fluid is used on a few older vehicle makes and models (Figure 10-1). An inattentive technician may assume the brake fluid is glycol-based, top it off without checking, and cause serious problems in the brake system. Using the wrong type of fluid will cause deterioration of the sealing components and degrade the braking operation. Even if the mistake is caught before the vehicle leaves the shop, the entire system must be taken down and all seals and flexible hoses must be replaced. This should be done at the shop's expense and the technician's expense. Always consult the service manual.

Glycol-based brake fluids are used in almost every brake system sold. Their popularity and frequent use are the reasons why mistakes, like the one listed above, occur. This brake fluid performs well and is fairly durable.

Brake fluids are classified by their resistance to boiling. The classes are DOT 3, DOT 4, and DOT 5 with the highest, 5, having the highest boiling point. DOT 3 is the most common brake fluid. If a system is being flushed, a lower-rated brake fluid should not replace a higher-rated one (DOT 3 *cannot* replace DOT 4), but a higher-rated brake fluid may replace lower-rated ones (DOT 4 may replace DOT 3). Consult the service manual for the proper type and classification.

The following precautionary statements should be as much a part of the technician's professionalism as the tools being used every day.

WARNING: Do not mix glycol-based and silicone-based brake fluids. The mixture could cause a loss of braking efficiency.

WARNING: Protect the vehicle's paint surface from glycol-based brake fluids. This fluid can damage and remove the finish very quickly. If brake fluid gets on the vehicle's finish, wash the area immediately with cold, running water.

WARNING: Store brake fluid in a sealed, clean container. Brake fluid will absorb water and its boiling point will be lowered. Exposure to petroleum products such as grease and oil may cause deterioration of brake components when the contaminated fluid is installed.

WARNING: Do not use petroleum-based (gasoline, kerosene, motor oil, transmission fluid) or mineral-oil-based products to clean brake components. These types of liquids will cause damage to the seals and rubber cups and decrease braking efficiency.

CAUTION: Wear eye protection when dealing with brake fluid because it can cause permanent eye damage. If brake fluid gets in the eyes, flush with cold, running water and see a doctor immediately.

Diagnosing Brake Systems by Inspection and Road Testing

Brake system failures can be classified as either hydraulic or mechanical failures. Both will create symptoms that can be readily connected to a component or assembly. Mechanical failures will be discussed by common symptoms and their probable causes.

Two conditions constitute hydraulic failure: external leaks and internal leaks. Although mechanical conditions will create the leaks, they will be addressed as hydraulic problems to clarify the operational symptoms. An external leak is defined as "brake fluid exiting the system completely." An internal leak is defined as "a fluid leak within the system where fluid does not exit the system."

Classroom Manual pages 251, 252

External Leaks

A driver may complain that when the brakes were applied, the pedal dropped drastically and it took much more distance to stop the vehicle. The most common cause of this problem is an external leak somewhere in the system. A complete leak where the fluid exits the system results in a complete loss of fluid and pressure. The pedal will drop much lower than normal or go to the floorboard immediately upon brake application. If a single-piston master cylinder is being used, the pedal will hit the floorboard and cause a complete loss of all brakes. Pumping the pedal in an effort to regain control only compounds the problem. On a dual-piston master cylinder, the pedal will drop much lower than normal, but the driver will have brakes on one of the split systems and can stop the car. Again, pumping the pedal will not improve braking. An external leak of this type is noticeable when performing a visual inspection. The brake fluid level will also drop in the reservoir.

If the master cylinder is leaking externally, brake fluid will be leaking between the rear of the master cylinder and power booster or firewall (Figure 10-2). Usually, an external

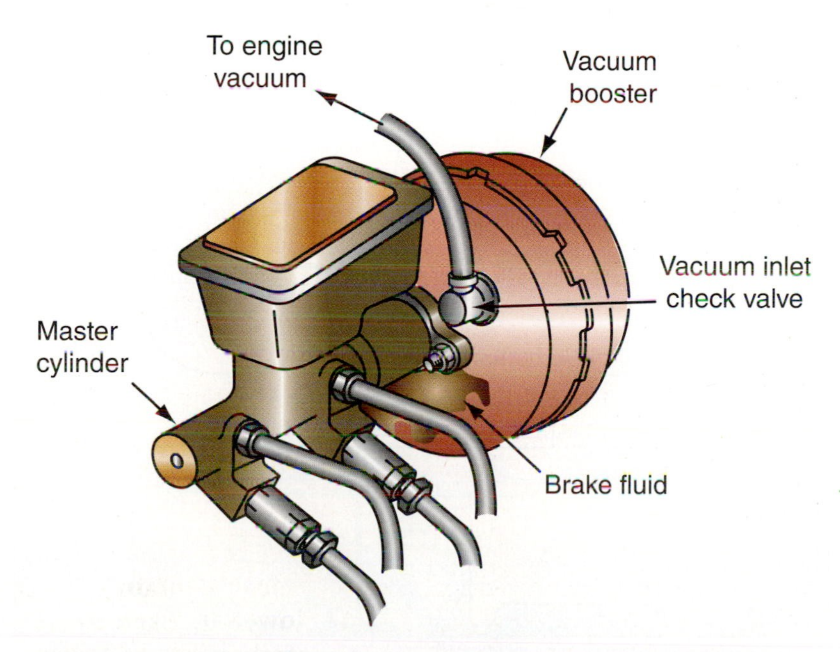

FIGURE 10-2 **Missing paint and bare metal on the booster or firewall is an indication of an external master cylinder leak.**

SERVICE TIP: At times, a customer will complain of the brake warning light flashing on when the vehicle is turning a corner or stopping. This is caused by the low fluid level and may be an indication of a leak or normal wear and tear. As the disc pads wear, the piston moves outward and pulls fluid from the reservoir. Before "topping off" the reservoir, check the wear on the disc pads. They may be near the replacement point.

master cylinder leak will not result in complete brake failure until the fluid in the reservoir is exhausted. An external leak at a line, wheel cylinder, or caliper will cause a total loss of brakes on that part of the split system. The leaking brake fluid will be visible on the inside of the wheel assembly and may be dripping from suspension or steering components. Small external leaks may be present but not readily visible. The leak is indicated when the reservoir must be topped off regularly. This type of leak is usually found at wheel cylinders during routine brake service. During inspection, the dust boots of the wheel cylinder are opened just enough to check for fluid inside the boots. The inside of the boots should be completely dry.

Internal Leaks

Drivers of older vehicles often complain of slow brake pedal drop while sitting at a stoplight. This situation is not as common on newer vehicles because of advances in machinery and materials, but it still occurs. The symptoms will usually occur as the car ages. Notice that the driver can stop the vehicle normally, but the braking action seems to lessen or disappear when the pedal is depressed for a time. Since the vehicle will stop with normal brake pedal movement, an external leak is eliminated. However, a pedal will drop when there is a loss of pressure, which points to an internal leak. An internal leak is almost always found in the master cylinder. Any other component that leaks will result in an external leak with the possible exception of ABS modulators. The modulators can leak internally, causing the same symptoms noted with a leaking master cylinder. However, because of regulations and new technology, this problem is usually found only on the isolation or dump solenoids on older rear wheel antilock (RWAL) systems.

An internal leak in the master cylinder allows fluid to flow from the compression chamber, past the seal or cup, and back into the reservoir (Figure 10-3). High pressure created during stopping may prevent the seals from leaking. However, as the driver reduces force to hold the vehicle in place, the pressure drops slightly. This pressure reduction is enough to relax the seals and allow fluid to seep. The fluid returns to the reservoir and the master cylinder pistons move forward, dropping the pedal. Eventually, the condition could cause the pedal to reach the floor if the brakes are applied long enough. The fluid level in the reservoir will remain the same. Usually, the cheapest, shortest, and best repair is the replacement of the master cylinder. There are rebuild kits for some master cylinders, but the additional labor cost may exceed the cost of a replacement component.

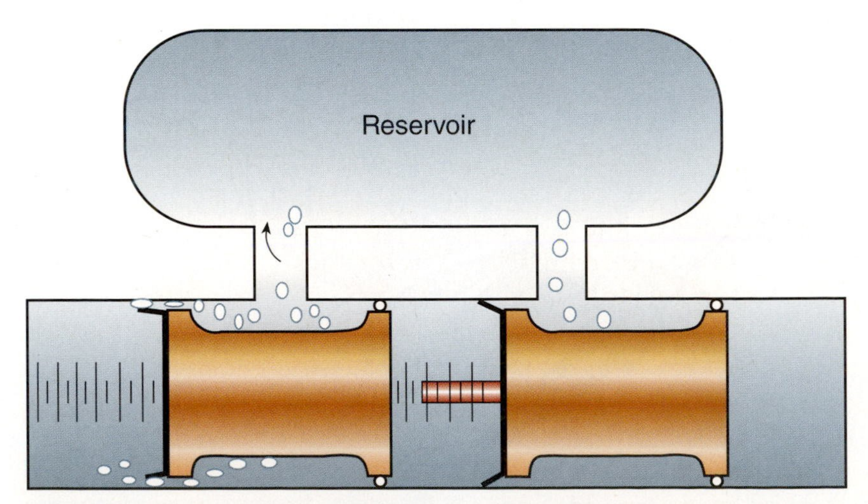

FIGURE 10-3 **If the piston cup is corroded or damaged, pressurized fluid can escape back to the reservoir, causing the brake pedal to drop.**

Grabbing or Locking

A **grabbing brake** occurs when one or more wheel brakes grab the rotor or drum and then quickly release it. The vehicle jerks as it slows. This situation could be caused by oil or brake fluid on the pads or shoes. **Hot spots** on the drum or rotor can also cause the problem as will worn or weak retaining hardware. Grabbing brakes are more common with drum brakes.

Grabbing brakes are usually caused by oil or brake fluid on the friction material.

A **locking brake** is one that will cause the wheel to stop turning and slide when the brakes are applied normally. This is caused by the friction material on that pad or shoe engaging with more friction than the other wheels. This can be caused by oil or brake fluid on the friction material or a brake that is adjusted too tightly. Brake fluid in particular will soften the friction material, which tends not to slide on the drum or rotor. Shoes and pads that have been contaminated by brake fluid or oil must be replaced and the rotor or drum cleaned. A second cause could be the opposing brake's failure to work correctly. For instance, if the left front wheel consistently locks, the first thing to check is the right wheel brake to make sure it is working. Locking brakes are more common with drum brakes because leaking brake fluid flows directly on the shoes. A locking disc brake can usually be traced to a failure of the brake on the opposite wheel.

Hot spots are formed by excessive heat on the weakest area of the disc or drum.

Locking brakes may be caused by mis-adjusted shoes, blockage in the brake line, or slick tires/road surface.

Pulling to One Side

In this instance, a steady, consistent, smooth pull caused by brake failure on the opposite side of the pull is being discussed. The driver must hold the steering wheel to one side to counteract the pull. Correcting the problem can be accomplished by adjusting the brake shoes or cleaning the slide pins, assuming there is no mechanical damage or bad parts. A pull to the side is often caused by a problem on the front axle, but an overly tight rear brake may cause a steady drift (light pull) to that side during slow to moderate braking. Excessive position camber may also cause pulling to one side during braking, but the driver will probably report this as a steering symptom since it will be present without braking.

Hose Damage

Another pulling condition can be caused by internal damage to a brake hose. The internal lining of the flexible hose can come apart, and every time the brake is applied, the torn lining moves and blocks the fluid flow (Figure 10-4). The result is good braking on one wheel and little or none on the damaged side. At times this same condition may exist, but the vehicle does just the opposite. The torn lining will allow fluid to flow to the brake component, but blocks its return when the brakes are released (Figure 10-5). This may cause a pull toward that brake during non-braking conditions. During the service, it will be noted that the linings for this wheel have worn much more than the other side, which is an indication that the caliper or wheel cylinder is stuck in the applied position. Usually, there are no external signs of the blockage. Over a period of time, the hose could become weak enough for a bulge to form during braking, but the driver will probably have the brakes checked before this happens. If a thorough inspection is completed without finding the cause of the locking brake, the technician should suspect the brake hose on the non-braking side. This hose can be removed and the interior can be inspected by straightening the hose and looking at a light through the hole.

Fading

Fading occurs when the brakes are applied for long distances such as driving down a long grade. The brakes will appear normal at first, but braking action will be reduced and pedal feel will become firmer the longer the brakes are applied.

When the brakes are applied for long distances, they heat and lose friction. As the friction between rotor or pad and drum or shoe decreases, the braking action decreases.

Fading is a condition where the brakes initially work but lose their stopping power because of heat or other conditions affecting the friction material.

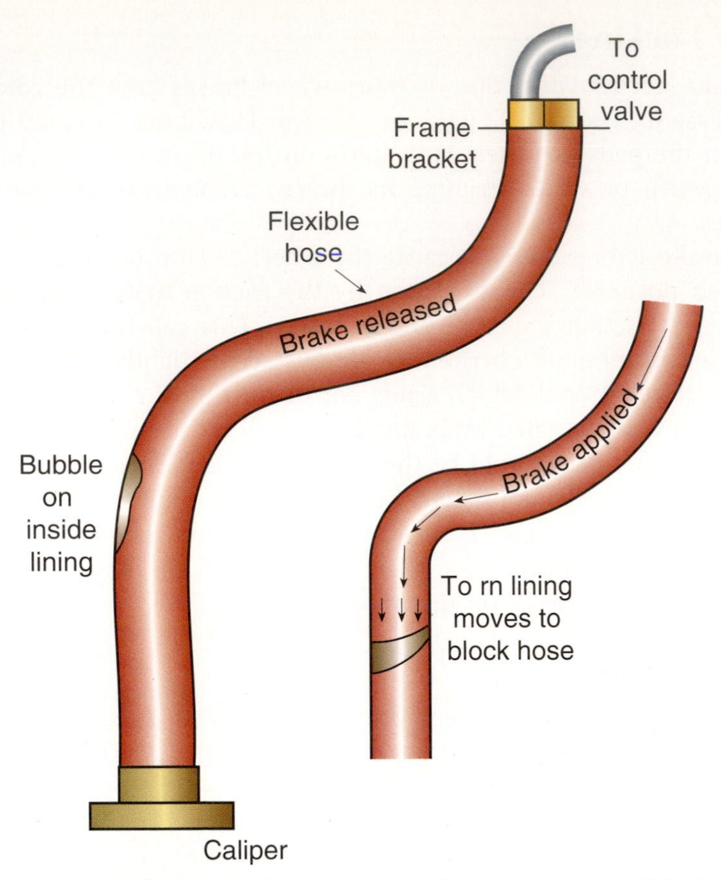

FIGURE 10-4 **A torn or damaged interior lining can move and block the hose during brake application.**

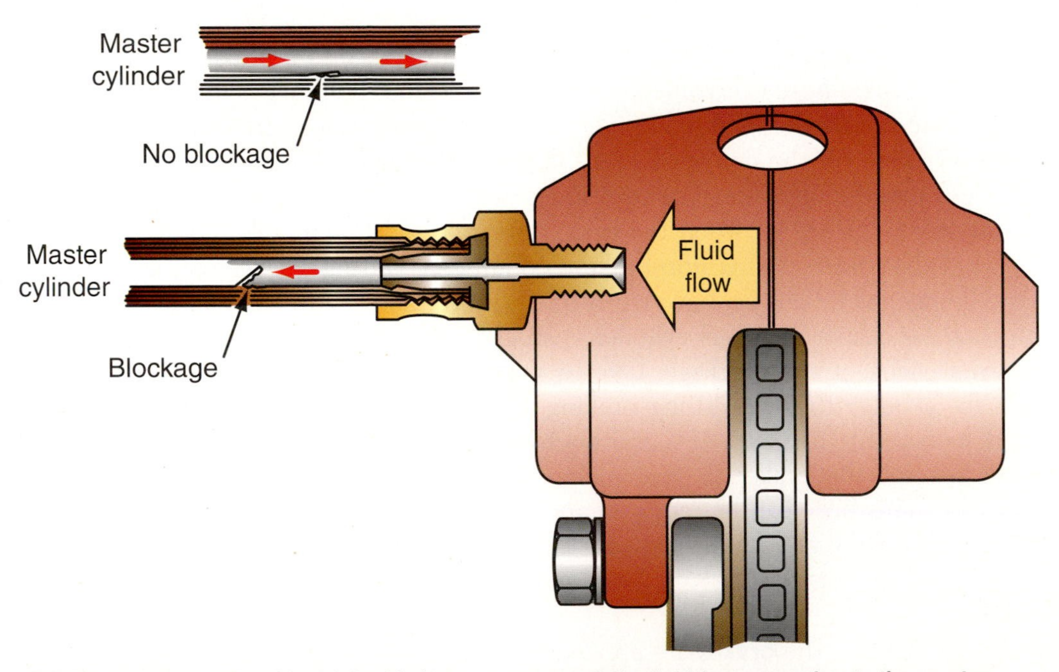

FIGURE 10-5 **The internal break in this hose may prevent the fluid from returning to the master cylinder, creating a constant brake application on this wheel.**

FIGURE 10-6 **This sign indicates that a slower gear should be selected before beginning descent. This reduces the need to apply the brakes during descent.**

A **hard pedal** condition is when high driver input braking force does not not sufficiently slow the vehicle.

Glazing is a condition when the friction material and rotor/drum developed a slick near non-friction surface. Usually caused by heat.

The driver applies more force and the heating becomes more intense. The eventual result is a complete loss of braking and a very **hard pedal**. Releasing the brakes will allow them to cool, but damage to the brake friction components may have occurred. When a light-vehicle driver complains of a hard pedal but poor braking condition, the technician should check for a trailer hitch. Pulling a trailer with a light vehicle can cause a fading condition.

Many times, signs are posted at the top of steep slopes warning drivers, particularly truck drivers, to select a lower gear before going down the hill (Figure 10-6). The vehicle's driveline and engine are used to slow the vehicle and reduce brake usage.

Hard Pedal but Little Braking

When a driver applies the brake and has a hard pedal, the vehicle may not slow as it should. Water between the rotor and pad or drum and shoes will cause this condition. The water will burn off quickly and the brakes will return to normal, but repairs are required if the same condition occurs every time the brakes are applied.

In most cases, the condition stems from brakes that have been overheated. Failure to completely release the parking brake, repeated hard or long braking, pulling a trailer, misadjusted drum brakes, or a stuck caliper will cause the friction material to glaze from heat. The result is loss of friction between the pads or rotor and the shoe or drum. The pad or shoe slides over the machined surfaces without gripping. The primary cause is a driver who consistently waits until the last moment to apply the brakes. Hand sanding the friction material might provide a quick fix to remove the glaze, but the correct repair procedure is to replace the shoes or pad and machine or replace the drums or rotors.

An overheating or **glazing** on one wheel may cause the vehicle to pull to one side during braking. Like other braking faults, pad or shoe glazing may not be obvious on first inspection. A glazed pad or shoe normally produces a very shiny, hard look to the rotor or drum. Bluish-colored spots on the machined surface are usually a sign that the brakes have been overheated and hot spots have formed on the braking surface.

SERVICE TIP:
The wrong shoes or pads can cause a hard pedal. They can also create almost all of the brake symptoms discussed in this section. A hard pedal is defined as a pedal that will not depress past a certain point, but the vehicle does not slow as it should.

Pedal Pulsation and Vibration

Pedal pulsation is a rapid up-and-down movement of the brake pedal during brake application.

A correctly machined drum and rotor, along with good, clean shoes and pads, will bring the vehicle to a smooth stop without vibration. During the life of the brake pad or shoes, heat attacks the working components during every braking action. This continuous heating and cooling, sometimes under poor weather and road conditions, causes the rotor or drum to warp or wear unevenly. Repeated hard braking increases the likelihood of damage. Disc brake systems are most susceptible to **pedal pulsation** and vibration during braking. Vibration on drum brake systems can be traced to hot spots on the drum, damaged shoes, or worn or damaged mounting components. For the most part, drum brakes do not produce noticeable pedal pulsation but they do tend to grab.

On a disc brake rotor, braking surfaces that are not parallel will cause pedal pulsation. The rotor has two machined braking or friction surfaces that should be parallel to each other. If one surface has a high or low area compared to its companion, or if the rotor is warped, the brake pad and piston will be pushed in and out of the caliper during braking (Figure 10-7). This action is transmitted back through the brake system, tries to kick up the brake pedal, and is repeated on every revolution of the rotor. In addition, there will be several points on the rotor that are not parallel or warped. The result is a light-to-heavy pulsation of the pedal and vibration of the vehicle. The damage could be severe enough to jerk the steering wheel back and forth as the two front brakes pulsate at different times.

Customers who are not familiar with the action of ABS may complain that at times the brakes seem to vibrate and shudder during heavy braking, or the vehicle feels like it is coming apart during a panic stop. Loose gravel, mud, wet leaves, or anything that makes the road surface slick can cause the ABS to function under light to moderate braking. A brake inspection and customer education may be the only repair needed.

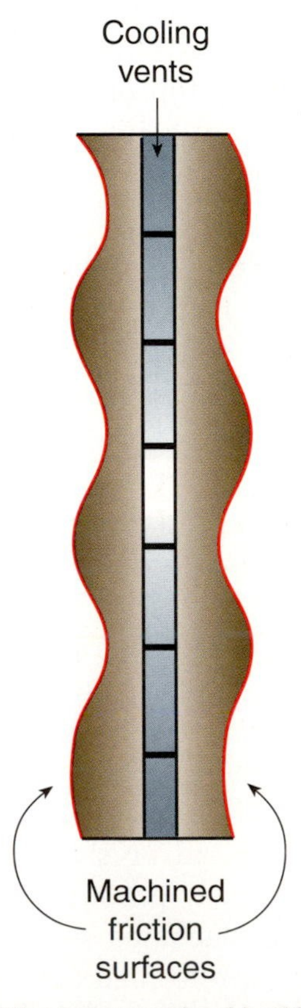

FIGURE 10-7 **If the two-machined braking surfaces of a rotor are not parallel, the brake pedal will pulsate during brake application.**

Spongy Pedals

A driver may complain that the vehicle stops quickly and quietly, but the pedal feels soft or bouncy. This is known as a **spongy pedal** and is caused by air in the brake system. As discussed in the section on hydraulic theory, liquids cannot be compressed. However, air can be compressed and used to perform certain types of work even though it does not transmit force well. When air enters the brake system, it forms an air pocket and compresses or expands every time the brakes are applied and released. What the driver feels through the pedal is the compression of the air pocket as the brake fluid tries to transmit the driver's force to the wheels. It will take more driver force to achieve the same braking action. The only solution is to locate and repair the air leak and bleed the entire brake system.

A **spongy pedal** is the result of air in the hydraulic system. Feels like stepping on an inflated balloon.

SERVICE TIP: Old brake fluid can cause an unsafe condition. Over time the fluid is heated, cooled, and reheated. Brake fluid is hygroscopic, which means it absorbs moisture. This lowers the boiling point of the fluid, and as it is heated or cooled, some of the fluid will evaporate, causing gases to be formed with the system. Muddy-looking brake fluid is a sign that the fluid is old and contaminated and that the system needs a complete flush and probably a lot of new parts.

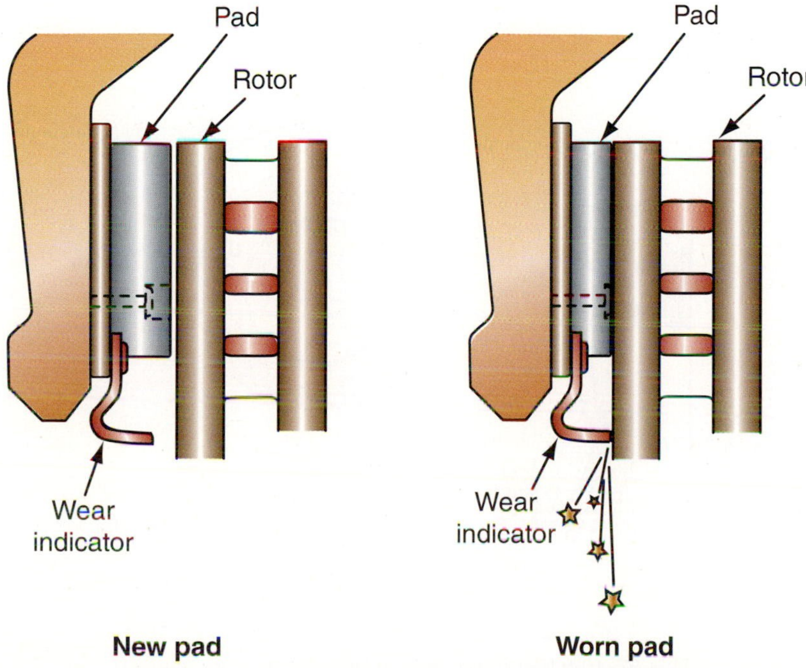

Grinding and Scraping Noises

Grinding and scraping noises are classic symptoms of completely worn out brake shoes and pads. The metal of the shoe web or pad mount comes into direct contact with the drum or rotor and causes the noise (Figure 10-8). The sound will become louder until it occurs even when the brakes are not applied. It may also be caused by other problems such as a shoe-return spring breaking and dropping loose from its mounts. However, the probable cause is a lack of friction material on the pad or shoe.

Another condition that can cause loud grinding or scraping noises is by installing the wrong type of pad or shoe. The friction material of the brake lining should always be suitable for the type of rotor used. With some composite rotors, a so-called "long-life pad" may cause very loud grinding noises during braking because metal particles or compounds in the pad friction material dig into the rotor surface and create a metal-to-metal noise. On a few vehicles, it may be necessary to install factory parts instead of aftermarket items to retain quiet braking.

Pad
Rotor
Wear indicator
New pad
Pad
Rotor
Wear indicator
Worn pad

FIGURE 10-8 **As the linings wear, the wear indicator eventually hits the rotor and creates noise to warn the driver that new pads are needed.**

Brake Booster Diagnosis, Inspection, and Replacement

Classroom Manual
pages 257, 258

WARNING: Before moving the vehicle after a brake repair, pump the pedal several times to test the brake. Failure to do so may cause an accident with damage to vehicles, the facility, or people.

CAUTION:
Do not work in or around the steering column without disarming the air bag systems. Serious injury could result from a deploying air bag.

WARNING: Before working on the brakes of a vehicle with ABS, consult the service manual for precautions and procedures. Failure to follow procedures to protect ABS components during routine brake work could damage the components and cause expensive repairs.

Master cylinders and boosters are easy to diagnose and repair. Listen to the operator's complaint, understand how the system works, and use the information presented on the last few sections to determine which component failed.

Brake Booster Operational Checks

To check the brake's power booster, turn off the engine and press down hard on the brake pedal several times to bleed (or exhaust) the booster's vacuum or hydraulic pressure. The pedal will rise with each stroke until it is firm and much farther from the floorboard than normal. Maintain force on the pedal and start the engine. When the engine starts, the pedal feel will become softer and the pedal will drop down as booster pressure builds.

Vacuum and Hydro-Boost

On vacuum and hydro-boost systems, bleed the boost pressure as previously discussed and start the engine (Figure 10-9). The pedal should immediately feel softer and drop lower. A firm, high, and steady pedal after engine start-up means there is no boost. On a hydro-boost system, attempt to turn the steering wheel. If power steering is present, the problem is in the valve mechanism mounted on the master cylinder. Consult the service

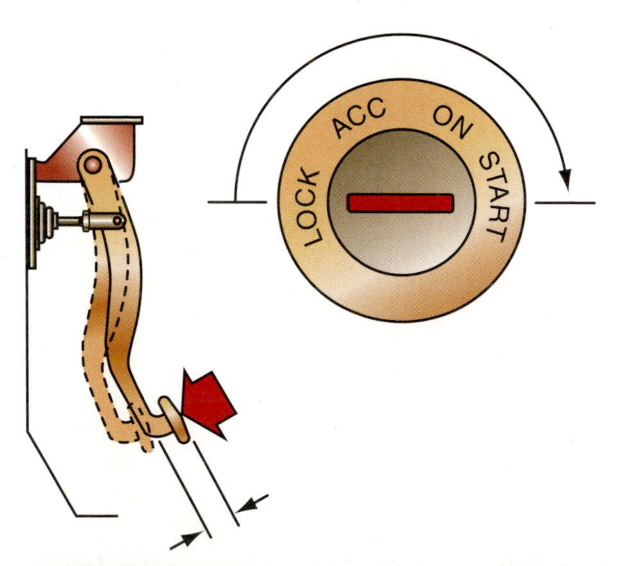

FIGURE 10-9 **Bleed the booster with the key off, hold the pedal down, and switch key to START. The pedal should fall away when the engine starts. (Reprinted with permission)**

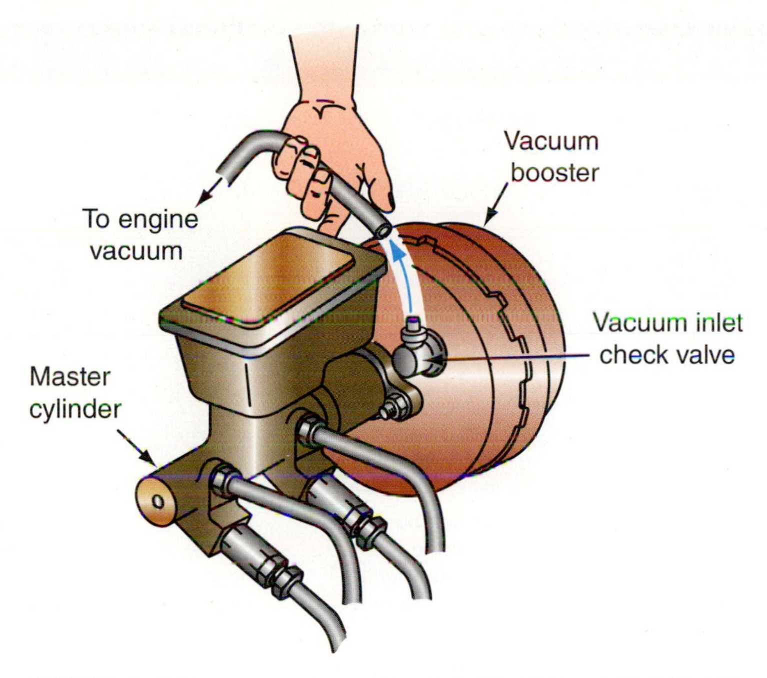

FIGURE 10-10 **Vacuum can be measured or felt at the booster end of the vacuum hose.**

Classroom Manual
pages 257, 258

CAUTION:
Do not work in or around the steering column without disarming the air bag systems. Serious injury could result from a deploying air bag.

A **vacuum pump** is a belt or gear driven pump to remove air from a system.

manual to determine if the valve mechanism can be replaced or serviced. Many manufacturers require that the master cylinder and valve be replaced together. A lack of power steering indicates that the power-steering fluid may be low, the power-steering pump is inoperative, or the power-steering belt is missing. Treat this condition as a steering component failure and perform the necessary repairs to that system.

A vacuum brake booster failure is caused by one of two things: booster failure or lack of vacuum. The first step is to disconnect the vacuum hose at the booster with the engine running (Figure 10-10). The engine will try to stall. Placing a thumb over the end of the hose will steady the engine and there should be a strong vacuum. A lack of vacuum usually means that the hose or check valve is closed or restricted. On a diesel engine, the **vacuum pump** and belt have to be checked if there is no vacuum at the brake booster hose. Photo Sequence 23 shows the typical procedure for vacuum booster testing.

SPECIAL TOOLS
Service manual
Wheel blocks

Vacuum Booster Replacement

If there is vacuum in the booster, consult the service manual for instructions on removing the booster and disarming the air bags. The general method of removing the booster is to disarm the air bag, disconnect the pushrod, disconnect and move the master cylinder, and remove the booster mounting fasteners (Figure 10-11). Do not disconnect the brake lines from the master cylinder. The master cylinder can be moved to the side without disconnecting the lines in most vehicles. Reverse the procedures to install the booster, vacuum hose, master cylinder, and the push rod. Check the service manual to determine if the push rod's length has to be measured and adjusted. The stoplight switch should be checked after installing the pushrod to the pedal.

SERVICE TIP:
An engine that operates poorly usually does not cause a total loss of vacuum in the brake booster. However, a tear in the booster's diaphragm can cause a good engine to stall and run roughly.

Test the new booster by starting the engine and applying the brakes. The pedal should be firm but easy to apply. The engine should not stumble or stall while the brakes are on. Stop the engine and pump the pedal to exhaust the vacuum from the booster. The pedal should rise in height above the floor and become very firm or hard to push. Hold the pedal down while starting the engine. The pedal should fall away as the engine is started. The brakes should be easy to apply with a soft but not spongy feel.

PHOTO SEQUENCE 23

Typical Procedure for Vacuum Booster Testing

P23-1 With the engine idling, attach a vacuum gauge to an intake manifold port. Any reading below 14 in. Hg of vacuum may indicate an engine problem.

P23-2 Disconnect the vacuum hose that runs from the intake manifold to the booster and quickly place your thumb over it before the engine stalls. You should feel strong vacuum.

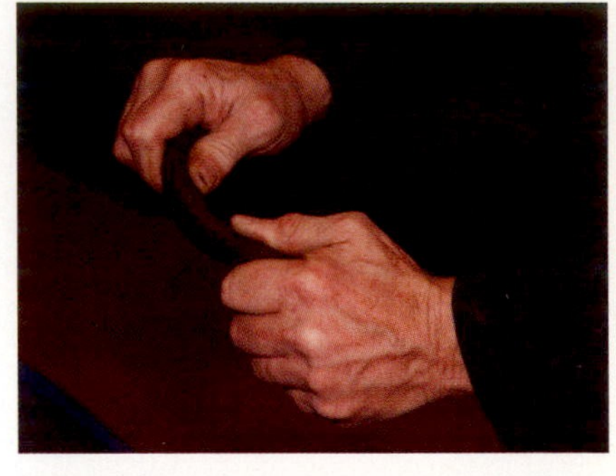

P23-3 If you do not feel a strong vacuum in step 2, shut off the engine, remove the hose and see if it is collapsed, crimped, or clogged. Replace it if needed.

P23-4 To test the operation of the vacuum check valve, shut off the engine and wait 5 minutes. Apply the brakes. There should be power assist on at least one pedal stroke. If there is no power assist on the first application, the check valve is leaking.

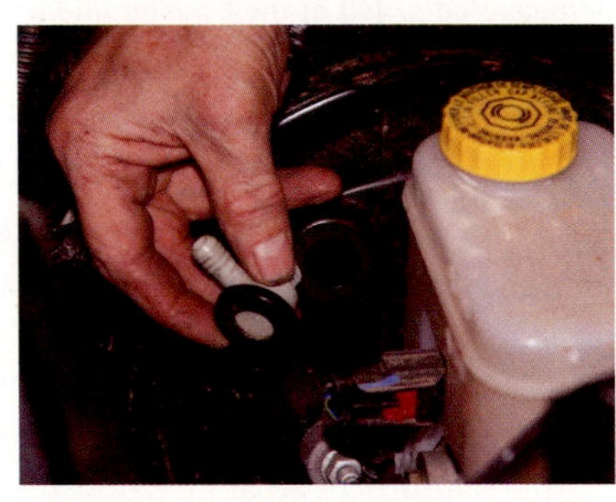

P23-5 Remove the check valve from the booster.

P23-6 Test the check valve by blowing into the intake manifold end of the valve. There should be a complete blockage of airflow.

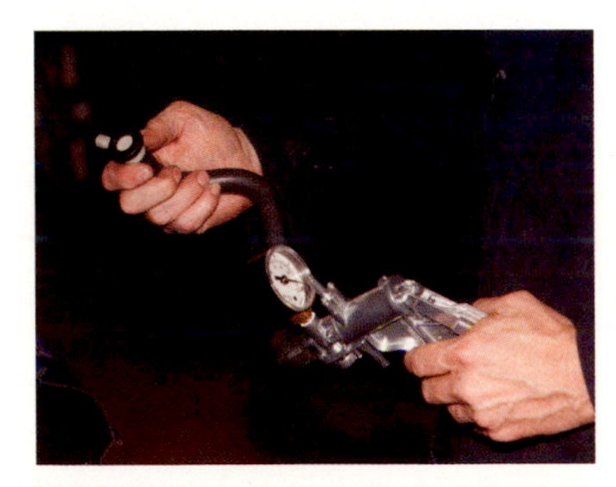

P23-7 Apply vacuum to the booster end of the valve. Vacuum should be blocked. If you do not get the results stated in step 6 and step 7, replace the check valve.

P23-8 Check the booster air control valve by performing a brake drag test. With the wheels of the vehicle raised off the floor, pump the brake pedal to exhaust residual vacuum from the booster.

P23-9 Turn the front wheels by hand and note the amount of drag that is present.

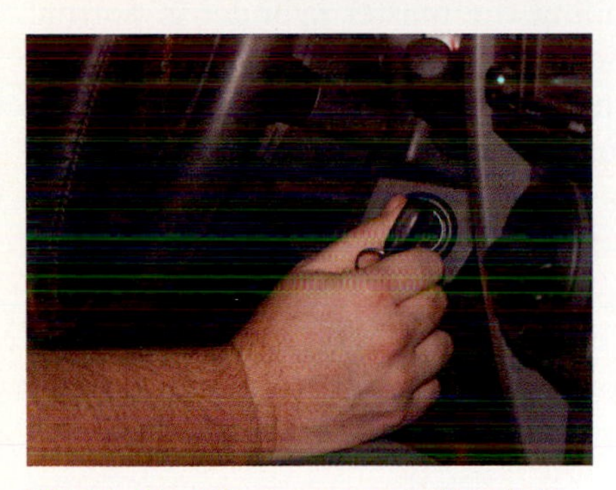

P23-10 **Start the engine and allow it to run for 1 minute, then shut it off.**

P23-11 **Turn the front wheels by hand again. If drag has increased, this indicates that the booster control valve is faulty and is allowing air to enter the unit with the brakes unapplied. Replace or rebuild the booster.**

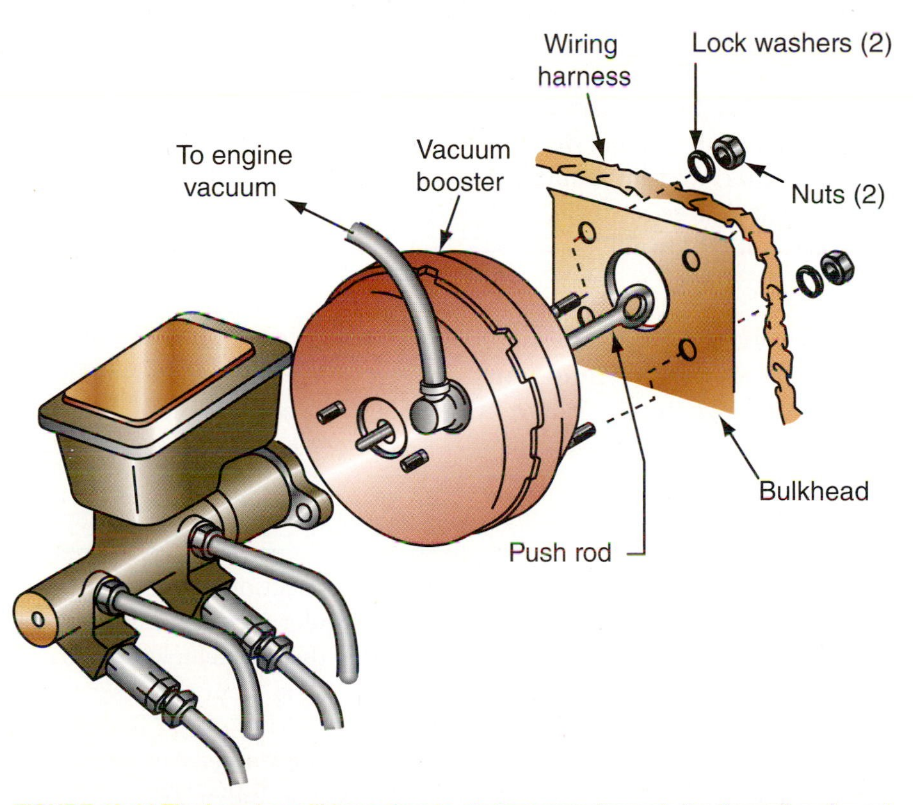

FIGURE 10-11 **The booster will have four to six fasteners through the firewall and must be removed from inside the vehicle.**

Classroom Manual
pages 259, 260, 261, 262

SPECIAL TOOLS

Wheel blocks

Vise

Master cylinder bleeder kit

Service manual

Master Cylinder Diagnosis, Inspection, and Replacement

WARNING: Handle master cylinders carefully to prevent spilling the fluid. Brake fluid is very corrosive and will cause paint and finish damage.

Master Cylinder Testing and Inspection

Check for cracks in the master cylinder housing, Look for drops of brake fluid around the master cylinder. A slight dampness in the area surrounding the master cylinder is normal. If the brake fluid is actually leaking, eventually the fluid level will be low enough and the warning lamp on the dash will illuminate. One of the best ways to check for leakage is to have a partner push on the brake pedal while you visually inspection the master cylinder and related components with a good lighting source.

Test driving the vehicle to determine a bad master cylinder can be difficult. You should always inspect the vehicle with the wheels off on a hoist before going for this test drive. Ensuring that the other mechanical components are not bad or causing issues will help determine and eliminate other components. If all other components check out to be good and you have a spongy brake pedal while braking, the master cylinder could have an internal leak.

Master Cylinder Replacement

The master cylinder creates and supplies the pressure for the brake system. An internal or external leak may affect both circuits of a dual system or just one circuit. In most cases, the master cylinder is fairly easy to replace. This discussion will cover the replacement of a master cylinder, with vacuum boost, on a non-ABS system. Consult the service manual before attempting to replace any brake components on an ABS system.

WARNING: Some ABS components may be located on or near the master cylinder. Consult the service manual on removing the master cylinder if any ABS or other system components present a problem to master cylinder replacement. Damage to the ABS or other systems may occur if proper procedures are not followed.

WARNING: Many ABS systems require special bleeding procedures. Do not remove a master cylinder on an ABS system before consulting the service manual. Damage to the brake system or loss of time could result if proper procedures are not followed.

Move the vehicle into the bay and block the wheels. Inspect the master cylinder mounting area. Remove the two or four metal brake lines attached to one side of the master cylinder body (Figure 10-12). An electrical connection for the fluid level sensor may be present. Four nuts should hold the master cylinder to the vacuum booster. Before loosening the nuts, place clean rags in the area under the master cylinder to capture any brake fluid that may leak. Remove as much fluid as possible from the reservoir.

Disconnect the electrical connection, if present. Select the proper line wrench and disconnect the brake lines. The flare nuts are loosened until the lines can be pulled out of the fitting. Bend the lines slightly so the master cylinder can be moved.

Select the correct wrench to remove the nuts. Generally, a ⅜-inch drive ratchet, socket and a 3- or 6-inch extension are the only tools needed to remove the cylinder's mounting nuts. With all of the nuts removed, slide the master cylinder toward the front of the vehicle until it clears the studs. Tilt the master cylinder slightly to one side with the line fittings up. This will reduce any leakage as the unit is moved.

Remove the reservoir cover and drain the remaining brake fluid into a suitable container. If necessary, remove the reservoir and install it on the new master cylinder using the new O-rings supplied (Figure 10-13). Place the old master cylinder in the box that the new one came in. It will be sent back to the parts vendor as a core for rebuilding. Secure a master cylinder bleeder kit or make one using two, short pieces of brake line and two fittings.

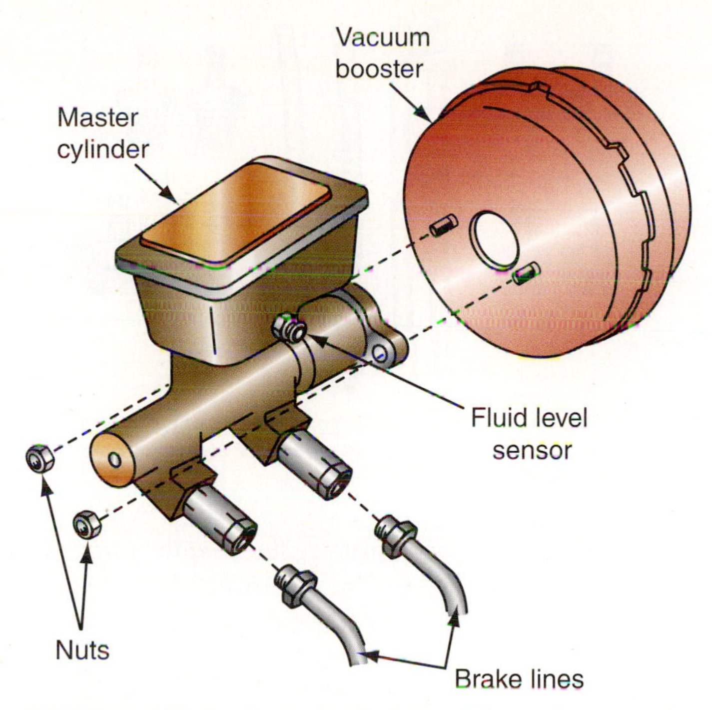

FIGURE 10-12 **Before loosening the brake fittings, place several cloths under them to catch any leaking fluid.**

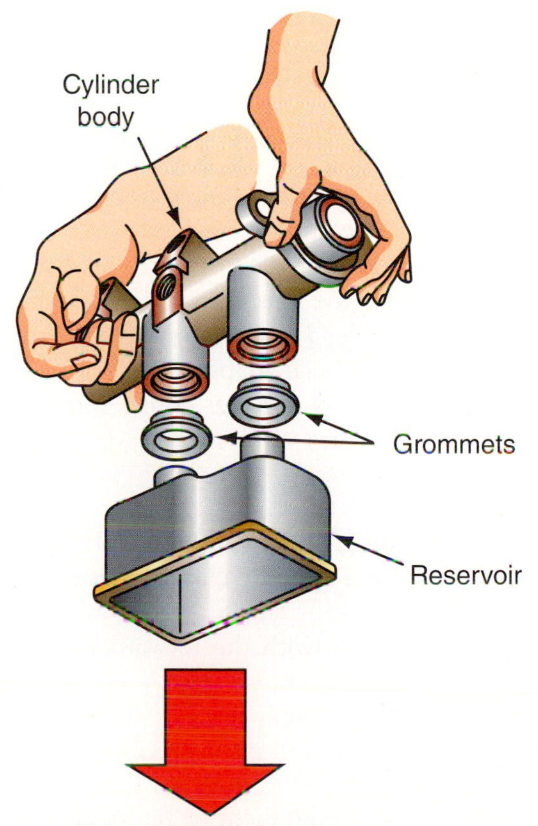

FIGURE 10-13 **Place the reservoir upside down, lubricate the new grommets, and press the reservoir straight down until it pops in place.**

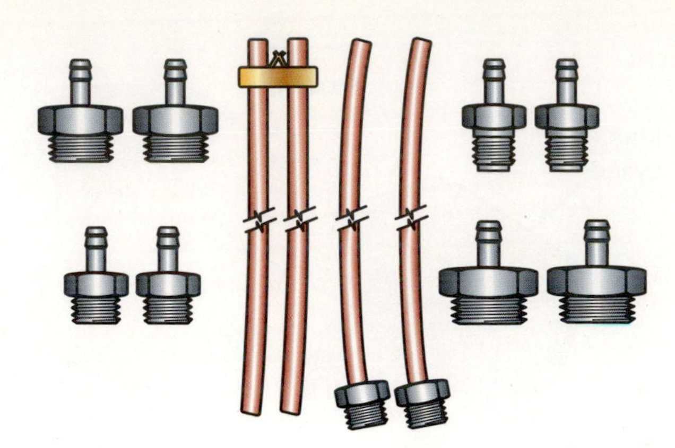

FIGURE 10-14 **This bleeder kit can fit several types and makes of master cylinders. (Courtesy of Snap-on Tools Company)**

The kit basically consists of two plastic lines with two plastic male fittings (Figure 10-14). The threads on the fittings are cut to fit several types of master cylinder threads and will flex to hold the fittings to prevent any leakage during braking. Usually, there are three types of threads on the fitting, each one a little larger than the one before it. The fittings are screwed into the line ports on the master cylinder. They do not need to be very tight because there is not much pressure applied during bench bleeding. Screw the fitting in by hand followed by only two or three turns with a wrench. Route the lines from the fittings back into the cylinder's reservoir, one per chamber. The lines should nearly touch the bottom of the chamber. The kit should include a small holding device to keep the lines in place during this procedure. Some master cylinders are packaged with a kit included, but most kits can be bought separately and kept on hand.

A bleeder kit can be made if necessary, but it is time consuming and can only be used for one type of master cylinder fitting thread. A brake line must be cut for two 6- to 8-inch pieces and bent to fit between the fitting and the bottom of the master cylinder's chamber. Select two male fittings that will slide over the line and screw easily into the master cylinder. The threads must be exact, since the fitting is either brass or steel and the wrong threads will cause problems. The fitting slides over the line and one end of the line is flared. The line is then bent to fit into the master cylinder chamber. Generally, a technician does not make a master cylinder bleeder kit unless one type of vehicle is being maintained or a special-purpose vehicle requires special adapters. For the most part, it's not worth the trouble to build a bleeder kit for each vehicle that comes into the shop when an inexpensive kit can be bought at almost any parts vendor.

WARNING: Before adding brake fluid, consult the vehicle's service manual. Most manufacturers require a specific classification of brake fluid to be used.

Mount the master cylinder in a vise with the forward end slightly elevated. Install the bleeder kit and remove the reservoir cover. Fill each chamber about half full with brake fluid. Use a punch or other rod to slowly apply force to the master cylinder's primary piston (Figure 10-15). Continue the stroke until both pistons bottom out. As the pistons are pushed in, air should escape from the bleeder tubes as evidenced by bubbles in the reservoir. Slowly release the force from the pistons and allow them to return to their resting position. Repeat the bleeding procedure until there are no bubbles visible in the reservoir during the pistons' inward movement. The rear or primary cylinder bore will probably bleed completely one or two strokes before the secondary chamber. During the bleeding process, always push both pistons to the bottom of their bore. Before removing the cylinder from the vise, close the

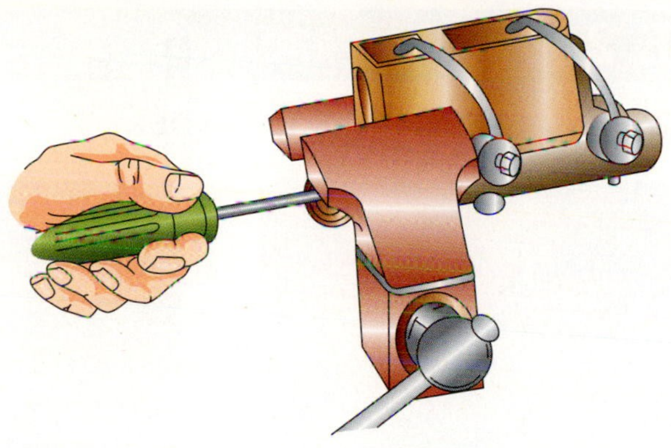

FIGURE 10-15 A screwdriver can be used to move the master cylinder pistons during bench bleeding.

two bleeder tubes with small paper clips and install the reservoir cover. The clips will prevent leaks and will prevent air from entering as the cylinder is installed on the vehicle.

Place the master cylinder in position on the brake booster and torque the fasteners. Position the two brake lines near their fittings. Remove one of the brake bleeder fittings and quickly insert the correct brake line into the master cylinder. Hand tighten the fitting enough to prevent leaking. Do not tighten completely at this time. Install the second brake line in the same manner. Drain and store the bleeder kit. See Photo Sequence 24 for details.

It is best to have a second person assist during the final stage of master cylinder bleeding. The next step is to remove any air that may have entered the line or cylinder fitting during the change over from bleeder to brake line. Ensure that enough rags are placed under the master cylinder brake lines to catch any fluid. Ask the second person to press slowly on the brake pedal. Both brake lines will be loosened and retightened on each stroke of the brake pedal. Loosen the rear line slightly as the pedal is being applied. Watch this small leak for signs of air. If no air is present, tighten the fitting and loosen the front line. Ensure that both fittings are tight before the brake pedal is released. Repeat this step until no air is being expelled past either fitting. Generally, one or two strokes of the brake pedal are sufficient. Complete the installation by tightening the fittings and topping off the fluid level. It will probably be necessary to bleed the entire brake system. Before moving the vehicle from the bay, apply the brakes and start the engine. If the vehicle has an automatic transmission, apply the brakes and put it in gear to see if the brakes will hold the vehicle. On a manual transmission, release the brakes and engage the clutch just enough to move the vehicle slowly. Apply the brakes to ensure that they will stop and hold the vehicle.

Classroom Manual
pages 273, 274, 275

Brake Flushings

There are other facts that must be considered during master cylinder replacement. If the vehicle is over 2 years old, or it has been more than 2 years since the last brake flushing, the old fluid should be flushed out with new fluid. This is done by bleeding each wheel until new fluid is visible. When a master cylinder fails because of age, it is probable that other components may require replacement. At the minimum, a thorough inspection of the brake system should be conducted.

SPECIAL TOOLS

Lift or jack and jack stands

Impact wrench

Torque wrenches

C-clamp

Disc brake silencer

Disc brake lubricant

Disc Brake Pad Replacement

WARNING: Before moving the vehicle after a brake repair, pump the pedal several times to test the brake and position the pads or shoes in place. Failure to do so may cause an accident that might damage the vehicle, the facility, or injure humans.

PHOTO SEQUENCE 24

Typical Procedure for Bench Bleeding the Master Cylinder

P24-1 **Mount the master cylinder firmly in a vise, being careful not to apply excessive pressure to the casting. Position the master cylinder so the bore is horizontal.**

P24-2 **Connect short lengths of tubing to the outlet ports, making sure the connections are tight.**

P24-3 **Bend the tubing lines so that the ends are in each chamber of the master cylinder reservoir.**

P24-4 **Fill the reservoirs with fresh brake fluid until the level is above the ends of the tubes.**

P24-5 **Using a wooden dowel or the blunt end of a drift or punch, slowly push on the master cylinder's pistons until both are completely bottomed out in their bore.**

P24-6 **Watch for bubbles to appear at the tube ends immersed in the fluid. Slowly release the cylinder piston and allow it to return to its original position. On quick take-up master cylinders, wait 15 seconds before pushing in the piston again. On nonquick take-up units, repeat the stroke as soon as the piston returns to its original position. Slow piston return is normal on some master cylinders.**

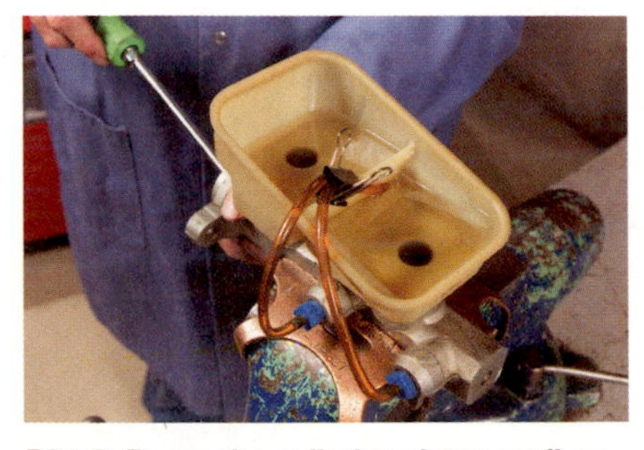

P24-7 **Pump the cylinder piston until no bubbles appear in the fluid.**

P24-8 **Remove the tubes from the outlet port and plug the openings with temporary plugs or fingers. Keep the ports covered until the master cylinder is installed on the vehicle.**

P24-9 **Install the master cylinder on the vehicle. Attach the lines, but do not tighten the tube connections.**

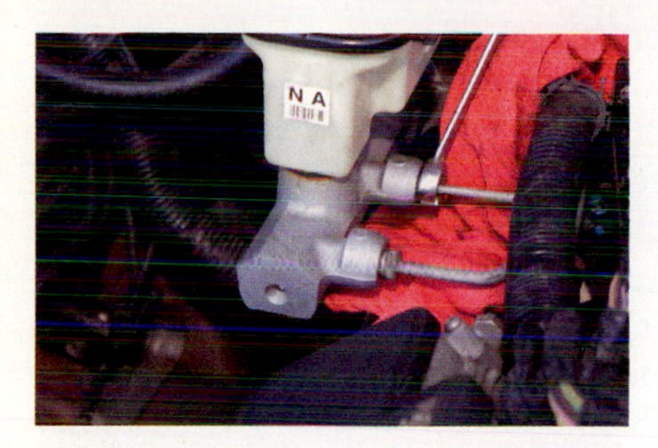

P24-10 Slowly depress the pedal several times to force out any air that might be trapped in the connections. Before releasing the pedal, tighten the nut slightly and loosen it before depressing the pedal each time. Soak up the fluid with a rag to avoid damaging the car finish.

P24-11 When there are no air bubbles in the fluid, tighten the connections to the manufacturer's specifications. Make sure the master cylinder reservoirs are adequately filled with brake fluid.

P24-12 After reinstalling the master cylinder, bleed the entire brake system on the vehicle.

WARNING: Before working on the brakes of a vehicle with ABS, consult the service manual for precautions and procedures. Failure to follow procedures to protect ABS components during routine brake work could damage them and require expensive repairs.

WARNING: Failure to lower the reservoir brake fluid level during brake repairs may result in spillover during brake repair. Damage to the vehicle's finish can occur. The fluid will overflow as the caliper pistons are pushed back into their bores and the fluid behind them is forced back into the master cylinder.

Disc brake pad replacement does not require any special tools except a **C-clamp** (Figure 10-16), although there are some specialty tools that will make the job go faster.

A **C-clamp** is so named because of its shape and function.

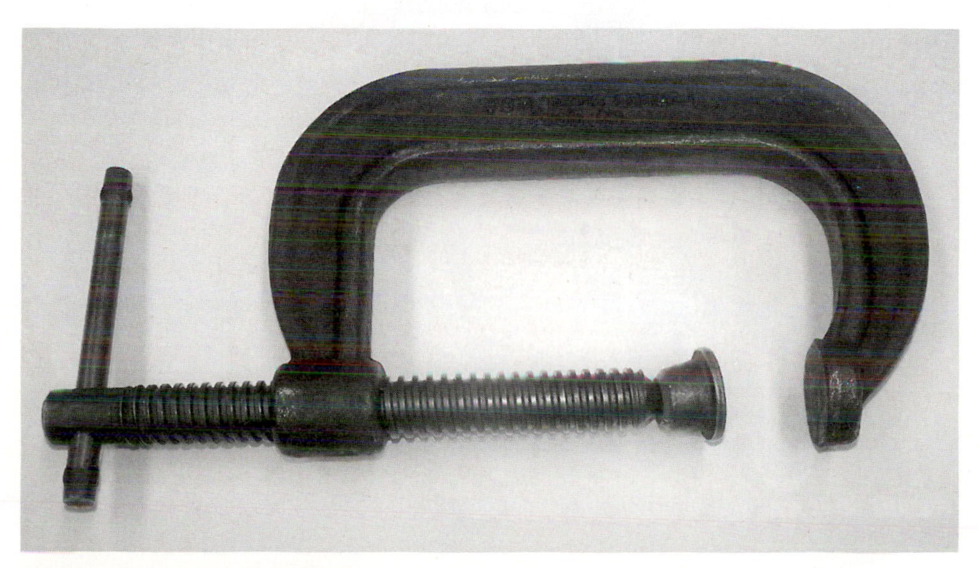

FIGURE 10-16 A typical C-clamp.

Before lifting the vehicle, inspect the fluid level in the reservoir. If the reservoir is more than half full, some of the fluid must be removed. A suction gun *should not* be used because the gun's rubber hose and piston seal will probably be damaged. A small, inexpensive syringe is the best choice. It can be discarded after several uses. Use cloth to capture any leaking fluid.

When replacing rear brake pads you need to determine what type of caliper is used. If it has a regular piston like the front, then the service is the same. Some rear disk brake calipers have an integrated parking brake. The caliper piston will not retract with a C-clamp. You will need a special tool (Figure 10-17) to turn the piston—which will reset it back into the caliper housing to make room for the new pads.

Each brake pad installation will be a little different. It is noticing this difference and then knowing what to do about it that makes the difference between having a customer come back because you didn't clean or install some small part that is causing a noise and having a customer that will have many miles of trouble free service.

On the back side of the brake pad, you must coat it with a noise suppression compound (Figure 10-18). Each technician has their own preference for what to use. Some brake pads have clips (Figure 10-19) and some calipers have support brackets that must be cleaned and lubricated (Figure 10-20). When servicing the brake pads you should always clean (or replace if badly corroded) the caliper slides (Figure 10-21).

WARNING: Consult the service manual before pushing the caliper's piston into its bore. Some ABS systems require that the caliper's bleeder screw be opened or they require other special procedures to prevent damage.

The next step is to compress the piston back into its bore. However, it is not desirable to push the old fluid back into the master cylinder. Instead obtain a small container and

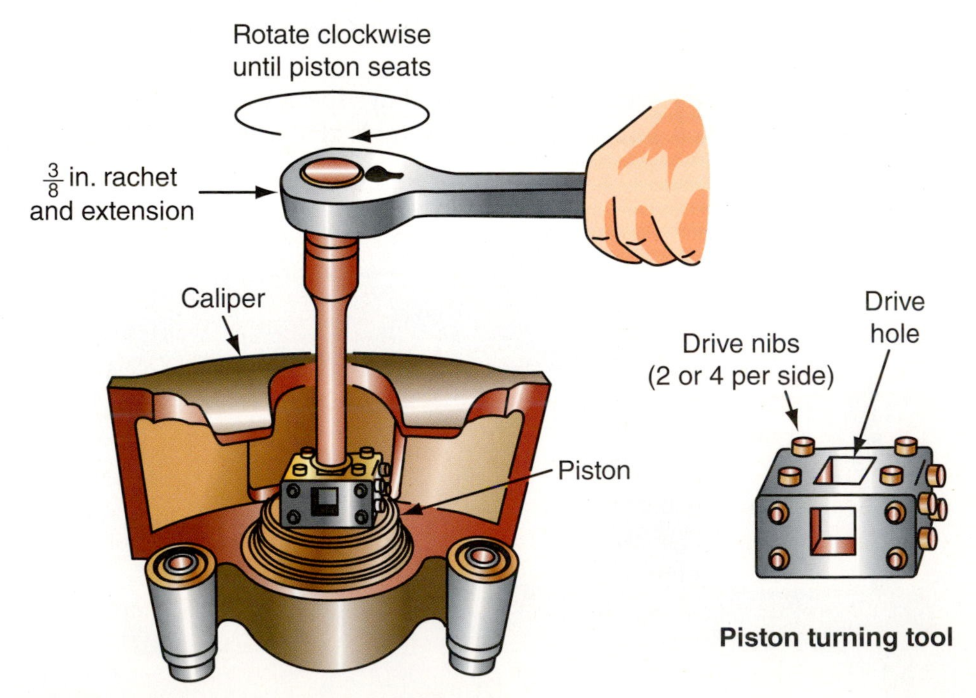

FIGURE 10-17 **One type of piston retracting tool.**

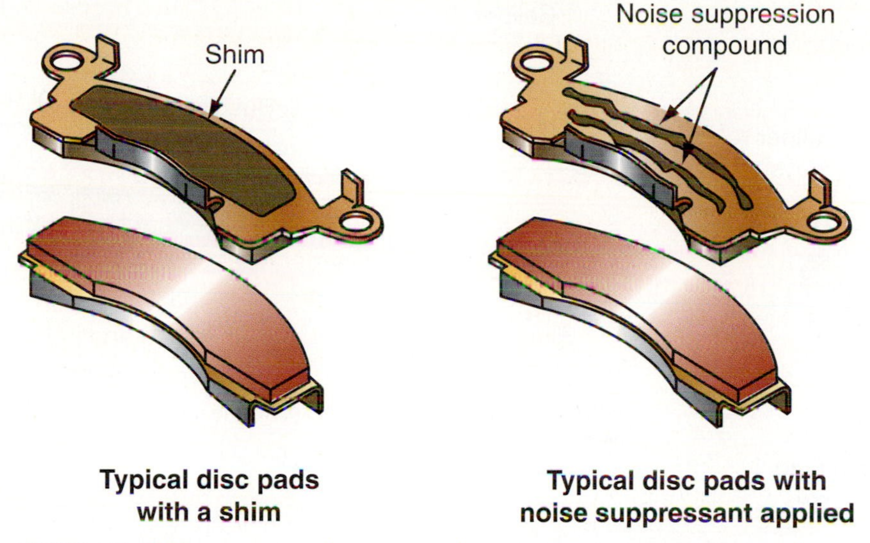

FIGURE 10-18 **You can use noise suppression compound or a soft shim on the back of a brake pad to help reduce noise, but do not use both on the same pad.**

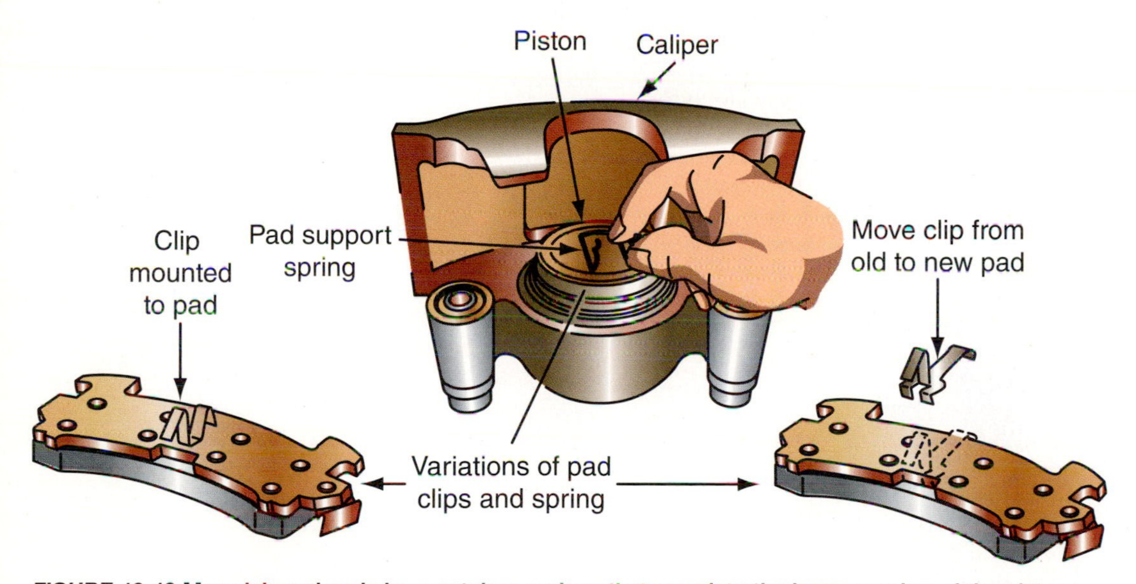

FIGURE 10-19 **Many inboard pads have retainer springs that snap into the inner opening of the piston.**

piece of flexible hose about 12 inches to 18 inches long with a diameter that will allow it to fit tightly over the bleeder screw. Position the container in a secure place on the lift or have someone hold it. Route the hose from the bottom of the container and fit the other end over the bleeder screw (Figure 10-22). Use a C-clamp and a block of wood or an old brake pad and apply a small amount of force to the piston (Figure 10-23). Loosen the bleeder screw and continue using the C-clamp to compress the piston into its bore till the piston is fully seated close to the bleeder screw. This prevents almost any of the dirty fluid from returning to the reservoir.

Remove the fasteners holding the adapter to the steering knuckle. With the fasteners removed, the adapter, caliper, and pads will slide off the rotor (Figure 10-24). Be cautious at this point because many rotors can drop off the hub once the caliper is removed. Once the assembly is removed, hang it with mechanic wire or other strong wire so the caliper's

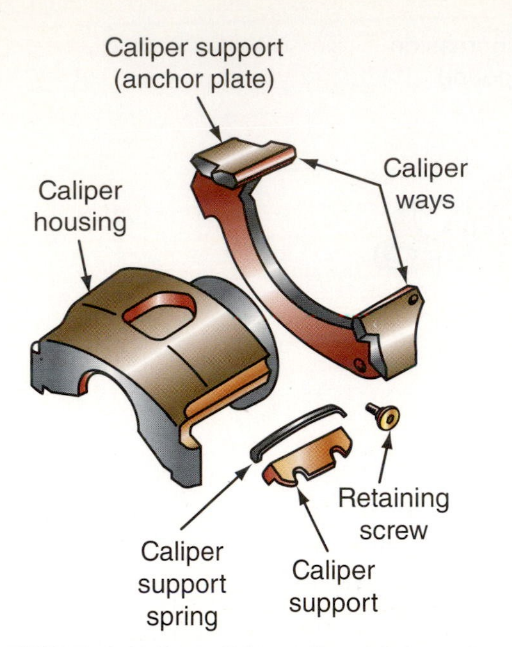

FIGURE 10-20 **For a sliding caliper, lubricate the caliper ways, support spring, and the mating areas of the caliper housing as highlighted here.**

FIGURE 10-21 **Lubricate the inside of the caliper bushing as highlighted here.**

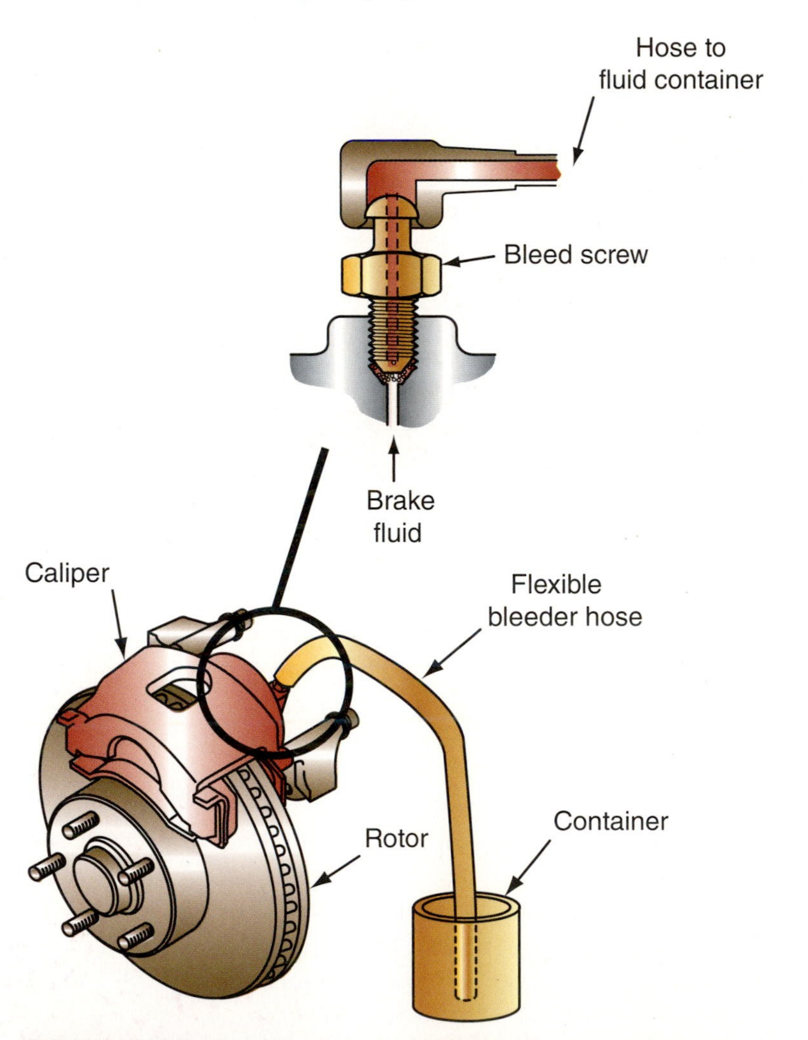

FIGURE 10-22 **Route a hose from the bleeder screw to a capture container. Just before the piston bottoms out, close the bleeder screw and remove the hose and container.**

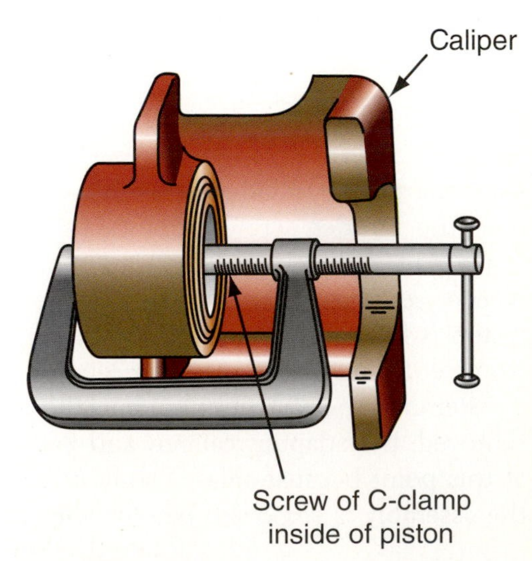

FIGURE 10-23 **Use a C-clamp to compress the piston into its bore.**

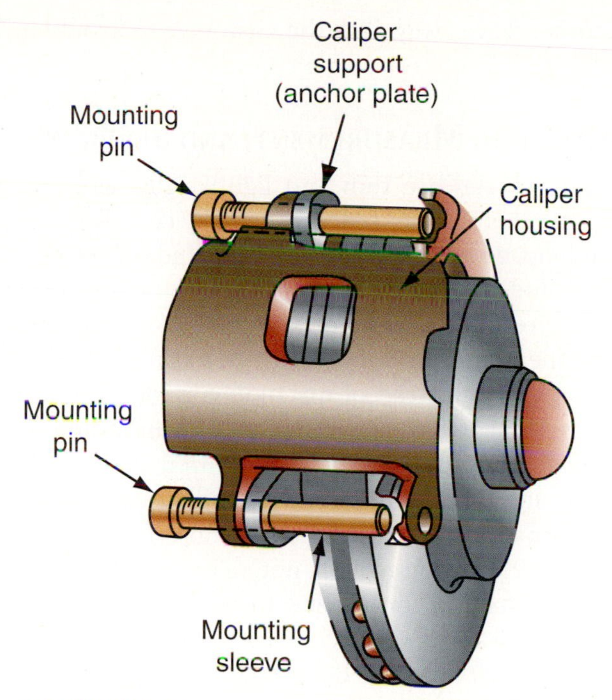

FIGURE 10-24 **Use care once the adapter fasteners are removed. The caliper and adapter can slide off and put unnecessary stress on the brake hose.**

weight does not hang on the flexible brake hose. Remove the pads, noting where they are located and what is used to hold them in place. Normally, antirattle clips keep the pads from clicking in the caliper when the brakes are applied and released.

If there is brake fluid present on the outside of the caliper, the caliper must be rebuilt or replaced. Like the wheel cylinder, it is usually best and cheapest to replace the caliper, but there are rebuilding kits on the market.

Before installing the pads, the slide pins or guides on the caliper and adapter must be lubed. Inspect the caliper and adapter to determine the slide points. Some systems have slide pins only, while others slide the calipers along rails on the adapter. Once located, clean the slides and then lube with a disc brake lubricant.

Assemble the pads into the caliper, inboard pad first. Ensure that the retaining clips or springs are in place before fitting the assembly over the machined rotor. Install and torque the adapter bolts, and install the wheel assembly.

When the vehicle is lowered to the floor and before the vehicle is moved, slowly pump the brake pedal several times to seat the pads against the rotor. The pedal should be firm and at a normal height above the floor before the vehicle is moved. Top off the brake fluid level as necessary and inspect for any overflow around the master cylinder.

Brake Pad Burnishing

Whenever new brake pads are installed, they need a short period of controlled operation that is called a **burnishing** or "break-in" period. This period of time polishes the pads and mates them to the rotor finish. Road testing the vehicle after installing new pads performs this procedure and verifies that the brakes are working properly.

Burnishing is an operation used to heat and seat the brake pads to the rotor.

To do this, drive the vehicle at 30 mph (50 kph) and firmly but moderately apply the brakes to fully stop the vehicle. Repeat this five times within 20 seconds of driving time between each application. Then drive the car to highway speeds and apply the brakes moderately to slow the car to 20 mph (30 kph). Then let the brakes cool down for 10 minutes before letting the customer drive away. Tell the customer to avoid hard braking for the first 100 miles.

Brake Rotor Inspection, Measurement, and Servicing

Rotors on modern light vehicles are thin and lightweight, when compared to the brake rotors of older vehicles. This design reduces vehicle cost and unspring weight; it also helps the vehicle's handling, adds to a smoother ride, and increases fuel economy.

Older-style, heavier, and thicker brake rotors (that are still found on heavier vehicles, light trucks, and SUVs) are often machined during brake pad replacement because they have more material on them and do not wear to the minimum legal thickness as quickly. You will find yourself replacing, instead of machining, a lot of brake rotors.

Brake rotors must be measured to see if they are causing some of the problems that the customer is experiencing. They must also be measured before a decision to refinish or replace them is made. Using a brake micrometer, measure the thickness of the rotor and compare it to the *machine to* and *discard* specifications. These specifications are found on the rotor itself (Figure 10-25) but are often not readable because of rust or dirt, so you need to refer to the service manual for specifications.

With the brake rotor still on the vehicle, use the brake micrometer and measure the rotor at several spots (Figure 10-26). The difference in these measurements is the thickness variation. A thickness variation may be the cause of a brake pedal pulsation that the customer is concerned about. The thinnest dimension that you measure should be compared to the machine to and discard specifications.

With the rotor still mounted, reinstall and tighten the lug nuts and mount a dial indicator on the rotor surface. Rotate the rotor and measure the runout of the rotor surface (Figure 10-27). Compare this reading to the specification; too much runout can result in brake pedal pulsation. After removing the rotor, measure the runout of the hub; although it is rare and usually only caused by a vehicle accident, if the hub is bent it will cause even a new rotor to pulsate the brake pedal and measure out-of-round. Excessive rust will also cause an incorrect measurement. Make sure to clean the rust from the surface (Figure 10-28).

After measuring the rotor, don't forget to visually inspect it. Look for grooves and a bluish glazed tint on the rotor surface. The glazing is caused by overheating of the rotor and can only be machined or resurfaced (Figures 10-29 and 10-30).

FIGURE 10-25 **The discard dimension, or minimum thickness specification, is cast or stamped into all disc brake rotors.**

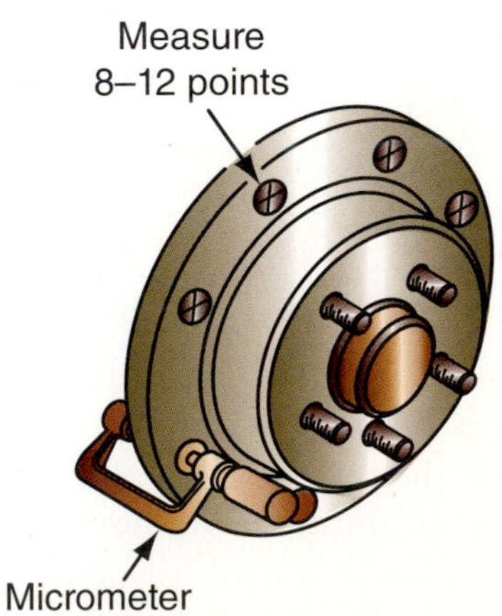

FIGURE 10-26 **Check for thickness variations by measuring at 8 to 12 points around the rotor.**

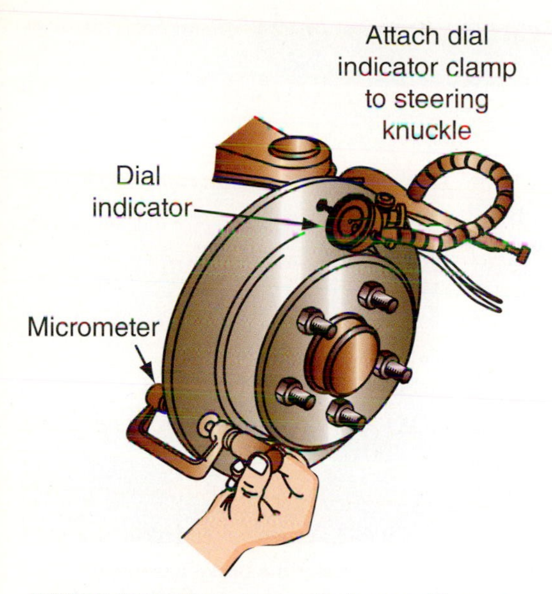

FIGURE 10-27 **The rotor and hub runout is measured with a dial indicator mounted to some fixed point on the vehicle (upper right). The rotor thickness and parallelism can be measured with a standard outside micrometer, as shown on the lower left, but is more easily done with an electronic brake micrometer.**

FIGURE 10-28 **Clean the dirt and rust from the hub-mounting area of the hubless rotor.**

FIGURE 10-29 **The groove in this rotor surface is machined to aid in cooling and noise reduction.**

FIGURE 10-30 **This rotor has a shiny glazed surface.**

BRAKE ROTOR REMOVAL AND RESURFACING

To remove the rotor, lift the vehicle to a good working height, set the locks, and remove the tire and wheel assembly. Inspect the disc brake area to determine the type of rotor or hub assembly and the method used to attach the caliper. Some systems require the adapter to be removed before the rotor is free. The rotor and hub may be one- or two-piece. If equipped, the wheel bearings and seal have to be removed on a one-piece assembly before the rotor can be machined. For this example, the rotor is separate and the adapter must be removed. Photo Sequence 25 shows the typical procedure for replacing the brake pads.

Most rotors can be removed and turned on the brake lathe (Figure 10-31), but there are portable lathes that can machine the rotor on the vehicle (Figure 10-32). They were originally sold to service one-piece hubs or rotors on front-wheel-drive (FWD) vehicles where the drive axle and the wheel bearing were press-fitted to the hub. This lathe requires some special setup procedures, but these can be learned quickly. The **on-vehicle brake lathe** can be

An **on-vehicle brake lathe** is designed to machine a brake rotor on the vehicle.

PHOTO SEQUENCE 25

Typical Procedure for Replacing Brake Pads

P25-1 **Begin front brake pad replacement by removing brake fluid from the master cylinder reservoir or reservoirs servicing the disc brakes. Use a siphon to remove about two-thirds of the fluid.**

P25-2 **Raise the vehicle on the hoist, making certain it is positioned correctly. Remove the wheel assemblies.**

P25-3 **Inspect the brake assembly, including the caliper, brake lines and hoses, and rotors. Look for signs of fluid leaks, broken or cracked lines or hoses and a damaged rotor. Correct any problems found before replacing the pads.**

P25-4 **Loosen and remove the caliper-to-knuckle bolts.**

P25-5 **Lift and rotate the caliper assembly up and off the rotor.**

P25-6 **Remove the old pads from the caliper assembly.**

P25-7 **To reduce the chance of damaging the caliper, suspend the caliper assembly from the underbody using a strong piece of wire.**

P25-8 **Check the condition of the caliper locating pin insulators and sleeves. Make sure all slide or guide hardware is lubricated with brake lubricate before assembly.**

P25-9 **Install a hose from the bleeder screw to the container. Open the bleeder screw.**

P25-10 Place a block of wood over caliper piston and install a C-clamp over the wood and caliper. Tighten the C-clamp to force the piston back into its bore.

P25-11 Remove the C-clamp and check the piston boot, then install new locating pin insulators and sleeves if needed.

P25-12 Install the new pads in the caliper. Then set the caliper with its new pads onto the rotor. Check the assembly for proper positioning.

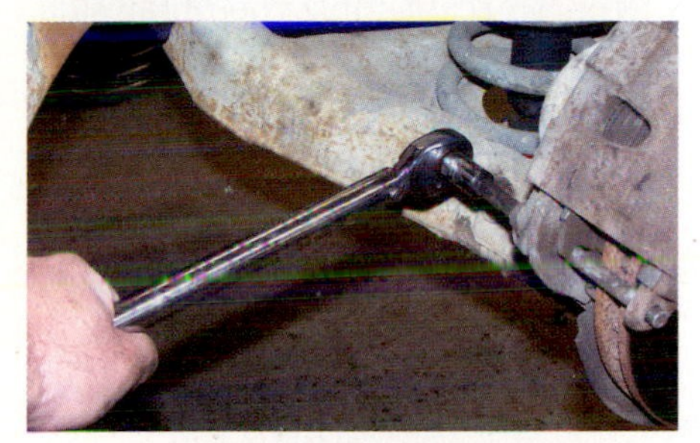

P25-13 Install and torque the caliper mounting bolts. Install the tire and wheel assembly and tighten to torque specifications. Then press slowly on the brake pedal several times to set the brakes.

FIGURE 10-31 AAMCO brake lathes like the one shown are found in most brake repair shops and are preferred by many older technicians.

FIGURE 10-32 This on-car lathe has its own drive motor.

used for machining other rotor setups as long as the rotor, machining head, and hub can be clamped tightly together with the lug nuts. Machining a rotor or drum should be done by an experienced technician following the lathe operating instructions and the procedures set forth in the vehicle's service manual. Photo Sequence 26 shows the typical procedure for mounting a brake rotor on the bench lathe.

Classroom Manual
pages 268, 269, 270, 271

Drum Brake Shoe and Wheel Cylinder Inspection, Measuring, and Replacement

Drum Brakes

Drum brake shoe replacement is a fairly simple process of removing the old and installing the new. However, drum brake shoes are held in place with return springs, retainers, parking brake cable, and adjusters, most of which can easily be installed incorrectly (Figure 10-33). As always, the service manual contains instructions for the repair. A quick method of learning drum brake assemblies is to lay out all parts in the order in which they were removed and never remove both sets of brakes at the same time. Many new technicians lose time because they pull both sides down at the same time and do not remember how the drum brakes were assembled. Keeping one set intact provides a model, or mirror, to reassemble the other side. Since the drum may have to be machined, no time is lost by keeping one assembly intact. It is suggested that this practice be followed during most drum brake repairs.

SPECIAL TOOLS

Brake spring pliers

Brake spring hold-down tools

Drum brake micrometer

Removing and installing drum brakes may also expose the technician to brake dust. Older brake materials contained asbestos. The brakes should be cleaned with a low-pressure aquatic cleaner to prevent dust (Photo Sequence 27). Another method is to use

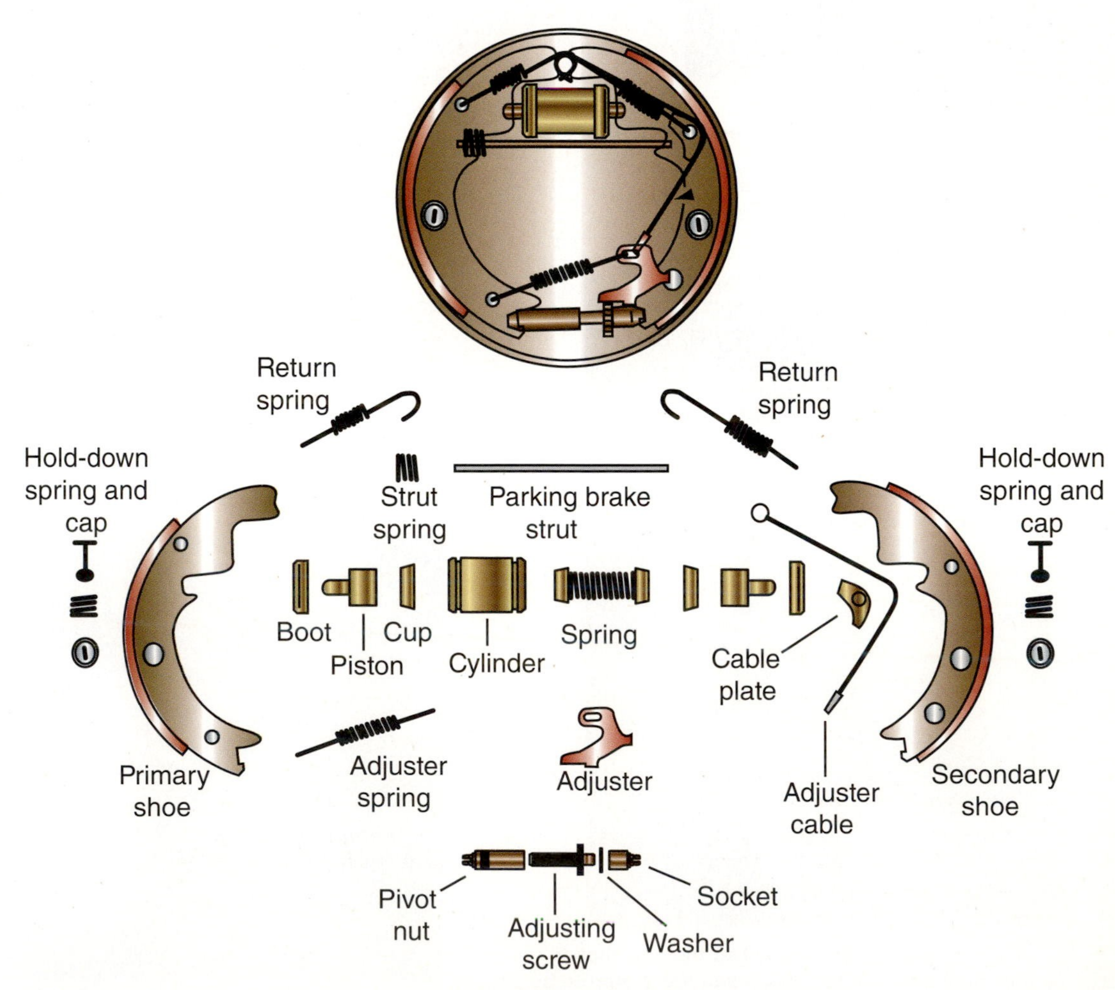

FIGURE 10-33 **Care must be taken when reassembling a drum brake.**

PHOTO SEQUENCE 26

Typical Procedure for Mounting a Brake Rotor on the Bench Lathe

P26-1 Clean the inside and outside of the rotor where the lathe's centering cone and clamps will be in contact with the rotor. Select a centering cone that fits into the rotor's center hole without protruding too much in either direction.

P26-2 Slide the inner clamp onto the arbor followed by the centering cone tension spring.

P26-3 Slide the cone onto the arbor followed by the rotor. The rotor is mounted in the same way as on the vehicle.

P26-4 Push on the rotor to compress the spring, and then hold the rotor in place as the outer clamp is fitted onto the arbor. The two clamps, inner and outer, must be of same size.

P26-5 Keep the pressure on the install parts as the spacers and bushing are installed on the rotor.

P26-6 Hold all parts in place as the arbor nut is installed and tightened to lathe manufacturer specifications.

P26-7 Install the vibration damper. The damper may be a flexible strap or spring that fits the outer circumference of the rotor, or two pad-like devices that are forced against each side of the rotors similar to installed disc brake pads.

P26-8 The rotor is now properly mounted and ready for the final check before machining.

PHOTO SEQUENCE 27

Typical Procedures for Wet-Cleaning Drum Brakes

P27-1 **After removing the wheel assembly, position the cleaner so the run-off waste is captured in the cleaner.**

P27-2 **Use a wet brush to remove any brake dust from the outside of the drum. This brush is not used to clean mud or accumulated dirt from the drum. If mud is present, use a dry brush to remove it.**

P27-3 **Remove the drum and use a wet brush to clean the inside of the drum. Once clean, set the drum to the side.**

P27-4 **Use a wet brush to wash the brake components free of dust. Ensure the waste material or liquid is draining into the cleaner's tub.**

P27-5 **Once the brake has been disassembled, wash the individual parts separately, holding them so the waste drains into the tub. Lay the parts on rags or a table for drying.**

P27-6 **After cleaning the various individual parts, complete the cleaning by washing down the backing plate.**

P27-7 **After cleaning is complete, check the cleaner's waste filter and replace it if necessary. The full filter must be treated as hazardous waste and disposed of accordingly.**

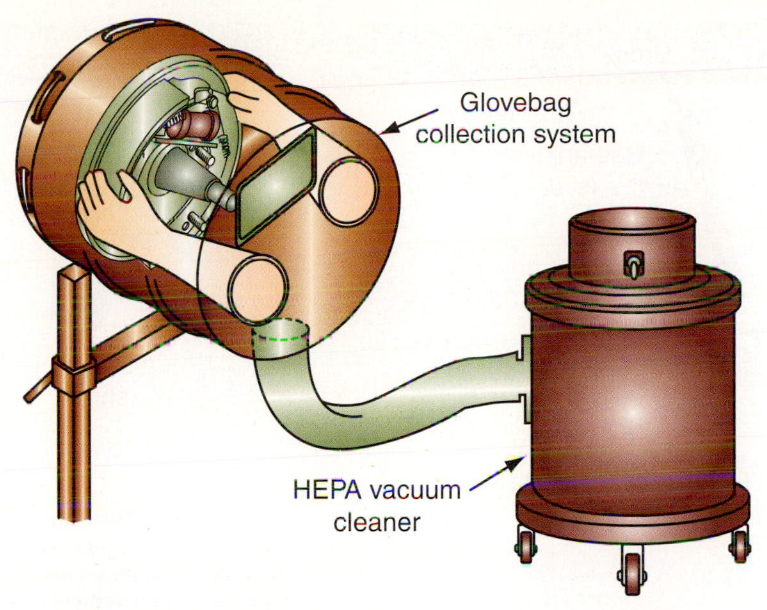

FIGURE 10-34 **This negative-pressure enclosure is used to contain brake dust during cleaning.**

a specially made high-efficiency particulate air (HEPA) vacuum cleaner (Figure 10-34). While there are several ways to clean the brake area and not spread brake dust, there is only one right way. The correct method is to use a washer or vacuum-type collector built for this purpose. One of the most common units is a portable, heated, low-pressure aquatic washer. This unit has a catch container that filters out the dust and shavings from the washer fluid and captures in a holding device. The waste filter is treated as hazardous waste when it is full. There are different brands and models available at nominal cost to the shop, which are much cheaper in terms of dealing with the EPA, OSHA, and, more importantly, employee safety in general.

Brake Shoe, Drum, and Related Components Inspection and Measurement

The first procedure for inspecting and measuring the brake shoes is removing the wheel and drum. Sometimes the drum will just come right off, sometimes you need to lightly hit it with a hammer to break up the rust on the hub, and sometimes you need to adjust the parking brake cable to ensure there is enough shoe clearance for drum removal (Figure 10-35). Be careful when removing the drum; sometimes adjustment of the

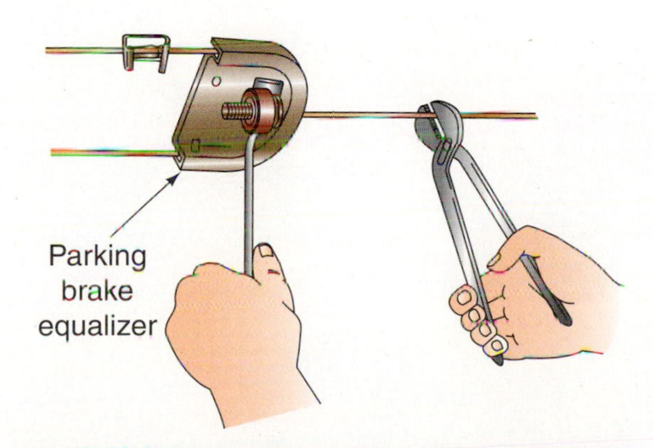

FIGURE 10-35 **Loosen the parking cables if necessary to ensure enough shoe clearance for drum removal.**

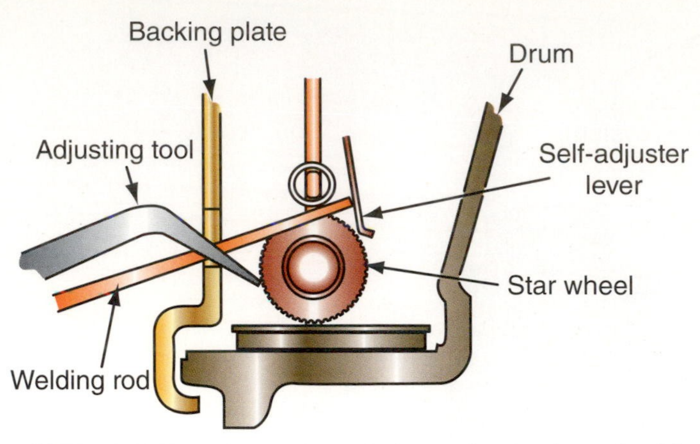

FIGURE 10-36 **Use a heavy piece of wire, such as a piece of welding rod, to move the adjusting lever away from the star wheel. Then rotate the star wheel to release brake adjustment for drum removal.**

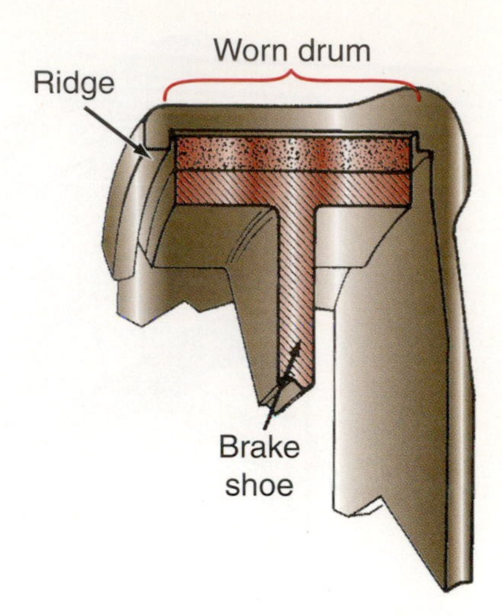

FIGURE 10-37 **These shoes have been adjusted to fit a badly worn drum, and the drum will jam on the lining if you try to remove it without backing off the brake adjustment.**

parking brake cable is necessary and other times you have to open the adjustment cover on the back side of the drum and adjust the shoes themselves (Figure 10-36). This may be because the drum is badly worn and the shoe is jammed into the groove (Figure 10-37).

Once you have the drum off and the brake shoes are cleaned, you can inspect the components. Use a depth micrometer (Figure 10-38) to measure the lining thickness. Inspect the shoes for cracks, glazing, and unequal wear (Figure 10-39). Pull the wheel cylinder boot slightly away and check for brake fluid leakage (Figure 10-40). Inspect the brake springs for cracks. Usually, when you install new brake shoes you will install new hardware, which will include new springs.

FIGURE 10-38 **Use a depth micrometer to measure lining thickness precisely.**

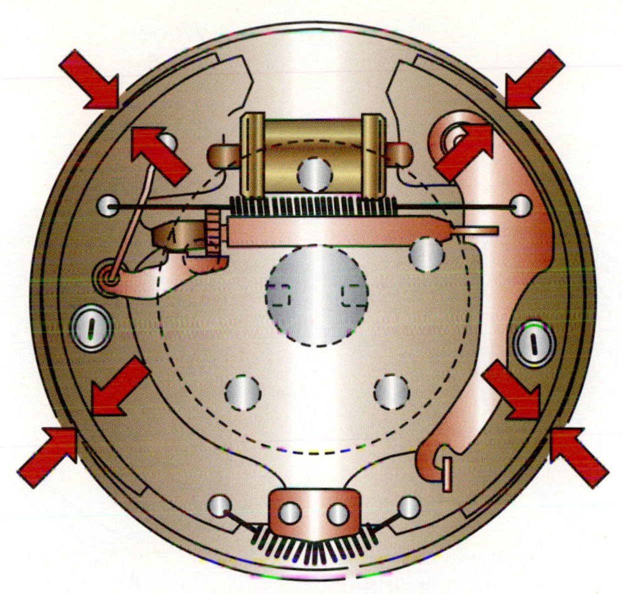

FIGURE 10-39 **Inspect the linings for unequal wear, as well as for thickness, cracks, fluid or grease contamination, and other damage. Unequal wear is most common at the indicated points.**

FIGURE 10-40 **Pull the boot slightly away from the cylinder and check for liquid or fluid within the boot.**

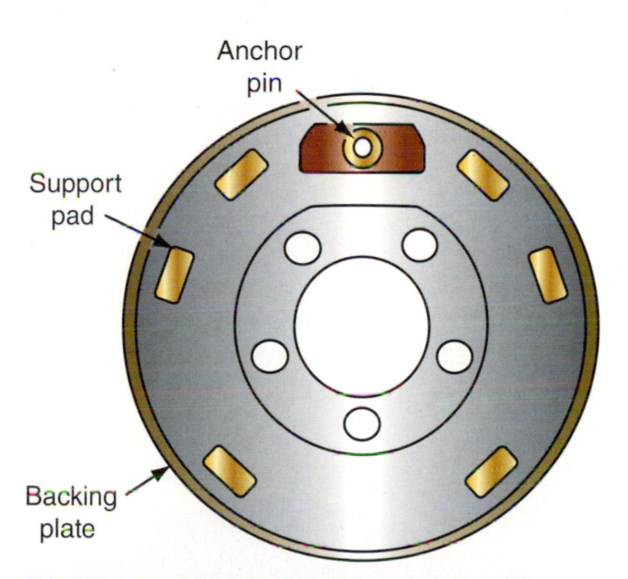

FIGURE 10-41 **Failure to lube the shoe or backing plate rub points will result in noise during brake application.**

Next, remove the brake shoes. Follow the steps in Photo Sequences 15 and 16 for drum brake disassembly and reassembly. After removing the shoes, inspect the backing plate and support pads (Figure 10-41). If needed, you should replace the wheel cylinder based on your inspection (Figure 10-42). Use caution, the brake line going to the wheel cylinder is often rusty and may break if you are not careful.

In general terms, the following steps for shoe removal and reinstallation may be followed and are illustrated in Photo Sequences 28 and 29.

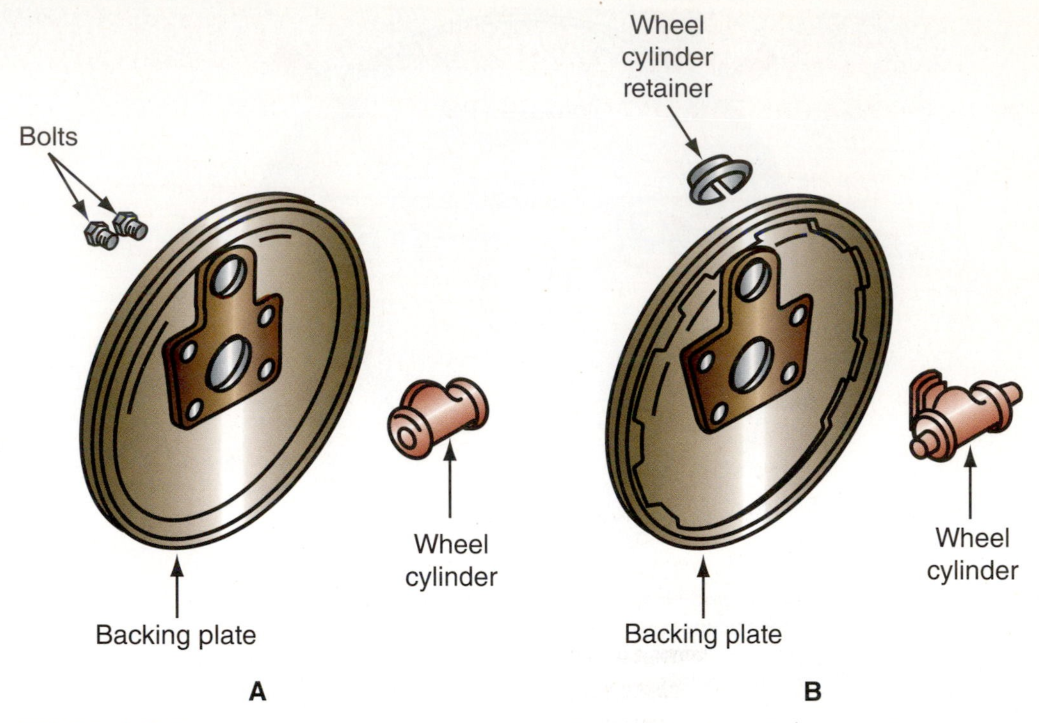

FIGURE 10-42 **Two common ways to mount wheel cylinders to backing plates are with small bolts or cap screws (A) or with a spring-type retainer ring (B).**

1. Remove both return springs.
2. Remove the retaining spring on one shoe and remove that shoe from the backing plate ensuring that the parking brake and self-adjuster components are held in place and then removed.
3. Remove the second shoe and disconnect the parking brake mechanism from the shoe to which it is attached.
4. Compare the new parts to the old ones and reverse the procedure to install.
5. Preset the shoe adjustment by measuring the drum and setting the brake shoes to that measurement (Figures 10-43 and 10-44).
6. After installing the drum, adjust the shoes by moving the adjuster with the drum on. Spin the drum until it catches on the brake shoes only slightly.

FIGURE 10-43 **Measure the inside diameter of the drum with the brake drum micrometer and tighten the locknut.**

FIGURE 10-44 **Flip the caliper over and place it over the widest point of the brake shoes. Adjust the brake shoes until they touch the caliper pads.**

PHOTO SEQUENCE 28

Typical Procedure for Disassembling Brake Drum Shoes

P28-1 Remove the top return spring from the anchor and shoe. Remember which shoe's return spring is on top. It must be reassembled in the same position.

P28-2 Remove the retaining spring from the secondary shoe. Do not remove the primary shoe retaining spring at this time.

P28-3 Rotate the secondary shoe forward and down. Catch the self-adjuster as the shoe releases the tension.

P28-4 Remove the parking brake strut and spring from the primary shoe. Withdraw it to the front. Remember which end holds the spring and which end faces forward.

P28-5 Remove the retaining spring on the primary shoe. Do not allow the shoe and self-adjuster mechanism to drop.

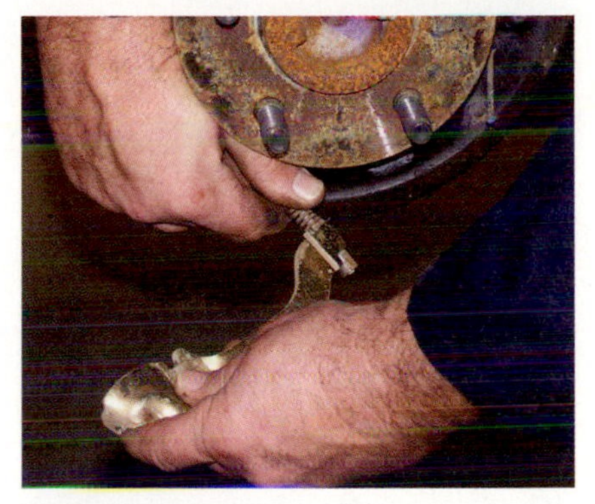

P28-6 There are two means to disconnect the parking brake from the primary shoe. One way is to remove the clip from the parking brake lever pin where it passes through the shoe. Do not lose the waved washer behind the shoe. The other way is to disconnect the cable from the lever as shown. The lever can be removed from the shoe later.

P28-7 Clean and lubricate the shoe support pads now. This is a common area that may be forgotten.

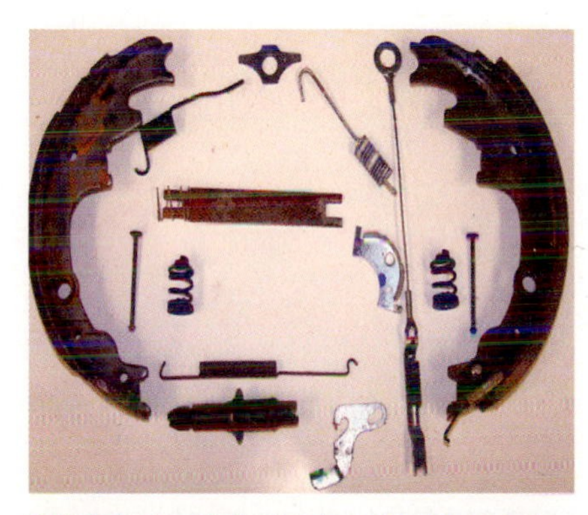

P28-8 Lay out the brake components in roughly the same position as they were on the vehicle will help in repositioning them correctly during assembly.

PHOTO SEQUENCE 29

Typical Procedure for Reassembling Brake Drum Shoes

P29-1 Compare the new parts with the old to ensure that the correct ones are available.

P29-2 Depending on how the parking brake was disconnected, there are two means to connect it: slide the lever pin with the washer installed through the primary shoe and install a new retaining clip, or connect the cable to the lever as shown.

P29-3 Hold the self-adjuster mechanism in place on the primary shoe as the retaining spring is attached.

P29-4 Pull the top of the secondary shoe forward and fit the parking brake strut between the two shoes. The large slot in the strut goes to the rear and the spring to the front.

P29-5 Pull the lower ends of the shoes apart and fit the self-adjuster between the two shoes. After releasing the shoes, check to make sure that the self-adjuster is correctly linked to the shoes and that the star wheel is to the rear.

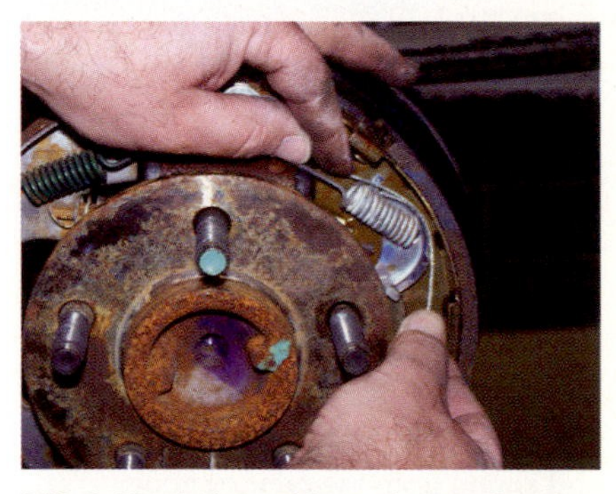

P29-6 Hook the cable to the pawl and push up on the pawl. Route the cable over the cam. After releasing the pawl, ensure that the cable stayed in place on the cam.

P29-7 Install the pawl and its spring(s).

P29-8 Pull the lower ends of the shoes apart and fit the self-adjuster between the two shoes. After releasing the shoes, check to make sure that the self-adjuster is correctly linked to the shoes and that the star wheel is to the rear.

P29-9 Hook the cable to the pawl and push up on the pawl. Route the cable over the cam. After releasing the pawl, ensure that the cable stayed in place on the cam.

P29-10 **Check each component of the assembled drum brake for proper positioning and anchoring.**

Parking Brakes

Parking brake repairs normally consist of cable replacements and adjustments or replacement of the ratchet locks on the pedal or lever (Figure 10-45). Most repairs are straightforward procedures laid out in the service manual. However, in most systems, some of the repairs require removal while some require partial disassembly of the rear brakes.

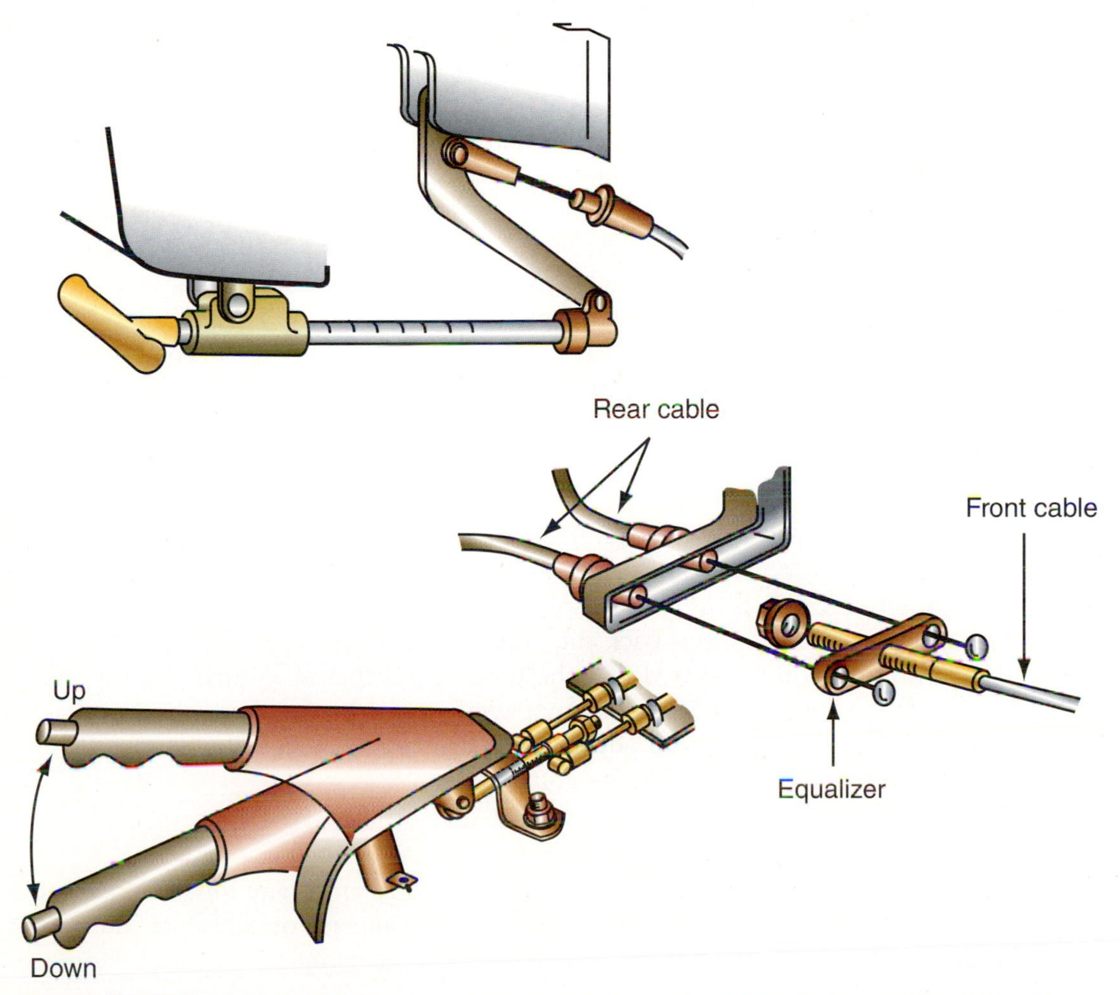

FIGURE 10-45 **The shaded points are the typical wear points on the parking brake actuation cable. They are also common lubrication points.**

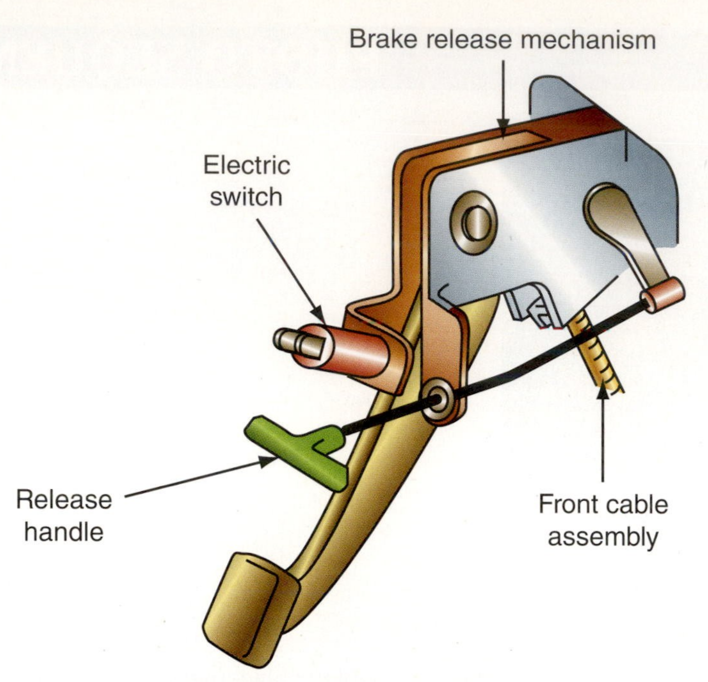

FIGURE 10-46 **Most parking brake warning light switches must be adjusted if the pedal is removed for any reason.**

Once the new brake parts have been installed, the rear brakes must be adjusted before the parking brake is adjusted. The parking brake should lock solidly about halfway along the pedal or lever's travel. Some manufacturers require that the parking braking locks after a certain number of clicks are heard or felt as the parking brake is applied. During the installation of any parking brake cable, ensure that the cable and sheaths are not clamped shut by fasteners. An inspection of the parking brake cable is done by moving it back and forth and making sure that the drum shoe moves as well. Some brake cables (especially those on automatic transmission vehicles) are rusty and will not move easily; they may break or get stuck.

The parking brake warning light switch may need adjustment after work is performed on the parking brake lever or pedal mechanism (Figure 10-46). The light should come on as soon as the parking pedal or lever is moved. It should not go off until the brake is completely released. After repairs are complete, turn the tires to make sure the parking brake is not causing the shoes or pads to drag.

Rear Disc Brakes

On vehicles with rear discs, the procedure for replacing the pads becomes a little more complicated. The rear disc may also be used as the parking brake or the disc rotor may have a small, internal drum brake whose sole function is for parking (Figure 10-47). In either case, compressing the piston back into the bore requires special tools and procedures. Normally, the piston is screwed into the bore with a special wrench (Figure 10-48). At times, even the correct procedures may not work properly and the technician has to spend time analyzing the problem. In addition, ABS presents special situations that must be dealt with before a proper repair can be completed.

Brake Bleeding

The brake system must be bled if the system was exposed to air at any time. This includes disconnecting the lines from the master cylinder, disc calipers, or wheel cylinders and external leaks at any point in the system. Bleeding the brakes completely is similar to flushing or replacing the brake fluid.

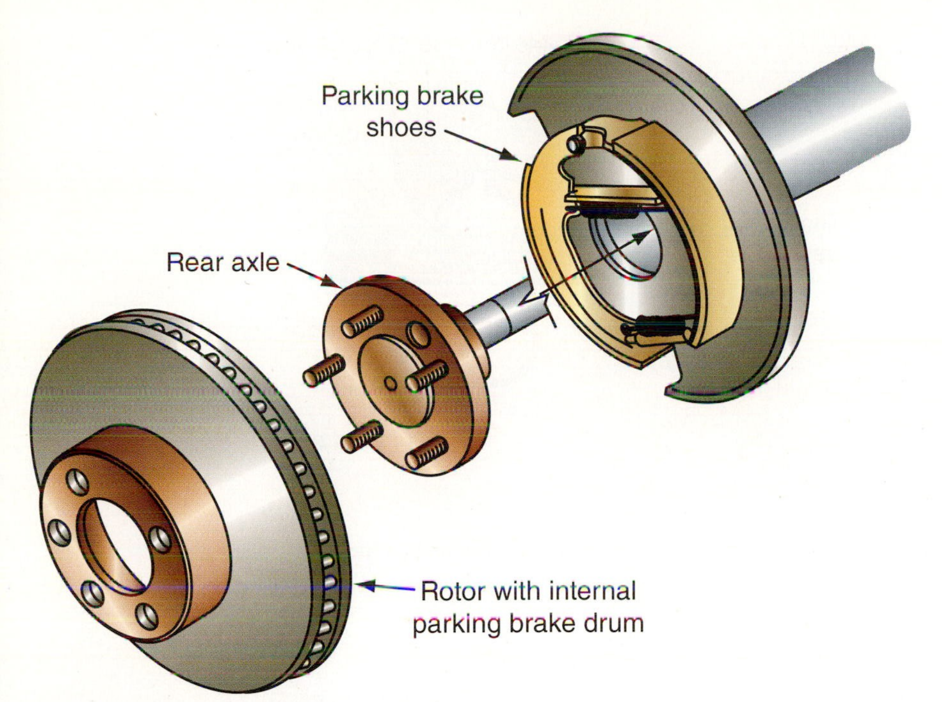

FIGURE 10-47 **This parking brake is located within a small drum built into the back of a rear disc brake rotor.**

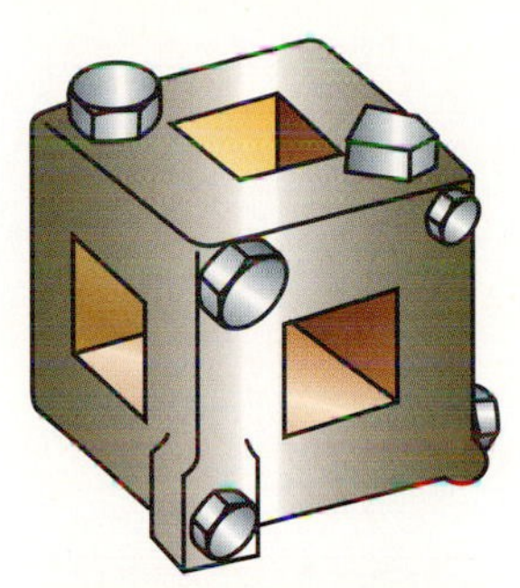

FIGURE 10-48 **It is not obvious in this picture, but this tool has six types of adapters to fit almost any rear-disc caliper piston. (Courtesy of MATCO Tool Company)**

FIGURE 10-49 **Bleeder screws, such as this one on a disc brake caliper, are also installed on wheel cylinders and some master cylinders and valves.**

A **pressure bleeder** is a small holding tank with a diaphragm between a compressed air chamber and a fluid chamber.

The key to brake bleeding is to locate the bleeder screws (Figures 10-49 and 10-50) and the bleeding sequence from the service manual (Figure 10-51). Bleeding can be done in several ways. Special suction equipment can be used as well as a **pressure bleeder** or two persons working together (Photo Sequence 30). In most systems, the wheels are bled starting with

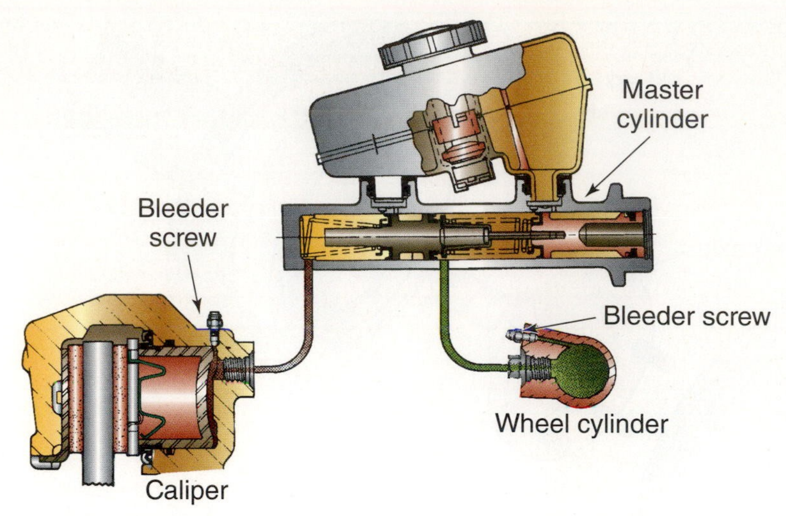

FIGURE 10-50 **Bleeder screws are located at the highest points on calipers and wheel cylinders.**

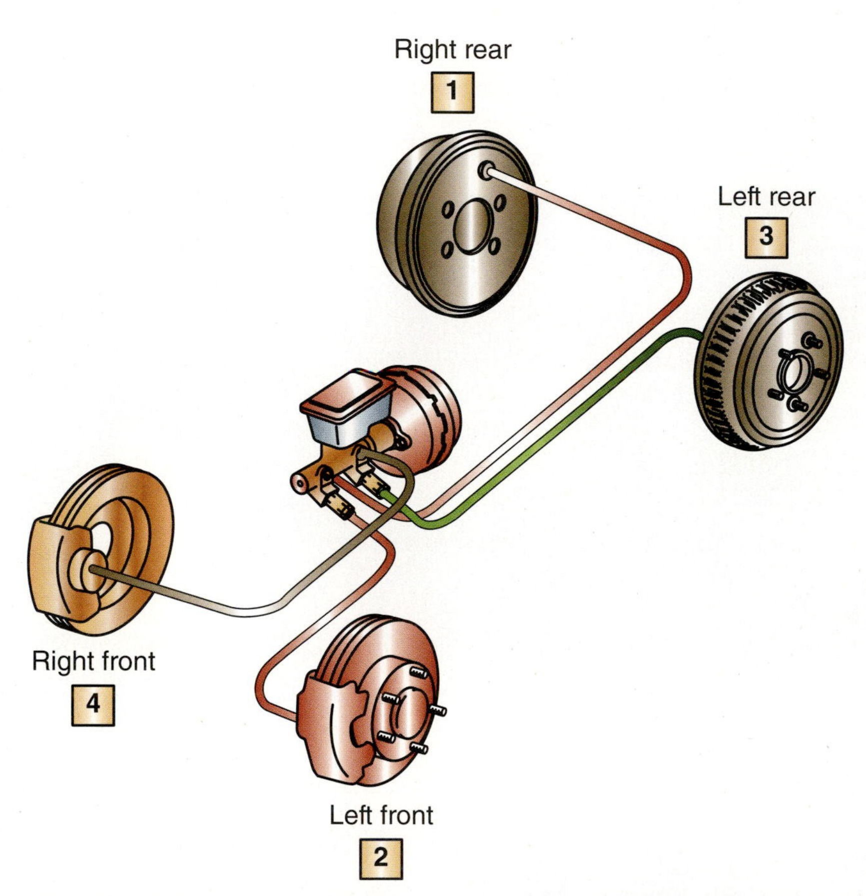

FIGURE 10-51 **Recommended bleeding sequence for a diagonally split brake system.**

the one farthest from the master cylinder. An example of a typical sequence is right rear, left rear, right front, and left front (Figure 10-51). Depending on the repairs accomplished, the master cylinder and some control valves may have to be bled before bleeding the wheels. Most ABS brakes require special bleeding procedures. Improper bleeding of the ABS may not lead to damage, but it could cost in labor time and frustration on the part of the customer, the technician, and the supervisor.

Typical Procedure for Manually Bleeding a Disc Brake Caliper

P30-1 Be sure the master cylinder reservoir is filled with clean brake fluid. Recheck it often to replace the fluid lost during the bleeding process.

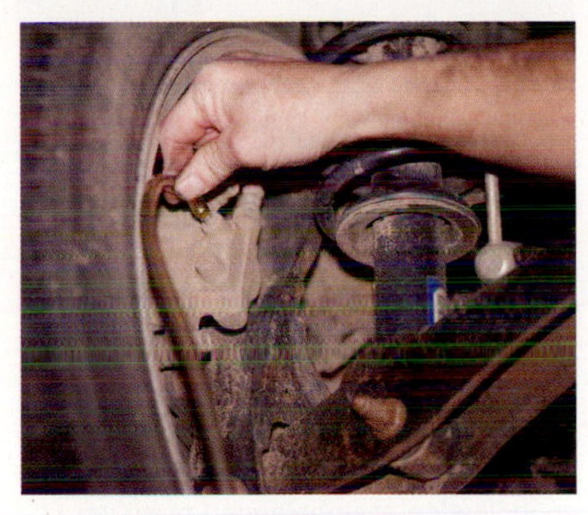

P30-2 Attach a bleeder hose to the bleeder screw.

P30-3 Place the other end of the hose in a capture container partially filled with brake fluid. Be sure that the free end of the hose is submerged in brake fluid. This helps to show air bubbles as they come out of the system and prevents air from being accidentally sucked into the system through the bleeder screw.

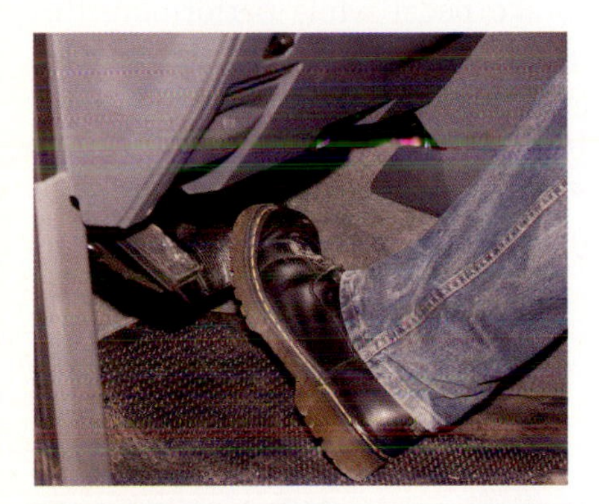

P30-4 Have an assistant apply moderate (40–50 pounds), steady pressure on the brake pedal and hold it down.

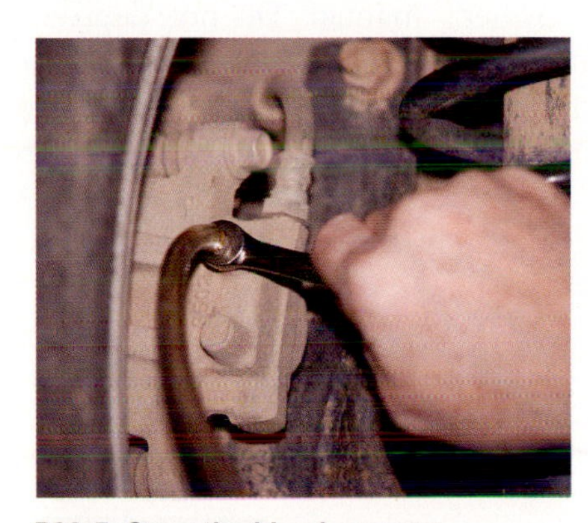

P30-5 Open the bleeder screw.

P30-6 Observe the fluid coming from the submerged end of the hose. At the start, you should see air bubbles.

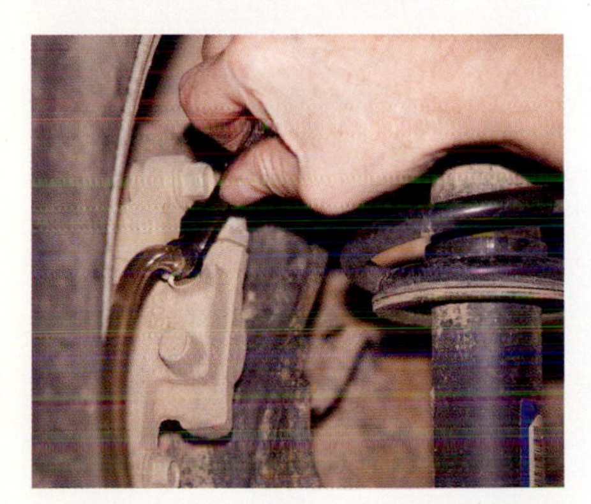

P30-7 When the fluid is clear and free of air bubbles, close the bleeder screw.

P30-8 Have your assistant release the brake pedal. Wait 15 seconds and repeat step 2 and step 3 until no bubbles are seen when the bleeder screw is opened. Close the bleeder screw at that wheel and move to the next wheel in the bleeding sequence. Bleed all four wheels in the same manner. Check the master cylinder fluid level often.

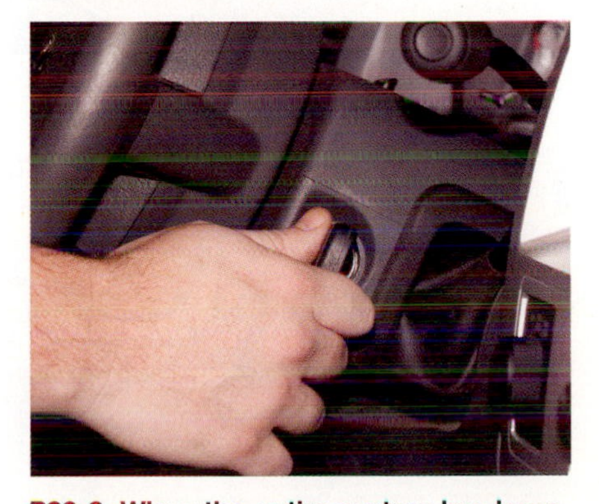

P30-9 When the entire system has been bled, start the engine and allow it to idle.

P30-10 **Check the pedal for sponginess.**

P30-11 **Check the brake warning lamp for an indication of unbalanced pressure. Repeat the bleeding procedure to correct the problem.**

P30-12 **Top off the master cylinder reservoir to the proper fill level.**

Figure 10-52 shows a typical pressure bleeder. To use it, connect the adaptor to the master cylinder reservoir (Figure 10-53), and pressurize it to the specifications outlined in the equipment's owner's manual. Do not depress the brake pedal while performing this procedure, as the fluid is already pressurized. Once pressurized, open the bleeders one at a time (and in proper sequence) and bleed the air out of the system.

Another brake bleeder device is shown in Figure 10-54. It can be connected to all four wheels at once as well as to the master cylinder. A reservoir on the device holds enough brake fluid to completely flush the system. Fluid is pumped into the master cylinder's

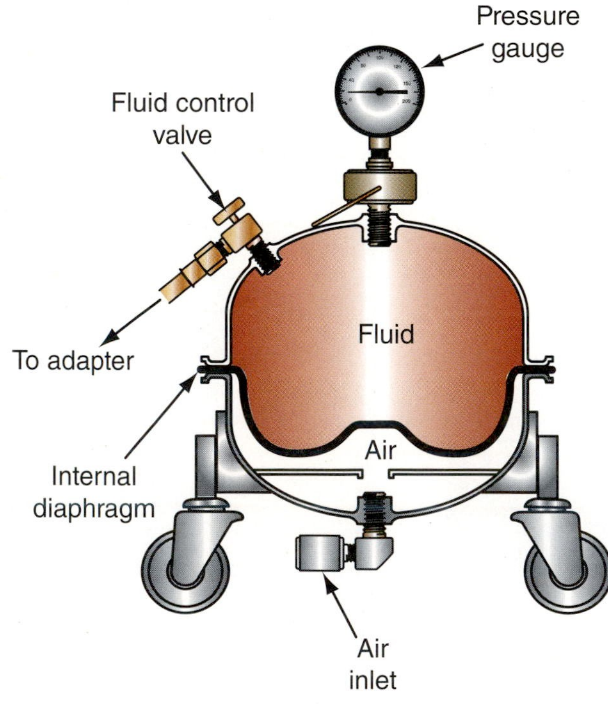

FIGURE 10-52 **The lower chamber is compressed air; the upper chamber holds brake fluid. The chambers are separated by a flexible, air-tight diaphragm.**

FIGURE 10-53 **Many different adapters are needed to properly pressure bleed the various master cylinder configurations.**

FIGURE 10-54 **A brake flushing machine.**

reservoir, through the brake system, and then returned to the brake bleeder device. Each wheel hose has a clear section near the wheel's bleeder screw. As new fluid is observed in these "windows," the bleeder screw is then closed for that wheel. A complete flush, using this device or a similar one, takes less than 10 minutes, even when performed by a newly trained technician or student.

Brake Lines and Hoses Inspection and Service

All manufactures include brake line and hose inspection in their maintenance schedules. Rigid hydraulic lines are made of double-wall, welded steel tubing that is coated to resist corrosion. Brake hoses are flexible and have multiple layers to withstand high pressures. Brake hoses will wear out due to movement and time. Brake lines will need to be replaced due to excessive rust that eats away the lines. Sometimes a hard line will have to be replaced because it was weak and brittle and the part it was connected to (like the wheel cylinder) was removed and moved the line enough to break it.

Start your inspection by looking for kinks in the lines and other damage due to rust and obvious leaks. Brake hoses can be inspected for cracks, leaks, and broken hold-down clips (Figure 10-55). If you have to remove a brake hose from a caliper, you have to install a new brake banjo washer. This washer is made of copper and creates a seal as you install it. This is a one-time-use component, so you must discard it and install a new one when removing a hose or caliper (Figure 10-56).

Brake hoses can be damaged on the inside due to a lack of fluid maintenance. To diagnose this you need to have a coworker pump the brakes while feeling the hose for swelling or bulging as pressure is applied internally (Figure 10-57). Ask the coworker to quickly apply and release the brakes and then quickly spin the wheel. If the brake at any wheel seems to drag after pressure is released, that brake hose may be restricted internally.

ABS Diagnosis and Service

Repairs to service brakes should not be started until the service manual has been consulted concerning cautions or warnings on the ABS system (Figure 10-58). Improper procedures or tools could damage wheel-mounted ABS components. In addition, some ABS systems have specific steps for bleeding or flushing the brakes. Though most "routine" repairs to the service brakes can be performed without fear of ABS damage, it is always too late to check the service manual after the damage is done.

Generally speaking, most shops will assign an experienced technician to diagnose and service an ABS system, but many of the simple steps can be performed by a new technician with adequate supervision. The first is the retrieval of diagnostic trouble codes (DTCs).

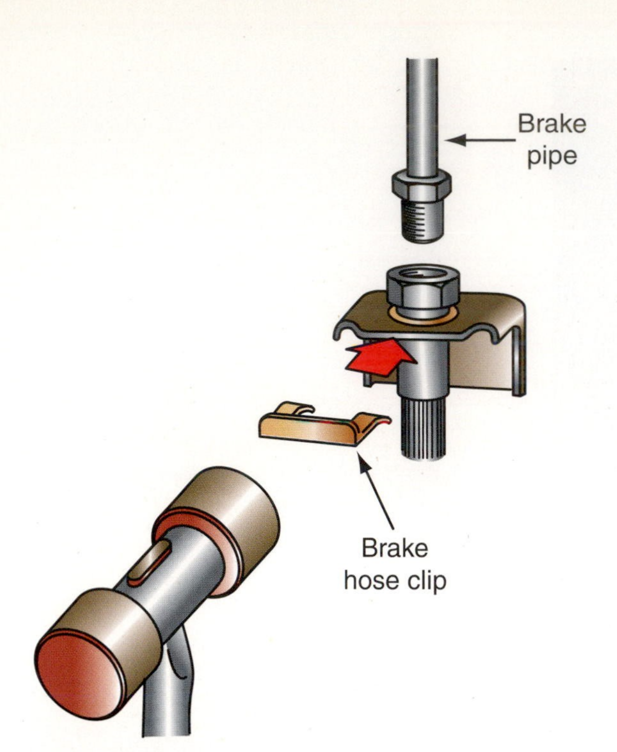

FIGURE 10-55 **The locking clip is tapped into place with a small hammer.**

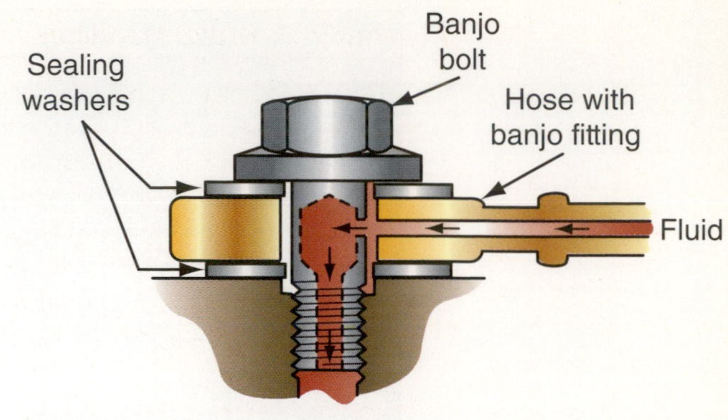

FIGURE 10-56 **Note the sealing washers on this banjo fitting. They should be replaced anytime the fitting is loosen.**

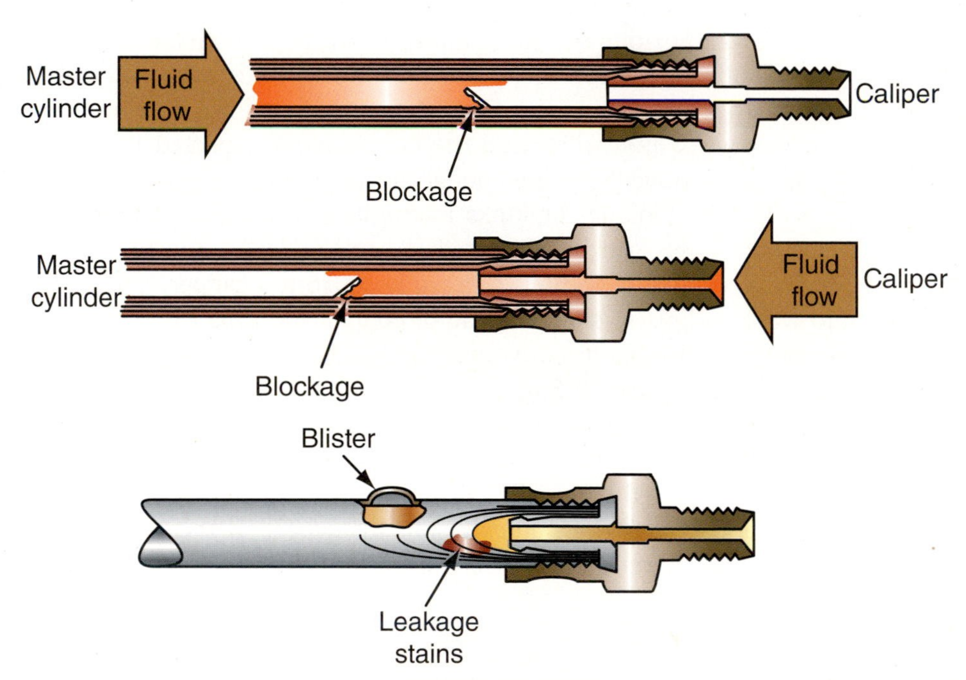

FIGURE 10-57 **Possible internal defects in a brake hose.**

When a customer complains that the ABS light (or other warning light) is on, the first tool used is the scan tool. All ABS DTCs can be retrieved with the vehicle's manufacturer-specific tool, and most can also be retrieved with an aftermarket scan tool like the ones sold by Snap-on, MasterTech, OTC, and others. In most instances, it is just a matter of

Antilock Brake Handling

CAUTION: Some components of the ABS are designed to be serviced as complete units, not separately. Removing or disconnecting certain components may damage the system, resulting in improper system operation and possible injury to passengers. Approved removal and installation procedures must be followed when servicing ABS components or their subcomponents.

FIGURE 10-58 **This is a typical type of warning for ABS systems. All manufacturers have cautions or warnings printed at the front of their manuals and in each section to which the caution or warning applies.**

identifying the vehicle to the tool and plugging the data link cable between the tool and vehicle. On-board diagnosis (OBD) I vehicles usually have separate data link connector and scan tool for ABS communication. Some of those vehicles required the DTC to be retrieved manually, and they will not communicate with the scan tool. OBD II vehicles have a more complex system for monitoring vehicle electronics. In most cases, but not all, the ABS computer can be accessed using the same OBD II Diagnostic Link Connector (DLC) and can sometimes be accessed with a generic (Global) OBD II scan tool.

Important Notes About Servicing Brakes on Hybrid Electric Vehicles

When servicing the brakes on a hybrid electric vehicle (HEV), make sure that you read all service precautions first. Some vehicles have an electronically controlled accumulator that pressurizes the brakes. This is unique to some hybrids because the engine will shut off when braking and the vehicle will mainly use the regenerative electric motor to slow the vehicle down. The brakes on these vehicles will typically last a little longer because they are not under the same braking stresses and conditions as non-hybrid vehicles. Follow the manufacturer's direction to depressurize or disable the system before servicing the brakes. Many times you will perform a brake inspection on a HEV and find that the brakes are not worn down. You should still inspect the caliper slides and moving parts. Those parts must be lubricated periodically as well, mostly from lack of use.

CASE STUDY

A student brought his car to our lab stating his brakes would not stop the car properly and he had added some brake fluid "a while back." A quick visual inspection showed the disc brake pads were completely worn out, to the point there was no lining left on any pad. If the brake fluid is low, obviously check for leaks but don't forget to check lining wear. This is particularly true of disc brake systems.

TERMS TO KNOW

Burnishing
C-clamp
Fading
Glazing
Glycol-based
Grabbing brake
Hard pedal
Hot spots
Locking brake
On-vehicle brake lathe
Pedal pulsation
Pressure bleeder
Spongy pedal
Vacuum pump

ASE-STYLE REVIEW QUESTIONS

1. Drum brake repairs are being discussed.
   Technician A says to keep one side intact while the other is disassembled to save some time.
   Technician B says wheel grease can be used to lube the rub points.
   Who is correct?
   A. A only
   B. B only
   C. Both A and B
   D. Neither A nor B
2. Pedal pulsation may be caused by any of the following EXCEPT:
   A. Warped rotors.
   B. Out-of-round drums.
   C. Broken or weak return springs.
   D. Damaged wheel bearings.
3. All of the following can be used to check a wheel speed sensor EXCEPT:
   A. Voltmeter.
   B. Test light.
   C. DMM on voltage.
   D. DMM on resistance.
4. The driver complains of the vehicle that front disc is pulling to the left during braking.
   Technician A says the left front brake is the cause.
   Technician B says the right front pads may be glazed from heat.
   Who is correct?
   A. A only
   B. B only
   C. Both A and B
   D. Neither A nor B
5. Diagnosing a disc or drum brake system is being discussed.
   Technician A says a bad hose on the rear brakes may cause the front wheels to lock too quickly.
   Technician B says pedal pulsation may be caused by incorrect pressure between the front and rear brakes.
   Who is correct?
   A. A only
   B. B only
   C. Both A and B
   D. Neither A nor B
6. Drum brake diagnosis is being discussed.
   Technician A says a grinding noise may be caused by broken hardware.
   Technician B says a firm pedal with little braking action could be caused by a parking brake misadjustment.
   Who is correct?
   A. A only
   B. B only
   C. Both A and B
   D. Neither A nor B
7. Rotor and drum machining is being discussed.
   Technician A says a drum can be machined on the vehicle.
   Technician B says a disc rotor must be removed from the vehicle for machining.
   Who is correct?
   A. A only
   B. B only
   C. Both A and B
   D. Neither A nor B
8. The customer complains of a slight pull to the left during braking. The technician finds the right inner pad worn out and the right outer pad with almost no wear.
   Technician A says the inner pad wear could be caused by dirty caliper slide pins.
   Technician B says the left pull could be a result of the right side brake failure.
   Who is correct?
   A. A only
   B. B only
   C. Both A and B
   D. Neither A nor B
9. The vehicle has very little brake boost and the engine stumbles severely during braking.
   Technician A says the hydro-boost power valve may be the problem.
   Technician B says the diaphragm in the booster is probably split.
   Who is correct?
   A. A only
   B. B only
   C. Both A and B
   D. Neither A nor B
10. A vehicle pulls slightly to one side during normal or slow braking.
    Technician A says a right rear drum brake shoe that is too tight could be the cause.
    Technician B says brake shoes soaked with brake fluid are the probable cause.
    Who is correct?
    A. A only
    B. B only
    C. Both A and B
    D. Neither A nor B

JOB SHEET 29

Name ______________________________ Date ______________

Replacing a Master Cylinder

NATEF Correlation

This job sheet addresses the following NATEF task:
Remove, bench bleed, and reinstall master cylinder.

Objective

Upon completion and review of this job sheet, you should be able to remove, bench bleed, and install a master cylinder.

CAUTION: Wear eye protection when dealing with brake fluid. Brake fluid can cause permanent eye damage. If brake fluid gets in the eyes, flush them with cold running water and see a doctor immediately.

Tools and Materials Needed

Vise
Oil drain pan and rags
Brake fluid
Torque wrench
Line wrenches
Fender cover
Service manual
Master cylinder bleeder kit

Procedures

Task Completed

1. Determine the following information from the service manual. 
   Vehicle Make ______________ Model ______________ Year ______________
   Master cylinder fasteners torque ______________________________
   Brake boost? ______________ Yes ______________ No
   If yes, what type? ______________________________
   Brake fluid type and classification ______________________________
   ABS cautions ______________________________
   ____
   ____
   ____
   ____

CAUTION: Do not work in or around the steering column without disarming the air bag systems. Serious injury could result from a deploying air bag.

WARNING: Do not mix glycol-based and silicone-based brake fluids. The mixture could cause a loss of braking efficiency.

WARNING: Before working on the brakes of a vehicle with ABS, consult the service manual for precautions and procedures. Failure to follow procedures to protect ABS components during routine brake work could damage the components and cause expensive repairs.

Task Completed

☐ **2.** Place thick wipe cloth under the master cylinder brake line connections.

☐ **3.** Disconnect the fluid level sensor if equipped and remove fluid from the reservoir.

☐ **4.** Loosen and remove the brake lines.

☐ **5.** Remove the master cylinder fasteners. Remove the master cylinder.

☐ **6.** Drain remaining fluid into an oil drain pan and dispose of the old fluid.

☐ **7.** Place the master cylinder body in a vise.

☐ **8.** Use a small pry bar to remove the reservoir from the master cylinder. Be careful not to crack the reservoir.

☐ **9.** Lube the new O-rings with brake fluid before installing the reservoir on the new master cylinder. The new master cylinder should have new O-rings.

☐ **10.** Install the reservoir to the master cylinder.

☐ **11.** Install the bleeder kit.

WARNING: Do not mix glycol-based and silicone-based brake fluids. The mixture could cause a loss of braking efficiency.

☐ **12.** Fill the reservoir chambers about half full with brake fluid.

☐ **13.** If necessary, reposition the master cylinder in the vise so the front (closed) end is higher than the rear.

☐ **14.** Use a punch or cross-tip screwdriver to push in the secondary piston. Push it in enough to bottom both pistons.

☐ **15.** Observe the reservoir for bubbles emerging from the bleeder hoses.

☐ **16.** Repeat steps 14 and 15 until there is no air coming out of the brake bleeder lines. Ensure that the line ends are kept below the level of the fluid.

☐ **17.** Close the lines with small paper clips.

☐ **18.** Install the reservoir cover.

☐ **19.** Install the master cylinder to the brake booster or firewall.

☐ **20.** Install and torque the master cylinder's fasteners.

WARNING: Protect the vehicle's paint surface from glycol-based brake fluids. This type of brake fluid can damage and remove the finish quickly. If brake fluid gets on the vehicle's finish, wash immediately with cold, running water.

☐ **21.** Remove the bleeder lines and attach the vehicle's brake line to the master cylinder. Do one line at a time and do not completely tighten the fittings.

NOTE: Step 22 should be done on a *single* stroke of the brake pedal.

Task Completed

22. Ask a second person to slowly depress the pedal while the secondary line fitting is loosened slightly to allow brake fluid to escape. Observe the fluid for air. Loosely tighten the fitting. Repeat with the secondary line. Ensure that both fittings are tightened before the pedal is released. ☐

23. Release the brake pedal and repeat step 22 as necessary to ensure that the lines are clear of air. If the procedure is not completed within two or three strokes of the pedal, the master cylinder is not bled properly and should be removed for bench bleeding. See steps 2 through 7 and 11 through 23. ☐

24. Test the system by applying the brakes with and without the engine running. The pedal should be firm and there should be no spongy feel. If the pedal is spongy, the complete brake system will have to be bled. It is suggested that the system should be bled any time the system is exposed to air. ☐

25. Remove the wipe cloth and ensure that the area is clean of any brake fluid. ☐

WARNING: Before moving the vehicle after a brake repair, pump the pedal several times to test the brake. Failure to do so may cause an accident with damage to vehicles, the facility, or it may cause personal injury.

26. Test the brakes before moving the vehicle from the bay. ☐

27. Clean the area and complete the repair order. ☐

Instructor's Response ______________________________

____

____

____

JOB SHEET 30

Name ______________________________ Date ________________

Replacing Disc Brake Pads

NATEF Correlation

This job sheet addresses the following NATEF task:

Remove caliper assembly from mountings; clean and inspect for leaks and damage to caliper housing; determine necessary action.

Clean and inspect caliper mounting and slides for wear and damage; determine necessary action.

Remove, clean, and inspect pads and related hardware; determine necessary action.

Disassemble and clean caliper assembly; inspect parts for wear, rust, scoring, and damage; replace seal, boot, and damaged or work parts.

Reassemble, lubricate, and reinstall caliper, pads, and related hardware; seat pads and inspect for leaks.

Clean, inspect, and measure rotor with a dial indicator and micrometer; follow manufacturer's recommendation in determining need to machine or replace.

Objective

Upon completion and review of this job sheet, you should be able to replace the brake pads on a front wheel disc brake and inspect the brake components.

Tools and Materials Needed

Lift or jack and jack stands
Impact wrench and socket
C-clamp
Disc brake silencer
Wire or cord
Torque wrench

Procedures

Task Completed

1. Determine the following information. ☐
   Vehicle Make ______________ Model ______________ Year ______________
   Caliper and caliper adapter torque ______________________________
   Brake boost? ______________ Yes ______________ No If yes, type?
   Brake fluid type and classification ______________________________
   ABS ______________ Yes ______________ No If yes, type? ______________
   Notes on ABS cautions ______________________________
   ____
   ____
   ____

WARNING: Before working on the brakes of a vehicle with an ABS system, consult the service manual for precautions and procedures. Failure to follow procedures to protect ABS components during routine brake work could damage the components and cause expensive repairs.

Task Completed

☐ **2.** Inspect and adjust the fluid level in the master cylinder so the reservoir is about half full.

☐ **3.** Lift the vehicle and remove the wheel assembly.

☐ **4.** Inspect the brake caliper mounting area for ABS components, caliper and adapter, and general condition of the steering, suspension, and brake components on each side of the vehicle. Record your observations.

____

____

WARNING: Do not use petroleum (gasoline, kerosene, motor oil, transmission fluid) or mineral-oil-based products to clean brake components. This type of liquid will cause damage to the seals and rubber cups and it decreases braking efficiency.

☐ **5.** Position a catch basin and spray down the braking components with a low-pressure washer.

☐ **6.** Remove the caliper mounting fasteners.

☐ **7.** Slide the caliper from the rotor. Use a large, flat screwdriver, if necessary, to pry the pads far enough to clear any ridge.

☐ **8.** Use wire or cord to support caliper weight from a vehicle component.

☐ **9.** Remove the pads and antirattle clips from the caliper or adapter.

☐ **10.** Remove the caliper adapter and rotor.

☐ **11.** Inspect the rotor for damage. Measure the thickness of the rotor and compare it to specification. If the rotor requires machining, consult with an experienced technician. Record your measurements and recommendation.

____

____

____

☐ **12.** Inspect the caliper piston boot, slide pins, and slide areas on the adapter and caliper. Record your observations. __________________________________

____

☐ **13.** Clean the slide pins or areas. Lube the pins or areas with disc brake lubricant.

☐ **14.** Coat the metal portion of each of the new pads with disc brake silencer. Allow them to set for about 5 minutes.

☐ **15.** Connect a flexible hose to the bleeder screw and route to a capture container.

☐ **16.** Loosen the bleeder screw and use a C-clamp to compress the caliper piston into its bore. Close the bleeder screw just before the piston bottoms out.

☐ **17.** Remove the hose and capture container.

☐ **18.** Install the inner pad into the caliper ensuring the pad mates and snaps into the piston. Install the antirattle springs. Install the outer pad.

Task Completed

19. Install the rotor onto the hub. Install and torque the adapter. ☐

20. If necessary, install the slide pins into the rotor. ☐

21. Slide the caliper or adapter with pads installed over the rotor and align the fastener holes. ☐

22. Install and torque the caliper fasteners. ☐

23. Install the wheel assembly and torque the lug nuts. ☐

24. Repeat steps 2 through 22 for the opposite wheel. ☐

25. Lower the vehicle when both wheel assemblies have been installed. ☐

26. Press the brake pedal several times to seat the pads to the rotor. ☐

WARNING: Before adding brake fluid, consult the vehicle's service manual. Many manufacturers require a specific classification of brake fluid to be used.

27. Check the brake fluid level and top off as needed. ☐

WARNING: Before moving the vehicle after a brake repair, pump the pedal several times to test the brake. Failure to do so may cause an accident with damage to vehicles and the facility, or it may cause personal injury.

28. Perform a brake test to ensure that the brakes will stop and hold the vehicle. Do this test before moving the vehicle from the bay. ☐

29. Clean the area and complete the work order. ☐

Instructor's Response ______________________________

____

____

____

JOB SHEET 31

Name ______________________________ Date ____________________

Diagnosing Drum Brake Hydraulic Problems

NATEF Correlation

This job sheet addresses the following NATEF task:

Diagnose poor stopping, noise, pulling, grabbing, dragging, or pedal pulsation concerns; determine necessary action.

Objective

Upon completion and review of this job sheet, you should be able to diagnose poor stopping, noise, pulling, grabbing, dragging, or pedal pulsation concerns.

Tools and Materials Needed

Basic hand tools

Protective Clothing

Goggles or safety glasses with side shields

Procedures

Task Completed

1. Determine the following information from the service manual. ☐

   Vehicle Make ________________ Model ________________ Year ________________

   Lug nut torque ____

   Brake boost ______________ Yes ______________ No If yes, type? ______________

   Brake fluid type and classification ________________________________

   ABS Cautions ____

   ____

   ____

2. Begin the inspection of the drum brake hydraulic system by checking the fluid level in the master cylinder reservoir. ☐
3. Check for leaks between the master cylinder and the power booster (bulkhead), between the reservoir and master cylinder, and the master cylinder outlet lines. ☐
4. Lift the vehicle to gain access to the undercarriage. ☐
5. Inspect for indication of fluid leakage down the inner side of each tire and wheel assembly. ☐
6. Inspect for leakage in and around the wheel cylinder area of each wheel. ☐
7. Inspect for leakage or damage on each line and hose with emphasis at each connection. ☐

Task Completed

☐ **8.** Road test the vehicle. As you apply the brake pedal, check for excessive travel and sponginess. What did you find?

____

____

☐ **9.** If the vehicle pulls or grabs to one side, raise the vehicle and remove the tire and wheel assembly.

☐ **10.** Remove the drum and inspect the brakes. Is there grease, fluid, or oil on the linings? Is the wheel cylinder firmly mounted without leaks? Are the wheel cylinder pistons froze in place?

☐ **11.** If the problem is not found, consult with your instructor.

Instructor's Response ____

____

____

____

JOB SHEET 32

Name ______________________________ Date ______________

Replace Drum Brake Shoes

NATEF Correlation

This job sheet addresses the following NATEF task:

Remove, clean (using proper safety procedures), inspect, and measure brake drums; determine necessary action.

Remove, clean, and inspect brake shoes, springs, pins, clips, levers, adjusters or self-adjusters, other related brake hardware, and backing support plates; lubricate and reassemble.

Pre-adjust brake shoes and parking brake before installing brake drums or drum or hub assemblies and wheel bearings.

Install wheel, torque lug nuts, and make final checks and adjustments.

Objective

Upon completion and review of this job sheet, you should be able to remove, inspect, install, and adjust drum brakes.

Tools and Materials Needed

Basic hand tools
Drum brake tools
Lift or jack and jack stands
Brake washer or vacuum cleaner
Brake lubricant

Protective Clothing

Goggles or safety glasses with side shields

Procedures

Task Completed

1. Determine the following information from the service manual. ☐
   Vehicle Make ______________ Model ______________ Year
   Is this a duo-servo or leading-trailing drum brake? ______________
   Lug nut torque ______________
   Brake boost ________ Yes ________ No If yes, type? ________
   Brake fluid type and classification ______________
   ABS Cautions ______________
   ____
   ____
   ____

2. Begin the repair of the drum brake hydraulic system by checking the fluid level in the master cylinder reservoir. ☐

Task Completed

☐ 3. Lift the vehicle to gain access to the undercarriage.

☐ 4. Inspect for indication of fluid leakage down the inner side of each wheel and wheel assembly.

☐ 5. Remove the wheel assemblies and drums. Inspect the brakes. Is there grease, fluid, or oil on the linings? Are the wheel cylinders firmly mounted without leaks? Are the wheel cylinder pistons free to move?

☐ 6. Use the proper cleaning procedures and equipment to clean the brake area including the backing plate.

SERVICE TIP:
Lay the brake components out in a manner similar to the way they are mounted on the backing plate.

☐ 7. Use the brake return spring tool to remove the shoes' return springs.

☐ 8. Use the brake retaining spring tool to remove one shoe's retaining spring. It can be either shoe.

☐ 9. On a leading-trailing system, remove the self-adjuster spring from the leading-trailing system.

☐ 10. On a leading-trailing system, rotate one shoe away from the bottom anchor and disconnect the self-adjuster and parking brake strut at the top.

OR

☐ 11. On a duo-servo system, rotate one shoe away from the wheel cylinder link and disconnect the self-adjuster mechanism at the bottom and the parking brake strut at the top. Remove the second shoe.

☐ 12. Disconnect the parking brake cable from the lever or the lever from the shoe.

☐ 13. Thoroughly clean and lubricate the six shoe rubbing pads on the backing plate and the thread portion of the self-adjuster.

☐ 14. Pull each wheel cylinder boot partially open and check for fluid inside the boot. Record your observations and recommendations.

____

____

☐ 15. If necessary, install the parking brake cable to the lever or install the lever to the rear shoe or both.

SERVICE TIP:
It does not matter which shoe goes on first. This job sheet will list the steps for installing the rear shoe on the right rear wheel. However, a right-handed person may find it easier to install the front shoe first.

☐ 16. Fit the top of the rear shoe web into the wheel cylinder link and install the retaining spring.

Task Completed

17. Install the return spring for the rear shoe. ☐

18. On a leading-trailing system, install the self-adjuster and parking strut onto the rear shoe. On a duo-servo system, install the parking brake strut onto the rear shoe. ☐

19. Align and fit the top of the front shoe web to the parking brake strut and self-adjuster, if necessary, and install the retaining spring, return spring, and self-adjuster spring (leading-trailing only). ☐

20. On a leading-trailing system, fit the bottom of the shoes into place on the anchor. ☐

21. On a duo-servo system, fit the self-adjuster spring into place on the shoes. ☐

22. If not already done, screw the self-adjuster to its minimum length. Place the front end of the self-adjuster into place on the bottom of the front shoe. Pull the bottom of the shoes apart and snap the rear end of the self-adjuster into place on the rear shoe (duo-servo only). ☐

23. Inspect the assembly for proper installation with emphasis on the springs and self-adjuster. ☐

24. Measure the inside of the drum and pre-adjust the shoes to match that measurement. Record the drum measurement. ☐

____

____

25. Slide the drum over the brake assembly. If it does not slide on easily, adjust the shoes in until the drum slides on with some drag. If it slides on too easy, adjust the shoes out. The drum should turn with a slight drag. ☐

26. With the drum installed, apply the brakes to center the shoes. Use a brake adjusting tool to make the final adjustment. ☐

27. Install the wheel assembly and torque the lug nuts. Lower the vehicle and top off the brake fluid as needed. ☐

CAUTION:
Before applying the brakes, ensure the brakes on any other wheels are installed and the drum(s) are installed.

28. Test the brake system by applying the brake pedal several times without and with the engine operating. Record your observations. ☐

____

____

Instructor's Response ____

____

____

____

JOB SHEET 33

Name ______________________________ Date ______________

Test and Replace Wheel Speed Sensor

NATEF Correlation

This job sheet addresses the following NATEF task:

> Test, diagnose, and service ABS speed sensors, toothed ring (tone wheel), and circuits using a graphing multimeter (GMM)/digital storage oscilloscope (DSO) (includes output signal, resistance, shorts to voltage or ground, and frequency data).
>
> Remove, clean (using proper safety procedures), inspect, and measure brake drums; determine necessary action.

Objective

Upon completion and review of this job sheet, you should be able to replace a wheel speed sensor for the ABS.

Tools and Materials Needed

Basic hand tools

Lift or jack and jack stands

GMM or DSO

Service manual

Protective Clothing

Goggles or safety glasses with side shields

Procedures

Task Completed

1. Determine the following information from the service manual. ☐

   Vehicle Make ______________ Model ______________ Year ______________

   What are the voltage or resistance specifications for the speed sensor? ____________

   ____

   General replacement procedures including fastener torque ______________________

   ____

   ABS cautions ____________________________________

   ____

   ____

2. Use the scan tool to retrieve and record any DTCs. ☐

3. Lift and support the vehicle to gain access to the sensor electrical connection. Unplug the sensor connection. What color are the conductors? ☐

   ____

Task Completed

Analog Meter

☐ 4. Select the AC volt position with the rotary switch.

☐ 5. Connect the meter leads to the terminals in the sensor plug.

☐ 6. Rotate the wheel while observing the action of the meter's needle. Record the needle action or inaction and make your suggestions on the sensor's serviceability.

NOTE: If there was no action, double check with the meter set in the DC volt position.

Digital Multimeter

☐ 7. Select the resistance or ohm position on the meter with the rotary switch.

☐ 8. Connect the meter's leads to the terminal in the sensor's plug. Record the resistance reading. Make your suggestion on the sensor's serviceability.

____

____

☐ 9. Follow the service manual's instruction to gain access to the sensor and its fastener.

☐ 10. Clean the area around the sensor and then remove the fastener.

☐ 11. Remove the sensor.

☐ 12. Install the new sensor following the service manual instructions. Install and torque the fastener.

☐ 13. Reconnect the sensor harness connection and any components removed for access.

☐ 14. Lower the vehicle, and, if not already done, clear any DTCs from the electronic control brake module (ECBM).

☐ 15. Start the engine. With the engine idling, pump the brake pedal several times. Did the ABS warning light stay off? If not, follow the scan tool's instructions and retrieve the DTC(s) and determine the necessary action(s).

Instructor's Response ______________________________

____

____

____

Chapter 11

Suspension and Steering Service

Upon completion and review of this chapter, you should understand and be able to describe:

- How to diagnose and inspect suspension and steering components.
- How to inspect and replace a shock absorber.
- How to replace inner tie-rod on a rack-and-pinion system.
- How to diagnose tire wear.
- How to remove and install wheel assembly.
- How to remove, balance, and mount tires on wheels.
- How to inspect and rotate tires.
- How to inspect and replace a TPMS sensor.
- How to perform a TPMS relearn procedure.
- How to repair a tire.
- How to replace a damaged wheel stud.
- How to perform a pre-alignment check and set up the alignment lift and machine.

Introduction

The driver may not realize that a steering or suspension problem exists until a part fails. The technician should make at least a thorough inspection of any vehicle being serviced. The inspection may only be a quick ride into the shop or a check of the thread on the tires. Spotting a potential hazard may bring business to the shop, but more important is the recognition and correction of a fault that could prevent accidents and injury.

Checking Frame and Suspension Mounting Points

Classroom Manual
page 287

As stated in the Classroom Manual, the condition of the frame affects the suspension and steering operations. A technician performing suspension and steering work should make an inspection of the frame. The frame should not show any unusual bends or cracks. Mounting fasteners that show any sign of movement indicate that a portion of the frame has shifted.

Attention should also be directed to the point where brackets are welded to the frame. Any break in a weld bead could indicate a twisted or bent frame. Abuse of the vehicle through overloading or driving can result in distorted frame members, particularly in unibody vehicles. Check the door, hood, and trunk alignment and hinge fittings. A misaligned door, hood, or trunk does not necessarily mean the frame is bent, but a bent or twisted frame could deform the body.

Diagnosing and Inspecting Suspension and Steering Problems

Classroom Manual
pages 287, 288, 289

At some time in their travels, everyone has seen obvious indications of poor suspension and steering components in vehicles on the road. The compact car with six passengers, the overloaded pickup, worn tires, and other indicators point to faults or potential faults.

Each could eventually result in a visit to a shop. Most suspension problems can be prevented with some maintenance and timely service. The components discussed in this section are control arm bushings, ball joints, springs, and shock absorbers. Torsion bars are checked in the same manner as springs, and struts are checked in the same manner as shock absorbers. Initial diagnosis of ARC systems will also be discussed as well as electric and 4WS systems.

CAUTION:
When placing the lift pads on a jack or lift under the vehicle, attempt to place the pad as close to the ball joint as possible. In some instances, the lower control can be bent by improper placement of the lift pads under the vehicle. Damage to the vehicle could occur.

Driver Complaints

The typical driver's complaint on suspension systems involves noise, swaying, and sagging of a corner or end of the vehicle. Most noise complaints are stated as a heavy thump, grating, or grinding noise as the vehicle wheels moves over the road surface. This may indicate a weak spring that would allow the control arm to bottom out against the rebound bumper on the frame, a bad ball joint, a missing or worn control arm bushing, or other component failure (Figure 11-1). The driver may also complain that the vehicle seems to roll or lean excessively during a turning maneuver. This could be caused by a weak spring, a worn or missing sway bar bushing, or mountings that are worn or missing. A technician should test drive the vehicle with the driver if possible. Test drive results along with the following inspections will isolate the problem component within a minimum time.

Studs on ball joints and steering linkages are tapered and are interference fitted into a tapered hole.

Control Arms

Most control arms do not suffer damage unless abused, but the bushings supporting the control arms do erode. The outer ends of the bushings are visible under the vehicle (Figure 11-2). The bushings should not be cracked, dry, or oil soaked. Pushing with a flat-ended rod or screwdriver can check the bushings' flexibility. The bushing material should flex and return to its original condition. If the material is very soft or does not return, it should be replaced. Replace all of the bushings on a control arm as a set. Worn bushings will allow the control arm to shift and may create a thud as the vehicle accelerates, brakes, or turns. Newer front-wheel-drive (FWD) vehicles have a lightweight lower control that can be bent by hard rearward shocks to the wheel or undercarriage.

SERVICE TIP:
Some vehicles have what is called a "rubberized" ball or steering linkage joint. This type of construction allows movement between the stud ball and its socket to reduce road shock to the vehicle. Consult the service manual for the specific amount of movement allowed. Some service manuals cite "excessive movement" without defining what that means. Typically, when specific measurements are not provided,

Ball Joints

Ball joints attach the control arm to the steering knuckle and must not have excessive movement between the **joint stud** and the steering knuckle. Excessive movement is any movement more than specified. There are several means to check ball joint wear. The service manual will provide exact procedures for performing this inspection. Some general guides are listed here.

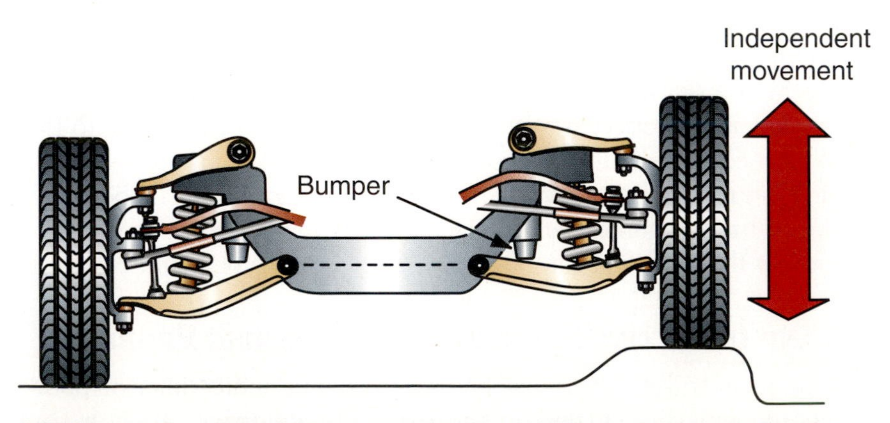

FIGURE 11-1 **The rebound bumper prevents the control arm from contacting the frame during hard shocks.**

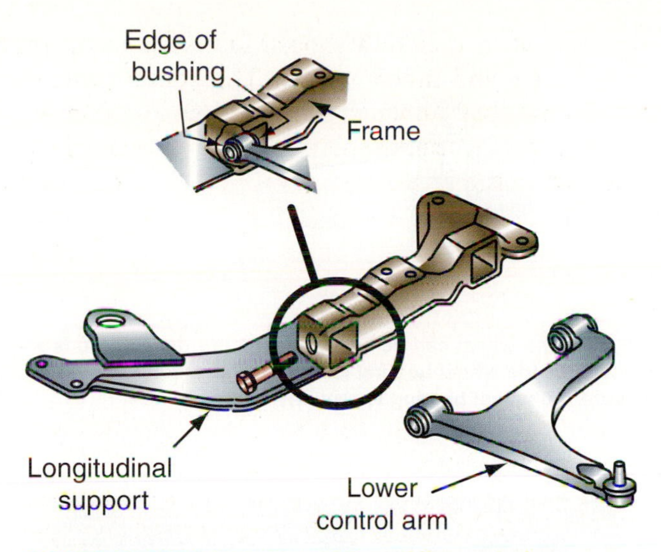

FIGURE 11-2 **The general condition of the control arm bushings can be checked by inspecting the visible ends of the bushing.**

Some ball joints have an indicator that is visible when the joint is worn. The grease fitting is designed and fitted into a floating or moveable retainer. As the joint wears, the grease fitting can be moved by hand. In some vehicles, any movement is cause for replacement of the ball joint. Some ball joints have the fitting's retainer extending above the ball joint. If the grease fitting shoulder should sink and become flush with the ball joint, the joint must be replaced.

WARNING: Do not replace a ball joint relying on hand and visual inspection alone. The quick test is used to make an initial diagnosis. Always refer to the service manual for specific inspection and tolerances.

Still other vehicles have ball joints that cannot be measured until the weight is removed from the joint. The weight of the vehicle must be on the control arm for a proper inspection. On a vehicle with the spring anchored on the lower control arm, place a jack under the control arm as close as possible to the ball joint (Figure 11-3). Operate the jack until the tire is an inch or two off the ground. Use a pry bar under the center of the tire and attempt to lift the tire (Figure 11-4). A good method is to place one hand at the top of the tire. This will give the technician a feel for any vertical movement of the wheel. If there is any movement, the ball joint needs to be checked further with a dial indicator or

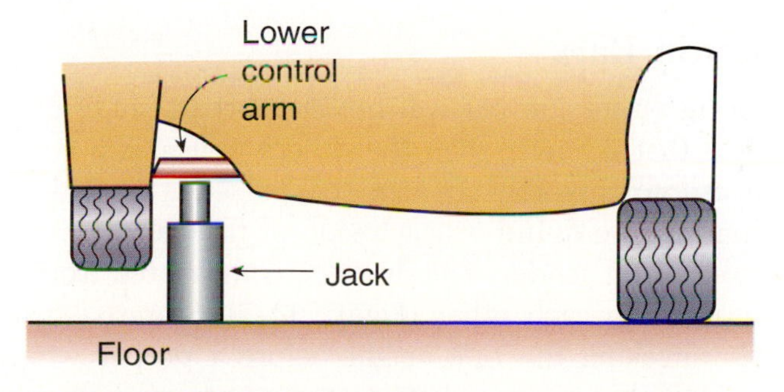

FIGURE 11-3 **The jack should lift directly under the ball joint or as near as possible.**

SERVICE TIP: *(continued)* and the technician feels the movement is excessive, it is best to replace the rubberized joint with a non-rubberized type. Before doing so, discuss the general technical issues and cost of changing to a non-rubberized joint with the customer. Technically, the only disadvantage to the technician and customer is the lack of "excessive movement" information and the fact that non-rubberized joint movement can be precisely measured. In fact, Ford, the primary user of rubberized joints, has never issued an exact movement specification for their rubberized joints and has discontinued their use. Both types have about the same life span, all things being equal. The cost of a rubberized versus a non-rubberized joint is about equivalent.

SPECIAL TOOLS

Pry bar

Jack

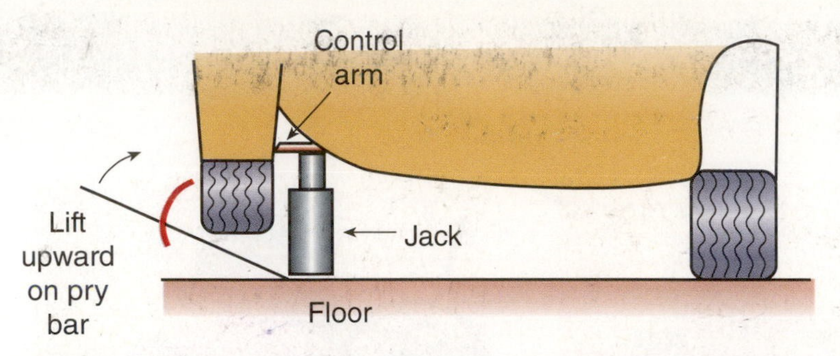

FIGURE 11-4 **The wheel can be shifted by the pry bar if there is wear in the ball joint or the wheel bearing is loose. Before condemning the ball joint ensure the wheel bearing is properly adjusted.**

FIGURE 11-5 **Dial indicator installed to measure horizontal joint movement.**

other procedures shown in the service manual (Figure 11-5). A typical ball joint measurement is outlined in Photo Sequence 31—typical ball joint measurement.

An short long arm (SLA) system with the spring on the upper control arm must have the upper ball joint checked in a similar manner. Systems that use torsion bars and struts also have specific inspection procedures to check the ball joints. Always check both wheels during the inspection.

Some ball joints have grease fittings that can indicate wear. Inspect these ball joints by inspecting the shoulder of the grease fitting for recession (Figure 11-6).

Springs and Torsion Bars

A worn or broken spring or torsion bar will not support its share of the vehicle's weight. A corner of the vehicle that is lower than the others indicates a weak or sagging spring. Some fairly quick measurements can confirm the presence of a weak spring or torsion bar. Locate a point that can be found on each side of the vehicle. An example could be a permanent point on each front fender. The distance is measured from each point vertically to the floor and compared to each other (Figure 11-7). Some vehicles require measurements to be made for checking the ride height before a proper wheel alignment can be accomplished. Figure 11-8 shows the measurement points on one vehicle. This diagnosis and test applies to all types of springs.

Typical Ball Joint Measurement

P31-1 Place a floor jack under a front, lower control arm, and raise the lower control arm until the front tire is approximately 4 inches (10 cm) off the floor.

P31-2 Lower the jack until the lower control arm is securely supported on a safety stand, and remove the floor jack.

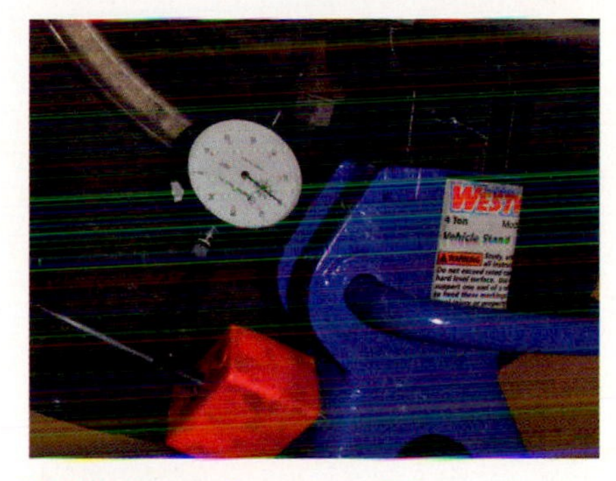

P31-3 Attach the magnetic base of a dial indicator for measuring ball joints to the safety stand.

P31-4 Position the dial indicator stem against the steering knuckle beside the ball joint stud.

P31-5 Preload the dial indicator stem approximately 0.250 inch (6.35 mm), and zero the dial indicator scale.

P31-6 Position a pry bar under the tire and lift on the tire-and-wheel assembly while a coworker observes the dial indicator. Record the ball joint vertical movement indicated on the dial indicator.

P31-7 Repeat step 6 several times to confirm an accurate dial indicator reading.

P31-8 Compare the ball joint vertical movement indicated on the dial indicator to the vehicle manufacturer's specifications for ball joint wear. If the ball joint vertical movement exceeds specifications, ball joint replacement is necessary.

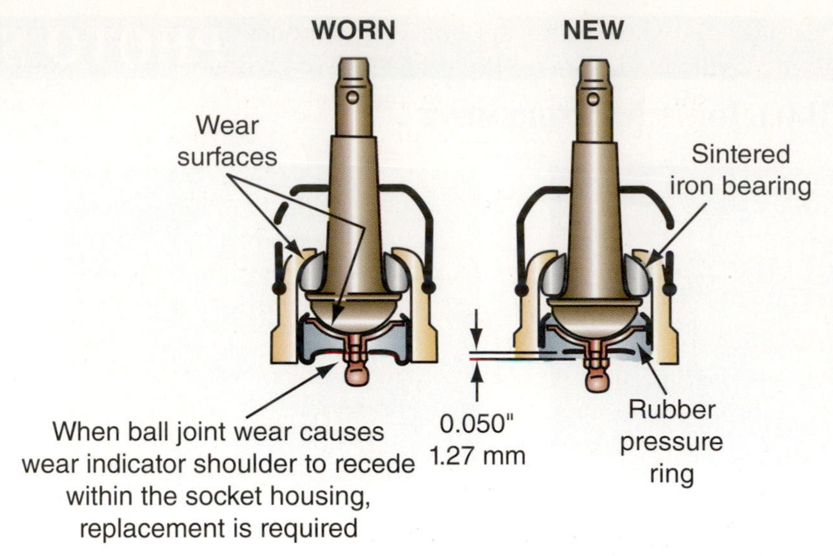

FIGURE 11-6 **Ball joint wear indicator with grease fitting extending from ball joint surface.**

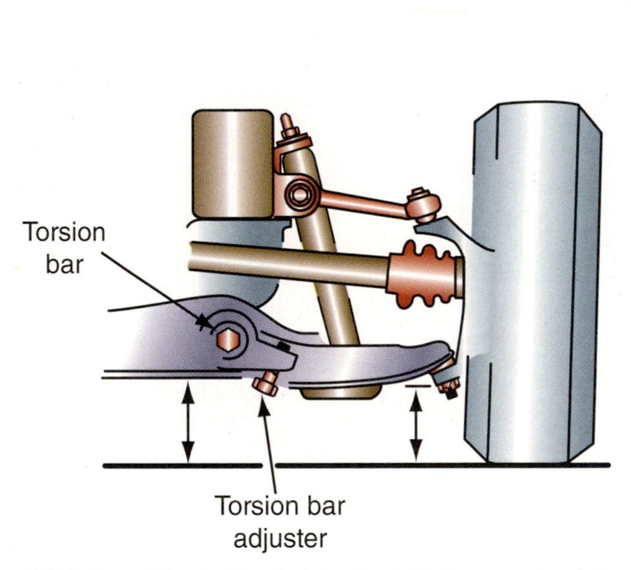

FIGURE 11-7 **Curb riding height measurement, torsion bar front suspension.**

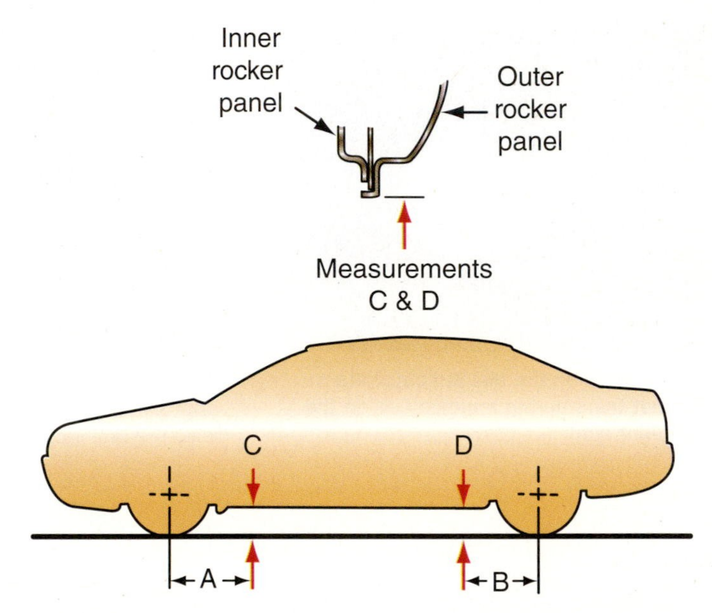

FIGURE 11-8 **Curb ride height measurements are made at specific points under the vehicles as determined by the vehicle manufacturer.**

Broken springs will cause the vehicle to drop in ride height, will not absorb shock, and will likely cause some noise. Any time the wheel with a broken spring encounters a dip or hump, the shock will be transmitted to the steering linkage and the vehicle frame. It will be readily noticeable by the passenger and particularly by the driver. A broken spring is replaced, not repaired. Some shops recommend that both springs on an axle be replaced together. That may be a valid recommendation in some cases, but it is not usually required by vehicle manufacturers and the makers of suspension components.

Shock Absorbers and Struts

Contrary to belief, shock absorbers and struts do not support the vehicle's weight. A weak spring does not necessarily mean the shock absorber is bad. However, a worn or broken spring may damage the shocks. Always check the shocks if a spring problem is found. Worn shocks and struts tend to make the vehicle **wave** or bounce as the wheels encounter road hazards. As the spring jounces and rebounds, the weak shock cannot control it. The end or corner of the vehicle will continue to move up and down until the spring's energy is exhausted.

As the vehicle **waves** from poor shocks or struts, the alignment angle changes on the front wheels and can cause steering control problems.

To test a shock absorber or strut, forcibly press downward on the vehicle corner with the suspected shock or strut several times. Release and watch for the vertical movement of the vehicle. A good, functioning shock or strut should allow the typical spring to jounce and rebound no more than once before stopping the vehicle at its ride height. Struts and shocks, like springs, can only be replaced.

Shocks or struts can also be broken. The broken part is usually the rod end (Figure 11-9). It will also produce a banging or knocking noise as the wheel encounters road irregularities. Naturally, there will be no shock absorber action. A more common condition with older shock or struts is oil leakage. There may be a very light indication of oil on the outside of the housing. This is considered to be **seepage** and is a normal condition. However, any visible liquid oil or indications of heavy leaking is a sign that the seals are bad and the shock or strut needs to be replaced (Figures 11-10 and 11-11). Again, some shops may recommend replacement of the shocks or struts on the same axle. With struts, it may be more desirable to replace both since front struts are part of the steering system also. It is the customer's choice, the overall condition of the vehicle, and the advice of the technician that determine the course of action.

Seepage is the very small amount of fluid that seeps past the seal. Most seals are designed to allow some lubrication seepage around the seal's contact point with the shaft.

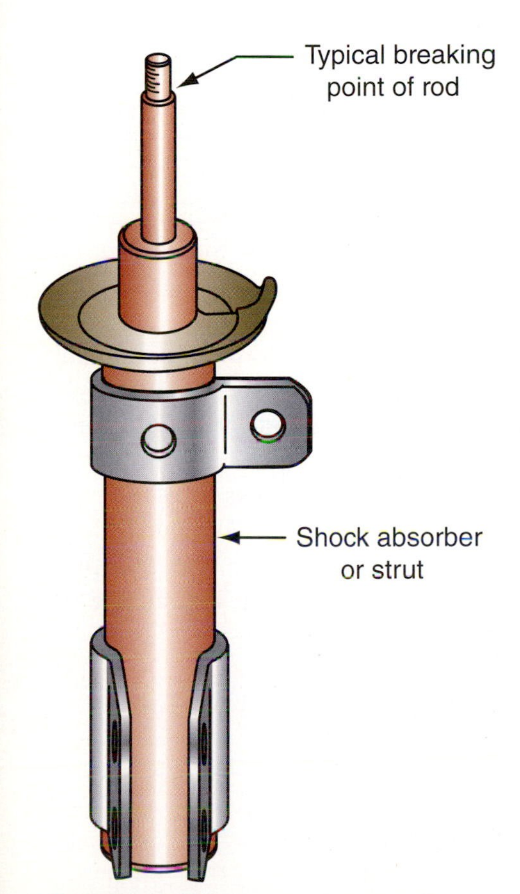

FIGURE 11-9 **The most common breaking point for a shock absorber or strut rod is at the stepped-down area near the top of the rod.**

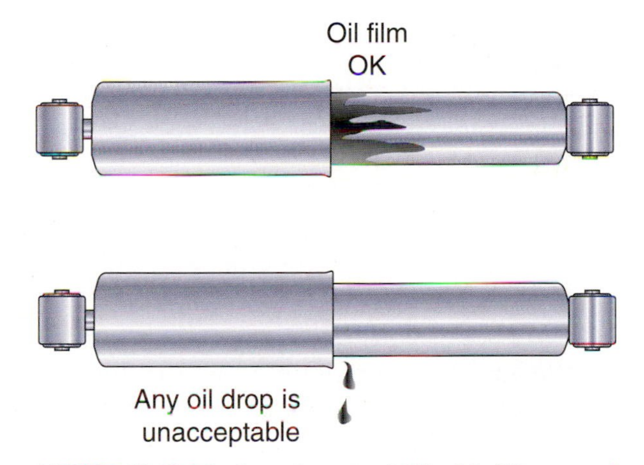

FIGURE 11-10 **A leakage is present if liquid oil is present along the outer shell of the shock absorber or strut housing.**

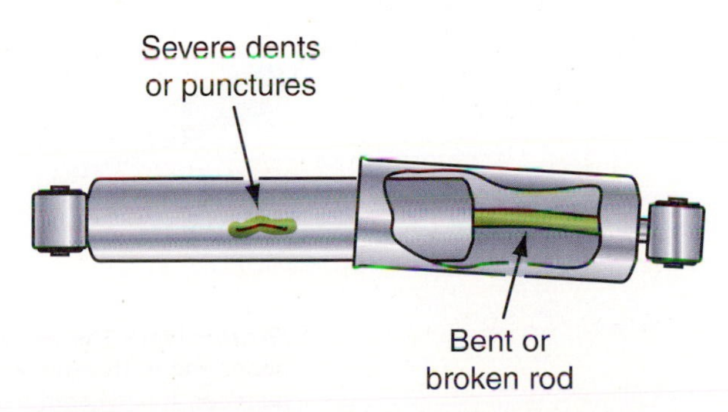

FIGURE 11-11 **Damaged shock absorber inspection.**

Isolators are made of rubber or rubber like material to reduce noise.

Gas-filled shocks have some nitrogen gas in them to increase the shock's action.

SERVICE TIP: Keeping the weight of the vehicle on the ground will help hold the shock's rod in as the upper fastener is removed. When installing a new shock, start the upper fastener on by hand and then lower the vehicle before completing the torque on the upper fasteners.

SERVICE TIP: On some vehicles the rear shock absorbers extend into the very rear of the passenger compartment. It may be necessary to remove the rear seat back, package tray, or radio speakers to access the top fasteners. Ensure hands and tools are clean before working inside the vehicle's passenger compartment. Care must be taken when dealing with fragile speaker wire connections.

Additional malfunctions of the suspension may initially be thought to be steering problems. Weak springs, shocks, or struts could result in poor steering and ride complaints. Poor springs on the rear axle can result in a vehicle that wants to wander from side to side.

Bumping or grinding noise can indicate poor spring **isolators.** The same is true of shock and strut mountings. Damaged upper strut mountings and bearings could result in crushing or grinding noises as the wheels are steered or a bump is encountered. Worn or missing body bushings can result in a noise that is similar to that of worn shock bushings. A thorough investigation must be conducted to locate noises that seem to be suspension problems. Another fact to consider is that worn or broken suspension components can damage other suspension and steering components.

Replacing Suspension Components

Front Shocks

Overall, shocks are very simple to replace. The wheel assembly may be removed for easier access. Remove the bolts or nuts that hold the shock in place. On some vehicles, the upper fastener is accessed through the engine compartment. Remove the top fastener on this type before raising the vehicle. The weight of the vehicle will help hold the rod while the nut is being removed. If the top nut and shock rod turn together, a special tool can be used to break off the nut (Figure 11-12), but in most cases, an impact wrench or air ratchet will remove the nut easily. Raise the vehicle and remove the lower fasteners. Remove the shock from the vehicle. Check to make sure all of the old bushings are also removed. The bushings on the rod type are two-piece with metal retainers.

WARNING: Do not use pliers or other unpadded tools to hold the strut or shock absorber rod still while the nut is being installed. A damaged or marred rod can damage the upper seal.

Gas-charged shocks hold the rod downward in the cylinder by a wire or lock. The rod must be released during installation. Inspect the mounting area and decide the best method of installing and releasing the rod. Many times, the best method is to fasten the lower end of the shock and then release the rod. The rod is not under a great deal of pressure, and extension is fairly slow and can be controlled. The rod's tip is guided into place. Other than the automatic extension of the rod, the installation is the same as non-gas shocks.

Install the rod's lower bushing retainer and bushing. Insert the rod into its mounting hole. Install the upper bushing, install the retainer, and loosely tighten the nut. Mount the lower end of the shock and install its bushings and fasteners. The shocks may have rod-type or plate-type mountings at each end or one of each (Figure 11-13). The rod-type mountings, where the nut screws directly onto the rod, always have bushings. The plate-type mountings usually do not have separate bushings. A special wrench is available that

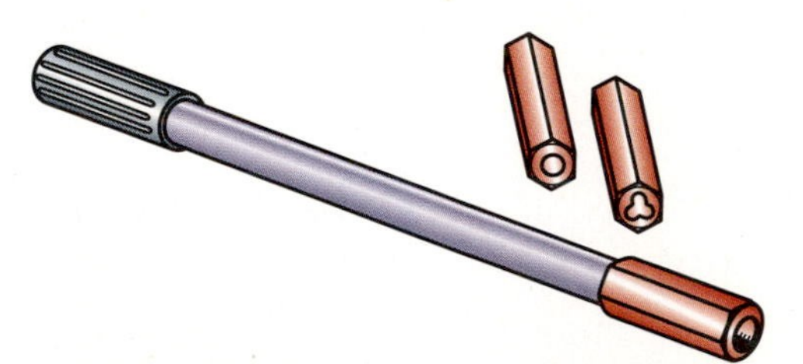

FIGURE 11-12 The end of the rod fits over the upper end of the shock absorber rod and its fastener. It is twisted back and forth until the rod breaks off.

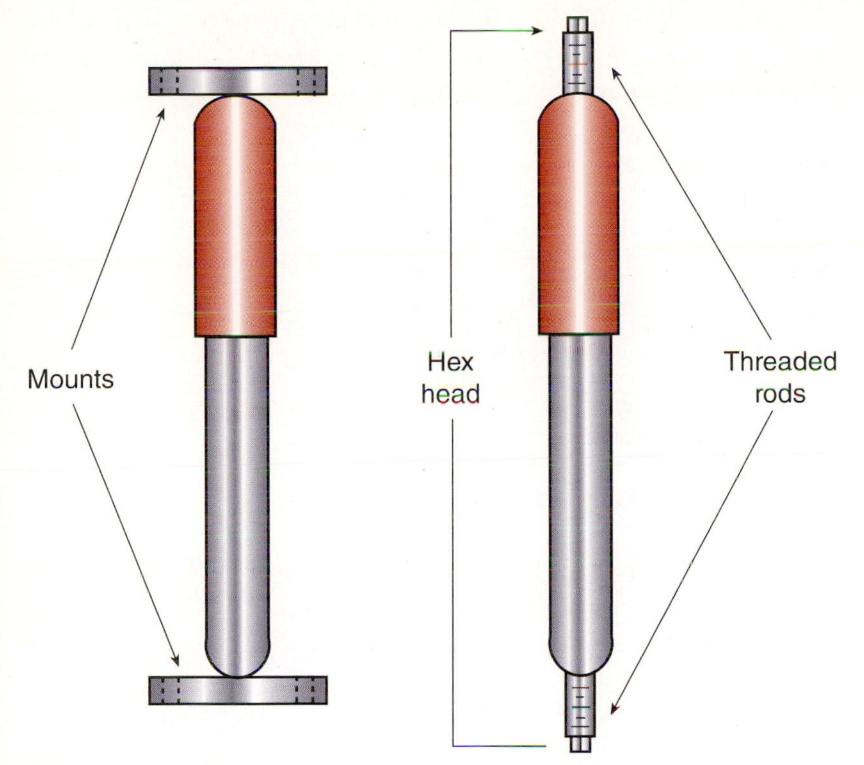

FIGURE 11-13 **A shock absorber may have the types of mount shown or a combination of the two.**

Flats for open-end wrenches

FIGURE 11-14 **This special socket is made of two connected pieces. The bottom fits over the fastener and the top fits over the hex end of the rod. The top is then held while the bottom is turned by an open-end wrench. Space around the shock may prevent using this type of socket.**

will hold the rod while the nut is being tightened (Figure 11-14). Torque all fasteners. Lower the vehicle, clean the work area, and complete the repair order.

Rear Shocks

Replacing the rear shocks is similar to replacing the front ones, and some rear struts can be replaced following the procedures for rear shocks. Do not remove rear struts if they are also used to mount coil springs. A service manual is required to remove this type of strut correctly and safely.

Before removing any fasteners, inspect the suspension to ensure that the spring will not come loose when the shock is disconnected. This is especially true of rigid axles with coil springs. The shock absorber helps hold the coil in place by preventing the axle from dropping too far. An axle with a leaf spring should also be supported. When the shock absorber is removed, the axle will drop and the new shock absorber can be connected without raising the axle. The upper shock absorber fastener on some cars is accessible through the trunk.

If the upper mounting is in the trunk, raise the vehicle enough to gain access to both ends of the shock. The vehicle can be raised to a good working height if both mounts are under the vehicle. Remove the upper fastener and then the lower ones. Remove the shock and bushing. Most rear shocks have bushings pressed into the shock absorber mounting. Others use bushings similar to the ones on the front (Figure 11-15).

If a gas-filled shock is used, decide on the best method to remove the hold-down wire and install the shocks. Usually, the easiest method is to place the upper end in position, loosely tighten the fasteners, and cut the wire. The shock absorber rod can be guided to the lower mount as it extends. Mounting the lower end first and then guiding the rod upward into place is another method. It is usually easiest to guide the shock into place instead of extending it and then forcing it into place.

Gas-charged shocks are a hydraulic shock absorber with nitrogen gas installed to pressurize the hydraulic fluid for quicker and more responsive shock action.

SPECIAL TOOLS

Jack stand, short or tall as needed

CAUTION:

Removing shocks on some solid rear axles will release their coil springs and drop the axle. Failure to support this type of axle can cause damage or injury.

SERVICE TIP: Before replacing some high-priced components, check the fuse for the air compressor and computer. Either one could cause the system to fail. Remember KISS ("Keep it simple, stupid."). If only one corner of the vehicle shows a possible suspension fault, then the problem is usually related to that wheel only. It could be an air spring, air valve, height sensor, or the air lines leading to the air spring. Assuming that the remainder of the system works, allow the compressor to switch on and listen for air leaks at the lowered corner. An air leak is indicative of a burst air spring, faulty air valve, or a burst or loose air line. A thorough visible inspection will pinpoint the malfunctioning component. If no air leak is heard, the air spring is not totally eliminated, but the fault is likely due to a height sensor or air valve. Shut down the system, find the air line connection to the air spring or valve and disconnect it.

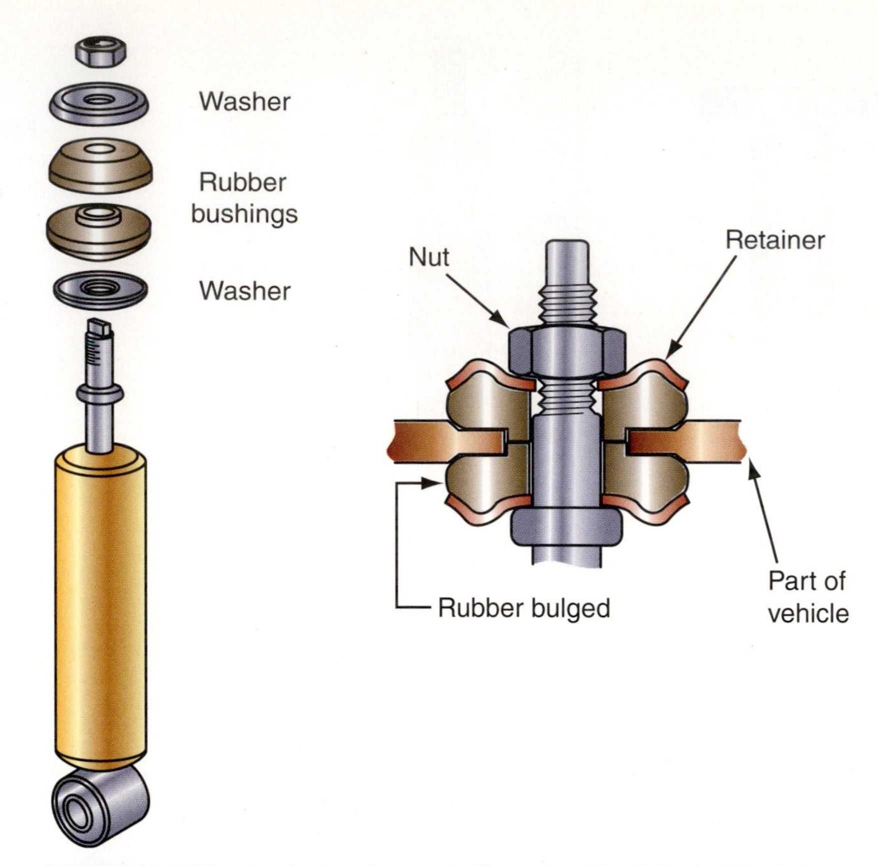

FIGURE 11-15 **The shock absorber or strut's upper end will be isolated from the body or frame by rubber-type bushings fitted over the rod. A shock absorber may have the same type of bushing at its lower end.**

Mount any necessary bushings and retainers. Install the shocks and torque the fasteners. Lower the vehicle, clean the work area, complete the repair order, and, if all repairs are complete, return the vehicle to the customer.

Automatic Ride Control (ARC)

Diagnosis of the mechanical components of automatic ride control (ARC) systems is the same as for true mechanical suspension systems. One common fault that is visible on many older ARC-equipped vehicles is the complete lowering of the vehicle. This can be so bad that the steering wheels cannot be turned without the tire rubbing the inside of the wheel wells. On a rear-load leveling system, the same fault may be visible by the extreme lowering of the rear and the high angle of the vehicle's front end.

In almost all cases like this, it is a sign that the entire ride control system is malfunctioning. Normally all four air springs, or four air valves, or four (or three) height sensors will not fail together, but a common component such as a faulty compressor or computer will cause the system to cease functioning.

Steering Component Diagnosis

Many times, the first indication of a steering component failure is the customer buying new tires or complaining of poor steering. The technician can identify some problems by doing a quick inspection while performing an oil change or some other repair work. Each item of the steering system should be checked each time the vehicle has any work performed on the steering or suspension systems, including tire replacement or repair. Most of the parts

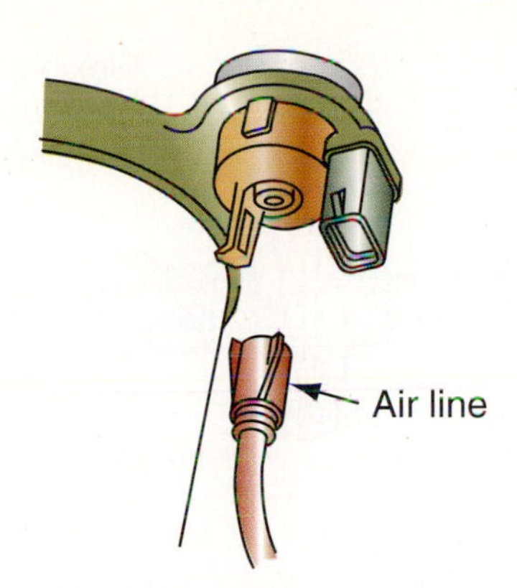

FIGURE 11-16 **Disconnecting the air line from the air spring solenoid valve.**

can be easily checked without any special tools except for lifting equipment. Tires will be addressed in a later section.

Driver Complaints

Like the suspension system, the customer may complain of noise or improper movement of the vehicle that may indicate a steering component failure. A worn linkage socket or inner tie-rod may produce a popping or snapping noise as the steering wheel is turned. A loose or worn idler arm may be stated by the customer as a "wander" on a straight road or excessive play in the steering linkage. Excessive steering wheel play may be the result of worn gears in the steering gearbox, but this is more common with recirculating ball and nut than with rack-and-pinion systems.

A symptom that came to be known as "morning sickness" may be present on older rack-and-pinion steering systems. This is a result of poor power-steering boost after the vehicle has been shut down overnight. The system leaks pressure until the fluid warms up and the rack's piston seals swell. The driver has to apply more force than usual to steer the vehicle during cold driveaway. The steering unit has to be replaced or rebuilt to correct this problem. Since rack and steering units are rebuilt by aftermarket vendors, it is usually best to replace the steering unit with a rebuilt model instead of trying to rebuild it at the shop. The overall cost will be about the same, but the time will be much less. Like suspension problems, the technician may have to road test the vehicle to further isolate the failed component.

Linkage

Before lifting the vehicle, check the linkage for **steering wheel free play.** With the engine off, move the steering wheel in each direction while watching the front wheels. Any movement of the steering wheel should result in an immediate corresponding movement of both wheels. If one wheel does not follow the steering wheel, there is probably a problem with the linkage to that wheel. A loose or worn steering gear will result in erratic action of both wheels. If there is excessive free play between the steering wheel and the steering linkage, the steering gear or **coupler** must be replaced (Figure 11-17). Neither can be repaired even though the steering gear can be resealed.

A more accurate method of checking free play is to measure the amount of travel of one spot on the steering wheel. Slowly, move the steering wheel in one direction until all of the free play is removed. Align a graduation on a rule or other linear measuring device with a spot on the wheel. Hold the measuring device in place and parallel to the wheel's direction

SERVICE TIP: (*continued*) Start the compressor again and listen for air leaking from the line. Escaping air from the lines indicates that air is available, but for some reason it is not being allowed into the air spring. This narrows it down to the air valve, height sensor, or the electrical wiring to and from the computer. The service manual will provide electrical information for using a digital multimeter (DMM) to test each electrical component and the wiring in the system. Some of the newest ARC systems will illuminate a warning lamp in the instrument panel. This alerts the driver and may provide some assistance in diagnosing the system. Still other systems allow the technician to retrieve diagnostic trouble codes (DTCs) from the system with a scan tool. As usual, when in doubt consult the service manual. This is especially useful if only one function of an ARC is not functioning

SERVICE TIP:
(*continued*)
(e.g., system will not lower the vehicle at cruise speeds). In some instances, the ARC computer uses data from the powertrain control module (PCM) or break control module (BCM) to properly determine when and what function to activate. Although ARC systems and their individual components can be expensive, diagnosing the system is usually straightforward as long as there is an understanding of the system's operation and the service manual is used.

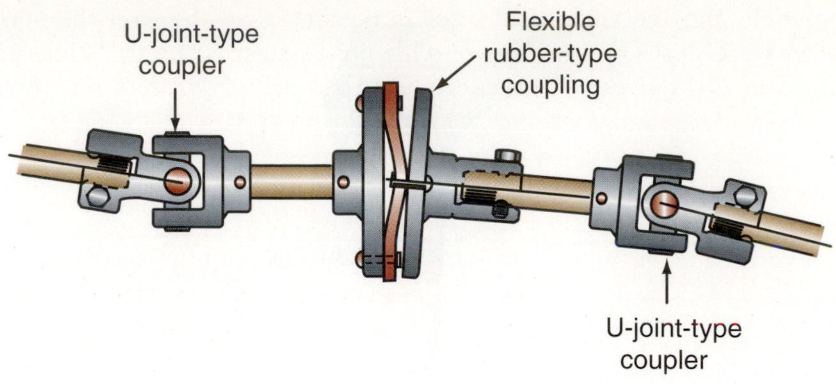

FIGURE 11-17 **The steering shaft coupling may be either a U-joint type or a rubber-type. A common symptom with the U-joint type is binding during a turn, whereas the rubber-type coupling tends to break loose or become chemically unstable.**

of rotation (Figure 11-18). Turn the wheel in the opposite direction to the first movement until the free play is removed again. Note where spot on the wheel aligns on the measuring device. The distance between the two graduations is the amount of free play.

Make a check of the power-steering system by starting the engine and turning the wheels completely from side to side. The complete movement should be smooth and continuous. The engine should not stall at idle, nor should there be any belt squealing, pump noise, or jerks of the steering wheel. Replace or adjust the belt if it is worn or loose. Add power-steering fluid if there is pump noise. If the power-steering pump emits a moaning sound, check its fluid level. If the level is correct, the pump may need to be replaced.

There are specific tests that can be used to diagnose the power-steering systems. One uses a beam-type torque wrench. Only an experienced technician should perform this test because the center pad of the steering wheel must be removed. The driver's air bag, if equipped, is part of the pad. Once the air bag is disarmed and the pad is removed, the torque wrench and socket are placed over the steering-wheel retaining nut (Figure 11-19). Lift the front wheels so they are free of the vehicle's weight. The steering wheel is turned using

SERVICE TIP:
Before lifting a vehicle equipped with some form of ARC, turn the system off before lifting the vehicle. The on/off switch is found in the trunk of many vehicles (Figure 11-16).

Classroom Manual
pages 304, 305, 306, 307, 308

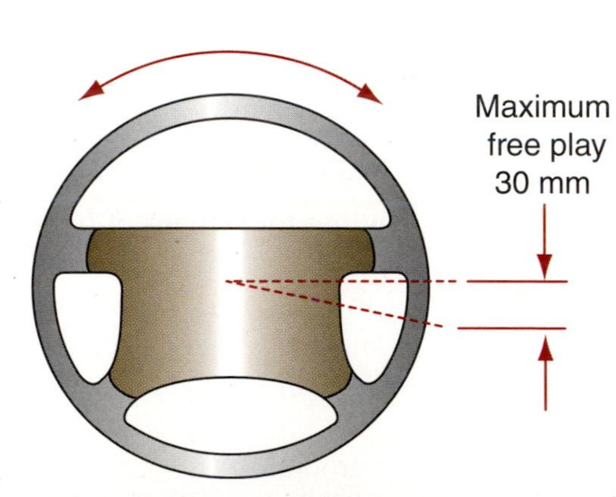

FIGURE 11-18 **Very few steering systems allow any free play at the steering wheel on modern vehicles. The free play can be measured with a ruler.**

FIGURE 11-19 **The amount of torque it takes to move the steering can be measured with a dial-type torque wrench. It should be checked any time a new steering gear is installed or new bearings are installed on the steering shaft.**

the torque wrench, and the amount of force is registered on the wrench. Another test is done with a pressure gauge set up for use with power-steering units. While any pressure gauge reading over 300 psi can be used, finding the correct adapters may be a problem. Most power-steering test gauges come in sets with the correct adapters and can be used fairly easily.

Most power-steering systems have a pressure switch. The switch sends a signal to the PCM to adjust idle speed during parking. A faulty switch may cause the engine to stall or stumble when parking. The engine may also have some problem in the fuel or air idle systems that may have caused it to stall during parking. The service manual lists the steps to test the switch and its circuits. However, the noted repairs do not have to be accomplished before the remainder of the inspection is performed.

The linkage has to be checked for worn or broken connections or parts. A drive-on lift is best suited for the inspection. It will keep the weight of the vehicle on the wheels and apply normal operating tension on the steering and suspension systems. Raise the vehicle to a good working height. With the vehicle lifted, apply linear force against the connections of each of the sockets in the linkage (Figure 11-20). Apply the force back and forth in line with the socket stud. There should be absolutely no movement between the stud and the hole it fits through. Any movement of the stud within its mating hole requires the component to be replaced. Do not confuse this with the side-to-side movement of the ball within the socket. The ball should rotate from side to side, and some manufacturers allow for some vertical movement of the ball within the socket. Consult the service manual for specific tests.

The pitman arm located at the bottom of a recirculating ball and nut system is splined onto the sector shaft and held in place by a large nut and lock washer (Figure 11-21). There should be absolutely no movement between the pitman arm and shaft and the nut should be completely tight. Here, any movement requires the replacement of the pitman arm, sector shaft, or both. If the sector shaft is the problem, it is recommended that the entire steering gear be replaced along with the pitman arm. It would be extremely unusual to find a problem at this connection. A nut that was not properly torqued is the only reasonable cause of slack between the arm and shaft. If there is a loose connection here, it advisable to closely inspect the entire system for similar loose fasteners or improperly installed components.

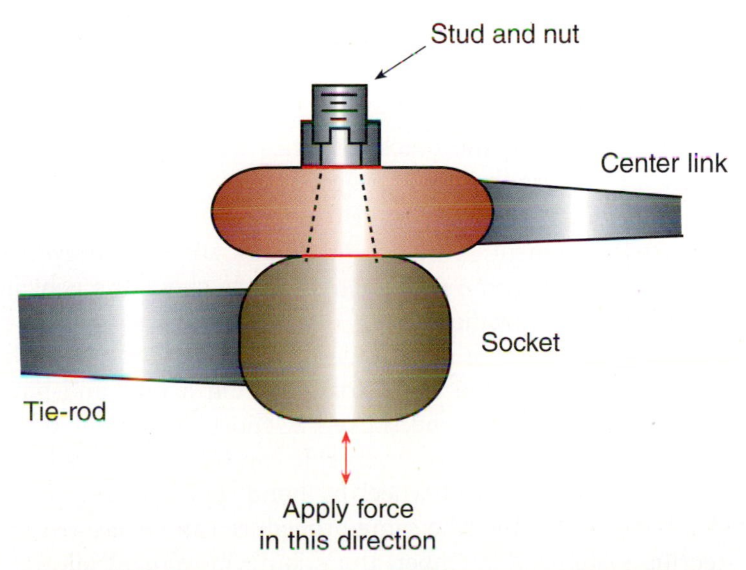

FIGURE 11-20 **There should be no excessive movement between the stud and the tapered hole.**

CAUTION:
Do not work on the steering column or steering wheel without disabling the air bag system. Serious injury or death can result if the air bag functions and the technician is working in the vicinity of the steering wheel or column. Consult the service manual for disarming procedures.

CAUTION:
Do not work on the steering column or steering wheel without disabling the air bag system. Serious injury or death could result if the air bag functions and the technician is working in the vicinity of the steering wheel or column. Consult the service manual for disarming procedures.

Steering wheel free play is the amount of steering wheel movement before the steering gear or linkage moves.

The **coupler** is a flexible connection between the steering column and the gear or the upper and lower shafts.

CAUTION:
Do not work on the steering column or steering wheel without disabling the air bag system. Serious injury or death could result if the air bag functions and the technician is working in the vicinity of the steering wheel or column. Consult the service manual for disarming procedures.

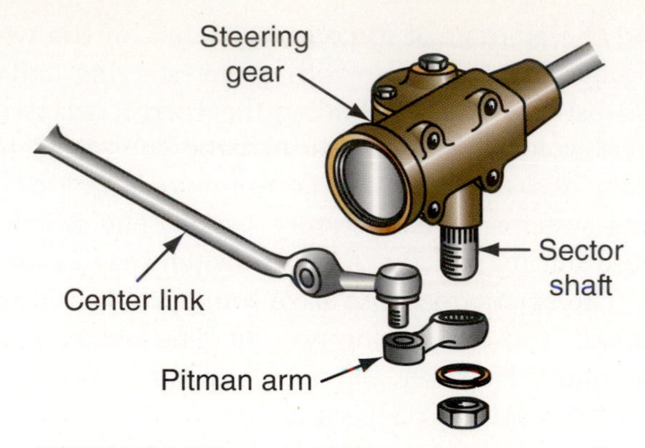

FIGURE 11-21 **There should be no movement at all between the pitman arm and the sector shaft. In almost all cases, excessive steering movement can be traced to other steering components.**

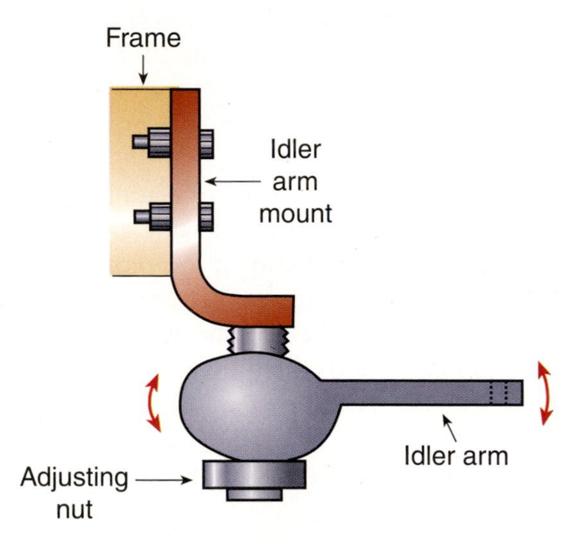

FIGURE 11-22 **Some idler arms have an adjustment to remove excessive movement.**

The idler arm, if equipped, must be checked for mounting and arm movement. The mounting bolts must be tight and the bracket flushes against its mount, which is usually the frame rail (Figure 11-22). As the connection between the center link and idler arm is checked, notice the movement of the idler arm on its pivot. Some manufacturers allow for a limited amount of movement at this point; others allow no movement. A few idler arms can be adjusted to remove some movement. Always check the vehicle's service manual for specific procedures and specifications.

Have someone move the steering wheel side to side while watching the movement of the linkage. The entire system should move together. On a rack-and-pinion system, one common problem is a worn inner tie-rod socket. A short movement of the rack with no movement of the tie-rod indicates a worn socket. It can be verified by locking the steering wheel and trying to move the affected wheel by hand. If there is a socket problem, the wheel and tie-rod will move slightly. The same procedure can be applied to test the linkage on all types of steering systems. Remember, there is no movement allowed at the steering linkage connections and the steering should be smooth from one side to the other.

Replacing Steering Linkage Components

WARNING: Do not apply heat to remove any steering linkage or fastener. Heat will change the temper of the component and could result in damage or injury.

Classroom Manual
Pages 304, 305, 306, 307, 308

The power-steering belt is replaced by loosening the tensioner. The pump may be driven by a single serpentine belt or it may have a separate belt. The replacement of a serpentine belt was covered in Chapter 8.

A single belt, or belts, used to drive the power-steering pump usually employs the pump as the belt tensioner. To replace the power-steering belts, it may be necessary to remove other belts. It is recommended that all of the belts be replaced at one time.

Some brackets have threaded adjusting bolts and brackets to adjust belt tension (Figure 11-23). The pump will have to be pushed or pulled on other types. Some pumps can be moved by hand while others have to be forced with a bar of some type. In most cases the bar will be a 1/2-inch or 3/8-inch breaker bar. The pump's bracket will have a square hole for the socket driver on the bar. Move the pump toward the engine until the belt is loose enough to be removed. Remove the belt and install the new one.

Position the bar to apply outward force to the pump. Do not pry on the reservoir. Pull the belt to the correct tension and hold in place. Tighten the easiest fastener to access. Release the bar and check the belt tension. If it is not adjusted properly, loosen the first fastener and tighten the belt again. If the tension is correct, torque all the fasteners.

Most steering linkage parts are replaced in similar methods by forcing the stud from its mounting hole. After lifting the vehicle, remove the cotter key, if used, and the nut from the stud. A special tool is then used to remove the stud from its mounting hole (Figure 11-24). Press the stud out and remove the steering linkage from the vehicle. In

CAUTION:
Do not work on the steering column or steering wheel without disabling the air bag system. Serious injury or death could result if the air bag functions and the technician is working in the vicinity of the steering wheel or column. Consult the service manual for disarming procedures.

SERVICE TIP:
Using a regular pry bar between the pump and a stationary component like the engine can damage the pump's reservoir. Use only the recommended tools to move the pump and tension the belt.

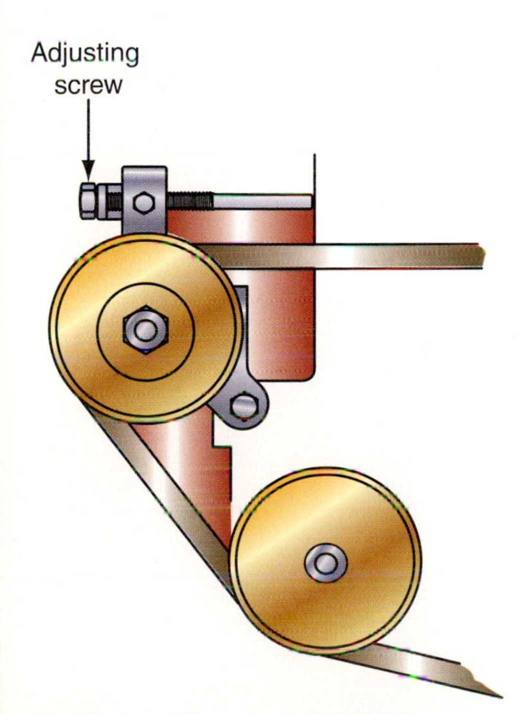

FIGURE 11-23 The screw-type adjuster can be found on many modern vehicles regardless of brand. The lock screw holds the pump in place after the belt is tensioned.

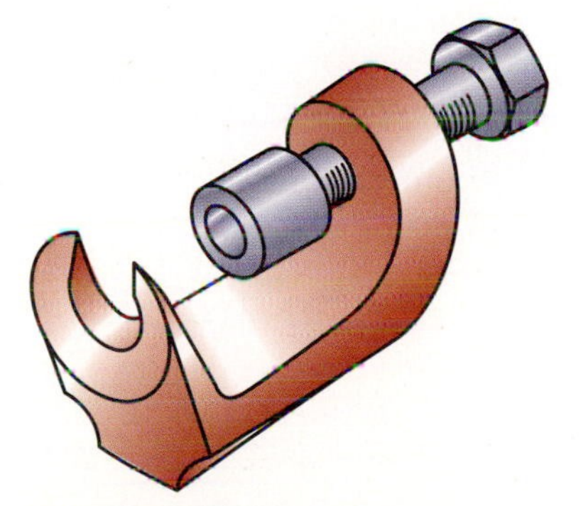

FIGURE 11-24 This tool works similar to a press.

CAUTION:
Do not overtension the belt. The belt will wear quicker and the pump's shaft bearings may be damaged.

SPECIAL TOOLS

Ball joint or tie-rod end press

Impact wrench

Torque wrench

A **castellated nut** is so named because of the turret-type projections on one side.

SPECIAL TOOLS

Inner tie-rod socket wrench

Ball joint or tie-rod end press

most cases, more than one connection will have to be removed to free the damaged components.

Tie-rod ends can be removed from the steering knuckle using the same tool. The end is then unscrewed from the tie-rod or adjusting sleeve. Count the number of turns as the end is being screwed off. Install the replacement end the same number of turns. This will approximately set the toe of the front wheels until an alignment is completed.

Installation is basically the reverse of removal. Before inserting the stud into its mating hole, ensure that the rubber boot is placed over the socket and a grease fitting is installed. Insert the stud and hand tighten the nut. Do the same with any other connection that needs to be done on this component.

The new part should come with a new nut and cotter key if used. (Do not reuse a locking nut.) The nut will be a self-locking type or a **castellated nut** (Figure 11-25). The castellated nut is used with a cotter key. A standard torque wrench and socket cannot be used because of space and the arrangement of the components. If special torque wrenches are not available, tighten the nut as much as possible using a standard box-end wrench. This is not really the proper procedure, but it is an accepted practice.

With a castellated nut, line up a slot in the nut with the cotter keyhole in the stud. Do not loosen the nut to align the hole or slot. Continue to tighten it a little at time until they are aligned. If the nut has to be loosened, the complete connection has to be broken and the installation must be started from the beginning. The nut will have to be removed and the stud will have to be forced from the hole and reinstalled. This is due to the tapered shape of the stud and hole. Once they are forced or pressed into place, they must be locked at that point. Loosing the nut relieves force on the taper, and the two mating components will not have the correct fit.

On rack-and-pinion systems, the inner tie-rod and socket are removed in a little different manner. Disconnect the tie-rod end from the steering knuckle as previously stated. On the inner end, remove both boot clamps and slide the boot outward toward the end of the rod. It may be necessary to turn the steering wheel until the boot and socket are clearly visible. Inspect the socket connection. There should be a small hole with the end of a pin

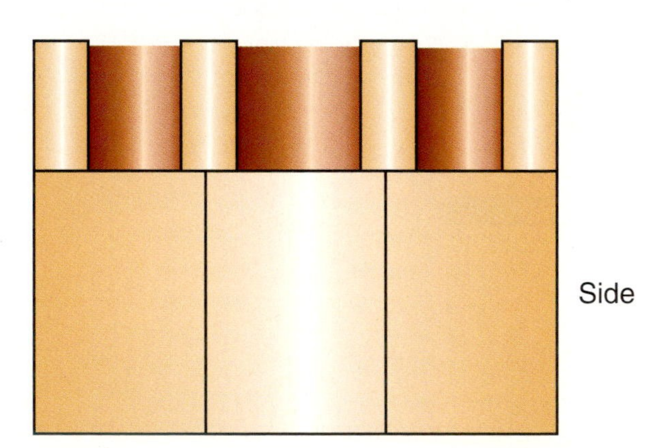

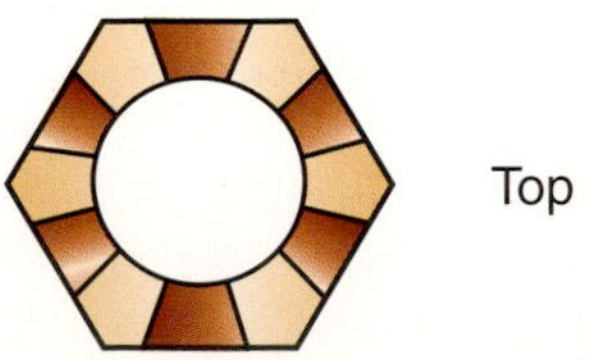

FIGURE 11-25 **The slots between the projections allow for the insertion of a cotter key.**

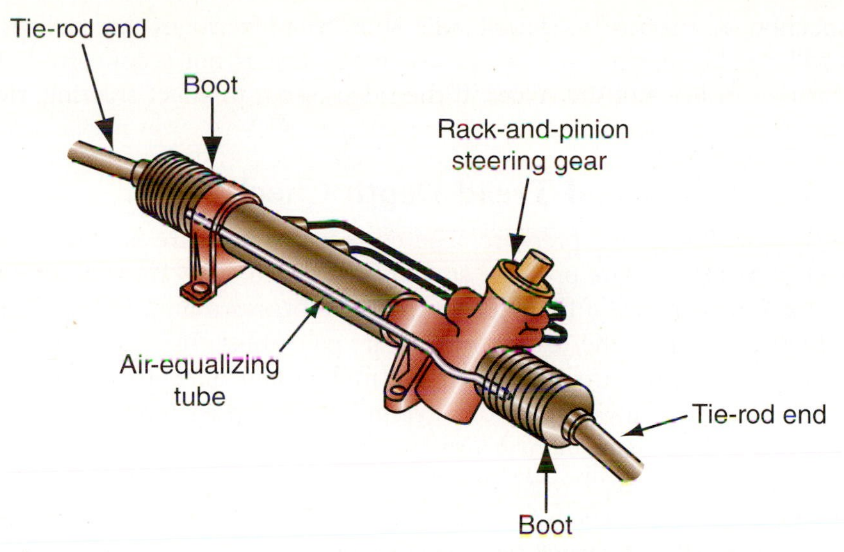

FIGURE 11-26 **The air-equalizing tube allows air trapped in the boots to move between the two as the rack is shifted left and right, collapsing and expanding the boots in the process. Loss or misplacement of this tube will result in grit and moisture entering the boots.**

Classroom Manual
pages 307, 308

visible toward the inner end of the socket (rack end). The pin locks the socket to the rack and needs to be removed. A pin punch and small hammer will accomplish the task. Other systems may use a setscrew or other device to lock the socket. They must be removed before proceeding.

There is an air-equalizing tube that runs between the boots on each end of the rack (Figure 11-26). As a boot is removed, ensure that the tube remains in place in the other tube. When the boot is installed, ensure that the tube is properly positioned in each boot.

With the lock removed, locate two or more flat sides on the socket. There are special extended sockets available, or an open-end wrench can be used to remove or install the inner tie-rod. Remove the tie-rod end, lock nut, clamps, and boot from the inner tie-rod. Remove the inner tie-rod from the rack. Slide the boot, clamps, and screw nut onto the new tie-rod. Screw and torque the inner tie-rod onto the rack and install any locking device. Move the boot over the rack housing and install the inner clamp. Leave the outer clamp loose for now since the toe must be checked and adjusted. Install the tie-rod end onto the tie-rod the same number of turns it took to remove it. Connect the end to the steering knuckle as previously stated. Hand tighten the lock nut until the toe is checked. The alignment can now be checked and the repair completed.

Pulling usually refers to a sudden movement of the vehicle to one side if the steering wheel is released.

Drifting usually refers to a slow movement of the vehicle toward one side if the steering wheel is released.

SERVICE TIP: **Pulling** or **drifting** can be caused by poor alignment, dragging or grabbing brakes, road conditions on older two-lane roads, or bad tires. In the absence of obvious tire faults, do not condemn the tires until all related systems are checked.

Tire Wear Diagnosis

Correctly interpreted, tire wear can point quickly to a steering or suspension malfunction. Abnormal tire wear is any one part of the tire that is worn more than a matching part of the same tire. It is primarily thread wear, but may include damage to the sidewall or bead. Also, individual tires may show different types of wear.

A driver may complain of the vehicle **pulling** or **drifting** to one side on a straight road. This is usually a result of poor wheel alignment, and corrective actions can be taken at a shop. Usually, a vibration felt at certain speeds can be traced to unbalanced tires, but it could possibly be a driveline problem as well. If tire rotation and balance have been recently performed and a pulling to one side starts after the work, radial tire belt movement is usually the problem. Switching the two front tires to opposite sides may correct the pull, but at times the tires must be placed back in their original condition to eliminate the pull.

An inspection of the tire's sidewall may show sloping ridges and valleys, and most radial tires will be affected to some extent. Normally, this is not a concern, but the driver may question the presence of the ridges. If the ridges seem to affect steering, ride, or durability, the tire needs to be replaced.

Tire Pressure Check and Tread Depth Check

Inspect the tire for the correct pressure. The correct tire pressure is always found on the tire placard (Figure 11-27). Tire pressure should be checked every month. Inform your customer to perform this service if they do not have tire pressure monitoring system (TPMS) sensors. When inspecting the tire tread wear you must use a tread depth gauge (Figure 11-28). Always measure the tire in the middle and on both edges. This will give you an early indication of any suspension, steering, or alignment problem. The legal minimum tread depth is 2/32 of an inch. A tire has wear indicators (Figure 11-29) that will be visible and flush with the tread when it is legally worn out. Most tires do not have good rain or snow traction below 4/32 of an inch. Most new passenger car tires start new with 12/32 to 13/32 of tread. Light truck tires have more.

Tire Pressure Wear

The tread should wear evenly around and across the tread area. The most common fault is incorrect tire pressure. An overinflated tire will ride on its center tread and not wear the other parts of the tire (Figure 11-30). An underinflated tire will ride on its edges with little

FIGURE 11-27 **Wheel weights are removed and installed with special wheel weight pliers.**

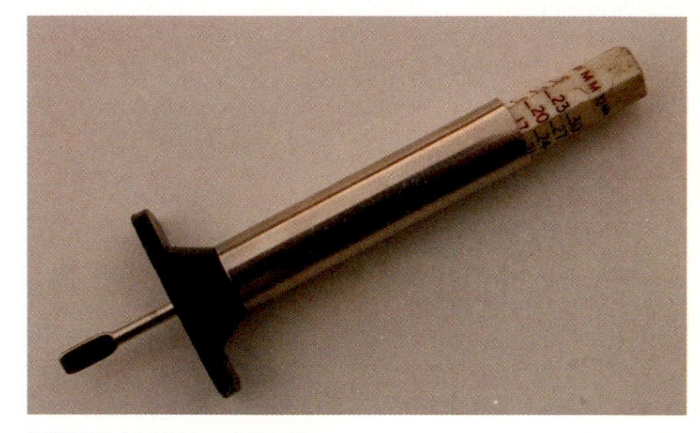

FIGURE 11-28 **Tread depth gauge.**

FIGURE 11-29 **Tire tread wear indicators.**

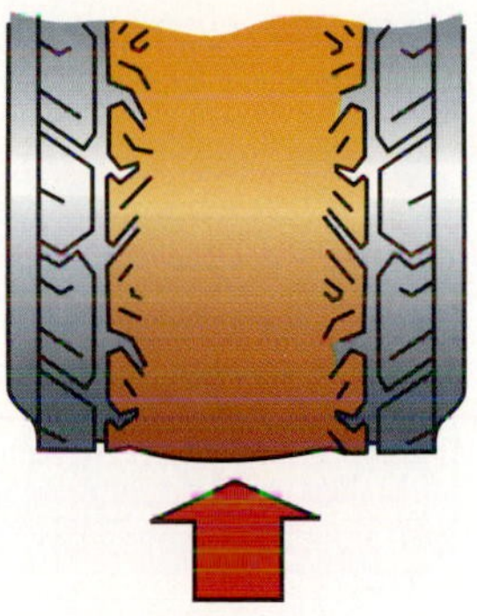

FIGURE 11-30 **Wear present on the center two or three treads is a sign of overinflation.**

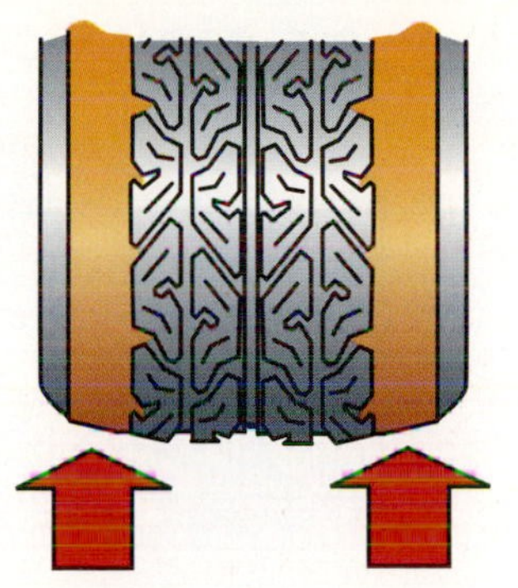

FIGURE 11-31 **Even wear on each outside tread indicates a tire driven while underinflated.**

wear to the inside area (Figure 11-31). Scuff marks on the sidewall may also be apparent on an underinflated tire. This results when the air pressure is not high enough to hold up the sidewall when cornering. The customer should be advised to check tire air pressure at least once a month or as the operator manual states. Underinflated tires also result in a loss of fuel mileage and poor steering control.

Camber Wear

Camber wear is indicated by excessive wear on the edge of the tire with little wear on the center or opposite edge (Figure 11-32). Too much **positive camber** results in wear on the outer three or four treads, and excessive **negative camber** shows up as inner tread wear. The remainder of the tread will show even wear. Camber is corrected using an alignment machine. Some vehicles do not have camber adjustment, while others have adjustments on each wheel. If there is no adjustment mechanism, the suspension, frame, and body have to be checked and damaged or worn components must be replaced.

Camber wear and **toe wear** can be confusing. Remember that toe wear will create a sawtooth pattern across all or most of the tread.

Toe Wear

Toe wear will also show the most wear at the tire's edge, but it will be spread over most of the tread width. Toe wear is indicated by a sawtooth pattern across the tread (Figure 11-33). The pattern is obvious when the technician's hand, palm down, is drawn across the tread

A **positive camber** is the inward tilt of the wheel at the top.

Negative camber is the outward tilt of the wheel.

FIGURE 11-32 **Camber wear affects the outer tread(s) of the tire. The wear will be fairly even across the affected treads.**

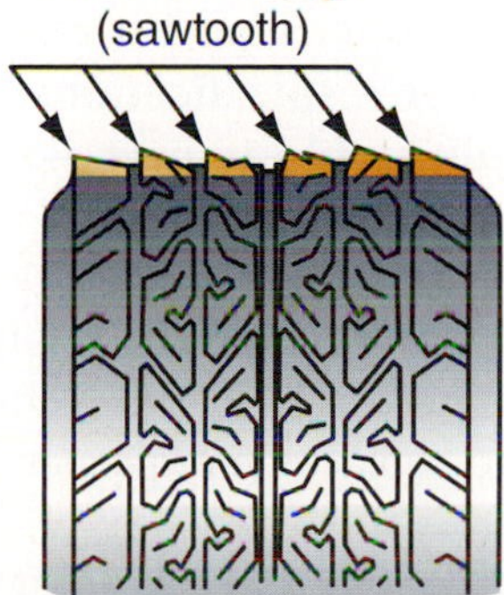

FIGURE 11-33 **Toe wear leaves a sawtooth effect with the outer tire tread being the most worn and the wear on the individual treads decreasing toward the center of the tire tread area.**

FIGURE 11-34 **Separating out-of-balance wear from lack of rotation wear sometimes is difficult to do. The best solution is to balance and rotate the tires.**

Cupping wear caused by balance or rotation is not as clearly distinguishable as other types of tire wear.

width. The edge of each tread will be sharpened as though it had been cut with a sharp object. The sharpened edges point in the direction of toe. If the edges point to the outside of the tire, the tire is toed-out too much. In this case, the greatest amount of wear will be on the inside treads, which is similar to camber wear, but camber wear will not create the sawtooth pattern.

CAUTION:
Before removing (or installing) a tire from a wheel equipped with an electronic tire pressure sensor, consult the service manual. There are specific steps in positioning the tire and wheel assembly on the tire machine. Failure to follow the specific steps for mounting the tire and wheel assembly on the tire changer could result in total destruction of the sensor.

Wear Caused by Balance and Rotation

Balancing and rotating tires are both important for long tire life. They should be part of routine tire maintenance and be performed about every 5,000 miles. Failure to do so could result in poor riding and steering comfort.

Tires that are not kept in balance may develop **cupping** of the treads at irregular intervals around the tire (Figure 11-34). The cupping occurs when the heavy spot of the tire hits the road, causing a scuffing of the tire against the pavement. Tires that have not been rotated may also cup the tread. When determining if balance or rotation causes the wear, remove the tire assembly and check its balance on a tire balance machine. Even at this point the technician may not be able to completely eliminate rotation as the cause of the problem, so the best solution is to balance and rotate all tires on the vehicle.

Wheel and Tire Service

WARNING: The lug nuts on a wheel should never be tightened with a straight impact wrench. They must be torqued according to the manufacturer's specifications using a torque wrench, a torque stick, or a reliable adjustable impact wrench. Damage to the wheel, studs, hub, or brake disc could result.

SPECIAL TOOLS

Jack and jack stands or lift

Impact wrench

Torque wrench or wheel torque wrench

Service manual

Some of the more common problems with wheels and tires can be traced to poor or no service. The driver or the technician may also cause a problem. Wheel and tire service is limited to replacement, proper torquing, or repairing a hole in the tire.

Wheel Assembly Removal and Installation

WARNING: Apply extra care when working with chrome or magnesium wheels. Improper torquing of the lug nuts could result in damage to the wheel.

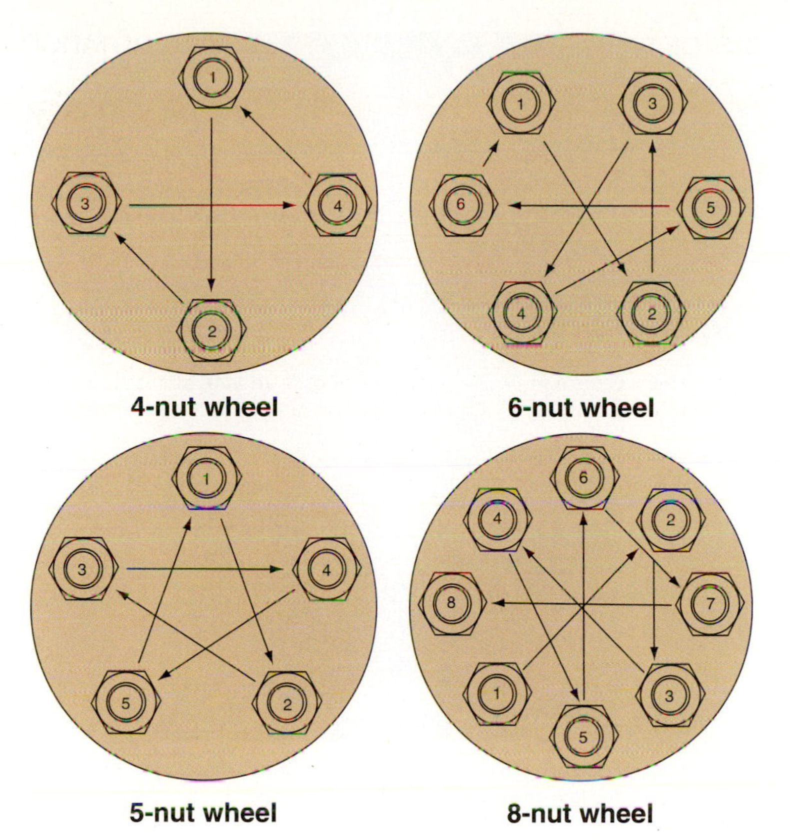

FIGURE 11-35 **Lug nuts should be tightened in a crosshatch pattern similar to the pictured sequence.**

SERVICE TIP:
When working with a TPMS valve, remember that a standard wheel rotation requires that a "relearn" procedure be performed on the TPMS. The relearn procedure is conducted after the tire and wheel assemblies have been rotated and remounted to the vehicle. The procedure is specific to the manufacturer and must be followed exactly. Consult the service manual for the instructions and use of special tools. Some relearn procedures have to be completed with a special tool, while others can be reprogrammed by the driver controls. Some vehicles have smart sensors that will automatically adjust to their new position.

Almost everyone at some point in their driving career must remove and install a wheel. Most flat tires do not occur at a shop but on a dark, lonely road or in the middle of rush-hour traffic. Drivers may carry some safety and repair equipment in their vehicles, but a torque wrench (available at the automotive shop) is the one thing they do not have. To change a tire, place a jack under the vehicle's lift point and raise it until the tire is clear of the ground or floor. If in a shop, a 1/2-inch impact wrench can be used to remove the lug nuts. Otherwise, the wrench that comes with the vehicle is used. Perform the necessary work on the wheel and tire or replace it with another assembly. Start the lug nuts on the studs and hand tighten them. Use a hand wrench to tighten the nuts as much as possible. If in a shop, use a torque wrench to tighten the nuts to specification. A cross pattern should be followed to torque or tighten the nuts so that the tightening process does not cock the rim on the hub (Figure 11-35). A cocked rim will prevent proper torquing of some of the nuts and will be obvious once the vehicle is on the road.

Tire Replacement

SPECIAL TOOLS
Tire changer
Service manual

There are several types of tire machines available. The following instructions are general guidelines for a tire changer with a side-bead breaker (Figure 11-36). The instructions for the tire changer must be followed. Remove the wheel assembly from the vehicle and release all of the air from the tire. Select the proper mounting tools to install the wheel assembly on the tire changer. Chrome and magnesium wheels require special mounts to prevent the wheel from damage or marring. Place the deflated tire and wheel assembly onto the bead-breaking portion of the machine. Ensure that the clamp does not lock over the rim but presses against the tire at a point near the rim. Operate the clamp until the tire bead breaks away from the rim. Turn the tire around and break the bead on the other side.

CAUTION:
Before removing (or installing) a tire from a wheel with an electronic tire pressure sensor, consult the service manual. Failure to follow the specific steps for mounting the tire and wheel assembly onto the tire changer could result in total destruction of the sensor.

CAUTION:
Keep hands and fingers clear of the bead area as the tire is being forced onto the wheel. Serious injury could result to hands or fingers if they get caught between the wheel and tire bead.

FIGURE 11-36 **A tire changer with adapters for low-profile and other special tires.**

WARNING: Inspect the tire for evidence that a quick-seal agent has been used. Any kind of liquid or gas other than normal air exiting from the tire after the valve has been removed may be an indication that a chemical commonly known as "fix-a-flat" has been used. This gas or liquid may be explosive. Serious injury could result should a spark occur during the tire removal from the wheel.

AUTHOR'S NOTE: The previous warning is given as a result of an accident in our town. A technician was removing a tire from a wheel when there was an explosion. The technician was seriously injured. Safety inspections later determined that the probable cause was that the customer had used a chemical for flat repair earlier that day and some of the chemical remained in the tire after deflation. A spark was generated from something (the exact source was never determined) that ignited this small amount of gas. Since about 1989, right around when this incident happened, manufacturers of tire chemical quick-seal kits have changed their propellant to an inert gas that is not flammable. However, exercising caution is always the best course of action.

Mount the wheel assembly onto the tire changer and lock it in place. Use the changer's bead guide to remove one side and then the other side from the rim. Before installing the tire, lube the beads with tire lubricant. Position one side of the tire on the rim and use the bead guide to install the tire. With the tire mounted, inflate it to the proper pressure. Some tires may require a band or air tool to initially seat the bead. Do not inflate the tire more than the maximum pressure noted on the sidewall. Once the tire is properly inflated, remove it from the changer. Photo Sequence 32 shows the typical procedure for dismounting and mounting a tire on a wheel assembly.

PHOTO SEQUENCE 32

Typical Procedure for Dismounting and Mounting a Tire on a Wheel Assembly

P32-1 Dismounting the tire from the wheel begins with releasing the air, removing the valve stem core, and unseating the tire from its rim. The machine does the unseating. The technician merely guides the operating lever.

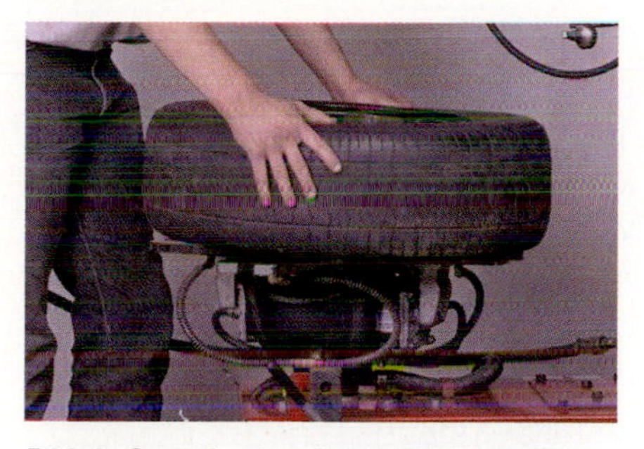

P32-2 Once both sides of the tire are unseated, place the tire and wheel onto the machine. Then depress the pedal that clamps the wheel to the tire machine.

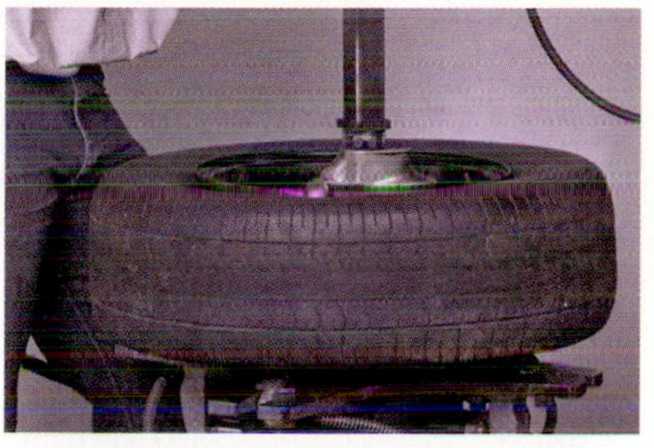

P32-3 Lower the machine's arm into position on the tire-and-wheel assembly.

P32-4 Insert the tire iron between the upper bead of the tire and the wheel. Depress the pedal that causes the wheel to rotate. Do the same with the lower bead.

P32-5 After the tire is totally free from the rim, remove the tire.

P32-6 Prepare the wheel for the mounting of the tire by using a wire brush to remove all dirt and rust from the sealing surface. Apply rubber compound to the bead area of the tire.

P32-7 Place the tire onto the wheel and lower the arm into place. As the machine rotates the wheel, the arm will force the tire over the rim. After the tire is completely over the rim, install the air ring over the tire. Activate it to seat the tire against the wheel.

P32-8 Reinstall the valve stem core and inflate the tire to the recommended inflation.

Diagnosing and Servicing TPMS Sensors and Tires with TPMS sensors

The first part of diagnosing and servicing a TPMS sensor is to inspect the warning light or warning message. During this inspection, make sure that the tires have the correct inflation by referencing the tire placard. Most systems are very accurate and a customer complaint may just be that their tire is actually low on air, even though it does not appear to be. Inspect the sensors for obvious damage and missing parts.

SPECIAL TOOLS

Tire balancer

Wheel weight hammer or tool

After rotating the tires, installing a new sensor, or installing new tires, you must perform a relearn procedure. A TPMS sensor can be inspected further with a scan tool. Specific software has to be used to access the TPMS data. When dismounting and mounting a tire with a TPMS sensor, you must take caution. First recognize the type of TPMS sensor used. If it is a valve-stem mounted sensor, then place the bead breaker of the tire machine 180 degrees away from it to minimize sensor damage. When removing the tire from the wheel, start with the arm of the tire machine 270 degrees away from the sensor. This will prevent sensor damage (Figure 11-37).

Wheel Balancing

WARNING: Before using any machinery, read and study the operating manual. Failure to do so could result in damage to the machine or injury to the operator.

To properly balance a wheel assembly it must be correctly mounted to the tire balancer (Figure 11-38). Various centering cones and fasteners are available on most tire balancers. Select a cone that fits approximately halfway through the wheel's large center hole. In most cases, all of the old weights are removed before mounting the assembly onto the balancer. The following instructions may differ between balancer brands and models.

Place the cone onto the balancer's spindle. There should be a spring already in place on the spindle. If not, install the spring and then the cone. The spring keeps the cone pressed into the wheel's center. Install the wheel assembly over the spindle and onto the cone.

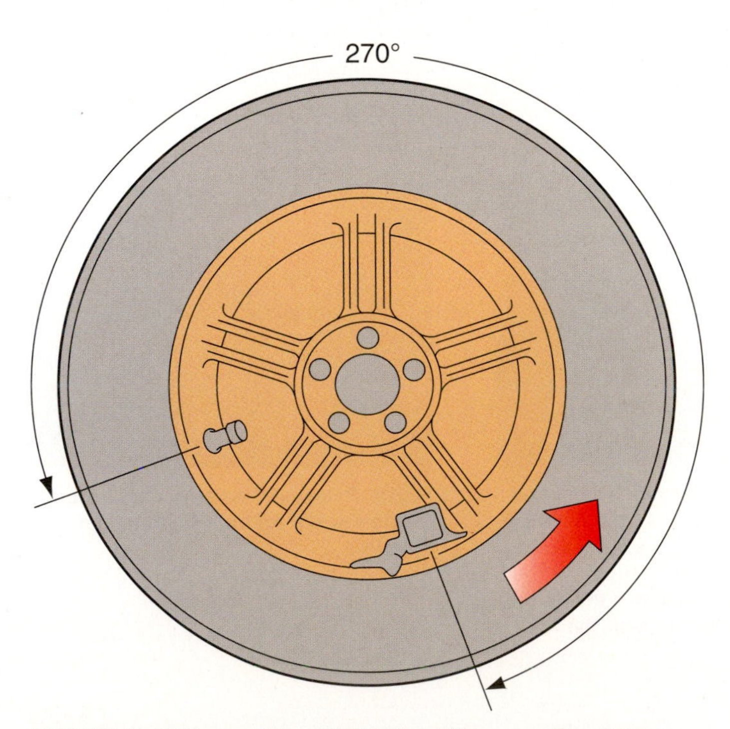

FIGURE 11-37 **Proper tire and wheel position on a tire changer.**

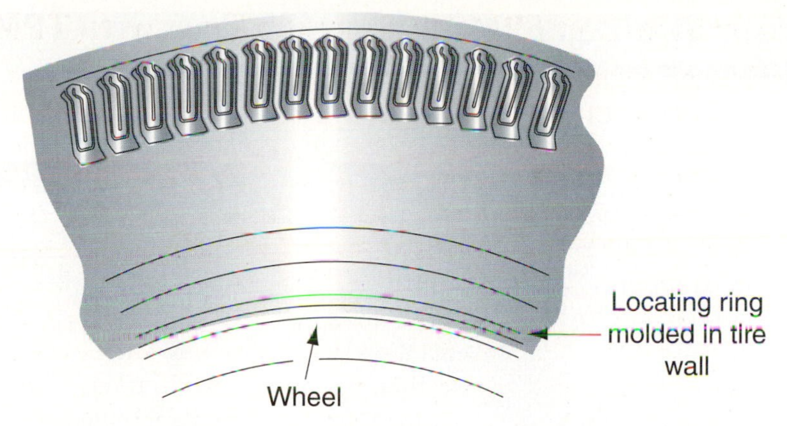

FIGURE 11-38 **The circular ring around the tire bead must be centered on the wheel rim.**

SERVICE TIP:
At times, the wheel assembly may not balance without adding several ounces of weight at various points. This may be corrected by removing all of the weight, deflating the tire, breaking the beads loose, and rotating the tire 90 degrees or 180 degrees on the wheel. The tire is inflated and balanced. This procedure may cause heavy spots on the rim and tire to counteract each other, which will require less weight to balance the wheel assembly. Follow the wheel balancers specific directions to achieve this. Larger tires typically require more weight.

Install the protection devices for magnesium or aluminum wheels, if required, followed by the retainer nut. There are usually three measurements that must be entered into the balancer's computer. They are rim diameter, rim width, and the distance from the edge of the rim to the balancer. There is a measuring bar that slides out of the balancer and the end is placed against the rim. This measurement is automatically entered into the machine. This is the distance between the wheel assembly and the balancer's point of reference.

With the measurements entered, close the hood. Most balancers will automatically spin. The balancer will spin the wheel assembly and bring it to a halt. Some balancers have a roller that will simulate a road force on the assembly (Figure 11-40). The display or scale will indicate where weights are to be placed and the size of each weight. Usually, the wheel assembly is rotated until the point of the new weight is straight up. A light or the alignment of two or more lines shown on the balancer's display will indicate that point. There will be two points selected for weights: one for the inside edge of the wheel and another for the outside edge. They are usually not directly across each other. Install the proper weights, close the hood, and respin the tire. If everything is correct, the balancer will indicate zero (0) or the letters "OK" to indicate that the assembly is balanced. If the assembly is still out of balance, additional weights will be indicated. Photo Sequence 33 shows the typical procedure for wheel balancing.

Some customers with magnesium or chrome wheels desire that no weights be shown on the outside of the rim. Most new balancers have procedures for measuring and marking points for adhesive weight on the inside of the rims. The balancer has to be set up differently. There are also special rim weights for chrome wheels. Follow the instruction manual for the special procedures.

FIGURE 11-39 **On-car wheel balancer.**

FIGURE 11-40 **The roller on the balancer senses force variation in the tire.**

PHOTO SEQUENCE 33

Typical Procedure for Wheel Balancing

P33-1 Complete all the preliminary wheel balance checks.

P33-2 Follow the wheel balancer manufacturer's recommended procedure to mount the wheel-and-tire assembly securely on the electronic wheel balancer. Some wheel balancers have a centering check in the programming procedure. If the tire and wheel are not mounted properly, tire and wheel balance will be inaccurate.

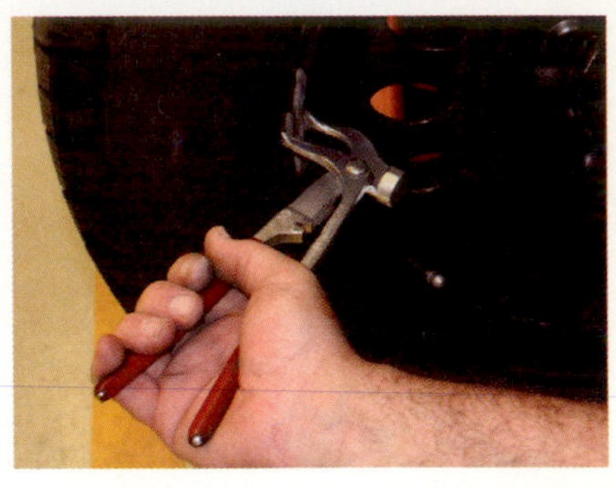

P33-3 Use a pair of wheel weight pliers to remove all the wheel weights from the wheel rim.

P33-4 Enter the wheel diameter, width, and offset on the wheel balancer screen.

P33-5 Lower the safety hood over the wheel and tire on the wheel balancer.

P33-6 Activate the wheel balancer control to spin the wheel-and-tire assembly.

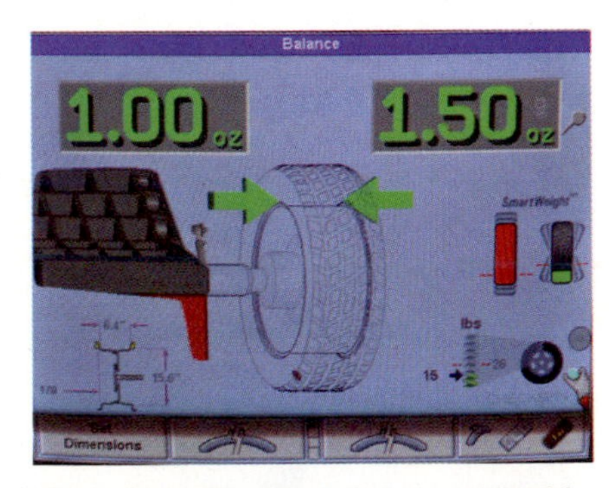

P33-7 Stop the wheel-and-tire assembly. Observe the wheel balancer display screen to determine the correct size and location of the required wheel weights.

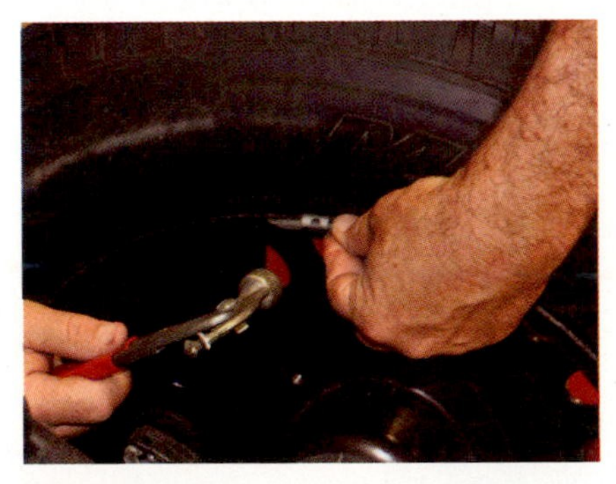

P33-8 Install the correct size wheel weights in the location(s) indicated on the wheel balancer screen.

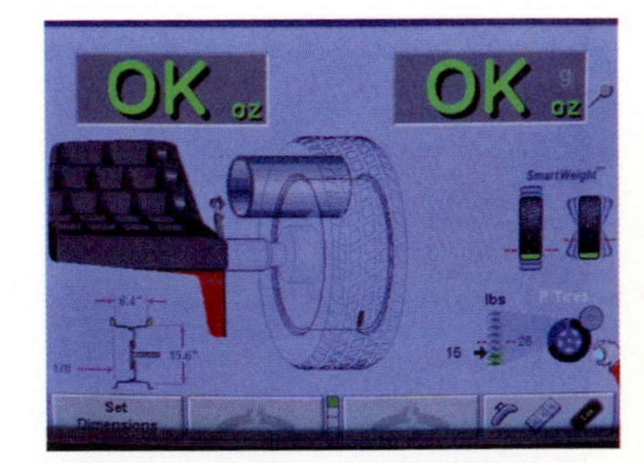

P33-9 Activate the wheel balancer control to spin the wheel again, and observe the wheel balancer display to confirm that it indicates a balanced wheel-and-tire assembly.

P33-10 **Stop the wheel, lift the safety hood, and remove the wheel-and-tire assembly from the balancer.**

Repairing a Tire

Tire repairs are a common procedure in automotive repair shops. However, the only repair that can be performed on a tire is to plug or patch a hole. This simple repair does have some drawbacks if not done correctly. The best method is to insert a plug patch, which is a combination of a plug and a patch (Figure 11-41). When inspecting a tire to repair, you must determine if the damage is in the repairable area (Figure 11-42). If the damage is outside of this area, you must install a new tire.

Once the hole is located, a plug can be inserted using a standard repair kit. Use the reamer in the kit to force the hole open. As the reamer is withdrawn from the hole, twist it in one direction. This forces the steel threads in the belt to align around the hole. Insert a plug into the plugging tool and coat the plug with vulcanizing cement. Push the plugging tool and plug through the hole. Do not twist the plugging tool as it is withdrawn. The plug should remain in the hole. Cut off any excessive plug material. Inflate the tire to its recommended pressure and check for leakage around the plug. The hole is now sealed and in an emergency the tire can be used. However, this is only a temporary seal and the following steps should be applied at the earliest time to complete the job.

Remove the tire from the wheel and trim the inside end of the previously installed plug. Install the tire on a bead-spreading tool if desired. The tire must be removed from the tool before the patch is installed. Buff on and around the plugged area with a fine buffing stone,

SERVICE TIP: A hole in the tire's sidewall should never be repaired. The tire must be replaced since any plugs or patches will work loose because of sidewall flexing.

SERVICE TIP: Installing a plug alone is a temporary repair. The only correct repair is to patch the hole with an internal patch supported by a plug. Inform the customer of this before accepting the job. Plugging a tire can be done with the tire mounted on the vehicle. Before doing any repairs, locate the leak and determine if the tire can be patched.

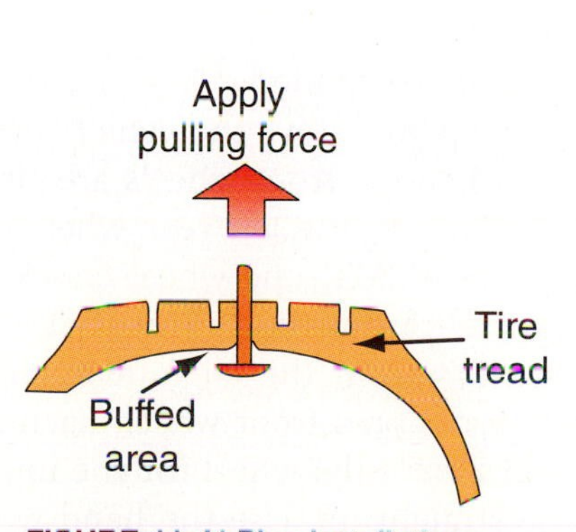

FIGURE 11-41 **Plug installation procedure.**

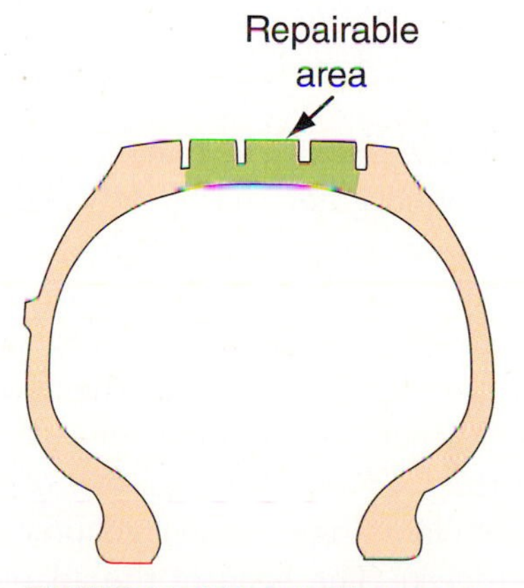

FIGURE 11-42 **Repairable area on bias-ply and belted bias-ply tires.**

SERVICE TIP:
(*continued*)
In most cases any hole over 1/4 inch in diameter cannot be thoroughly sealed. There are plugs and patches to cover larger holes but the results are not always satisfactory. Small leaks can be found by submersing the portions of the tire into a tire vat. A tire vat can be built by cutting a 50-gallon drum in half lengthwise and filling with water. Any hole will be evident by tracing the air bubbles.

but care must be exercised to prevent removing the tire's inner lining. The buffed area should be about twice as large as the patch being installed. Clean the inside of the tire of all buffed material and dust with a vacuum cleaner or a damp cloth. Remove the tire from the bead spreader if used. Apply a film of vulcanizing cement over the entire buffed area. Once the cement is dry, select a patch and remove the film covering the adhesive side of the patch. Do not touch the adhesive side. Locate the arrows on the patch and install the patch with the arrows pointing toward the tire's beads. Center the patch over the hole and press it firmly into place. Roll the patch in place with a corrugated tire stitcher moving from the center of the patch outward. Once the patch is rolled in, remove the plastic shield from the back of the patch. Mount the tire to the rim and inflate to the recommended pressure. Check for leaks around the sealed hole with a soapy solution.

Replacing a Damaged Wheel Stud

Many times the wheel studs are damaged by overtightening the lug nut or cross-threading the nut onto the stud. Once in a while the stud will just be stripped or break because of old age, but this is very unusual. Replacing a stud can be easy on some wheel assemblies, whereas others require removal of the hub from the vehicle. The general procedure is the same in either case.

Remove the tire and wheel assembly and inspect the area on and around the stud being replaced. Determine if the stud can be replaced on the vehicle as it now sits or whether additional components require removal to fully access the stud. Once a sequence of activities is established, remove any component necessary to gain clearance for removing and installing the stud.

If the hub is still on the vehicle, one of the quickest ways is the use of washers and a lug nut. Select three or four washers with an inside diameter slightly larger than the knurled area of the stud. Insert the new stud in from the backside of the hub and slide the washers over it. Try to keep the washers centered on the stud as much as possible. Select an old lug nut without the chrome covering and screw it onto the stud, flat side first. Use a pry bar between two other studs to help keep the hub from rotating. While holding the hub stationary, use a box-end wrench or ratchet and socket to tighten the nut. This will draw the stud through the hole until the inside end of the stud is flat against the back of the hub.

SERVICE TIP:
Very small holes will sometimes not leak except when the vehicle is operating and the tire is flexing on the road surface. The customer may complain of having to add air every 2 or 3 weeks. The hole can sometimes be found by applying a soapy solution over the tire tread and moving the vehicle slowly while an assistant watches for air bubbles. This technique may require some extra time.

Setting Up an Alignment Machine

There are several brands and models of alignment machines in use today. They range from units that measure a white light (older models) to fully computerized units that do practically everything except turn the wrench. The following section covers a typical setup for a computerized system. Most alignment brands are very similar in their setup procedures, but you should always go through the manual or a training course before using it.

Switch on the alignment machine so that it warms up while the vehicle is positioned. Then use a ground guide mirror or another person to help you move the vehicle onto the lift. Keep the vehicle centered on the runway and stop when the front wheels are centered on the turntables. The ground guide should chock (i.e., block) the left rear wheel on the front and back prior to the driver getting out of the vehicle. With the wheel chock firmly in place, the driver can now exit the vehicle, leaving the transmission in neutral and the ignition key turned off. Remove one of the large sensors from the alignment rack and install it on the rear wheel. Install one of the smaller sensors on a front wheel. Each sensor is installed by clamping its wheel adapter over the outer rim of the wheel (or the inner rim on some vehicles). The head of the adapter should be straight up. Use the hand wheel to tighten the adapter onto the wheel and do not overtighten.

The alignment machine should be warmed up by now and show a graphic display screen. On the alignment machine, select the make, model, and year from the vehicle selection screen. At times, it may be necessary to input additional data such as ride height and number of drive wheels. The machine will prompt the technician for the necessary information. Note that the brand of car or truck must be selected on this screen (e.g., Chevrolet or Chevrolet truck). Also, note that full-size vans and some sport utility vehicles (SUVs) are listed as trucks under the appropriate brand.

The next screen on the alignment machine is the sensor calibration screen. Each of the four sensors on the vehicle should be highlighted. If not, the sensor will be shown in red and the machine will not proceed. With all sensor signals received by the machine, a large green arrow will appear on-screen between the four wheels. This indicates that the vehicle is to be rolled backward until the arrow becomes a wide green line. There will also be a loud tone indicating the vehicle has been moved sufficiently for the machine to receive a calibration reading. The sensors will be calibrated at approximately 45 degrees to their installed position.

To properly move the vehicle for such a calibration reading, the technician should place the left hand on the front edge of the left rear tire without obstructing the sensor or receiver's line of sight. Move the rear chock about 20 inches away from and in line with the tire. Placing the right hand opposite the left, manually rotate the tire rearward until the arrow changes or the chock is reached. Do not rush the movement. Try to make the tire and vehicle move at a very steady slow rate until the arrow changes. If the arrow has not changed when the tire hits the chock, then the chock will need to be moved several more inches. It will take some practice over several alignments to get the distance and movement right the first time.

Once the arrow on the screen changes, hold the wheel (and vehicle) in place until the machine sounds a tone or changes color and the arrow changes back into an arrow pointing to the front of the vehicle display. Use the same procedures in the previous paragraph to roll the wheel forward until the arrow again changes into a straight line. Hold the vehicle until you hear the alignment machine's tone and then slide the rear chock back behind the wheel. Make sure it is firmly in place. The vehicle should be just about in the same position as when the calibration was started. The machine will automatically reset to the "Measure caster screen." Remove the locking pins from the turntables and slide plates.

The easiest way to measure caster on the Snap-on John Bean alignment machine is to sit in the driver seat of the vehicle, set the transmission to PARK, and apply the parking brake on firmly. On the alignment machine, there will be a center drawing of the steering wheel and the angle at which it is now positioned. On-screen, near the top of the graphic display, will be a small red dot. Turn the steering wheel slowly to the right while noting the red dot. When it reaches about 5 degrees of rotation, the ball will turn yellow. Continue guiding left until the ball turns green and the matching ball just under it also turns green. This will happen at about 10 degrees into a left turn. Hold the steering wheel in place until the upper ball turns red again. Steer to the right, past the center, until the upper and lower balls turn green, approximately 10 degrees right. Again, hold the steering wheel steady until the upper ball turns red and then steer left until the ball turns green over the green center ball. Adjust the steering wheel slightly until the degree reading in the center of the display screen shows zero-degree center. The caster has now been measured, recorded, and the machine will display the actual alignment angles for all four wheels. Turn the ignition key off and exit the vehicle.

Most alignment machines have software to enter customer, vehicle, and other repair order information. They can also print out the data and include some diagnostic and repair information. The amount, style, and display methods differ on each brand and model. The machine's operating instructions will detail the exact procedures.

SERVICE TIP: If the hub is removed from the vehicle to gain clearance, it will be necessary to support it in some manner. A large vise will work or blocks of wood under the edges of the hub will provide a steady base. With the necessary clearance achieved, use a steel flat-nose punch of approximately the same diameter as the stud and a hammer to drive the stud out. There are two ways to install the new stud. If the hub is installed in a vise or supported by blocks, flip the hub over and drive in the stud with the hammer. Drive it in until the back of the stud is flush flat with the inside of the hub.

Classroom Manual
pages 323, 324, 325, 326, 327

SPECIAL TOOLS

Alignment machine

Alignment lift

Alignment procedures taken after the setup will not be covered here, but for the reader's general information, the following steps are taken to complete an alignment:

- Sensors are compensated individually.
- Locking pins are removed from turntables and skid plates.
- The vehicle is lowered to lift.
- Caster measurement is taken.
- Rear caster, camber, and toe adjustments are made.
- Front caster, camber, and toe adjustments are made.
- Heads are removed and stored.
- The vehicle is removed from the lift and the turntable and skid plate pins are reinstalled.

Detailed information on suspension, steering, and alignment components and procedures can be found in *Today's Technician: Automotive Suspension and Steering Systems*, 5th Edition, Delmar Cengage Learning. Other sources include school, local, and Internet libraries. Suspension and steering parts manufacturers like MOOG and TRW have Internet sites with information on design and function of their products. Photo Sequence 34 shows the typical procedure for setting up an alignment machine.

PHOTO SEQUENCE 34

Typical Procedure for Setting Up an Alignment Machine

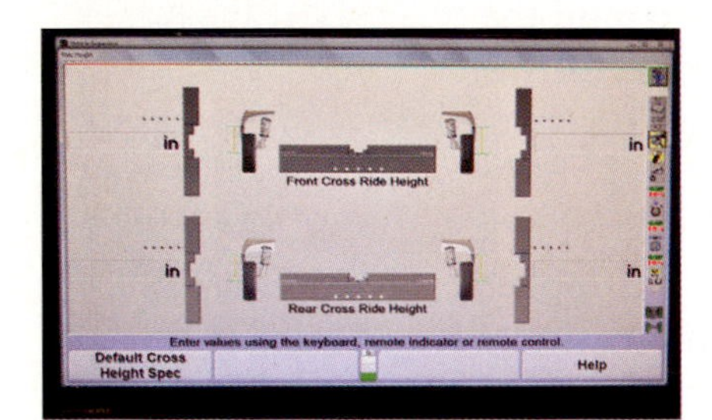

P34-1 **Display the ride height screen. Check the tire condition for each tire on the tire condition screen.**

P34-2 **Position the vehicle on the alignment rack.**

P34-3 **Make sure the front tires are positioned properly on the turntables.**

P34-4 **Position the rear wheels on the slip plates.**

P34-5 **Attach the wheel units.**

P34-6 **Select the vehicle make and model year.**

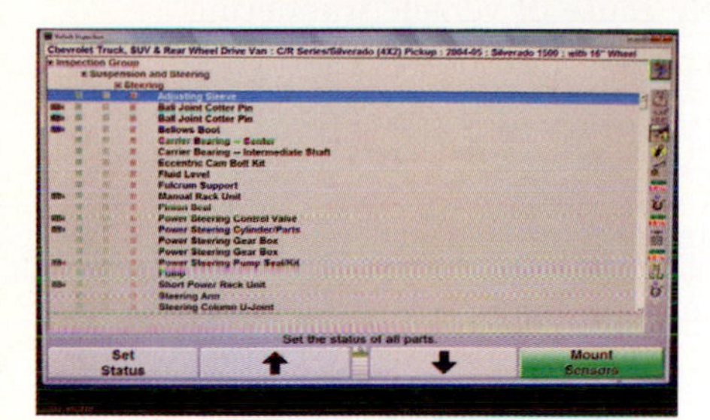

P34-7 **Check the items on the screen during the preliminary inspection.**

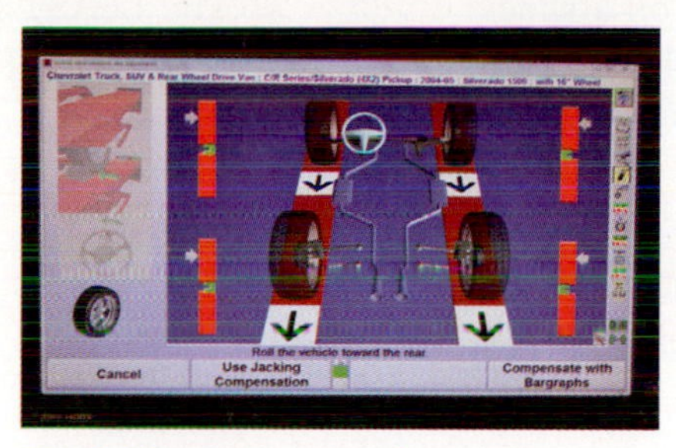

P34-8 **Display the wheel runout compensation screen.**

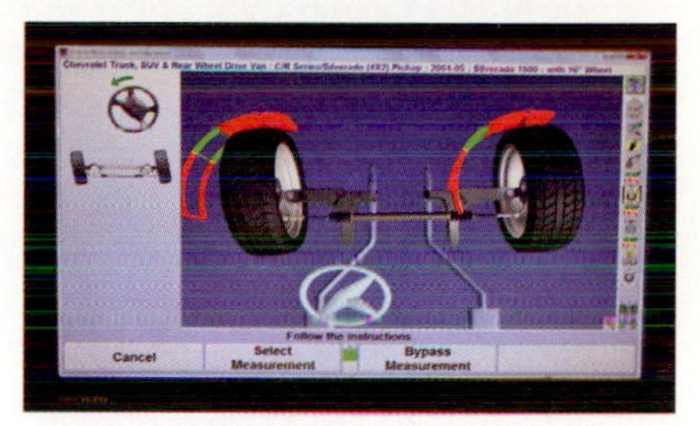

P34-9 **Display the turning angle screen and perform the turning angle check.**

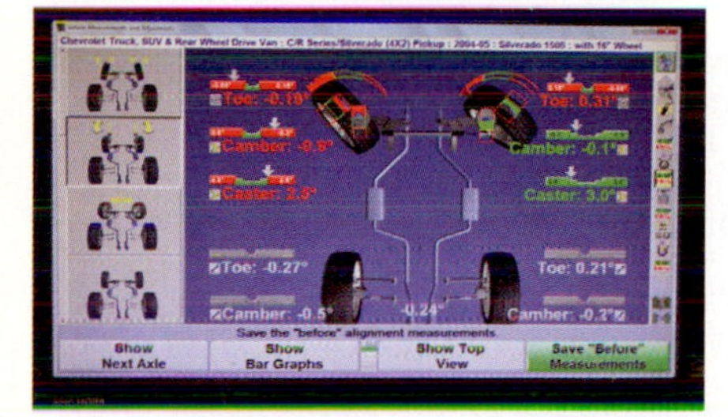

P34-10 **Display the front and rear wheel alignment angle screen.**

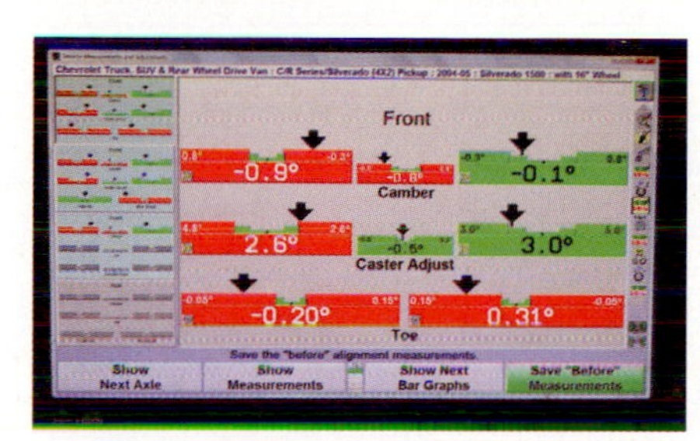

P34-11 **Display the adjustment screen.**

CASE STUDY

A technician was performing a routine tire rotation and balance along with other repairs. One tire proved difficult to balance and required excessive weights in the technician's opinion. The tire was deflated, broken down, and slid 180 degrees on the wheel. The amount of weight was reduced and the tire finally balanced out. Due to the other repairs, the shop foreman tested the vehicle and it had a pronounced vibration similar to an unbalanced tire. The technician rechecked the difficult tire and found it to be over two ounces off. The technician removed the tire from the rim. Inside were several large lumps of chassis grease and a new patch. Based on a conversation with the owner, it was determined that the tire had been patched several days earlier. The owner also stated that the vibration started after the tire was patched. The tire was a low-profile tire mounted onto a wide rim. It was deduced that the shop performing the patch could not seat the tire and had used chassis grease to fill the large gap between the tire and rim so air could be held long enough for the tire to seat. Apparently, some of the grease got into the tire. The technician cleaned the interior of the tire and balanced it with no further problems.

TERMS TO KNOW

Camber wear
Castellated nut
Coupler
Cupping
Drifting
Gas-charged shocks
Isolators
Joint stud
Negative camber
Positive camber
Pulling
Seepage
Steering wheel free play
Toe wear
Wave

ASE-STYLE REVIEW QUESTIONS

1. *Technician A* says the upper bearing on a shock absorber can cause a grinding noise during turns.
   Technician B says a broken coil spring may allow the frame to connect the bumper on the lower arm.
   Who is correct?
   A. A only
   B. B only
   C. Both A and B
   D. Neither A nor B

2. Coil springs are being discussed.
   Technician A says a broken spring may affect steering angles.
   Technician B says a weak spring on an SLA system may affect steering angles.
   Who is correct?
   A. A only
   B. B only
   C. Both A and B
   D. Neither A nor B

3. A rack-and-pinion system is being discussed.
   Technician A says a worn pinion gear may cause erratic movement between the steering wheel and wheels.
   Technician B says a worn inner tie-rod socket will cause both wheels to respond slowly to steering wheel movement.
   Who is correct?
   A. A only
   B. B only
   C. Both A and B
   D. Neither A nor B

4. The steering system is being discussed.
   Technician A says any movement in the socket or ball requires a replacement of that steering component.
   Technician B says any movement between the stud and its mating hole requires a replacement for that component.
   Who is correct?
   A. A only
   B. B only
   C. Both A and B
   D. Neither A nor B

5. A vehicle with a lowered corner is being diagnosed. This condition could be caused by any of the following EXCEPT:
   A. worn shock absorber.
   B. split air spring.
   C. broken leaf spring.
   D. broken torsion bar mount.

6. The LEAST likely cause of vehicle wandering is:
   A. worn recirculating ball nut.
   B. loose shock absorber mounts.
   C. broken rear springs.
   D. worn rack-and-pinion bushings (mounts).

7. Inspection of the steering linkage is being discussed.
   Technician A says sideways movement of the socket is normal.
   Technician B says force should be applied against the centerline of the socket stud to check for movement between the stud and hole.
   Who is correct?
   A. A only
   B. Both A and B
   C. B only
   D. Neither A nor B

8. A tire with abnormal wear is being discussed.
   Technician A says cupping of the tread could be a sign of poor maintenance.
   Technician B says balancing and rotation may help correct the problem.
   Who is correct?
   A. A only
   B. B only
   C. Both A and B
   D. Neither A nor B

9. *Technician A* says a sawtooth wear pattern across the tread and excessive wear on a tire's edge may indicate that both the toe and camber are incorrect.
   Technician B says excessive wear on the outer edge of a tire indicates the caster is too negative.
   Who is correct?
   A. A only
   B. B only
   C. Both A and B
   D. Neither A nor B

10. Wheel alignment is being discussed.
    Technician A says a cable is used to align the sensor with the wheel assembly.
    Technician B says that vehicle and customer data can be printed out by most computerized alignment machines.
    Who is correct?
    A. A only
    B. B only
    C. Both A and B
    D. Neither A nor B

JOB SHEET

Name ______________________ Date ____________

Lubricating Suspension and Steering Systems

NATEF Correlation

This job sheet addresses the following NATEF task:

Lubricate suspension and steering systems.

Objective

Upon completion and review of this job sheet, you should be able to lubricate the various components of the steering and suspension systems.

Tools and Materials Needed

Service manual

Lift or floor jack with stands

Grease gun, powered or hand-operated

Procedures

Task Completed

1. Raise the vehicle to a good working height. ☐
2. Locate each grease fitting on the suspension. ☐

Component	Fitting (Y/N)
Upper ball joint	________
Lower ball joint	________
Steering knuckle (used on solid front axle without control arms; one top, one bottom; four total)	________
Tie-rod ends	________
Tie-rod or center link	________
Idler arm mount	________
Center link or idler	________
Center link or pitman arm	________
Front mount of rear leaf spring (rarely used, but there may be two)	________
If the vehicle is RWD, check drive shaft U-joints, usually included as part of routine service (one per joint).	________
Inspect other components for fittings.	________
Are there additional fittings?	________
If so, list. (Possible locations: trailer hitch, U-joint on rear-drive axle like some Corvettes; power takeoff and shaft; mounts on front leaf spring; U-joint-type coupling on steering shaft; other visible points where two or more components move or work together.)	________

Task Completed

☐ 3. Lube each fitting in turn, starting at one front wheel and moving across the vehicle to the other wheel. It may be necessary to turn the wheel left or right to access the fittings.

☐ 4. Do not overlube. Stop lubing when a very small amount of grease is forced from the boot or a slight "pop" is heard.

☐ 5. Lube any fittings at the rear of the vehicle.

☐ 6. Lube any other fittings found under the vehicle.

☐ 7. Lower the vehicle.

☐ 8. Lube any suspension or steering components not accessible from under the vehicle.

☐ 9. Check the power-steering fluid level.

☐ 10. Clean the work area, complete the repair order, and return the vehicle to the customer.

Instructor's Response ______________________________

JOB SHEET 35

Name ______________________________ Date ______________

Dismounting and Mounting a Tire

NATEF Correlation

This job sheet addresses the following NATEF task:

Dismount, inspect, and remount tire on rim.

Dismount, inspect, and remount tire on wheel equipped with tire pressure sensor.

Objective

Upon completion and review of this job sheet, you will be able to dismount, inspect, and remount tire on a wheel equipped with a tire pressure sensor or without a sensor.

Tools and Materials Needed

Tire changer

1/2-inch drive impact wrench

Tire valve removal tool

1/2-inch drive torque wrench

CAUTION:
Treat the TPMS cap and valve with care. The cap is aluminum and the valve is nickel-plated to prevent corrosion. Failure to protect them may cause damage to the monitor.

CAUTION:
Use caution when the pry bar is inserted between the tire and guide or wheel while the turntable is rotating. Guide the bar until it can be lifted from the tire. Injury can occur if the pry bar jerks loose or becomes dislodged and hangs in the wheel assembly.

CAUTION:
Keep hands and fingers clear of the bead area as the tire is being forced onto the wheel. Serious injury could result if they get caught between the wheel and tire bead.

CAUTION:
Do not allow tire pressure to exceed the maximum pressure listed on the sidewalls. Serious injury or death could result from an exploding tire.

CAUTION:
Treat the TPMS cap and valve with care. The cap is aluminum and the valve is nickel-plated to prevent corrosion. Failure to protect them may cause damage to the monitor.

Task Completed	Procedures
☐	1. Locate the tire data and lug nut torque. Tire size ______ Tire type ______ Recommended tire inflation ______ Maximum pressure on sidewall ______ Lug nut torque ______
☐	2. Study all operating instructions for the tire changer.
☐	3. Inspect the tire changer and operate each of the control valves. The valves should be spring loaded to the OFF position. If any valve does not return automatically to the OFF position immediately after being released, do not use the machine until repairs have been completed. **NOTE:** Special instructions for a wheel with TPMS are listed after each appropriate step.
☐	4. Lift the vehicle and remove the wheel assembly.
☐	5. Remove the valve from the tire's valve stem. Place the TPMS cap and valve on a clean and dry surface.
☐	6. Once the tire is completely deflated, remove all wheel weights.

WARNING: Do not place a magnesium or chrome rim onto the tire changer without using special adapters. Damage to the wheel may result if protective adapters are not used. Consult the machine's operating instructions.

☐ 7. Place the tire in the bead breaker.
   A. If the breaker is on the side of the changer, move the tire until the clamp reaches the *edge* of the rim, but is not directly on the rim. On a TPMS-equipped rim, position the bead breaker 90 degrees from the TPMS valve.
      1. Guide the clamp to the tire and operate the power valve.
      2. Maintain the valve in operating mode until the tire breaks from the bead. Release the valve.
      3. Turn the tire until the other side is under the clamp.
      4. Repeat steps 7.A.1 and 7.A.2 until the second side breaks loose.
   B. If the breaker is mounted on a turntable, place the tire on top of the turntable.
      1. Operate the power valve to lock the tire assembly to the platform.
      2. Position top and bottom clamps, as required, near the rim's edge.
      3. Operate the clamp-operating valve until the tire breaks loose from the rim on both sides.

Task Completed

8. If the tire is not on the turntable, place it there and clamp it in place using the changer's clamping device. ☐

9. Position the tire removal or installing guide and lock it in place. On a TPMS-equipped rim, position the guide so the pry bar can be inserted a little counterclockwise of the TPMS sensor when prying the tire over the dismounting guide. ☐

10. Use the changer's pry bar to lift a section of the tire onto the guide. Do only one side of the tire at a time. ☐

11. Operate the turntable rotation valve until the pry bar can be removed and this side of the tire is separated from the rim. Turn off the rotation valve. On a TPMS-equipped rim, operate the turntable in a CLOCKWISE rotation only. ☐

12. Pull up on the tire near the guide. On a TPMS-equipped rim, position the guide so the pry bar can be inserted a little counterclockwise of the TPMS sensor when prying the tire over the dismounting guide. ☐

13. Insert the pry bar between the lower side of the tire and the wheel. Lift the tire over the guide. ☐

14. Operate the turntable rotation valve until the pry bar can be removed and this side of the tire is separated from the rim. On a TPMS-equipped rim, operate the turntable in a CLOCKWISE rotation only. Turn off the rotation valve. ☐

15. Move the guide to the side and remove the tire. ☐

16. Lightly lubricate the bead area of the new tire with tire lubricant. ☐

17. Place the tire over the wheel. On a TPMS-equipped rim, position the mounting guide 180 degrees from the TPMS valve stem. Position the bead or transition at 45 degrees before (counterclockwise of) the valve. The transition area is the place where the mounted portion of the tire crosses from the rim's center up over the rim's lip to the nonmounted portion of the tire. ☐

18. Position the tire guide within the circle of the tire's beads and lock in place onto the rim. The tire should circle the guide. ☐

19. Start the front of the tire's lower bead down over the edge of the rim and over the guide. ☐

20. Operate the rotation valve and apply downward pressure on the tire as the guide forces one side of the tire onto the wheel. On a TPMS-equipped rim, operate the turntable in a CLOCKWISE rotation only. ☐

21. Turn off the rotation valve. ☐

22. Position the top of the tire over the edge of the rim and the guide. On a TPMS-equipped rim, position the mounting guide 180 degrees from the TPMS valve stem. Position the bead or transition at 45 degrees before (counterclockwise of) the valve. The transition area is the place where the mounted portion of the tire crosses from the rim's center up over the rim's lip to the nonmounted portion of the tire. ☐

23. Operate the rotation valve and apply downward pressure on the tire as the guide forces one side of the tire onto the wheel. On a TPMS-equipped rim, operate the turntable in a CLOCKWISE rotation only. ☐

24. Turn off the rotation valve. ☐

Task Completed

☐ **25.** Connect the changer's tire inflation line to the tire's valve stem.

☐ **26.** Inflate the tire until the bead "pops" into place on the wheel.

☐ **27.** With the tire inflated, disconnect the air line and release the changer's locking clamps.

☐ **28.** Install the valve into the valve stem.

☐ **29.** Reconnect the air line to the valve stem and inflate it to specified pressure.

☐ **30.** Disconnect the air line and install the valve stem cap. On a TPMS-equipped rim, install the special aluminum cap and nickel-plated valve in place. Do not use caps made of any material other than aluminum.

NOTE: Wheel assembly should be balanced before completing step 31.

☐ **31.** Install the wheel assembly onto the hub; install and torque lug nuts; and calibrate the TPMS if applicable.

Instructor's Response ____

____

____

____

JOB SHEET 36

Name ______________________________ Date ______________

Balancing a Wheel Assembly

Objective

Upon completion and review of this job sheet, you should be able to properly balance a wheel assembly.

Tools and Materials Needed

Tire balancer

Wheel weight hammer or tool

Procedures

Task Completed

1. Study the balancer's operating instructions. ☐
2. Select special adapters if the wheel is made of magnesium or chrome. ☐
3. As required, select and install the centering cone onto the balancer's spindle. The cone may be installed on the outside of the wheel on some balancers or certain types of wheels. ☐
4. Position the wheel assembly onto the cone or spindle as required. ☐
5. Install the lock nut onto the spindle and screw it into place. ☐
6. If the balancer is not powered, turn on the power switch. ☐
7. Select the type of balancing procedures desired: static or dynamic. Some balancers have special procedures that can be used. ☐
8. Measure the distance from the balancer to the edge of the wheel. Measure at the part of the rim where the weights are installed. Use the measuring device attached to the balancer. ☐
9. Enter the measurement using the buttons or knobs on the balancer. ☐
10. Enter the wheel's width into the balancer. ☐
11. Enter the wheel's diameter into the balancer. ☐
12. Ensure that all wheel weights are removed from the wheel. ☐
13. Close the hood and operate the balancer to spin the wheel assembly. ☐
14. After the wheel assembly has stopped, observe the reading. If it is not balanced, rotate the wheel by hand until the balancer's alignment marks are matched for the inside of the wheel. ☐
15. Attach the proper size weight, as determined by the balancer, to the inside of the wheel. ☐
16. With the balancer's marks aligned, install the weight onto the rim. Use the weight hammer. ☐

Task Completed

☐ **17.** Rotate the wheel by hand until the marks are aligned for the outside of the wheel.

☐ **18.** Select and install the weight on the outside of the wheel.

☐ **19.** Close the hood and operate the balancer.

☐ **20.** If the wheel assembly is balanced, remove the wheel assembly from the balancer.

☐ **21.** If the wheel assembly is not balanced, repeat steps 13 through 19 as necessary to complete balancing.

NOTE: If the wheel assembly cannot be balanced on the second attempt, recheck the measurements and the mounting of the wheel assembly on the balancer. The tire may have to be removed from the wheel and repositioned.

☐ **22.** Once the tire is balanced, remove the wheel assembly from the balancer and install it on the vehicle.

Instructor's Response ____

____

____

____

JOB SHEET 37

Name ______________________________ Date ______________

Repair a Tire

NATEF Correlation

This job sheet addresses the following NATEF tasks:

Inspect tire and wheel assembly for air loss; perform necessary action.

Repair tire using internal patch.

Objective

Upon completion and review of this job sheet, you should be able to inspect a tire and repair the tire.

Tools and Materials Needed

Tire vat

Plugging tool

Patching tool

Protective Clothing

Goggles or safety glasses with side shields

Describe the vehicle and tire being worked on:

Year ______________ Make ______________ Model ______________

VIN ____________________ Engine type and size ____________________

Tire size ________________ Recommended tire pressure ________________

CAUTION: Plugging alone is a temporary repair. The plug should be replaced or supported by an internal patch. Failure to patch a tire properly could result in a swift loss of tire air pressure and may cause damage or injury.

Procedures

Task Completed

1. Remove the tire and wheel assembly from vehicle. ☐
2. If necessary, inflate the tire to recommended pressure. Inspect the tire for damage and leaks. Record your findings for each area. ☐
   a. Tire sidewall ________________________________
   b. Tire thread area ________________________________
   c. Tire bead ________________________________
   d. Leaks. Where are they located on the tire? Is the hole small enough to be plugged and patched?

   ____

   ____
3. If there is no damage other than a repairable leak, is the hole readily identifiable? ☐
4. If the leak is not easily located, place the tire assembly into the tire vat and force the lower rim under water. Check for bubbles. Rotate the tire assembly enough to place another portion of the tire under water. Continue until bubbles are seen and the leaking hole is located. Where is it located on the tire? ☐
5. Insert a plug through the tines of the plugging tool. Apply vulcanizing cement to the plug. ☐

Task Completed

☐ 6. Force the reamer through the hole until it is completely through the tire fabric. Rotate the reamer in one direction only to align the strands of the tire's steel belt in one direction.

☐ 7. Remove the reamer.

☐ 8. Using the plugging tool, push the plug through the hole until less than half of the plug is showing.

☐ 9. Remove the plugging tool and trim the plug even with the tread.

☐ 10. Inflate the tire to recommended pressure, if necessary. Place soapy water or other liquid around the plug and check for air bubbles. The plug is sufficient if no air bubbles are present. Record your findings.

____

____

☐ 11. If plugging only, advise the customer that the plug should be replaced with a patch as soon as possible.

To complete the repair, perform the following steps.

☐ 12. Remove the tire from the wheel.

☐ 13. Install a plug as directed in steps 5 through 13. This time also trim the plug flush with the tire's inner lining.

NOTE: A bead-spreading tool may be used during buffing, but the tire MUST be removed from the tool before applying the cement and patch.

☐ 14. Use a fine buffing stone to lightly clean the area around the plug. The cleaned area should be slightly larger than the patch to be applied and should not go through the lining.

☐ 15. Use a vacuum cleaner to clear the tire's interior of buffing dust.

NOTE: Remove the bead-spreading tool, if used.

☐ 16. Apply a film of vulcanizing cement over the entire buffed area.

☐ 17. Allow the cement to dry completely. Do not apply flame to quick-dry it.

☐ 18. Remove the backing from the adhesive side of the patch. Do not touch the adhesive side.

☐ 19. Position the patch so the bead arrows on the patch point to the tire beads.

☐ 20. Center the patch over the hole and press the patch into place.

☐ 21. Use a corrugated tire stitcher to roll the patch in place. Roll outward from the center with as much hand force as possible.

☐ 22. After completely rolling the patch, remove the plastic film from the top of the patch.

☐ 23. Mount the tire to the wheel and inflate to recommended pressure.

☐ 24. Check for leaks around the plug or patch. Record your findings.

☐ 25. Mount the wheel assembly on the vehicle.

Instructor's Response ____

____

____

____

JOB SHEET 38

Name ______________________________ Date ______________

Inspect and Replace a Wheel Stud

NATEF Correlation

This job sheet addresses the following NATEF task:

Inspect and replace wheel studs.

Objective

Upon completion and review of this job sheet, you should be able to inspect a tire and replace a wheel stud.

Tools and Materials Needed

Basic hand tools

Protective Clothing

Goggles or safety glasses with side shields

Describe the vehicle and tire being worked on:

Year ______________ Make ______________ Model ______________

VIN ____________________ Engine type and size ____________________

Procedures

Task Completed

1. Lift the vehicle sufficiently to remove the wheel assembly. ☐
2. Inspect the stud for damage. If not broken off, note any damage to the threads and determine the probable cause for the damage. ☐
3. Is this a drum or disc brake system? ______________ If drum, remove the drum. ☐
4. Inspect the rotor and hub or axle to determine if additional components must be removed before removing the stud. If yes, consult the service manual for proper removal of those components and any special instructions for removing or installing a stud. Record you findings. ☐

   ____

   ____

5. Remove any additional components as needed. List the components removed. ☐

   ____

   ____

6. Use a hammer to drive the old stud from the hub or axle. ☐

   NOTE: Installation of the stud may be done in several ways. If the hub or axle is removed from the vehicle, a hydraulic or arbor press can be used to press-fit the stud into the hub or axle (best method). A hammer may be used if space permits to allow the stud to be squared in the hole during installation. In cases where the hub or axle remains on the vehicle and space is not sufficient for the use of a hammer, use the method in step 6 to install the stud.

Task Completed

☐ 7. Select three or four washers or spacers with a hole that allows enough room for the stud's locking ridges to clear when it is pulled through the hub or axle.

☐ 8. Insert the stud from the rear of the hub or axle and slide the washers over the stud.

☐ 9. Screw a standard lug nut, with cone side out, onto the stud. Hand tighten sufficiently to hold stud and washers in place.

☐ 10. Ensure the stud is square in its hole and the washers are centered around the stud.

☐ 11. Tighten the lug nut with a wrench or socket. Do not use an impact wrench.

☐ 12. Use the wrench (socket) to draw the stud completely through the mounting hole until its head is completely flush against the rear of the hub or axle.

☐ 13. Remove the lug nut and washers.

☐ 14. Inspect the stud carefully to ensure:

a. The head of the stud is absolutely flat against the rear of the hub or axle. It must be completely flat without any gaps.

b. There is no damage to the stud's threads. The best method to do this is to screw the lug nut onto the stud. It should screw on without any dragging or binding using hand force only.

c. If the stud head is not flat or the threads are damaged in any way, the stud must be replaced.

Record your inspection results. ______________________________

____

Instructor's Response ______________________________

____

____

____

Chapter 12

Transmission and Driveline Service

Upon completion and review of this chapter, you should understand and be able to describe:

- How to diagnose driveline noise.
- How to adjust a clutch linkage.
- How to inspect and service a manual transmission or transaxle.
- How to inspect and service an automatic transmission or transaxle.
- How to inspect and service a drive shaft or axle.
- How to inspect and service a differential.

Introduction

Like the engine, most of the powertrain components need routine servicing for providing the best performance. Many times a driver will routinely change the engine oil and filter but fails to do the same for the transmission or transaxle and differential. Noise is sometimes the first indication that the driveline is failing. In some cases, the customer will not bring the vehicle in for transmission or driveline service until it is too late. At this point, major repairs or replacement are sometimes the only option. It is your job as a technician to help prevent this by inspecting the transmission and driveline properly and helping the customer (or service advisor) make informed decisions about the maintenance.

Driveline Diagnosis by Noise

Classroom Manual
pages 337, 338, 339

Most driveline noises are easy to separate from the noise of an engine failure. However, the noise can echo or vibrate through the vehicle, leading the technician to look in the wrong place. If the noise can be associated with a type of component, it can help eliminate some of the possible sources. Some of the driving conditions needed during a test drive can be produced on a lift in the shop. Only an experienced technician operating within the shop's policy should perform or supervise a test drive on the road or on the lift.

WARNING: Care must be taken when operating the vehicle in an attempt to isolate a driveline noise. The vehicle has to be operated to load, unload, and stress the driveline. It is best to use an empty parking lot or a low-traffic road. Failure to pay attention to other motorists or traffic conditions can result in serious consequences.

WARNING: Do not conduct a test drive without permission of the owner and managers, and only in accordance with shop policy. An experienced technician should perform or supervise any test drive. Failure to follow proper procedures could result in damage to the vehicle, injury, and legal actions.

WARNING: Use extra care in placing lift pads and lifting and operating a vehicle while raised on the lift. Do not raise the vehicle any more than necessary to clear the wheels of the floor. It is best to keep the nondrive wheels chocked on the floor. Clear the areas of equipment and people in front of and behind the vehicle. Damage to the vehicle or injury could occur if the operation is not done in a safe and cautious manner.

Manual Transmissions or Transaxles

Noises from the transmission and transaxle usually include growling from the gears or whining from the bearings. Gear noise will change or disappear when another gear is selected. Sometimes a damaged gear, synchronizer, or maladjusted clutch will result in grinding or grating noises as the damaged teeth engage teeth on the mating gear (Figure 12-1). Bearing whine will change in pitch as the vehicle speed changes.

Noises can be further diagnosed by operating the vehicle under different conditions. Usually, transmission noise will not change because of road condition or the turning of the vehicle. It will appear as the load is applied and released. If the clutch is disengaged as the vehicle is driven and the noise does not change, it is probably caused by driveline components behind the transmission.

Disengaging the clutch unloads the driveline and the noise will change in pitch, volume, or both. Placing the transmission in neutral as the vehicle is moving will cause most transmission noise to disappear or at least change drastically. A vibration may be felt in the shift lever as a certain gear is engaged.

All of the internal components of an automatic transmission or transaxle are heavy, continuously lubed, and tend not to fail unless abused.

Automatic Transmissions or Transaxles

Normally, automatic transmissions or transaxles do not produce major noises. The most common noise is a whine from the hydraulic pump or possibly the torque converter. If there is a clunking or other noise coming from the transmission area, the flexplate may be cracked (Figure 12-2). A cracked flexplate will also result in noise during cranking as the starter tries to turn the flexplate and torque converter. Any other noise from the automatic transmission or transaxle usually means it has to be replaced or rebuilt.

FIGURE 12-1 Gears can make noise if they are chipped or missing parts. Proper lubrication and maintenance are crucial.

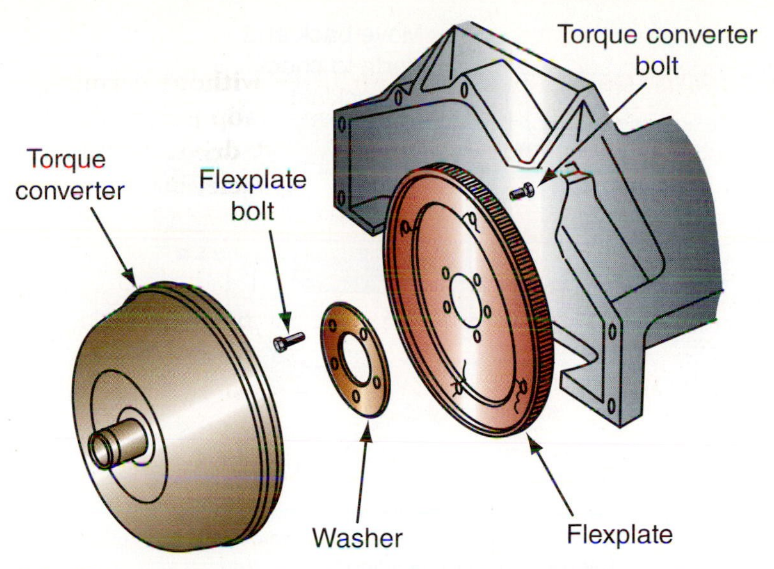

FIGURE 12-2 **Inspect the flexplate (or flywheel) for cracks around the mounting bolt holes.**

Drive Shafts

A bent or unbalanced drive shaft will cause a continuous vibration, which gets worse as the shaft's speed is increased. The vibration may lessen but will not go away completely as the load is changed. This vibration should not be confused with bad or unbalanced tires. Tires will usually produce a vibration on an axle. Front tires will cause a vibration or jerking of the steering, while rear tire vibration is felt in the seat. Drive shaft vibration may resound throughout the vehicle.

Most original equipment U-joints do not have **grease fittings**. Replacement U-joints will include a fitting with the joint.

U-joints can create a clunking noise as the drive shaft is loaded and unloaded (Figure 12-3). This very closely resembles noise from worn differential gears. Closer inspection is required. A squeak at speeds of 3 to 6 mph usually means the U-joints are dry. If a **grease fitting** is present on the joint, lubricating it may help. It is best to replace a

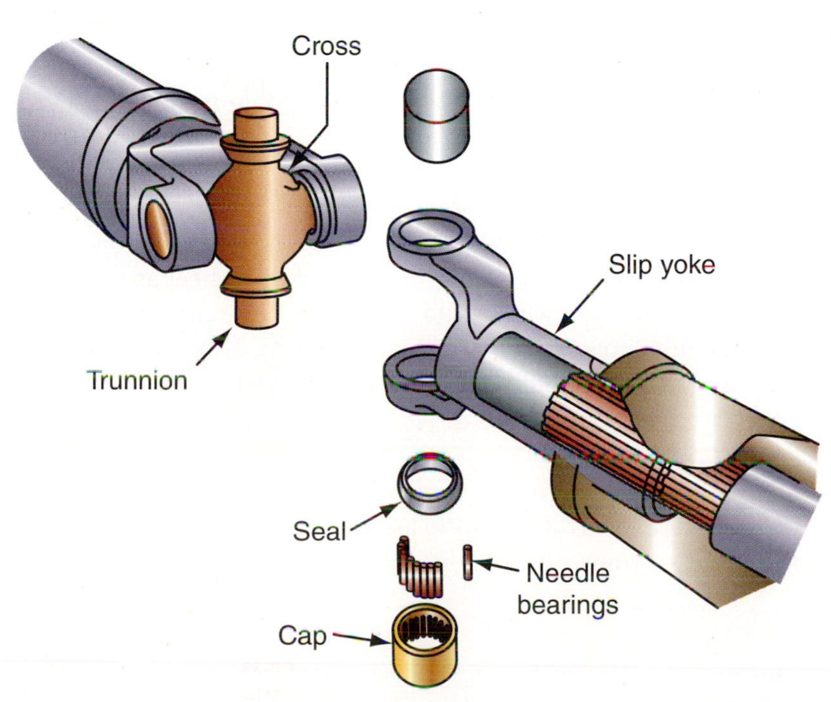

FIGURE 12-3 **A common customer complaint is a squeak at low speeds. This noise indicates a dry (not lubricated) U-joint(s).**

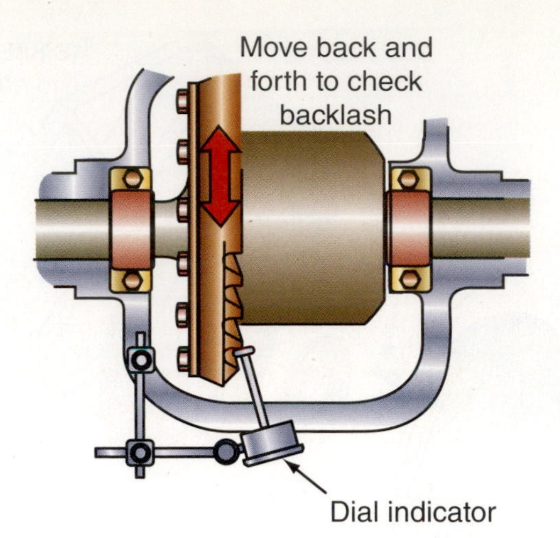

FIGURE 12-4 **A roaring or whining sound traced to the differential could indicate that the gear backlash is incorrect.**

squeaking U-joint as soon as possible. If the joint breaks apart during vehicle operation, the drive shaft assembly can come through the floorboard, causing injury to the passenger. A drive shaft that drops to the road and digs into the road surface can flip a car over.

Differentials

As stated earlier, clunking noise from under the car may be caused by worn U-joints or worn differential gears. Other differential noises can be just as difficult to isolate. A high-pitched whine from the differential can mean the pinion gear and ring gear are set too tightly or one of the bearings may be going bad (Figure 12-4). Operating the vehicle in tight, slow turns can isolate axle bearing noises. As the vehicle leans, its weight loads and unloads the bearings sited near the drive wheels. The bad bearing will make more noise when it is on the outside of the turn and loaded. Rear-wheel-drive (RWD) axles usually do not make any noise until they break. Then there is no drive or noise.

FWD Axles

Most front-wheel-drive (FWD) axle bearings are one-piece, double-ball bearings that are permanently lubed.

The noise from front-wheel-drive (FWD) axles will originate in the constant velocity (CV) joints. They are fairly easy to diagnose and isolate. From a dead stop, accelerate hard straight ahead. A fast, clicking noise is usually related to a bad inner CV joint. Operating the vehicle in a tight, slow circle will produce a slow-to-fast clicking noise from bad outer CV joints. The noise will be worse when the bad joint is loaded or on the outside of the turn. Bad outside CV joints may also be felt in the steering wheel as a turn is made.

A bearing mounted in the wheel hub supports the outer end of the drive axle (Figure 12-5). A common noise from a damaged bearing is a loud crushing or cracking sound. This may also produce a noticeable feel in the steering wheel and increases when the bearing is loaded.

Clutches

Clutch problems can be easy to diagnose because the system is so simple. Clutch noises have to be separated from transmission noises.

Diagnosis

A rattling or whining sound with the clutch disengaged can be traced to the release bearing. If the noise is audible at all times but changes in pitch or volume as the clutch is

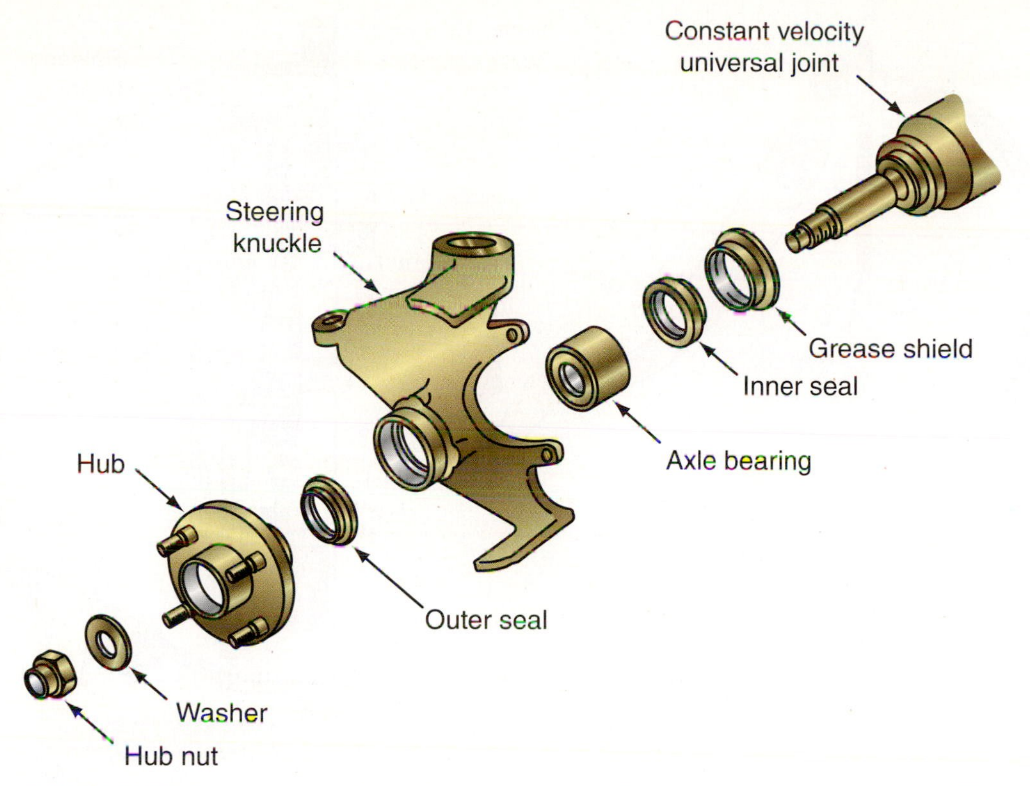

FIGURE 12-5 **To isolate axle bearing noise (roaring), test drive and change vehicle direction. The noise will change as the bearing on the outside of the turn is loaded and unloaded.**

operated with the transmission in neutral, the noise points to a continuously running release bearing that is dry of lubrication. A transmission input shaft's bearing noise will be most prominent with the clutch engaged and the transmission in neutral (Figure 12-6). It will disappear within a short time when the clutch is disengaged and held that way for a few seconds. A chattering noise and jerking as the clutch is being engaged usually indicates a bad clutch disc, pressure plate (clutch cover), or flywheel. A leak in the engine's rear main or transmission input seal will also cause chattering and jerking during clutch operation. The clutch assembly must be removed for an exact diagnosis.

A clutch friction disc can get overheated and have hot spots similar to those found in brake friction material. It usually results in chattering or grabbing during clutch engagement.

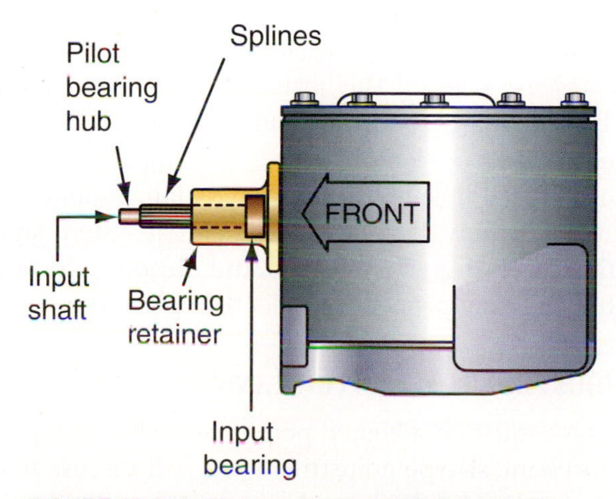

FIGURE 12-6 **Noise from the input shaft bearing is most noticeable when the transmission is under load. It will cease when the clutch is disengaged (pedal depressed) and a few seconds allowed until the shaft stops spinning.**

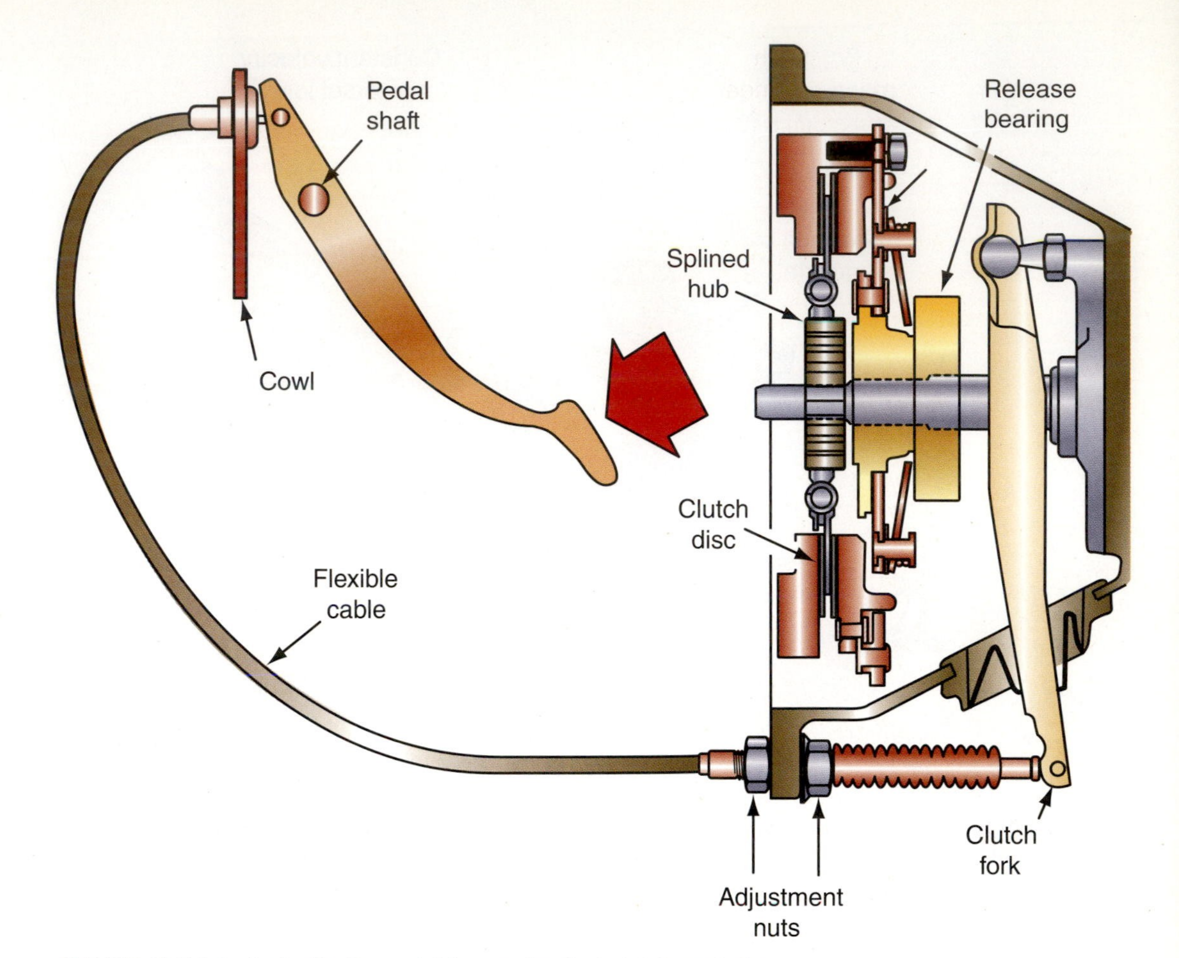

FIGURE 12-7 **A typical adjusting point for mechanical clutch controls.**

SERVICE TIP:
Problems engaging first or reverse gears may also be caused by low fluid level in the transmission. Low fluid may cause the gears and synchronizers to bind. They can completely block the first or reverse shift particularly after the vehicle has sat idle for a time. Check the fluid before beginning removal of the transmission and clutch. A new customer may be found when informed that the suspected $300.00 clutch replacement will only cost $40.00 or less.

A worn clutch disc will slip during load conditions, most notably in the higher gears. A racing engine as the engine load goes up in high gear is the first sign of a worn clutch disc. Mechanical clutch linkage that has not been properly adjusted will not allow the pressure plate to clamp correctly and will cause the same problem. This type of linkage requires periodic adjustment as the clutch disc wears (Figure 12-7). An early sign of pressure plate wear or maladjusted linkage is a difficulty in shifting into first or reverse gears while the vehicle is halted. The pressure plate in this instance is not completely releasing the clutch disc. There may not be a slipping condition, but other gears may also be hard to select when the vehicle is under load.

Since most hydraulic clutch linkages are self-adjusting, slipping normally indicates bad components in the clutch assembly. If the hydraulic clutch system should leak, the pressure plate will not release the clutch disc. Topping off and bleeding the system may temporarily correct the problem, but the only proper repair is to replace the leaking component.

Classroom Manual
pages 346, 347

Inspecting and Adjusting the Clutch Linkage and Fluid

Mechanical clutch linkages must be adjusted periodically due to normal clutch lining wear. Before discussing the mechanical-type adjustment, we will discuss inspecting and changing the fluid in a hydraulic clutch. The fluid can be replaced easily by attaching a rubber hose to the slave cylinder bleeder screw and draining the old fluid into a container, preferably a clear one (Figure 12-8). Add fluid to the master cylinder and operate the pedal until clean fluid comes out of the slave. Keep the end of the hose below the level of fluid in the

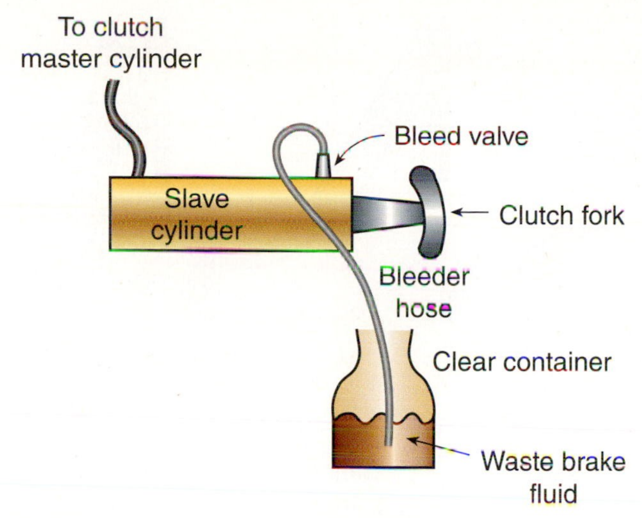

FIGURE 12-8 **The hydraulic clutch system can be bled using a rubber hose and a catch container.**

SERVICE TIP:
Bleeding a hydraulic clutch can be aggravating on some vehicles. If the system cannot be bled using the manual method use a pressure bleeder.

SPECIAL TOOLS
Lift *or*
Floor jack and jack stands

Very few hydraulic control clutch systems have pedal **free travel**. Almost all manual clutch linkage systems have a specific free travel.

container and watch for air bubbles as the system is bled. Bleed the air out of the system by closing the bleeder screw and pumping the pedal several times. Hold the pedal to the floor and slightly open the bleeder. Continue this process until there are no air bubbles in the fluid. This process should be done at about the same interval as that for the coolant and brake fluid, about every 2 years. Consult the service or operator manual for the maintenance schedule. This process is very similar to brake bleeding. When completed make sure to check the level of fluid in the clutch (Figure 12-9). A low level may indicate a leak. Clutch fluid is similar to brake fluid, so it must be inspected in a similar manner.

Many newer models of cars and light trucks have hydraulically controlled clutches, but there are many older vehicles still on the road. A heavy-truck or equipment fleet technician will perform repairs and adjustments as a routine maintenance service on heavy vehicles (on- and off-road).

Adjusting the pedal's **free travel** sets the mechanical linkage engagement point for the clutch. Free travel is the distance the pedal moves before the release bearing

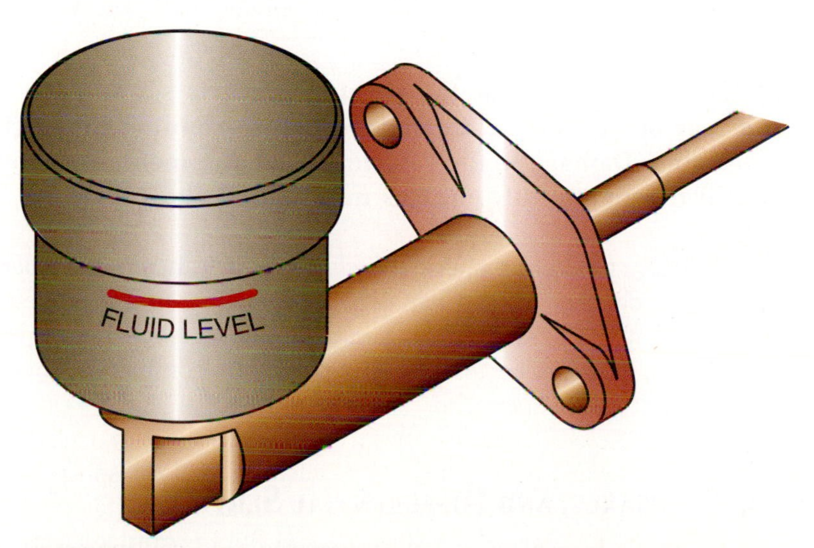

FIGURE 12-9 **Fluid for a hydraulic clutch is stored in a small reservoir that may be attached to the clutch master cylinder or remotely mounted on the firewall.**

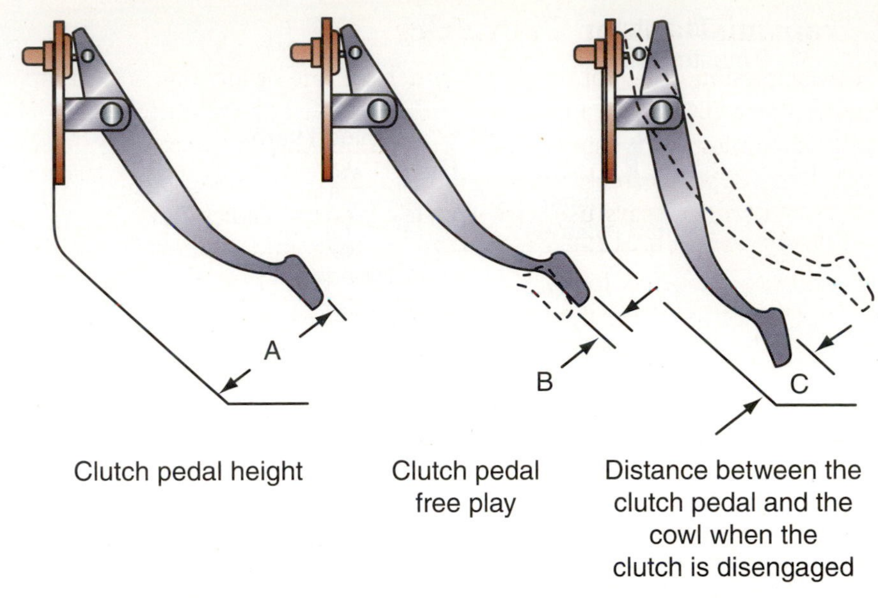

FIGURE 12-10 **The adjustment shown is for adjusting pushrod length on a hydraulic-controlled clutch. Remember to check the brake switch operation whenever the clutch linkage, mechanical or hydraulic, is adjusted.**

SERVICE TIP: Always make sure to check the manufacture's Technical Service Bulletins for issues related to the transmission servicing, maintenance, and inspection. Some manufactures use a dye in their automatic transmission fluids and that dye will change color as it ages. The color is misleading and can cause an improper maintenance suggestion for service.

contacts the release levers of the pressure plate (Figure 12-10). The vehicle will probably have to be lifted some amount to gain access to the adjustment. If the vehicle is lifted to a normal (stand up) working height, a second person will have to ride the vehicle up the lift to operate the clutch and measure the pedal free travel. If only one technician is available, it is probably quickest and easiest to jack and block the front of the vehicle. This will allow the technician to use a creeper to make the adjustment, slide out, measure the free travel, and if necessary slide back under the vehicle to make additional adjustment.

The adjustment device will be located on the clutch linkage at the left side of the bell housing near the release bearing's fork. It usually consists of one or two nuts that are screwed on a threaded rod. Adjust the linkage using the adjusting nuts until the pedal free travel is the correct distance. Most clutches have a free travel of 3/4 inch to 1 inch. Consult the service manual for the correct specifications.

WARNING: Ensuring that the parking brake is set and the areas to the front and rear of the vehicle are cleared before running the engine to check clutch operation. Damage to the vehicle, shop equipment, or injury to persons may result if the clutch fully engages unintentionally.

With the adjustments completed, check the clutch and transmission operation on the lift. All gears should be checked at the halt with the engine running. First and reverse should be selected easily at the halt. All other gears should be smooth with no grinding or hanging. The vehicle will have to be tested on the road to completely check the adjustment. Consult the instructor or supervisor.

Classroom Manual
pages 347, 348, 349, 358, 359, 360, 361, 362, 363, 364

Transmission, Transaxle, and Differential Service

Both manual and automatic transmissions and transaxles need routine servicing and maintenance. Most of the time it is only a fluid and filter change. The service manual specifies the interval, type of fluid and capacity.

Manual Transmissions or Transaxles

This type of transmission does not need very much service other than a fluid change. A few manual systems use a filter. The fluid can be drained through a drain plug located at the bottom of the transmission housing. Manual transmission fluid comes in different blends. Older vehicles use a 85W-90 thick oil with few additives, but newer systems require lighter, thinner fluid. Some smaller cars use an automatic transmission fluid. The current classification is listed as GL-4 or GL-5 (Figure 12-11). GL fluids may have additives that lubricate, clean better, and are designed for specific temperature ranges. Many manufacturers also used special lubricants for their manual transmissions. Most newer light trucks require a special fluid for their manual transmissions. Other manufacturers use automatic transmission fluid or a lightweight engine oil for their manual transmission or transaxle systems. Most manual transmissions require 3 to 5 quarts of fluid.

SPECIAL TOOLS

Drain pan

Air- or hand-operated pump *or* suction gun

To change the fluid in a manual transmission, lift the vehicle to gain access to the drain plug (Figure 12-12). The transmission should be hot enough for the thick fluid to flow freely. Cold transmission fluid is very slow to flow and will not remove all of the debris that may be inside the bottom of the transmission. The transmission should be at the same angle as it is when the vehicle is setting on its tires. Remove the drain plug, capture the old fluid, and dispose of it according to the shop's policies. While the fluid is draining, check for leaks at the transmission's extension housing seal where the drive shaft mounts or the drive axle seals on a transaxle. Repair as necessary.

SERVICE TIP: Attempting to service a transmission that is not at the same angle as an operating vehicle may not allow all of the fluid to drain.

Reinstall and torque the drain plug once all of the fluid has drained. Fluid can be installed with an air-operated pump or using a hand **suction pump**. The suction pump is slow and can be messy.

The **suction pump** is used to draw fluid from a container and force the fluid into the transmission or transaxle by hand.

FIGURE 12-11 **Manual transmission and differential fluids range from a thick 85W-90 to thin automatic transmission fluid or a special blend fluid.**

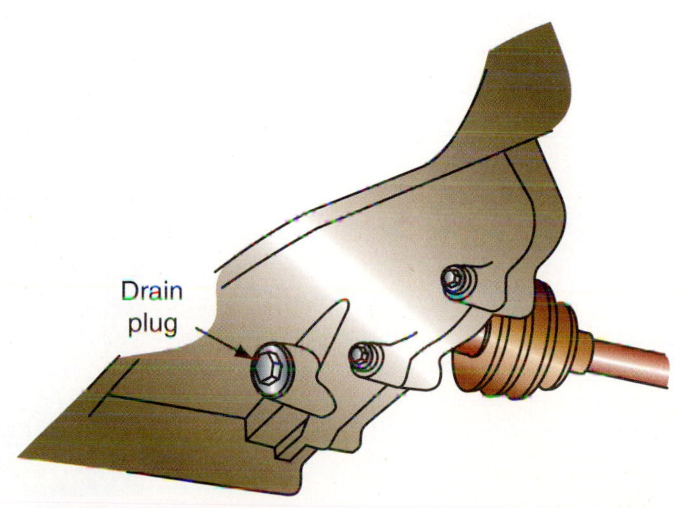

FIGURE 12-12 **A typical drain plug for a manual transmission or transaxle.**

WARNING: Overfilling a transmission could cause high pressure within the housing and rupture a seal or gasket. Damage could result from an overfilled transmission.

SERVICE TIP:
Always refer to the service manual when checking the fluid level on a manual transmission, transfer case, or differential. Some manufactures call for a special tool to check the fluid level and the level is not supposed to be all the way up to the fill plug.

Locate the fill plug. It is normally about halfway up on the side of the transmission. Remove the plug and insert the pump's hose tip into the hole. If an air-or hand-operated pump is used, the amount being installed is measured with a dial on the pump (Figure 12-13). Suction pumps will hold about a pint of fluid and will have to be refilled until the transmission is full (Figure 12-14). Most manual transmissions do not have a dipstick to check fluid level. The most common method of checking the fluid involves the use of the technician's finger (Figure 12-15). Insert the little finger into the fill hole to the first knuckle. Bend the fingertip downward inside the transmission. If the fluid touches the finger, the level is correct. Special tools can be purchased or made to check the fluid level. A quick tool can be made from a welding rod. Bend the rod 90 degrees, about the length of the last little finger joint from one end (Figure 12-16). About an inch from this bend, curve the rod so the bent end can be fitted into the fill hole. Place the bent end into the fill hole and the rod against the bottom of the hole (Figure 12-17). If the fluid level is correct, there will be fluid on the end of the rod inserted into the fluid. A tool is not an absolute necessity unless the technician performs this type of service several times a day or week.

During filling, the fluid is added until it barely begins to run from the fill hole. If it is overfull, drain and capture the excess. Reinstall and torque the fill plug.

FIGURE 12-13 **This type of lubrication gun is attached to a large bulk container of lubricant.**

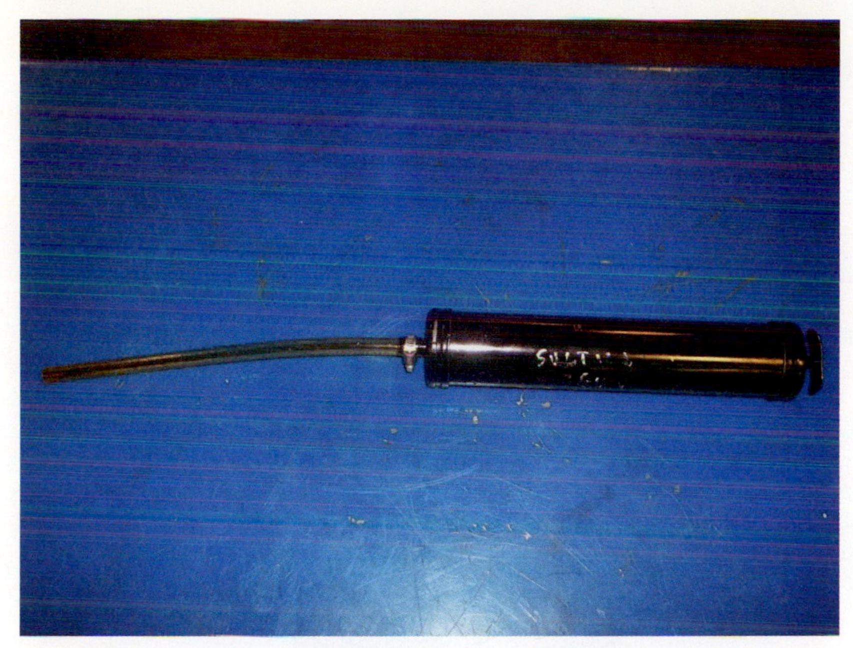

FIGURE 12-14 **A suction gun is a slower method of transferring fluids, but it still works.**

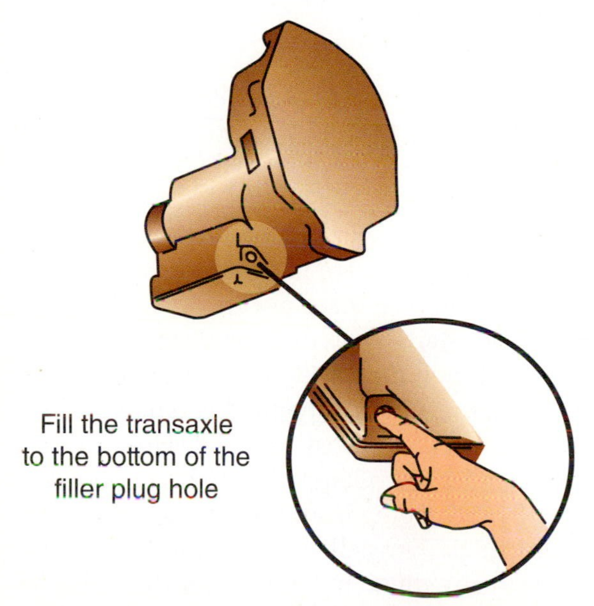

FIGURE 12-15 **Typically the transmission or transaxle/differential check plug is on the side of the housing. The fluid level can be checked with a finger.**

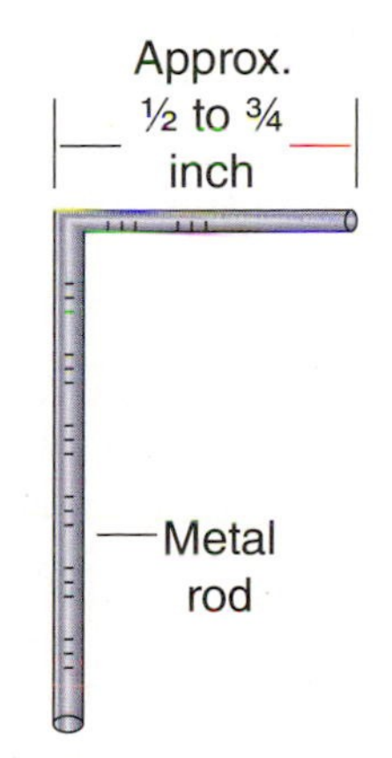

FIGURE 12-16 **A straight piece of small bar stock can be made to check the fluid level in a manual transmission or differential.**

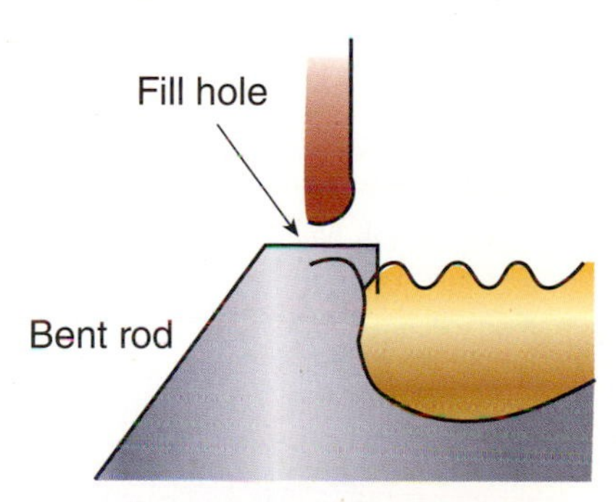

FIGURE 12-17 **Place the rod on the bottom of the fill hole and allow the short portion to extend down into the fluid.**

Automatic Transmissions or Transaxles

Automatic transmission or transaxle service includes a fluid and filter change. The fluid removed will not be all of the fluid in the system. A typical fluid change will only remove the fluid in the transmission pan. The torque converter and pump will retain some fluid. Before starting the service, check the manual for the correct amount of fluid used for a filter change service. Ensure the amount used is not for a full flush. It is best if the transmission is warm. Transmission and transaxle services are performed using similar procedures.

SPECIAL TOOLS

Large drain pan

Funnel

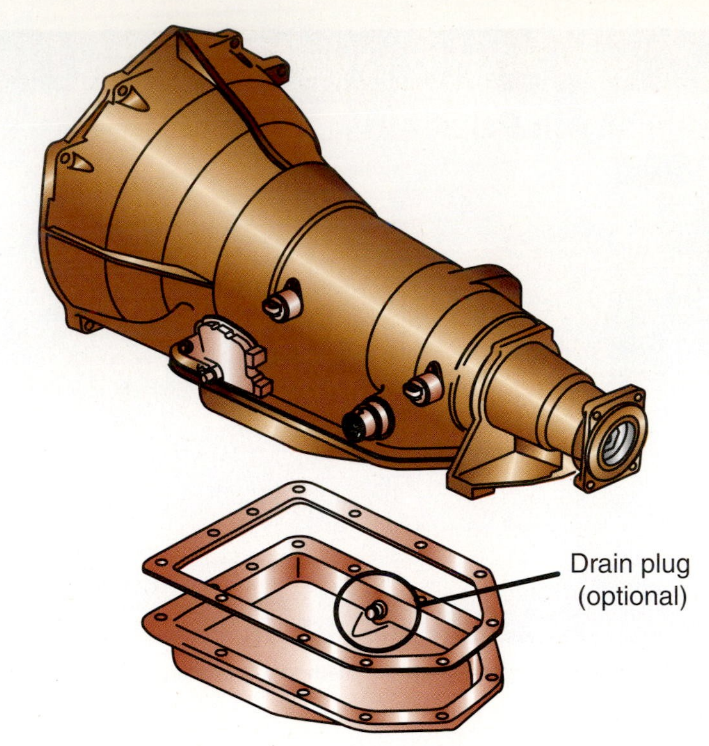

FIGURE 12-18 **A drain plug for an automatic transmission or transaxle is uncommon but very handy if present.**

SERVICE TIP:
Check the used transmission fluid using feel, sight, and odor. If the fluid is burnt or very dirty, changing the filter and three quarts of fluid will not help the transmission. Many times a full-system flush will help restore most of the transmission's performance, but a flush may not help and may even decrease performance more if there is a lot of mileage on the fluid. Photo Sequence 35 highlights the procedures for changing the filter and gasket.

Before proceeding, compare the pan gasket in the filter kit with the pan. There are many different pan shapes, none of which are particular to a certain brand or model. If the pan is removed and the gasket is incorrect, the lift and bay will be tied up until another gasket is available.

Lift the vehicle to gain access to the transmission's oil pan. Most automatics do not have drain plugs. The pan has to be removed (Figure 12-18). Care must be taken to prevent spills. Remove the pan bolts from the lowest side of the pan. Place a wide drain pan under the transmission and remove bolts along each side, starting at the point where the first bolts were removed. At some point, the pan will tilt and fluid will begin to drain. Loosen the other pan bolts, but do not remove all of them until most of the fluid is removed. Keep the last two or three bolts loose until the pan can be supported by hand. Hold the pan against the transmission and remove the last several bolts. Control the oil pan and pour the remaining fluid into the drain. There will be fluid dripping from the filter and transmission. Keep the drain pan in place until the oil pan is reinstalled. Note how the filter is installed. Automatic transmission filters come in many different shapes and their installation location may not be readily apparent once the old one has been removed. Remove the filter fasteners and filter.

Usually, there is a small magnet at the bottom of the oil pan (Figure 12-19) with a few metal shavings on the magnet. This is normal. Excessive shavings indicate gear or bearing damage. Look for black friction material at the bottom of the pan as well. There will probably be some, but like the metal shavings, this is normal. Pieces or chucks of friction material indicate that the hydraulic clutches are bad. The lower side of the filter needs to be checked for the same items. Excessive shavings or friction material normally requires a transmission overhaul or rebuild. Consult the instructor, supervisor, or service manual for advice as needed.

Clean the oil pan of all materials, including any gasket material. The oil pan mounting area on the transmission also has to be cleaned of all gasket materials. Be sure to check the bolt holes in the pan and transmission for damage or gasket material. Some manufacturers

Typical Procedure for Performing an Automatic Transmission Fluid and Filter Change

P35-1 To begin the procedure for changing automatic transmission fluid and filter, raise the car on a lift. Before working under the car, make sure it is safely positioned on the lift.

P35-2 Before continuing, compare the new pan gasket with the shape of the oil pan.

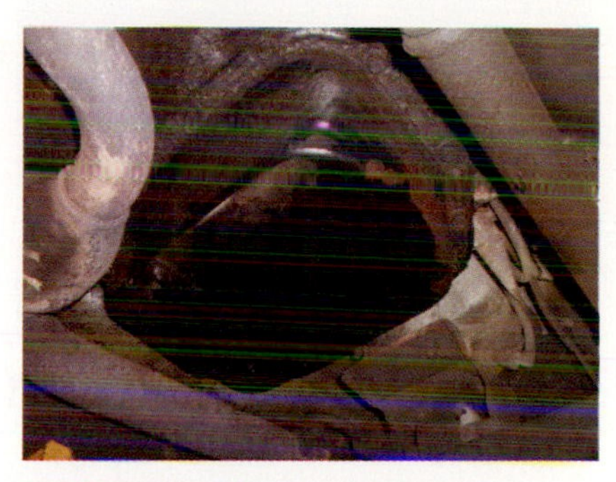

P35-3 Locate the transmission pan. Remove any part that may interfere with the removal of the pan.

P35-4 Position the oil drain under the transmission pan. A large-diameter drain pan helps prevent spills.

P35-5 Loosen all of the pan bolts and remove all but three at one end. This will cause fluid to flow out around the pan into the drain pan.

P35-6 Supporting the pan with one hand, remove the remaining bolts and pour the fluid from the transmission pan into the drain pan. Be careful, there may still be up to a quart of fluid left in the pan.

P35-7 Carefully inspect the transmission pan and the residue in it. The condition of the fluid is often an indication of the condition of the transmission and serves as a valuable diagnostic tool.

P35-8 Remove the old pan gasket and wipe the transmission pan clean with a lint-free rag.

P35-9 Unbolt the fluid filter from the transmission's valve body. Keep the drain pan under the transmission while doing this. Additional fluid may leak out of the filter or the valve body.

P35-10 **Install the new filter onto the valve body and tighten the attaching bolts to proper specifications. Then lay the new pan gasket over the sealing surface of the pan and move the pan into place.**

P35-11 **Install the attaching bolts and tighten them to the recommended specifications. Carefully read the specifications for installing the filter and pan. Some transmissions require torque specifications of inch-pounds rather than the typical foot-pounds.**

P35-12 **With the pan tightened, lower the car. Pour automatic transmission fluid (ATF) into the transmission. Start the engine. With the parking brake applied and a foot on the brake pedal, move the shift lever through the gears. This allows the fluid to circulate throughout the entire transmission. After the engine reaches normal operating temperature and with transmission in park, check the transmission fluid level and correct it if necessary.**

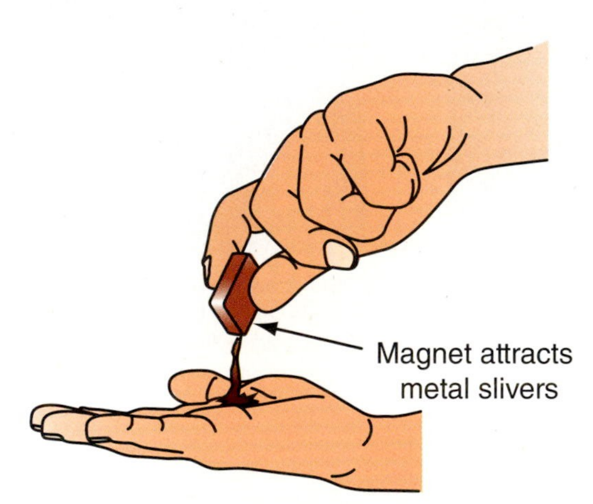

FIGURE 12-19 **The magnet located in the transmission oil pan will attract metal shavings. Check it to help determine the mechanical condition of internal components.**

use room temperature vulcanizing (RTV) instead of a gasket to seal the pan, but RTV can enter and block the hole. Like the manual system, take the time now to check the seals at the drive shaft or drive axles.

WARNING: Overfilling a transmission could cause high pressure within the housing and could rupture a seal or gasket. Damage could result from an overfilled transmission.

Install the filter and torque the filter bolts (Figure 12-20). If RTV was the original sealing material but an aftermarket gasket is available, use the gasket. Place the gasket on the

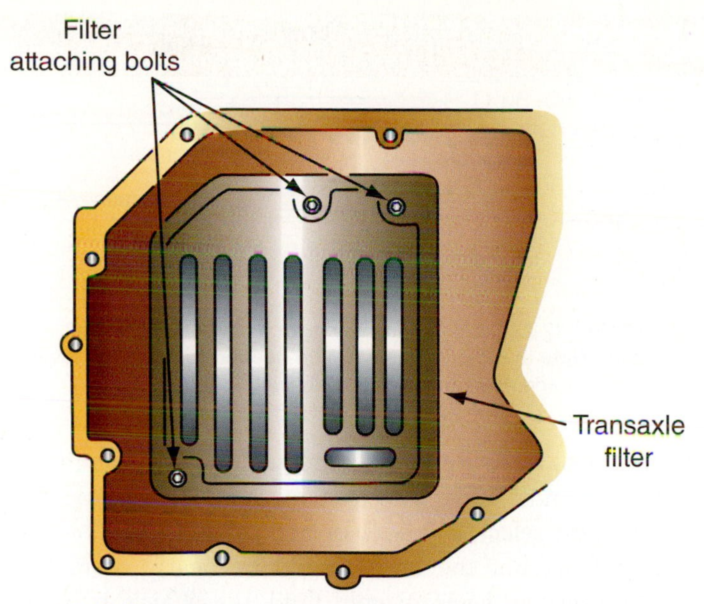

FIGURE 12-20 **Always torque the filter fasteners. The filter can allow air to enter the hydraulic system if not properly torqued.**

oil pan. Do not use RTV on the gasket unless the gasket manufacturer recommends it. If RTV has to be used in place of a gasket, it should be laid in a 1/8-inch bead and applied evenly around the pan. Remember to lay the bead on the fluid side of the bolt hole and *do not* allow RTV to cover the hole. Ensure that the RTV meets the vehicle manufacturer's specifications and is allowed to cure for 10 to 15 minutes before mounting the pan to the transmission. Before installing the oil pan, wipe the fluid from the mounting area on the transmission. Place the pan in place and start the bolts. Overtightening the bolts can strip the hole threads very easily. It is important to use a good torque wrench and tighten the bolts, in sequence, to torque specification (Figure 12-21). With the pan installed, lower the vehicle and add the proper amount of fluid. The fluid should register at the cold level mark or slightly below the full mark on the **dipstick**.

In automotive applications a **dipstick** is a thin flexible rod used to check fluid levels in certain components.

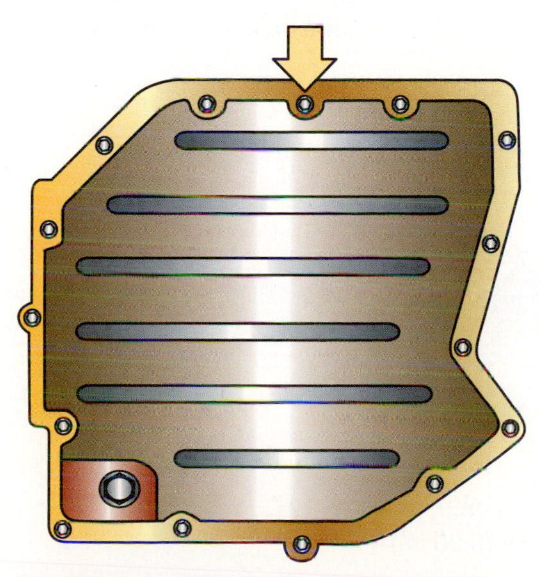

FIGURE 12-21 **Always torque the oil pan fasteners. A too-tight condition will cause just as much leakage as a too-loose condition.**

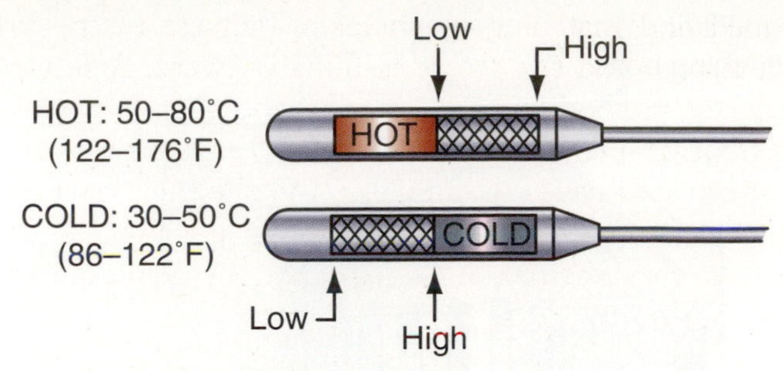

FIGURE 12-22 **Dipsticks usually have a "hot" and a "cold" mark. Note that the amount of fluid needed to move the level from the add mark to full is usually just 1 pint.**

SPECIAL TOOLS

Drain pan

Air- or hand-operated pump *or*

Suction gun

With the fluid installed, start the engine and select each gear in turn to fill the pump and clutches. The fluid in most automatic transmissions or transaxles is checked with the engine running and the gear selector in park or neutral and hot. The first check after a service will be with cold fluid, but the level should be nearly correct but not be overfull (Figure 12-22). With the proper level for cold fluid, the vehicle may be driven to heat the fluid so it can be checked at the hot fluid level. Next, check the oil pan for leaks. Remove any fluid over the full mark and repair any leaks if necessary.

Differentials

Classroom Manual
pages 367, 368, 369

The differential is serviced by changing the oil. Some differentials have a drain plug whereas others have a rear cover that must be removed. The fill plug is usually about halfway up the side, similar to the manual transmission. The fluid is drained, refilled, and checked the same way as a manual transmission and uses the same type of fluid on older differentials (Figure 12-23). Lockup or **limited slip** differentials use special additives for the clutches. Using regular fluid can seriously damage a lockup differential. Newer differentials use special blended fluids. Final drives on transaxles are serviced as part of the transaxle.

Drive Shaft and Drive Axle Repair

Classroom Manual
pages 367, 369, 370

Drive shaft repair usually consists of replacing a U-joint or replacing the entire drive shaft assembly. On the FWD axle, CV joints or boots can be replaced, but the most common task is to replace the entire axle. CV joint boots can be replaced, but because of the CV

A **limited slip** differential is designed to reduce wheel spin using some type of clutch mechanism. Using incorrect differential fluid will damage the clutches.

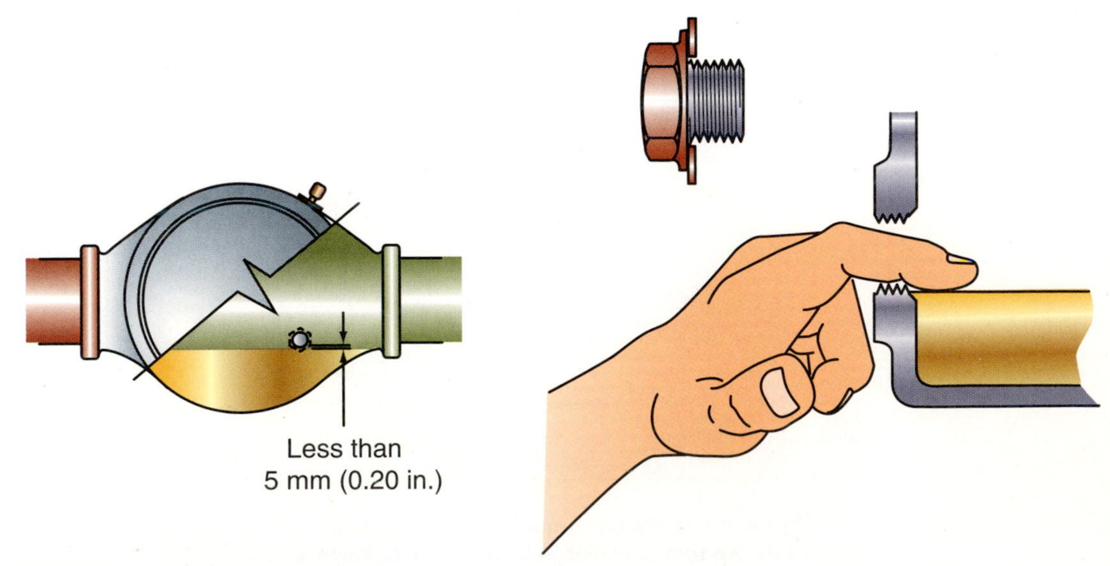

FIGURE 12-23 **The differential fluid level is checked the same way as a manual transmission.**

joint's highly machined and matching mechanism, damage to the joint can occur very quickly when a boot is torn and the grease is flung outward. Axle replacement is usually the cheapest and quickest method for the customer and shop.

Replacing just a boot or one joint may save some time or the cost of the part, but the labor will offset most of that savings. In addition, the shop will have to warrant the repair component and the labor. Replacing the entire drive axle allows the shop to give the customer a new axle or one that was rebuilt in a specialty driveline shop at roughly the same total cost and a better warranty.

SPECIAL TOOLS

Vise

Marker

U-joint bearing press

Grease gun

Drip pan

Drive Shafts

We will cover a basic U-joint replacement. Some vehicles may have a **center bearing** on the drive shaft and more than two U-joints. The procedures will be the same except for removing the center bearing. Before removing the drive shaft, mark the position of the shaft to its companion flange on the differential (Figure 12-24). The companion flange is the part where the drive shaft's rear U-joint mounts to the differential. There may be another companion flange at the center bearing or the transmission depending on vehicle design. Use a **sealing tool** to slide over the transmission's output shaft once the drive shaft is removed. This will stop any fluid from leaking.

The **center bearing** is used on two-piece drive shafts.

A **sealing tool** is a pliable cover shaped like the slip joint. The plug slides over the output shaft to prevent oil from leaking out during servicing.

WARNING: Do not overtighten the vise on the drive shaft and do not place the vise over the balancing weight. It may bend or dent the tube, knock a weight off, and unbalance the assembly.

WARNING: Do not use tools for purposes for which they were not designed. Damage to components, tools, or injury to a person may occur.

Place the drive shaft in a vise near the U-joint being replaced. Remove the rear U-joint by first removing the retaining clips and pressing the bearing cups out. The clips may be internal or external. Some original U-joints do not use clips but have plastic retainers that will be sheared when the cups are pressed out. The replacement U-joints for this type will have internal clips. When pressing the cups, use an adapter to support the drive shaft's yoke. This prevents the **yoke** from spreading. A common practice is to use two sockets, one slightly larger than the cup hole and the other slightly smaller than the cup itself. The smaller socket is used to push the cups away from the yoke while the larger socket

The **yoke** of a light drive shaft (small car) can be easily distorted. The correct use of the proper tool will help prevent this distortion.

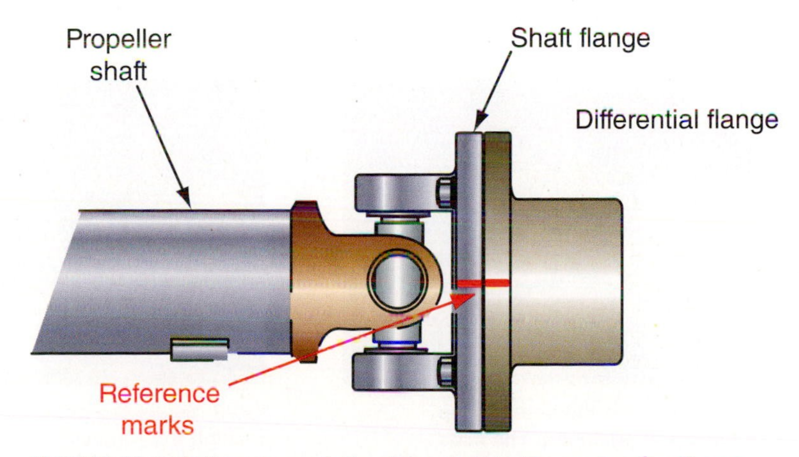

FIGURE 12-24 **Failure to mark the drive shaft to the companion flange could result in an out-of-phase drive shaft with vibration problems.**

FIGURE 12-25 **Special tools are available for pressing the U-joints in and out of the drive shaft yoke.**

acts as a standoff between the yoke and the vise. However, in Chapter 3, "Tools and Equipment Usage," there is no mention of using sockets as pressing or standoff tools. Another common practice is to use a hammer and punch. Though the hammer and punch can be used as drivers in some instances, this is not one of them. There are tools designed for removing and installing U-joints (Figure 12-25). See Photo Sequence 36 for U-joint replacement with the correct tools.

Pressing on one cup will force the opposite cup out of the yoke. With the first cup out, try to remove the U-joint cross by moving it into the empty bore and twisting it to the side. If the space is not sufficient, the yoke will have to be rotated and force applied to the end of the cross. This will move the second cup outward enough to remove the cross. The second cup can be knocked out with a rod installed through the empty bore once the cross is removed.

PHOTO SEQUENCE 36

Replacing a Single Universal Joint

P36-1 **Ensure the vise is not placed over a drive shaft weight and is not overtightened.**

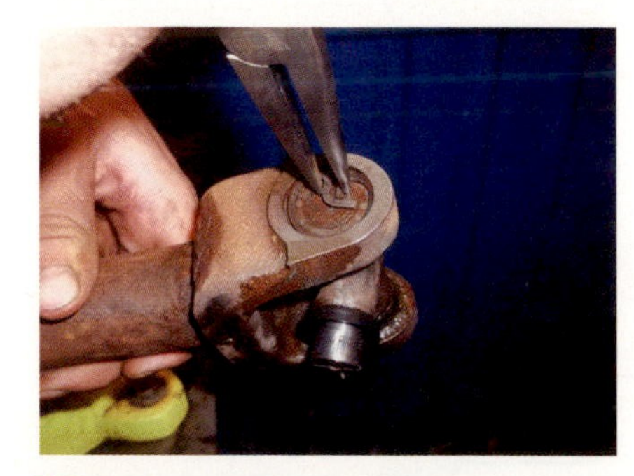

P36-2 **Remove the bearing cup retainers, if equipped. The clips may be internal or external on the yoke.**

P36-3 **Select the proper adapters and position the removal clamp over the adapters.**

PHOTO SEQUENCE 36 (CONTINUED)

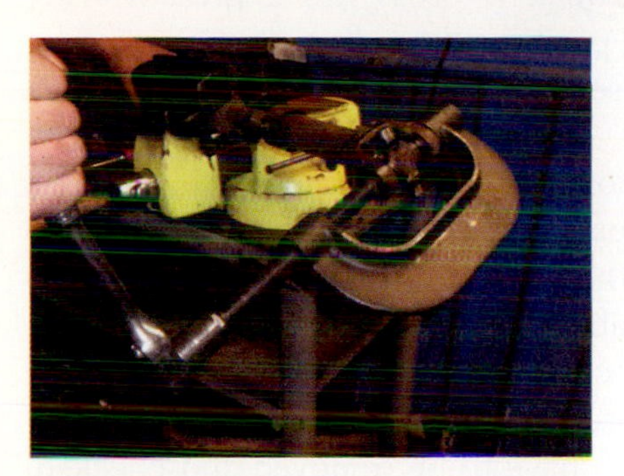

P36-4 Use a box-end wrench to turn the screw that applies force to the adapters. The adapters will have to be switched from one side to the other to press the second cup out.

P36-5 When installing the cups, first position one cup hand tight into the cavity.

P36-6 Install one end of the cross into the partially installed cup, then maneuver the cross so the other end fits into the opposite cavity. The hole for the grease fittings should face toward the drive shaft.

P36-7 Hold the cross in place to retain the needle bearings in place of the first cup as the second cup is fitted over the cross end and hand-pressed into the cavity.

P36-8 Position the clamp and adapters over the cups, ensuring the cross does not slide out of one of the cups. This could cause a bearing to drop inside the cup.

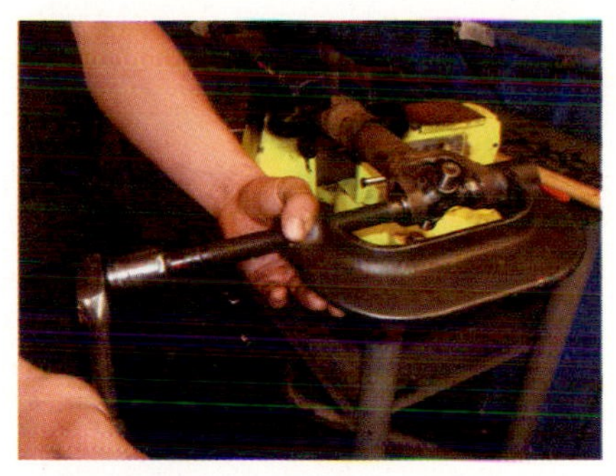

P36-9 Use the box-end wrench to press the cups in place. At about three-fourths of the way in, check to make sure the cross is centered and free in the bearing cups. If not, remove the cups and cross, inspect the bearings, and reinstall.

P36-10 Once the cups are completely seated, install the cup retainers. Sometimes the aftermarket retainers may be installed differently from the original parts.

P36-11 Install the grease fitting. Do not lube until all four cups are fitted and locked in place.

The **grease fitting** on some aftermarket U-joints are placed in the bearing cup instead of the joint cross.

During the installation of the U-joints, the bearing cups must be kept squarely in the bore as force is being applied. The needle bearings can be displaced if the cup is dropped or is not aligned correctly on the cross (Figure 12-26). The cross has a grease fitting hole that must be placed correctly as well. The hole is positioned to allow the insertion of the grease fitting and allow the connection of a grease gun with the drive shaft on the vehicle.

Place the cross in a bore and align a cup on the cross to keep the bearings in place as the cup is pressed. Press the first cup just enough to position the other end of the cross to align the second cup. With both cups being guided by the cross, apply force to each side so the cups are pressed into the yoke. One cup will probably be in position before the other. Install the retaining clip on the cup and then press and clip the second cup into position. If the second cup does not go all the way in, one or more of the needle bearings may have dropped to the bottom of one of the cups. The joint will have to be removed and checked. Install the grease fitting.

Some rear U-joints have the second set of cups in a flange, which bolts to the companion flange (Figure 12-27). Replace the cups on this type of arrangement at this time. Many systems have outer cups that connect to the companion flange by U-bolts. If this is the case, move the two new cups out of the way. They will not be needed until the drive shaft is reinstalled. Rotate the U-joint to ensure that it is free. It should be stiff but can be moved by hand.

Mark the position of the slip joint to the drive shaft. This U-joint is removed and installed the same as the rear one; however, it may require several attempts to clamp the shaft in the proper position in the vise. Ensure that the slip joint is positioned correctly as the U-joint is being installed. Install the grease fitting and grease the joint. If all four cups are installed on the rear joint, it can also be greased. If only two cups are installed, the joint cannot be greased correctly until the drive shaft is installed into the vehicle.

Remove the sealing tool from the transmission and slide the slip joint into position. Place the two rear bearing cups on the cross and in position on the companion flange. Ensure that the marks are aligned. Install and torque the bolts. If necessary, grease the rear

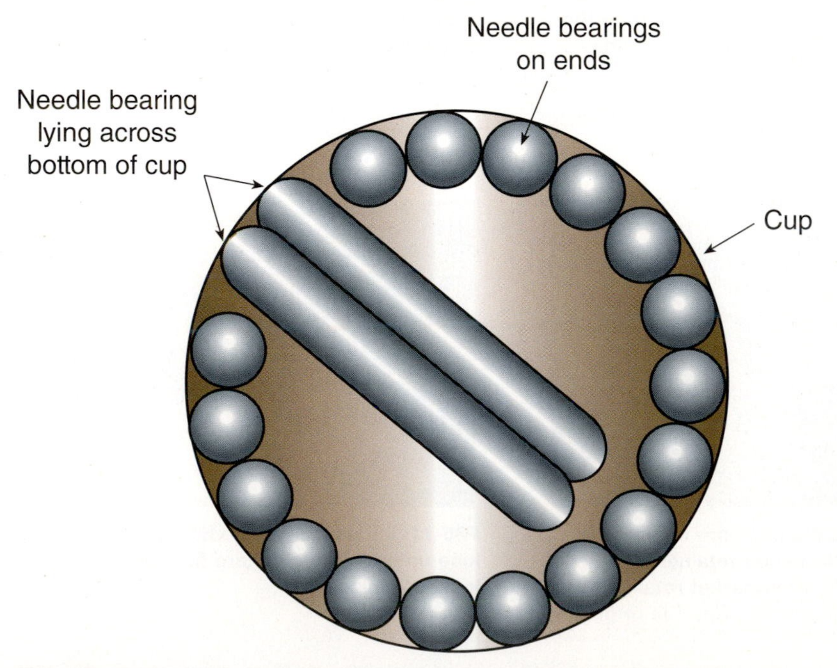

FIGURE 12-26 **Needle bearings stand on end and can be easily knocked over during installation.**

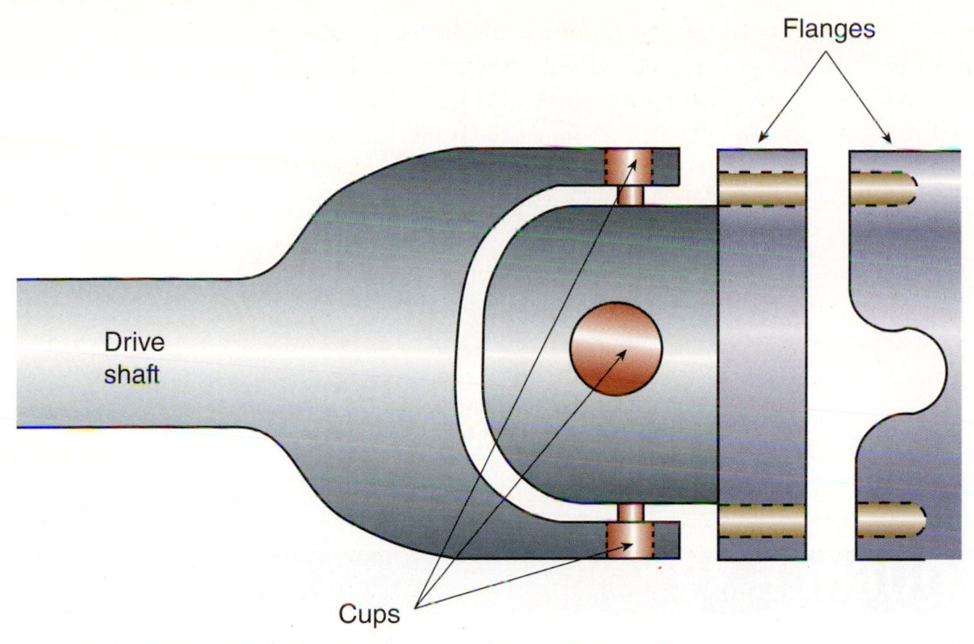

FIGURE 12-27 **Flanges like the two shown must be bolted together.**

U-joint. The vehicle can now be lowered, the area cleaned, the repair order completed, and the vehicle returned to the customer.

Drive Axles

Replacing a FWD axle is not normally assigned to an entry-level technician, but he or she may be asked to assist in the replacement. The replacement process is fairly simple and can be learned quickly.

Front-wheel-drive axles consist of two CV joints and a shaft. If the shaft is bent, the entire axle is always replaced. If a CV joint is damaged or the boot is torn, either can be replaced, but as mentioned earlier, it is usually cheaper in the long run for the customer to replace the axle as an assembly. The replacement of a typical drive axle will be discussed here. The following is typical of manual or automatic transaxles and should be supervised by the instructor or an experienced technician until proficiency is achieved by the new technician.

WARNING: Use extreme care when working on vehicles with antilock brake systems (ABSs). Damage to the system or components may occur if improper tools or procedures are used.

SPECIAL TOOLS

Torque wrench, 3/8 and 1/2 inch

Small pry bar

Lift the vehicle to gain access to the inner end of the shaft. A drain pan should be available to catch any fluid that may drain from the transaxle's final drive housing. Remove the wheel assembly. Locate, remove, and hang the brake caliper and rotor out of the way. Before removing any other components, inspect the area around the steering knuckle. Closely inspect for antilock system components. If a sensor is present, consult the service manual for procedures to protect it during axle replacement. Also check the outer CV joint for a tone ring (Figure 12-28).

There are several methods used to attach the hub to the suspension. If a wheel alignment is performed at the strut, the vehicle will have to be checked for alignment before returning it to the customer. Photo Sequence 37 shows the typical procedures.

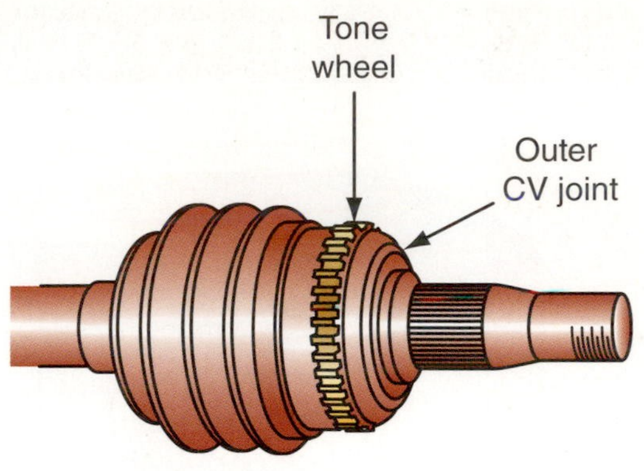

FIGURE 12-28 **This tone ring on the CV joint can be damaged if care is not taken as the axle is removed and installed.**

PHOTO SEQUENCE 37

Typical Procedure for Removing and Installing Drive Axles

P37-1 **With the tire on the ground, remove the axle hub nut and loosen the wheel nuts.**

P37-2 **Raise the vehicle and remove the tire and wheel assembly.**

P37-3 **Remove the brake caliper and rotor. Be sure to support the caliper so that it does not hang by its brake hose.**

P37-4 **Remove the bolts that attach the lower ball joint to the steering knuckle. Then pull the steering-knuckle assembly from the ball joint.**

P37-5 **Remove the drive axle from the transaxle by using the special tool.**

P37-6 **Remove the drive axle from the hub and bearing assembly using a spindle remover tool. Most times the drive axle can be pushed out by hand.**

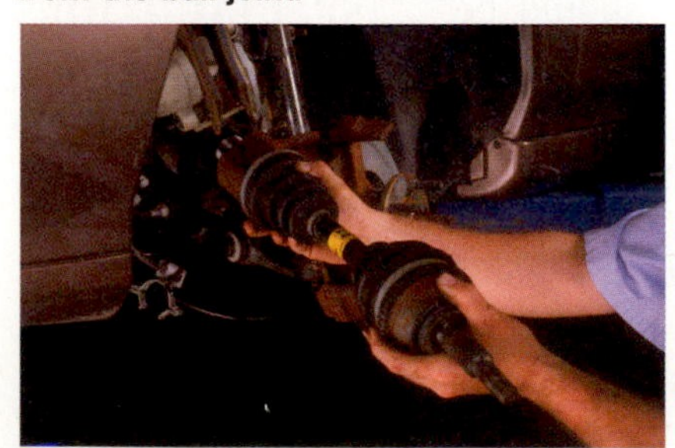

P37-7 **Remove the drive axle from the car.**

P37-8 **Install joint in a soft-jawed vise to make necessary repairs to the shaft and joints and prepare to reinstall the shaft.**

P37-9 **Loosely install the axle shaft into the transaxle.**

P37-10 Tap the splined end into the transaxle or install the flange bolts and tighten them to specifications.

P37-11 Pull the steering knuckle back and slide the splined outer joint into the knuckle.

P37-12 Fit the ball joint to the steering-knuckle assembly. Install the pinch bolt and tighten it.

P37-13 Install the axle nut and tighten it by hand. Do not reuse the old axle nut.

P37-14 Install the rotor and brake caliper.

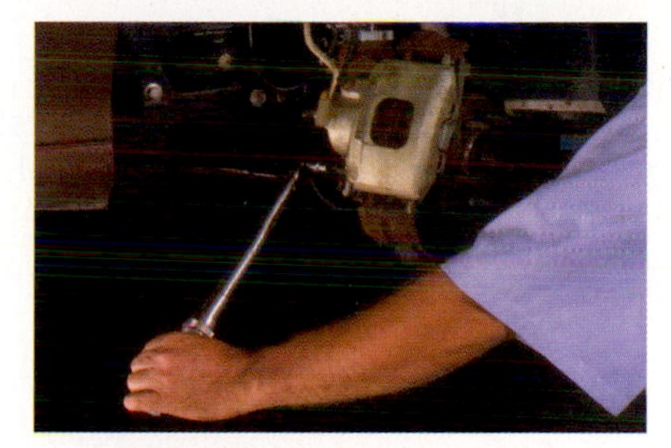

P37-15 Torque the caliper bolts to specifications.

P37-16 Install the wheel and tire assembly.

P37-17 Lower the car.

P37-18 Torque the axle nut to the specified amount.

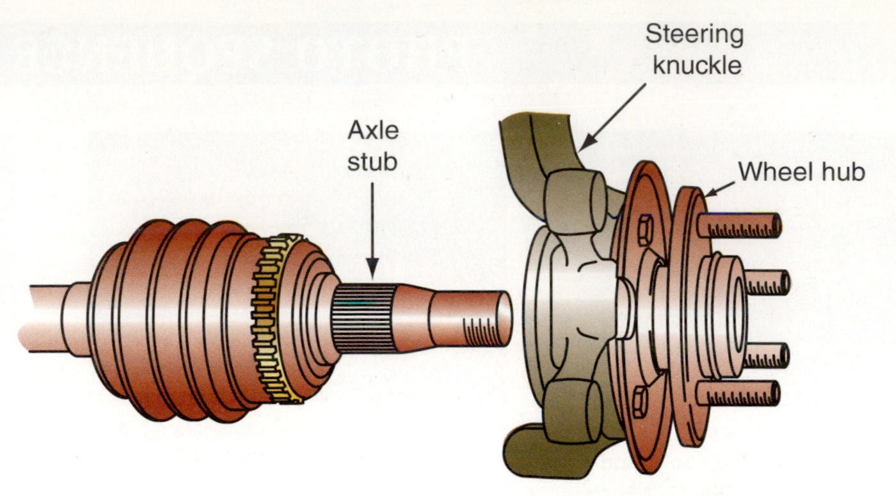

FIGURE 12-29 **With either the top or bottom steering knuckle's fasteners removed, the steering knuckle can be twisted to one side so the axle stub will slide out. The bolts removed here will require the wheel alignment to be checked after axle installation.**

Axle stubs are the outer ends of the axle.

The first step after inspection is to remove the axle nut. The cotter key and locking plate, or tab, are removed followed by the nut. There are two ways to set up the knuckle for removal of the axle. The first is to remove the steering linkage from the knuckle and then remove the strut or knuckle mounting bolts. The freed knuckle, swinging on the lower ball joint, can be swung out of the way as the **axle stub** is pushed from the hub (Figure 12-29). The wheel alignment will probably have to be checked in this instance.

A second method is to separate the steering linkage from the knuckle and unbolt the lower control arm. When this is done, the knuckle will hang from the strut (Figure 12-30). This is the best method because it will not change the wheel alignment.

WARNING: Do not drive directly on the end of the axle. Damage to the axle threads will occur. Use the axle nut or a piece of wood between the driver and the axle.

Separate the axle from the hub by pulling out the steering knuckle to the side. In most cases, the axle will slide from the hub. If the axle will not slide out, press the axle stub from the hub using a special puller tool. Once loosened, the axle should slide out. Allow the hub and knuckle to hang to one side as the axle is separated from the transaxle. Do not just pull on the axle shaft. This will cause the inner CV joint to separate.

WARNING: Always use a drive axle that has the correct locking device on the transaxle end of the axle. An undersized clip may allow the axle to come out of the transaxle during operation. An oversized clip may damage transaxle gears during axle installation and hinder removal if it does lock in place.

WARNING: Do not pry or hit the CV boot. Damage to the boot may occur and allow the CV grease to be flung out during operation.

FIGURE 12-30 **Separating the steering knuckle from the lower control arm prevents misalignment of the steering following a drive axle or CV boot installation.**

A small pry bar is fitted between the inner end of the inner CV joint and the transaxle housing to unlock the inner end of the axle (Figure 12-31). A special tool is also available to unlock the axle. Slide the axle out. Check the end of the axle's inner stub for a retaining clip or lock. Ensure that the replacement axle has the same type and size of lock or clip.

Place the new axle into the transaxle until it locks or is fully positioned against the transaxle housing. The axle must not slide out when using only hand force. A brass hammer can be used to tap the axle in place. If it does not lock in, inspect and correct the

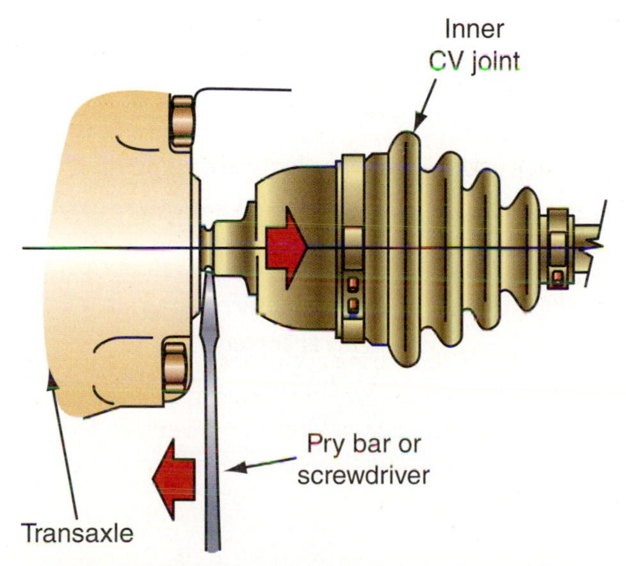

FIGURE 12-31 **A flat or curved pry bar can be used to "pop" or unlock the inner axle stud from the transaxle's side gear. Do not pull on the axle; the CV joint could be pulled apart.**

problem. Line the outer axle stub with the hub and slide the hub onto the axle. Install the washer and loosely tighten the axle nut. Install the lower ball joint and the steering linkage. As the ball joint and steering linkage nuts are tightened to torque, line up the nuts to the cotter key's hole, if used. If the nut passes the hole, do not loosen the nut. Tighten it until the hole aligns again. If the nut has to be loosened, it must be removed completely and the two components broken apart before reinstalling.

Install the brake components and wheel assembly. Lower the vehicle until the tires are resting on the floor. Torque the axle nut and install the locking collar and cotter key. If locking tabs are used, tap or crimp the tabs in place. Operate the brakes to reseat the pads. Finish the repair by completing the repair order, cleaning the area, storing the tools, and returning the vehicle to the owner.

CASE STUDY

An automatic transaxle was rebuilt and installed in a vehicle. A rebuilt pump and torque converter were also installed. During the test drive, a serious fluid leak was detected at the bottom of the bell housing. Knowing that the only possible transaxle components that could leak in this area were the pump gasket, pump seal, or torque converter, the technician removed the transaxle. Inspection of the seal and torque converter showed no damage to either part. Fluid was present around the inside of the bell housing and over the torque converter. The technician removed the pump and found the gasket to be dry. A closer inspection of the suspected components was conducted. When the pump seal was removed, it was found that the pump bushing behind the seal cavity was missing. This allowed the converter pump lug to move within the pump and break contact with the seal. A replacement pump seal and bushing were supplied and the transaxle reinstalled. A test drive revealed no leaks and the vehicle was returned to the customer.

TERMS TO KNOW

Axle stub
Center bearing
Dipstick
Free travel
Grease fitting
Limited slip
Sealing tool
Suction pump
Yoke

ASE-STYLE REVIEW QUESTIONS

1. Drive axles are being discussed.
   Technician A says some drive axles have center bearings.
   Technician B says the CV joint or boot may be replaced individually.
   Who is correct?
   A. A only
   B. B only
   C. Both A and B
   D. Neither A nor B

2. Drive shafts are being discussed.
   Technician A says the drive shaft must be marked before removal.
   Technician B says the U-joint does not need to be marked before removal.
   Who is correct?
   A. A only
   B. B only
   C. Both A and B
   D. Neither A nor B

3. Transmission fluid is being discussed.

   Technician A says most manual transmissions use automatic transmission fluid.

   Technician B says the level of fluid in an automatic transaxle should be checked while hot for an accurate measurement.

   Who is correct?

   A. A only C. Both A and B
   B. B only D. Neither A nor B

4. Transmission servicing is being discussed.

   Technician A says the oil pan bolts must be torqued.

   Technician B says the fill plug is on the side of a manual transmission.

   Who is correct?

   A. A only C. Both A and B
   B. B only D. Neither A nor B

5. A vehicle has a rattling noise from the area of the clutch. The noise goes away when the clutch is disengaged. This noise could be caused by any of the following EXCEPT:

   A. dry release bearing.
   B. input shaft bearing.
   C. worn friction (clutch) plate.
   D. worn input gear.

6. A whining noise is heard as the engine is running and the transmission is in neutral.

   Technician A says this may be caused by a torque converter.

   Technician B says the output shaft bearing may be the cause.

   Who is correct?

   A. A only C. Both A and B
   B. B only D. Neither A nor B

7. During servicing of an automatic transmission, metal shavings are found on the magnet.

   Technician A says a small amount of shavings is normal.

   Technician B says excessive shavings may indicate a problem with the overrunning clutch.

   Who is correct?

   A. A only C. Both A and B
   B. B only D. Neither A nor B

8. A whining sound is present at all times when the engine is operating. The vehicle is equipped with an automatic transmission.

   Technician A says this could be caused by low transmission fluid.

   Technician B says the transmission pump could be worn.

   Who is correct?

   A. A only C. Both A and B
   B. B only D. Neither A nor B

9. U-joint replacement is being discussed.

   Technician A says the bearing cups are pressed out.

   Technician B says a cup that fails to seat completely may be caused by a misplaced needle bearing.

   Who is correct?

   A. A only C. Both A and B
   B. B only D. Neither A nor B

10. A clunking noise is heard during gear changing.

    Technician A says the noise is probably in the differential or U-joints on an automatic-transmission-equipped vehicle.

    Technician B says the noise may be caused by a bad inner CV joint.

    Who is correct?

    A. A only C. Both A and B
    B. B only D. Neither A nor B

JOB SHEET 39

Name ________________________________ Date ______________

CHANGING FLUID IN AN AUTOMATIC TRANSMISSION

NATEF Correlation

This job sheet addresses the following NATEF tasks:

Drain and replace fluid and filter(s).

Objective

Upon completion and review of this job sheet, you should be able to change the fluid in an automatic transmission or transaxle.

Tools and Materials Needed

Drain pan

Hand tools

Service manual

Torque wrench 1/4-inch drive, inch-pounds

Procedures

Task Completed

1. Identify the transmission and specifications: ☐
   Vehicle make____________ Model ____________ Engine ____________
   Transmission ____
   Type of fluid ____
   Amount of fluid ____________________ L/Qt
   Fluid to be checked _____________ hot/cold _____________ park/neutral
   Filter bolt torque ________________________ inch/foot-pounds
   Pan bolt torque __________________________ inch/foot-pounds
2. Lift the vehicle to gain access to the oil or fluid pan. ☐
3. Position the drain pan and remove the lowest pan bolts. ☐
4. Remove other bolts slowly until the fluid begins to drain. ☐
5. Support the pan and remove all bolts. Drain any remaining fluid. ☐
6. Inspect the magnet. Describe any materials that adhered to the magnet. __________ ☐
   ____
   ____
   ____
7. Remove the filter. Remove any filter gasket or O-ring if present. ☐
8. Remove all gaskets or RTVs from the pan and pan-mounting area on the transmission. ☐
9. Install the filter and gaskets or O-ring. Torque the filter bolts. ☐
10. Place the gasket or RTV on the pan. Wipe the pan mounting area on the transmission clean. ☐

Task Completed

☐ **11.** Position the pan and gasket and install the bolts.

☐ **12.** Torque the pan bolts in a sequence.

☐ **13.** Lower the vehicle. Keep the wheels off the floor.

☐ **14.** Add the specified type and amount of fluid. Conduct an initial fluid level check. Top off as needed.

☐ **15.** Set the parking brake if it has not already been set. Start the engine and engage each gear for 5 to 10 seconds. Ensure that the transmission engages reverse and drive.

☐ **16.** Select the park (neutral) gear.

☐ **17.** Check the transmission fluid for cold level. Top off as necessary.

☐ **18.** Test drive the vehicle until the engine is at operating temperature.

☐ **19.** Check the fluid level. Is the fluid hot? ________________ Cold? ________________

☐ **20.** Top off as necessary.

Instructor's Response ____

____

____

____

JOB SHEET 40

Name ______________________________ Date ______________

Servicing a Differential

NATEF Correlation

This job sheet addresses the following NATEF tasks:

Drain and refill differential housing.

Objective

Upon completion and review of this job sheet, you should be able to service a differential on an RWD vehicle.

Tools and Materials Needed

Hand tools

Service manual

Torque wrench, 3/8-inch or 1/4-inch drive

Drain pan with screen or magnet or wide flat pan

Air- or hand-operated pump *or* suction gun

Procedures

Task Completed

1. Identify vehicle and specifications: ☐
   Vehicle make__________ Model __________ Engine __________
   Transmission ____________________manual/automatic
   Type/amount of fluid ____________________
   Cover bolt torque ____________________
2. What is your reason for servicing this differential? (diagnosis, symptoms) __________ ☐
   ____
3. Lift the vehicle to a comfortable working height. ☐
4. Position the drain pan under the differential. If possible, use a pan that will capture and hold metal shavings with a screen or magnet or use a wide, flat pan in which fluid can be examined after draining. ☐
5. If the differential has a drain plug, remove it and drain the fluid. ☐
6. Once the fluid is completely drained, install and torque the plug to specifications. Go to step 21. ☐
7. Remove the lower bolts on the cover. Loosen the side bolts starting at the bottom on each side. Do not remove bolts. Fluid may start to drain as the bolts are loosened. ☐
8. If necessary, use a screwdriver to pry the bottom of the cover from the housing. Use caution since fluid will begin to flow as soon as the seal is broken. ☐
9. Allow the fluid to drain. ☐

Task Completed

☐ **10.** With most of the fluid drained, remove the remainder of the bolts and the cover.

☐ **11.** Examine the waste fluid for any metal shavings. There should be none. Inspection results:
____
____

☐ **12.** If shavings are found, further diagnosis of the gears and bearings is required.

☐ **13.** Clean the cover and housing of all sealing material. Clean the inside of the housing as much as possible.

☐ **14.** If a gasket is available, position it on the cover and insert a bolt through the cover.

☐ **15.** Place the gasket and cover on the housing. Start two bolts to hold them in place. Start the remaining bolts and hand tighten all of them.

☐ **16.** Torque the cover bolts to specifications. Go to step 21.

☐ **17.** If a gasket is not used, lay a bead of RTV on the cover. Ensure that the bead is about 1/8-inch thick, evenly applied around the cover edges, and attached to the fluid side of the bolt holes. Let it stand for 10 to 15 minutes.

☐ **18.** Align the cover's bolt holes with the holes in the housing.

☐ **19.** Align the cover with the housing. Ensure that the cover does not slide as the bolts are started.

☐ **20.** Start and hand tighten all bolts. Torque all bolts to specifications.

☐ **21.** Remove the fill plug. It is normally about halfway up on the side of the housing just behind the drive shaft connection.

☐ **22.** Secure the correct fluid and install it through the fill plug. Use the available fluid pump to install the fluid.

☐ **23.** As the fluid nears specified capacity, reduce the fluid flow.

☐ **24.** Cease filling just as soon as the fluid begins to flow from the fill hole.

☐ **25.** Check the fluid level using the first joint of the little finger or a measuring tool. Add or drain fluid as necessary to the correct level.

☐ **26.** Check the cover for leaks.

☐ **27.** If the repair is complete, clean the area of tools and waste fluid and lower the vehicle.

☐ **28.** Complete the repair order and return the vehicle to the owner.

Instructor's Response ____
____
____
____

JOB SHEET

Name ______________________________ **Date** ______________

Replacing a U-Joint

NATEF Correlation

This job sheet addresses the following NATEF tasks:

Diagnose universal joint noise and vibration concerns; perform necessary action.

Objective

Upon completion and review of this job sheet, you should be able to replace a U-joint in a drive shaft assembly.

Tools and Materials Needed

Lift *or* floor jack and stands
U-joint bearing press
Grease gun
Marker
Transmission sealing plug
Drain pan
Vise

Procedures Task Completed

1. Identify the vehicle: ☐
   Vehicle make __________ Model __________ Engine __________
   Transmission type ____________ U-joint bolt torque____________
   Are all of the U-joints being replaced?____________________
   Is there a center bearing on this shaft?____________________
   Is it being replaced?________________________
   If yes, what is the bearing bolt torque?____________________
2. Lift the vehicle to a good working height. ☐
3. Compare the replacement U-joint with the installed joint. ☐
   Are they the same in size of bearing cup? ________________
   Do they have the same locking devices? _________________
4. Mark the rear end of the drive shaft to the differential companion flange. ☐
5. Position a drip pan under the rear of the transmission. ☐
6. If present, remove the center bearing fasteners. Support the shaft. ☐
7. Remove U-joint or flange fasteners at the differential. Support the shaft. ☐
8. Slide the slip joint from the transmission. Set the shaft aside. ☐
9. Install the sealing plug. ☐

Task Completed

☐ **10.** Position the shaft in a vise with the vise jaws near the U-joint being removed.

☐ **11.** Remove all locking clips, if present, from the joint.

☐ **12.** Position the U-joint bearing press.

☐ **13.** Tighten the press until one bearing cup is forced out.

☐ **14.** Remove the press, twist the joint cross sideways, and remove the cross. If it will not clear, position the press in an opposite direction to the first setup. Force the second bearing cup out. Remove the cross.

☐ **15.** Rotate the shaft in the vise to more fully expose the second set of bearing cups.

☐ **16.** Repeat steps 12 through 14.

☐ **17.** Remove the bearing cups from the replacement joint and place them aside.

NOTE: Read steps 18 through 25 before starting the installation.

☐ **18.** Start one cup into one hole of the yoke.

☐ **19.** Insert the cross into the yoke by placing one end of the cross into the seated cup.

☐ **20.** Twist the cross until the opposite end is fitted into the second and opposite hole of the yoke. Hold in place.

☐ **21.** Start a second bearing cup over the exposed cross end and into the yoke.

☐ **22.** Position the cross so the two ends are started slightly into their respective cups and the grease fitting is facing in the correct direction. Hold in place.

WARNING: Do not apply excessive force while pressing the cups. They will require some force, but if it becomes difficult, the bearing cups will have to be removed to check for dropped needle bearing. If a bearing is dropped, excessive force could damage it, the cup, and the yoke.

☐ **23.** Position the bearing press and slowly press the cups into the yoke. Rotate the cross as the cups are pressed.

☐ **24.** At about the halfway point, stop pressing the cups. Check the position of the cross within the cups. If the cross is positioned properly, continue pressing the cups.

☐ **25.** When the cups are in position, move the cross as far as it will go in each direction. Install the locking clips if the cross can be moved by hand. If it will not move, the cups and cross will have to be removed to check for dropped bearings.

☐ **26.** Install the grease fitting.

☐ **27.** Repeat steps 18 through 25 to remove and install the other cups on this joint.

☐ **28.** Repeat steps 8 through 25 to remove and install the other U-joints.

☐ **29.** Grease all U-joints that have all bearing cups installed and locked. The other U-joints will be greased after the drive shaft has been installed on the vehicle.

☐ **30.** Move the shaft to the vehicle.

☐ **31.** If required, replace the center bearing onto the shaft.

Task Completed

32. Remove the sealing plug and slide the slip joint into the transmission. ☐
33. If this is a two-piece drive shaft (with a center bearing), support the shaft as the rear bolts are being installed. ☐
34. Install the bearing cups, if necessary, onto the cross. ☐
35. Position the rear U-joint into the companion flange and start the bolts. Align the marks by turning one rear wheel to rotate the flange. ☐
36. Torque the fasteners. ☐
37. Install the center bearing, if present, and torque the fasteners. ☐
38. If necessary, grease the U-joints that have not been lubricated. ☐
39. If necessary, top off the transmission fluid at this time (manual) or after the vehicle has been lowered (automatic). ☐
40. Secure all tools and clean the area of any oil spots. ☐
41. Lower the vehicle, complete the repair order, and return the vehicle to the customer. ☐

Instructor's Response ____

____

____

____

Chapter 13

Auxiliary System Service

Upon completion and review of this chapter, you should understand and be able to describe:

- The proper method to disarm the air bag system.
- How to inspect and repair the warning light systems.
- How to conduct a performance test of dash-mounted instrument circuits.
- How to diagnose, inspect, and repair or replace the interior lighting.
- How to diagnose and inspect the power windows and door locks.
- How to conduct a performance test of the climate control system.
- How to perform a function test of the antitheft system.

Introduction

Most passenger comfort and convenience features will fail like any other component on the vehicle. Most of the failures will be electrical in nature, either a failure of the wiring or the electrical components. Since many of the accessories on today's vehicles require skills gained from many hours of experience and training, this chapter will cover some of the common tests and repair that could be performed by entry-level technicians.

Disarming an Air Bag

Many of the repairs to the interior of the vehicle present a danger to the technician from the vehicle's supplemental restraint system (SRS). Before attempting any repairs in or around the steering column, the SRS should be disconnected from the battery.

A **memory saver** is a small battery-powered device used to save the vehicle's various memories, such as radio settings.

WARNING: Do not use a memory saver when the battery is disconnected for interior service. Even a small 9-volt battery is sufficient to charge the air bag capacitor. Serious injury could result if the air bag deploys while a person is leaning over or is near the bay mounting site.

Classroom Manual
pages 407, 408, 409, 410

Confirm the customer's complaint that indicates a repair is needed within the passenger compartment. If repairs or testing are deemed necessary in that area, disarm the air bags by first disconnecting the NEGATIVE cable from the battery. Consult the service manual for the required waiting period needed for the capacitor to naturally discharge. Typical waiting time is at least ten minutes, but some manufacturers recommend or require a longer time. Always consult the service manual before assuming that the capacitor is discharged.

Lamp check is a program where certain dash indicator lights will illuminate for only a few seconds so that the technician can check their operation.

SPECIAL TOOLS

Service manual

Vehicle's diagnostic routine

KOEO means ignition key on, engine off, a test condition required to performed certain tests.

KOER means ignition key on, engine running, another test condtion required for certain tests.

CAUTION:
The following tests are for nonelectronic circuits only. Sensor circuits can be determined by observing the number of wires attached to the sensor or sending unit. True sending units will have only one wire, whereas sensors will have two or more wires. Grounding a sensor circuit could damage the computer or the circuit.

In many instances, it will be necessary to have electric power to test components located in or near the air bag areas. There are two ways to do this safely:

1. Use an ohmmeter (a digital multimeter or DMM) to measure resistance of the component and compare it to manufacturer specifications after it has been disarmed and removed from the vehicle.
2. Remove the component(s) from the vehicle and test it at a test bench equipped with the proper power supply and tools.

WARNING: Do not reconnect the battery again until all of the service work is completed, including reassembly of components and their mounting. A quick touch of the cable to the battery will instantly recharge the capacitor. This would supply the power needed to fire the air bag and serious damage and injury could result.

Warning Lights

Many of the warning lights are simple, electrical circuits. A good understanding of their operation will give the technician a head start on basic repairs. Most tests have to be run when the key is on with engine off (KOEO) or key on with the engine running (KOER). You may have to turn the key off and back on several times to read and understand all of the lights. A light that goes on for only a few seconds and then goes off does so because of the program designed to work with it. This is known as **lamp check**. Lamp check is a program in which the bulbs will illuminate for only a few seconds. This is done so that the technician (or driver) can check if the bulb is working or not. If the oil pressure warning light bulb is burnt out, then the driver would never know when or if there is a problem with the oil pressure.

KOEO and **KOER** running are two test conditions that are associated with powertrain control module (PCM) test routines, but they appear in other system test procedures as well.

Older vehicles use a mechanical switch to turn the light on when certain conditions are present. Other vehicles use a sending unit for the warning light and gauge. One basic function occurs on both types: the ground to the warning light is completed.

WARNING: Do not grab hoses or other components under the hood if an overheated engine is suspected. Even an engine that is operating at normal temperature can cause serious burns.

WARNING: Never remove the radiator cap on a warm or hot engine. Always let the engine cool down. Extremely serious burns and other injuries can occur when removing a hot radiator cap.

If the red engine light is lit, the problem indicated is either an overheated engine or low oil pressure. On some vehicles, there is a light for oil and one for overheating (Figure 13-1). Before starting work on the electrical circuit, do the obvious and check the level and condition of the coolant and oil. For a cooling problem, look for missing or damaged fan belts. The coolant level on a hot engine can be checked in the overflow reservoir. The reservoir may not contain a true representation of the coolant level, but it will be an indication. Note

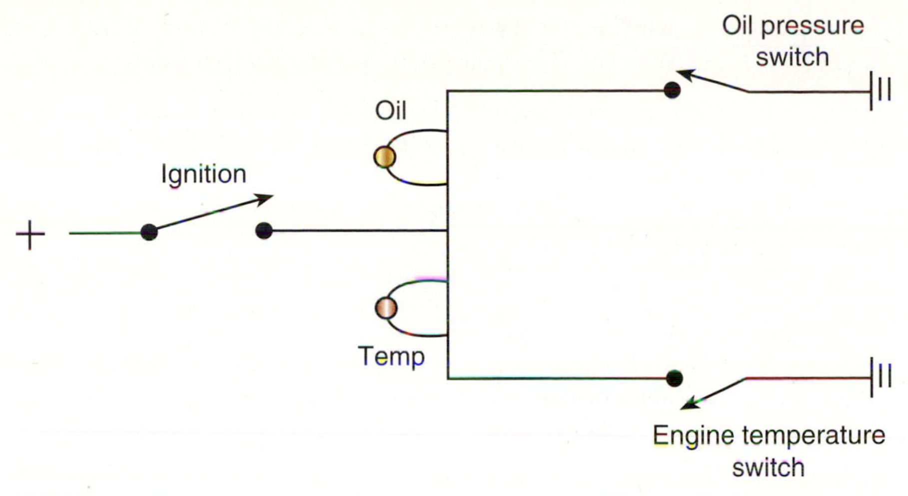

FIGURE 13-1 **Two warning lights require two completely different circuits powered through the ignition switch.**

that on some cooling systems, the reservoir may become pressurized. Notice the condition of the engine by checking for oil or coolant leaks. The hoses should be inspected for swelling. Do not touch hot hoses. In many cases of low oil pressure or overheating, the condition will be found during the inspection.

SERVICE TIP: Before starting the engine, check the repair order or with the customer to determine if there were unusual noises coming from the engine prior to shutdown. Noise is usually a sign of internal damage. Further operation of the engine may not give the technician additional information for diagnosis, but it will probably cause even more damage.

WARNING: Do not perform a KOER test if there are any visible or audible signs that the engine is damaged or could possibly suffer damage when running. Doing so could cause more damage.

WARNING: Never apply 12-volt direct current (VDC) to a sending unit unless directed to do so by manufacturers' test routines. Sending units are almost always on the negative side, may be electronic, and will be damaged by direct 12-VDC, positive current.

Assuming that the problem was not apparent under the hood and the engine operates normally, run the engine just long enough to confirm the customer's complaint and determine if there is more than one warning light for the engine. Before starting the engine, switch the ignition to run and observe the lights. There should be several red lights and possibly an amber light. If all of the red warning lights are lit, the circuit to each light is good.

Do not run the engine any longer than it takes to check the lights. If the warning light stays on during engine idle, speed up the engine to see if the light goes out. As the engine speed increases, the oil pump produces more flow and pressure. If the light goes out, the likely problem is in the oil system. In this case, the oil pump and engine could be worn or sludge is developing in the oil pickup or passageways. Generally, an overheated engine will not cool down enough to turn off the light by simply speeding up the engine. Note the matching gauges, if equipped. When a red light turns on but the gauge reads normal, the problem is usually in the light circuit.

A bulb test or proof-out test of the warning light is done every time the ignition key is switched to ON. Other warning lights are tested while the key is in the START position.

For a vehicle with a shared light for oil pressure and temperature warning, shut down the engine and locate the oil pressure and temperature sending units (Figure 13-2). Perform a KOER test by disconnecting the temperature unit. It usually is the easiest unit to reach because it is on or near the thermostat housing in most vehicles. Start the engine. If

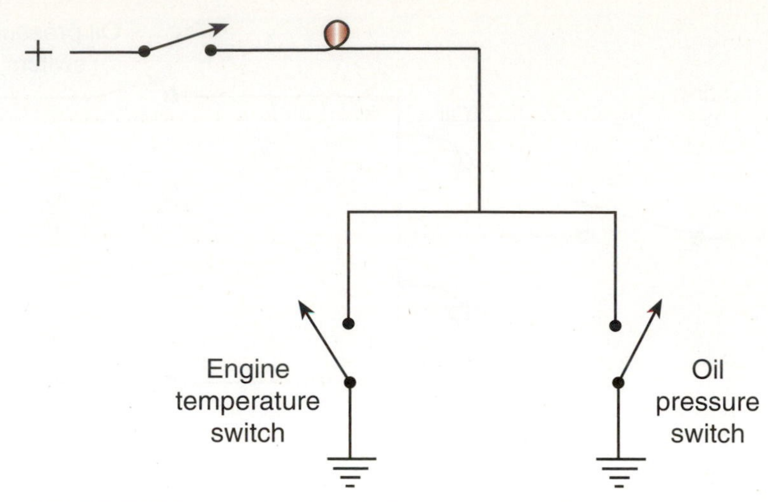

FIGURE 13-2 **One warning light for two systems requires parallel circuits from each sending unit that are powered from the ignition switch.**

the light goes off, the problem is either overheating or circuit trouble. A light that stays on may indicate an oil pressure or circuit problem. Repeat the test with the oil pressure sending unit. With the engine running and both circuits unplugged, the red light should go out. If it does not, there is a problem within the circuit. With the light out, connect one of the two sending units. Now, the appearance of a red light indicates a failure in that system. Lack of a red light indicates that the problem is in the disconnected system.

SPECIAL TOOLS

Service manual

Jumper wire

If there are separate lights for temperature and oil, disconnect the sending unit for the affected circuit and test that circuit only (Figure 13-3).

On a dual light system, check the electrical circuit from the sending unit to the light by disconnecting the wire at the sending unit. Run a KOEO test. Use a jumper to ground the sending unit's wire to a metal part with ignition switch in ON (Figure 13-4). Each time the wire is grounded, the warning light should come on. If the light and circuit function properly, the sending unit is bad. On a temperature warning system, let the engine cool down, if necessary, and replace the sending unit. Do not remove the temperature sending unit on a hot engine. If the lubrication system is being checked, it is possible to manually check the oil pressure before replacing the sending unit. This would also be a direct test of the oil pump output.

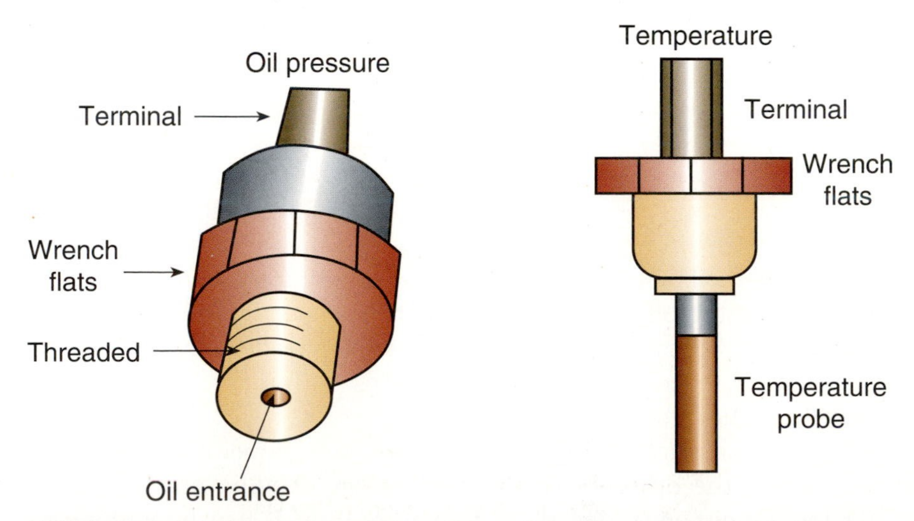

FIGURE 13-3 **The oil pressure sending unit is usually different in appearance and operation from the temperature sending unit or switch.**

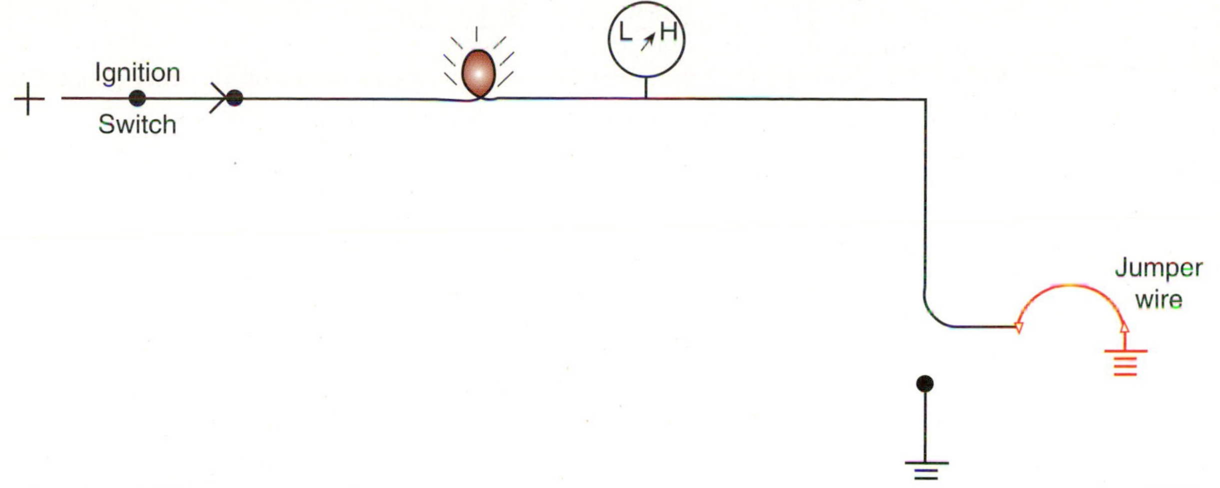

FIGURE 13-4 **This test will ground the circuit and it should turn on the warning light or make the gauge read at its maximum.**

WARNING: Do not start the engine if the oil level is low or the engine is making unusual noises. Serious damage could result if there is not enough lubrication on the bearings and journals.

Test Oil Pressure Warning Circuit

Remove the oil pressure sending unit and install a manual oil pressure gauge (Figure 13-5). Start the engine and record the oil pressure. Consult the service manual for exact specifications. Generally, pressures over 20 psi at idle are acceptable. Shut down the engine immediately if oil pressure does not reach 20 psi within a few seconds of engine start-up. If the

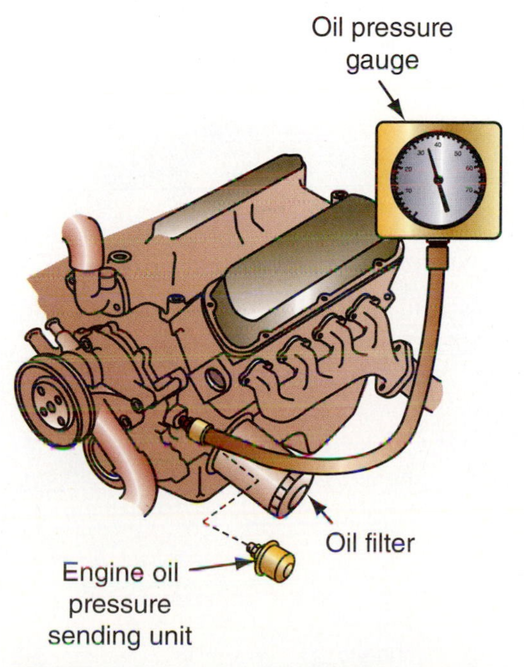

FIGURE 13-5 **The gauge for this test is a dash-mounted aftermarket gauge fitted with a long tube or hose for attachment to the engine's block.**

SPECIAL TOOLS

Manual oil pressure gauge

pressure is at least 20 psi, speed up the engine. There should be an increase in pressure. If the pressure is low or does not increase, the probable cause is a worn oil pump or a worn engine. However, each of these conditions occurs over a period of time, depending on the condition of the engine. A discussion with the operator can verify this. A broken oil pump will suddenly turn on the light, cause almost immediate engine failure, and the condition will be obvious to the technician.

Test Engine Temperature Warning Circuit

CAUTION: Electric cooling fans may turn on at any time when the temperature sending unit is connected. Observe safety procedures by remaining clear of an electrical cooling fan during testing.

If the temperature light or gauge circuit proves correct during the test, the engine's cooling system must be checked. That includes the fan, water pump, thermostat, and coolant condition and level. A detailed diagnosis routine will be set forth in the service manual. This is particularly true of computer-controlled engines. Many computer-controlled systems use the PCM to turn on the warning lights based on input signals from its sensors. Diagnosing computer systems is best left to experienced technicians, but check the basics such as belts, oil and coolant level, leaks, or anything else that may point to a mechanical failure. If the top of the engine shows signs of recent repairs, check to make sure all connectors are, in fact, properly connected and tight.

Test Alternator Warning Circuit

CAUTION: Some electric cooling fans with a temperature switch (not to light or gauge) will turn on when the switch is disconnected with the ignition key on. Observe safety procedures by remaining clear of an electrical cooling fan during testing.

Low voltage or current in the charging circuit will switch on the alternator warning lights. A loose alternator drive belt may slip and not allow the alternator to produce the required voltage or current. Always check the belt and all battery and alternator connections before testing the charging system or the battery. If the inspection does not indicate a mechanical problem, the battery and charging system can be checked using a VAT-40 or VAT-60 as outlined in Chapter 8, "Engine Preventive Maintenance." Do not use a VAT-40 on a computer-controlled alternator.

The voltage or current sensor is usually in the regulator and, depending on the system, the alternator may have to be replaced or rebuilt or an external regulator may need to be replaced (Figure 13-6).

Other warning light systems can be checked in the same manner. Before condemning the electrical part, always check the mechanical portion of the system. Most of the time,

SERVICE TIP: If the alternator has been replaced, it is possible that the wrong alternator pulley has been installed. An oversized pulley will slow down the speed of the alternator, thereby reducing its output. It will be most notable at idle speeds.

FIGURE 13-6 **The voltage regulator on some model vehicles is internal to the generator.**

the warning system is doing what it was designed to do. The most common failure in a warning system is the switch or sending unit. Disconnecting the switch or sending unit should turn off the light.

GAUGES

Classroom Manual
pages 377, 378

WARNING: Do not check electronic circuits using a test light, grounding a conductor, or jumping a connection unless specifically directed to do so by the service manual. Incorrect test equipment and procedures can damage electronic circuits and components.

Before proceeding, determine the type of gauges used on the vehicle. If the instrumentation is electronic and is operated by a computer module, do not use the test procedures suggested here. Most sending units with one wire are not considered to be electronic while two or more wires are generally part of an electronic circuit in some manner. Also consider the customer's complaint. If only one gauge is not working, then the problem is in that system. If all gauges are not working, it is probable that the fault lies in some shared component such as a fuse or common conductor. Always consult the service manual to determine the system's operation and the basic diagnostic test procedures. The tests discussed here are very similar to or the same as the ones for warning lights. Some vehicles may use separate circuits for each but share a common sending unit.

Sending units for analog gauges use a material that changes resistance based on the amount of heat or pressure being applied, or they use some variable resistor (Figure 13-7). Both types can fail at OFF, ON, or any point in between. If there is a one-wire connection at the sensor, grounding the wire in the KOEO mode will cause a maximum gauge reading. A disconnected wire will allow the gauge to drop to its lowest limit. If the gauge works in this test, then the sending unit is bad. A gauge that does not move when the wire is disconnected or grounded (KOEO) indicates a bad circuit or gauge. Consult the service manual for test procedures for the circuit and gauge.

Variable resistor sending units are used for fuel gauges. A float in the tank moves a wiper arm over a very thin resistor wire (Figure 13-8). The position of the arm on the wire determines the amount of resistance between the gauge mechanism and ground. The test for this circuit is similar to the one for one-wire sending units on the engine. Lift the vehicle and locate the connection for the sending unit. It will be near or on the tank. On older vehicles without in-tank electric fuel pumps, there is usually one wire leading to the tank. On systems with in-tank pumps, there may be several electrical connections. Before disconnecting any of them, consult the service manual. Once the sending unit wire is

SPECIAL TOOLS
Service manual
Jumper wire

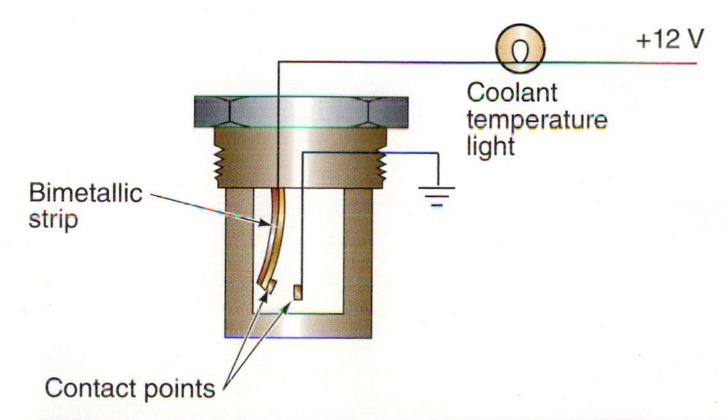

FIGURE 13-7 **A cutaway view of a typical temperature warning light switch.**

SERVICE TIP:
Like in all electric circuits, when nothing works, start with the most common point if none of the circuits work. Common points for gauges and warning lights are fuses, switches, integrated circuits, and the instrument voltage regulator. Also understand that the fuse for the instrument panel light is not the one for the warning lights and gauges.

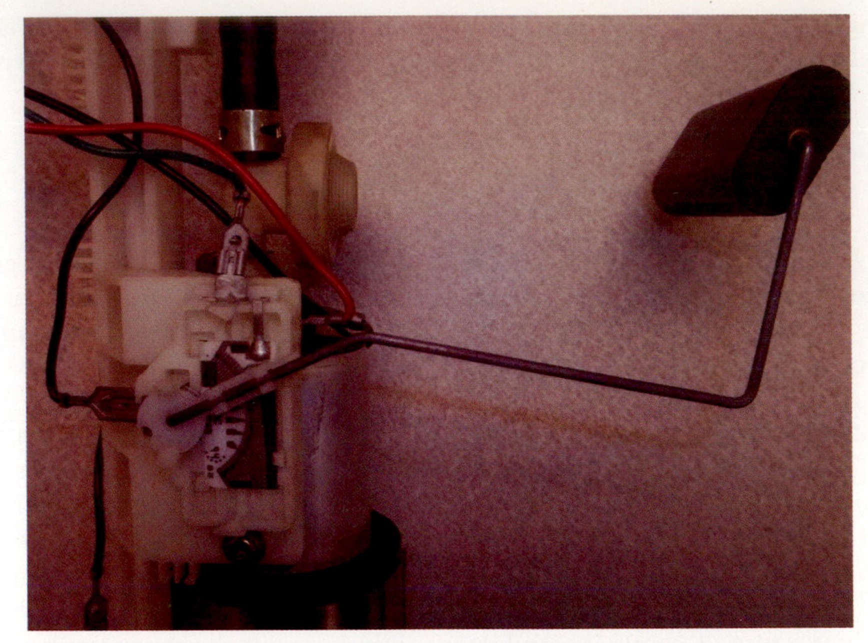

FIGURE 13-8 **The wiper arm sliding across the resistor wire changes the resistance between the signal wire and ground.**

located and disconnected, perform a KOEO test. Turn the ignition key on while the wire is grounded. A good circuit will cause the gauge to read full. Removing the wire from ground should cause the gauge to drop to empty. If either condition is not met, then the circuit or gauge is bad. An operating circuit means the fuel tank has to be dropped and a new sending unit installed.

Here is a word of caution before leaving this section. Most instrument panels use some type of voltage limiter to regulate the amount of voltage being supplied to the various gauges and sometimes to the entire gauge circuits (Figure 13-9). It is normally referred to as an **instrument voltage regulator (IVR)**. Jumping 12 VDC to a gauge or circuit may damage them and cover up the original problem. The tests listed here are simple and cover most initial testing of analog, nonelectronic gauges and circuits. If the test indicates that the circuit or gauge is inoperative, the test procedures to be followed from that point are found in the service manual. Testing of the actual instrument panel components should always be done according to the manufacturer's procedures and performed or supervised by an experienced technician.

An **instrument voltage regulator (IVR)** limits the voltage supplied to the gauges to 12 VDC so the gauges are more accurate.

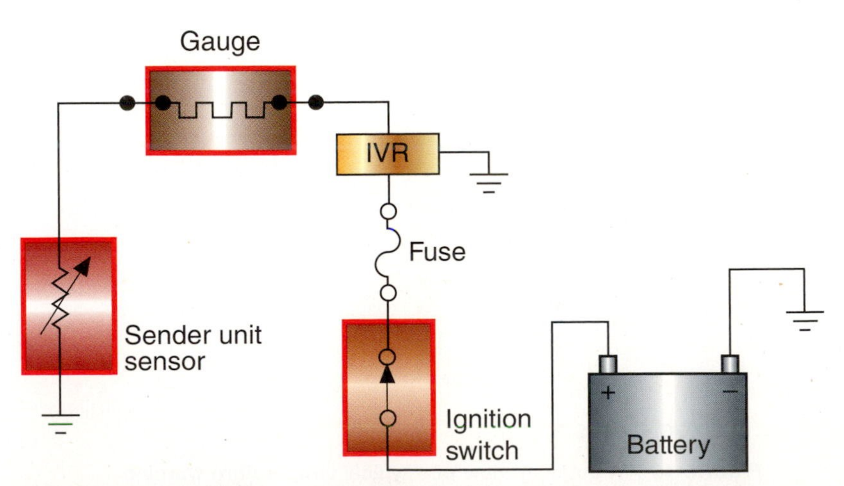

FIGURE 13-9 **Most IVRs can be individually tested and replaced from the instrument panel or any gauge or light circuit.**

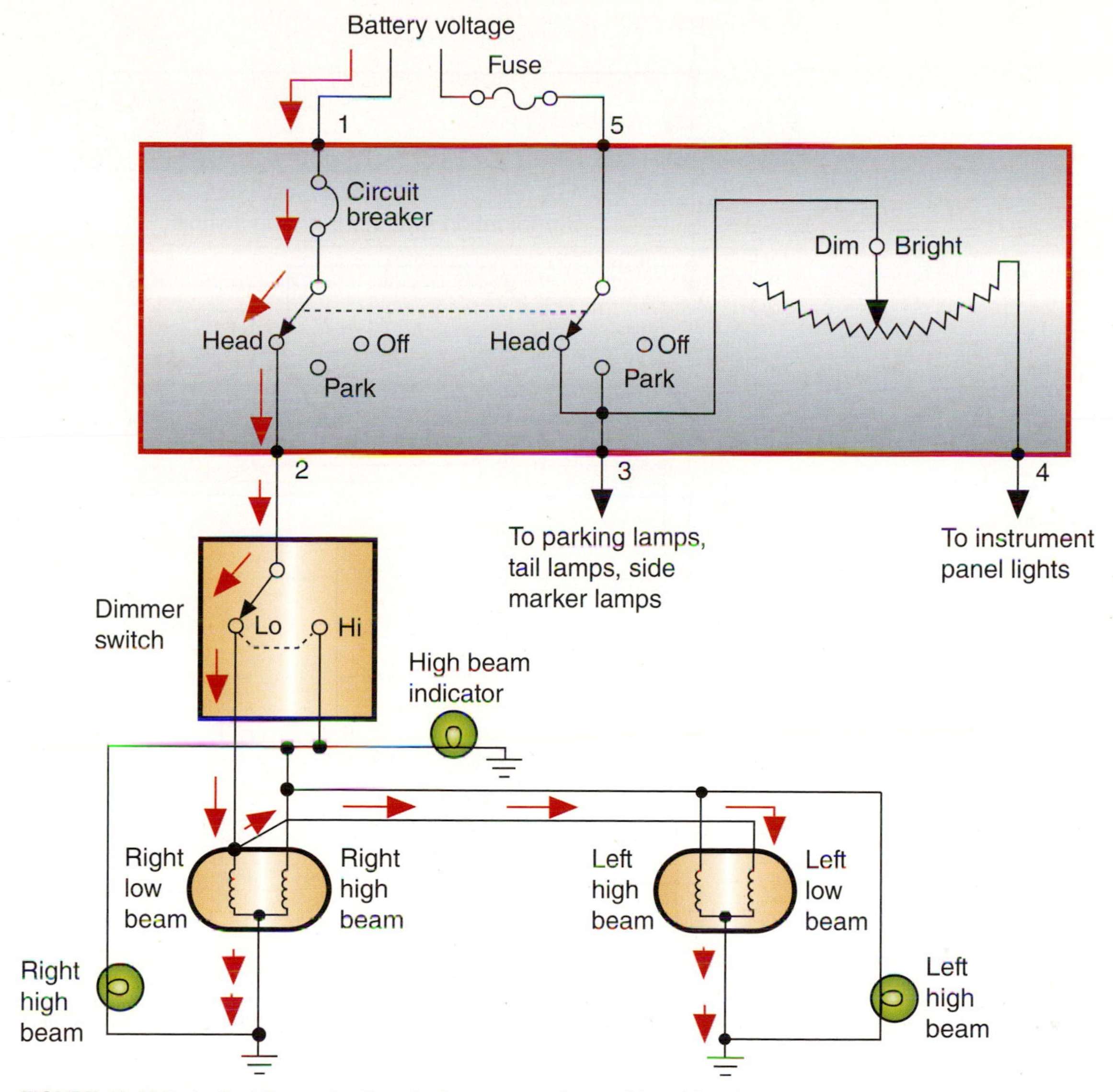

FIGURE 13-10 **Note that the protection devices are on the positive side of this light system.**

Lighting Systems

Automotive lighting systems are very simple electrical systems. They have a positive side and a negative side. Almost all systems are series-parallel circuits with the controls and protection devices located in the positive or hot side (Figure 13-10). After the last control, the circuit parallels out to two or more lamps. The initial diagnosis is easy. If one lamp on a system does not light but the others do, the circuit, controls, and protection devices are operating properly. The usual problem would be a burned out lamp. However, there are times when all the lamps on a system go out, indicating a control or protection device failure. It is possible, but not probable, that all the lamps on a system burn out at once. Even then the root cause is probably somewhere else in the circuit, such as a protection device not opening when the circuit is overloaded with current or there is a circuit defect. Diagnosing the headlight and marking light circuits and the turn signals will be discussed, covering the most probable causes of complete system failure.

Classroom Manual
pages 383, 384, 385, 386, 387

Headlights and Marking Lights

Each of the two systems is turned on by different contacts within the same switch (Figure 13-11). This multiposition switch may be found on the dash or steering column

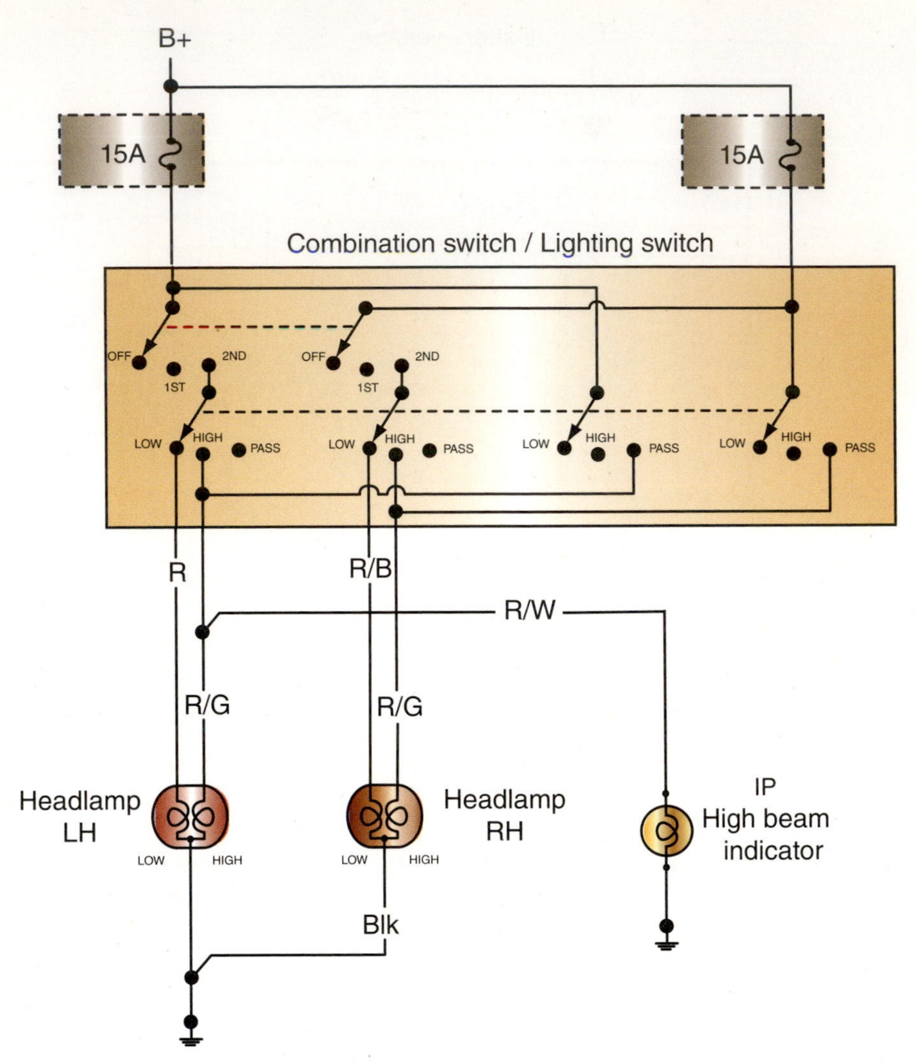

FIGURE 13-11 **The headlight switch has several different contacts that feed parallel circuits.**

SERVICE TIP:
A test light could be used in the following tests, but the amount of voltage will not be registered. It is best to use a multimeter when testing for voltage.

and has OFF, PARK (marker), and headlight positions. The switch is normally connected to one or more relays on the grounded side of the circuit. The relays are electrical/mechanical or electronic switches that connect the battery to the light when the switch is on. A fuse protects power to the marking side while a circuit breaker is normally used on the headlight feed. With all of the marking lights (or daytime running lights) and headlights inoperative, the most probable cause is a bad headlight switch or total loss of power from the battery. However, always check the fuse and circuit breaker before proceeding further.

Remove the switch or disconnect the switch from the harness, whichever is the easiest. If the headlight switch is mounted on the steering column or near the steering wheel, ensure the air bag is disarmed. Locate the two heavy conductors feeding (input power) the switch. They are usually red or orange and will be 10- or 12-gauge wires (Figure 13-12). If the wires are much smaller, then the switch is probably controlling the headlight relay and not the headlights directly. For the purpose of the discussion, we will assume there is no relay. The two feed conductors should register within 1 volt of battery voltage. On some vehicles, it is necessary for the ignition switch to be in the RUN position before the marker lights or headlights or both will have electrical power. If necessary, turn the key to RUN and retest the connections for voltage. This confirms the presence or absence of voltage at this connection. If neither conductor shows the presence of voltage

SPECIAL TOOLS
Multimeter or voltmeter
Service manual
Jumper wire

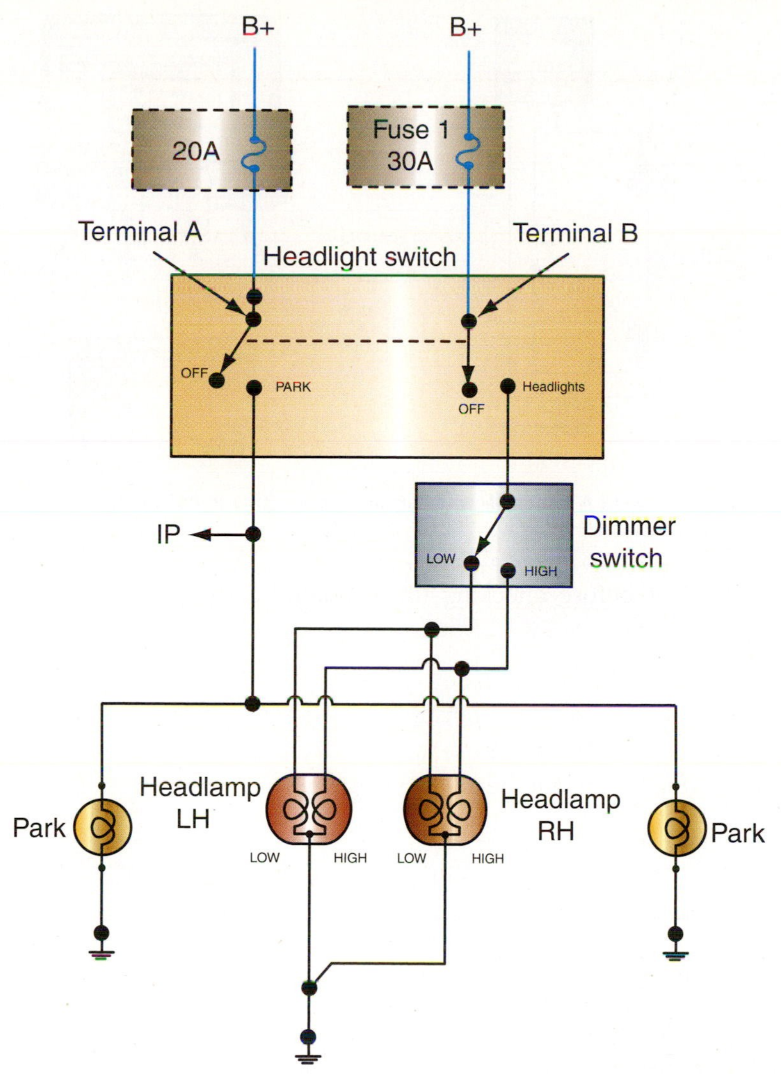

FIGURE 13-12 **The red or orange wire from the fuse(s) is the power feed to the headlight switch. There should be battery voltage (within 1 volt) available at terminals A and B in this illustration.**

or has low voltage and the circuit protection devices are good, then the problem is between the switch and battery. Remember that a weak or dead battery will cause problems with the other vehicle circuits. If only one conductor registers a voltage, then a problem exists in the other. In many cases, failure to investigate thoroughly may not reveal the presence of a relay or other electrical device that may be controlling or protecting the dead circuit. Check the circuit layout carefully using the wiring diagram(s) and component locator. If there are none installed, it is obviously a break between the battery and the switch.

In most cases where only one circuit controlled by a gang switch is inoperative, the switch has failed in one area or one set of contacts. However, the first thing to check is the protection device for the failed circuit. Do not just replace the protection device and return the vehicle to the customer. The device failed for a reason. It could be old age, a common reason on older vehicles, but the circuit should always be checked for damage or excessive load. Replacing the device without checking the circuit may result in customer comeback and loss of profit and time.

Many headlight systems have a relay that routes a large current flow directly from the battery to the headlamps (Figure 13-13). The relay is operated by a set of contacts in the headlight switch. If the relay, normally under the hood, is easier to access, it may be better

A **gang switch** is a device with multiple switches opened or closed by one mechanical lever, arm, or button. A turn signal switch is a gang switch that connects two different contacts to switch on a front and rear marker lamp.

Some vehicles, especially imports, have more than four terminals on the **relay**. Relays of this sort have more than one switch within them.

SERVICE TIP: Most switches located in the passenger compartment are switches on the ground side of the circuit. Many times you will not find a "positive" feed at the switch connection. When studying the wiring diagram, determine the position of the switch in the circuit with regard to the fuse, load, and grounding connection. This will give an indication of whether positive voltage will be present at the switch.

CAUTION:
Many new vehicles actually use computer modules that turn on lights upon receiving a ground (negative) signal from a specific switch. Do not use a standard test light on this type of circuit. A DMM or LED test light is the only acceptable tool to use when checking an electronic circuit. But remember that a LED test light will show the presence of voltage, but not the amount. Damage to the computer and circuit could result from using incorrect test procedures or tools.

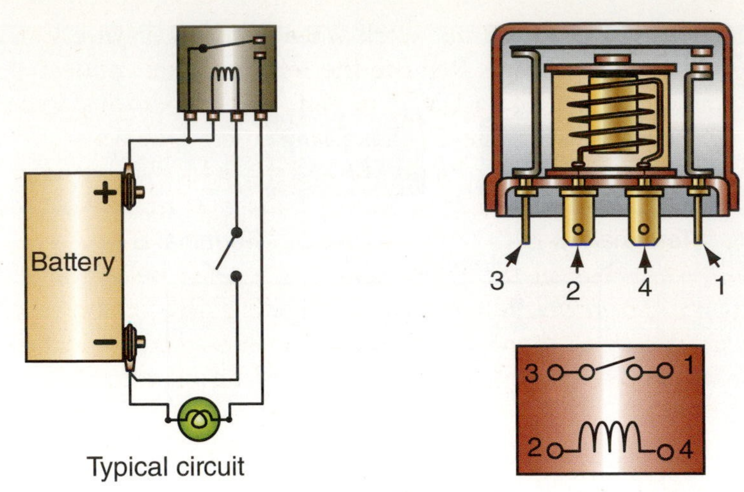

FIGURE 13-13 **A typical 4- to 5-terminal relay. This is not considered to be an electronic relay.**

and quicker to check it before checking the headlight switch. A typical simple relay has four terminals:

- One for battery feed to the relay's coil
- One for battery input feed to the lamps
- One for output feed to the lamps
- One for ground feed from the headlight switch

One of the most common failures in any electrical circuit is a loose ground or a completely missing ground. Before testing the relay, use the wiring diagram to determine which terminal is positive and which is negative. For this exercise, the headlight switch provides the ground for the relay's coil. Once the terminals and conductors have been positively identified, a quick test with a jumper wire can be performed.

Locate the terminal that is connected to the headlight switch. Connect a fused jumper wire from the terminal to a good ground (Figure 13-14). If the system works, the problem is either in the switch or in the conductors between the switch and relay. Since the system(s) worked in this test, it proves the relay coil is working. The next "easier" thing to check will probably be the switch.

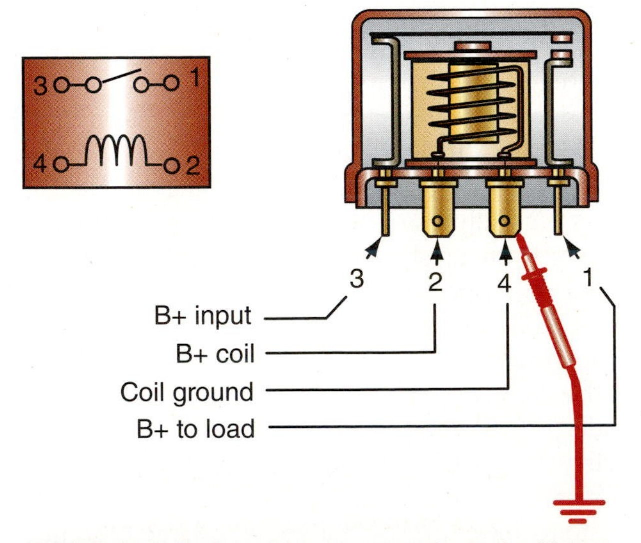

FIGURE 13-14 **Back-probing a jumper wire to the #4 terminal creates a good ground for the relay coil. If the lights function, then the headlight switch is probably bad.**

Assuming that the system(s) did not work when the jumper wire was installed, more extensive testing must be performed. Remove the relay from its connection to expose the connection's internal terminals. Use a DMM to perform the following steps in sequence:

Step 1. Select DC voltage on the DMM. Connect the DMM's black lead to a good ground and probe the two power terminals (#3 and #2) with the red lead (Figure 13-15). The meter should register within 1 volt of battery voltage. If not, repair the circuit to the battery. If voltage is present on both terminals, go to step 2.

Step 2. Select resistance on the DMM and turn on the headlight switch. Leave the black lead at ground and move the red lead 12 (Figure 13-16) to the terminal connected to the headlight switch. There should be continuity between the terminal and the headlight switch with little or no resistance. If not, repair the circuit or switch. If continuity is correct, go to step 3.

SERVICE TIP: A common symptom of a loose ground connection is apparently impossible actions with the various circuits on that ground. A customer may complain that one day this works, the next day that one doesn't but this one does, sometimes nothing works, and sometimes everything works perfectly fine. Look for a loose ground that is common to all the strange-acting circuits or several ground connections in the same area of the vehicle. Failure to recognize the symptoms of a loose ground can be frustrating.

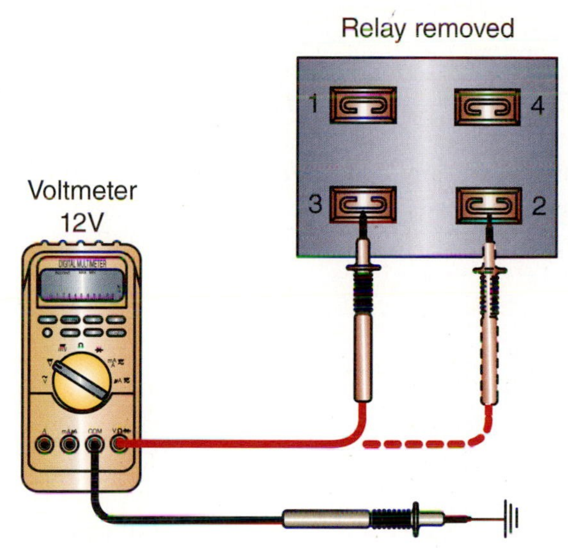

FIGURE 13-15 **Remove the relay for the tests. Use a voltmeter or a DMM set on VDC to measure power at terminals #2 and #3. It should be within 1 volt of battery voltage.**

SPECIAL TOOLS

Jumper wire

Multimeter or voltmeter

Relay removed

Voltmeter 12V

FIGURE 13-16 **Test for continuity between terminal #4 and ground with the headlights switch ON. There should be continuity and little resistance.**

Step 3. Turn off the headlight switch and move the red lead to the output terminal of the headlights (Figure 13-17). The meter should register continuity and a resistance reading dependent on the type and size of the lamps. If an open circuit is indicated or the resistance is very low, then the problem exists either in the circuit or the headlamps. Note that a positive result on each step thus far indicates the relay is probably at fault, but step 4 will definitely confirm the suggested diagnosis.

Step 4. By this time, the problem should be isolated but one final test can be performed if desired. Locate the headlight output terminal (#1) and the power input terminal (#2). Install a fused jumper wire between the two terminals (Figure 13-18). The lights should illuminate. If not, there is a problem between the battery and lights.

The previous tests seem to be more difficult than the ones discussed earlier. However, many times the most difficult part of electrical testing on a vehicle is locating and accessing the conductor or the component being tested. Experience, wiring diagrams, and electrical component locator manuals will show the technician how to conduct the same tests at different and easier-to-access points in the circuit (Figure 13-19). The preceding tests on the headlight circuit can be used to test almost any electrical, but not electronic, system on the vehicle by understanding the circuit's operation and tracking the circuit and components back to the voltage source.

Two tools can assist the technician in testing electrical circuits. One is known as a short finder. It is connected to the circuit and a power source. The short finder indicates the presence of a magnetic field. When electric current flows through a conductor, a

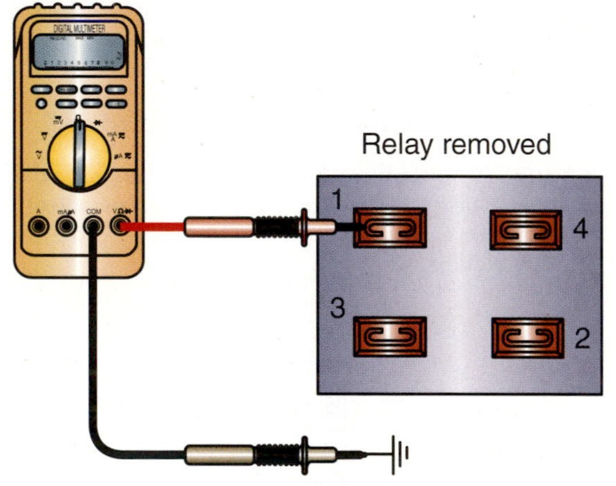

FIGURE 13-17 **Test for continuity between terminal #1 and ground. There should be continuity and a resistance value dependent on the size of the lamps.**

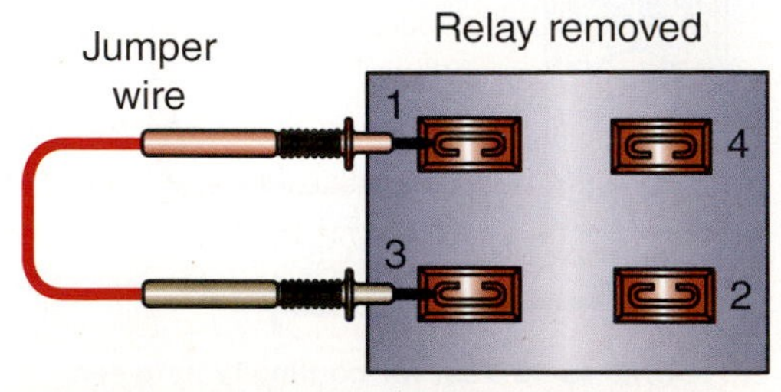

FIGURE 13-18 **Placing a fused jumper wire between terminals #1 and #3 should activate the headlights. If not, there was an error at the initial diagnosis or during the test procedures.**

SERVICE TIP:
Many relays have a small diagram of their internal wiring printed on the outside of the relay. Use the diagram to determine exactly which terminal connects to the relay coil and internal switch. On relays with more than five terminals, it will be necessary to apply extra diligence to the wiring diagrams before testing.

SERVICE TIP:
The power terminals on most four-terminal relays are #30 and #86 with #85 and #87 as the output and ground terminals, respectively. In the illustration, #2 and #3 are the power with #1 and #4 as outputs.

SERVICE TIP:
If the fuse in the jumper wire opens (blows), there is a short-to-ground somewhere in the circuit being tested. Some probable locations for short-to-ground in the light circuits are under the battery tray, in and around moving components, at the connections, or at the place where the harness is routed through the bulkhead.

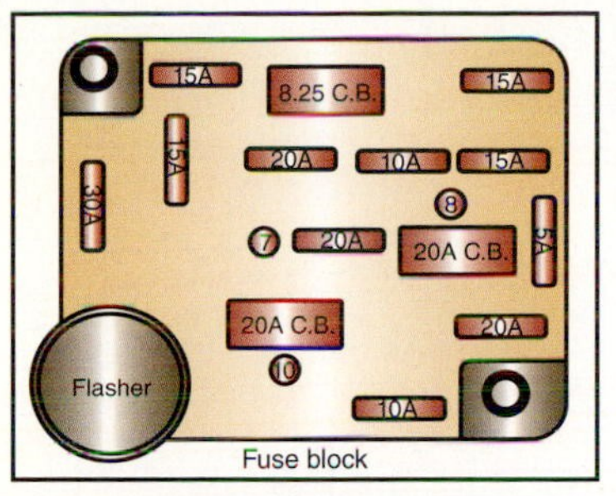

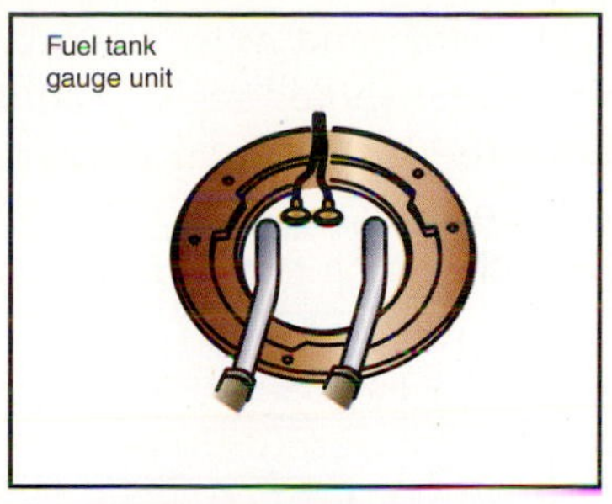

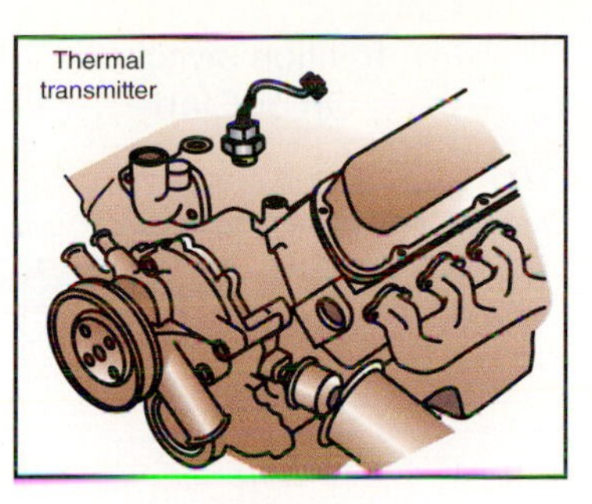

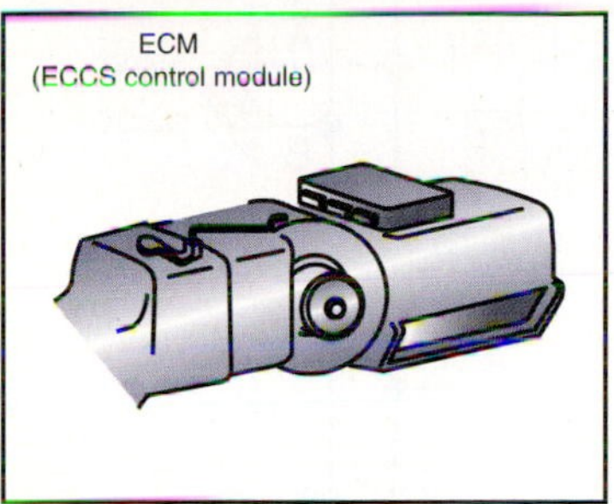

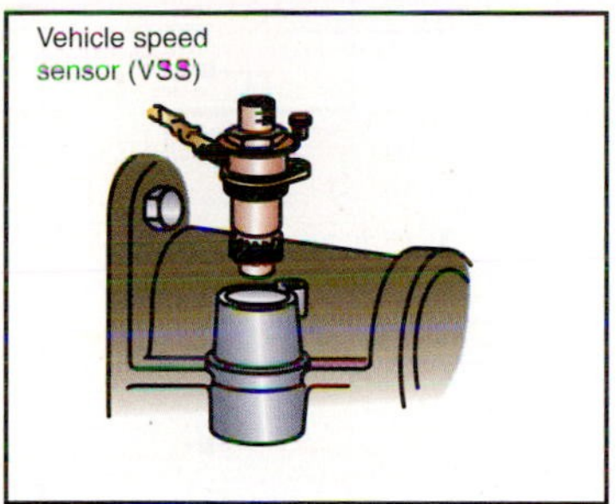

FIGURE 13-19 **Component locators show wiring routing and components within a circuit. This can help make it easier to find accessible test points.**

SERVICE TIP: A quick check used by many experienced technicians is to locate a working relay exactly like the one under testing. Use the known good relay to replace the "bad" one and switch on the circuit. If it works now, then the relay is bad. However, don't use this method until you are absolutely certain of which relays can be used for this swap test.

magnetic field is established around that conductor. A needle on the face will move from side to side as the finder is moved over a conductor. When the needle stops moving, there is no current in the conductor at that point.

A second tool is designed for testing relays. The tool has a female plug for connection to the suspected relay. A short **pigtail** is plugged into the connection where the relay is usually inserted. Operating the tester and observing the tester's meter and the relay operation can pinpoint the electrical or mechanical problem.

SPECIAL TOOLS

Jumper wire

Multimeter or voltmeter

Turn Signals

The turn signal circuits work like most other lighting systems except that they use a fast-acting circuit breaker called a flasher to blink the lamps. Testing of this system is done in a manner similar to other lighting systems.

When the turn signal switch is moved to either the left or right turn position, current flows from a fuse through the flasher and switch to a front and rear lamp. When one lamp burns out, the flasher will do one of three things: blink the other lamp faster, blink the other lamp slower, or turn on the other lamp continuously. Replacing the nonoperative lamp will usually repair the system.

A second problem may be indicated by signals working on one side that do not turn on or do not blink on the other side. Because of the design of the signal switch, one side of the turn signal system may work well, but the other side may not receive any power (Figure 13-20). Another condition that could cause only one side to work is the installation of the wrong lamp or excessive or extremely low resistance in the nonworking side. Resistance determines the amount of current needed for operation, and the proper amount of current is needed to make the flasher open and close.

SERVICE TIP: The most common cause of brake, turn, and marking light failure on a trailer is its grounding wire. If all trailer lights blink or become very bright when the turn signal or brake is used, suspect the trailer's master ground. Usually, it is a white wire connected to the trailer frame near the hitch. The problem may also affect some or all of the towing vehicle lights.

The easiest method of checking a turn signal with one inoperative side is to try the hazard warning system. If all the lights work, the circuits are good and the turn signal switch is probably bad. Sometimes, replacing the turn signal flasher with a new one will make the system work, but the flasher will not be the cause of the failure. A new flasher will temporarily make the circuit a little better and allow the current to flow. Corrosion or looseness of the switch contacts would be the most likely cause, and the switch must be replaced.

When the turn signals, marking lights, or brakes work incorrectly, one thing to check is the addition of a trailer electrical connector on the vehicle (Figure 13-21). When additional

A **pigtail** is a replacement connector with a short length of the wiring for connection to the vehicle wiring.

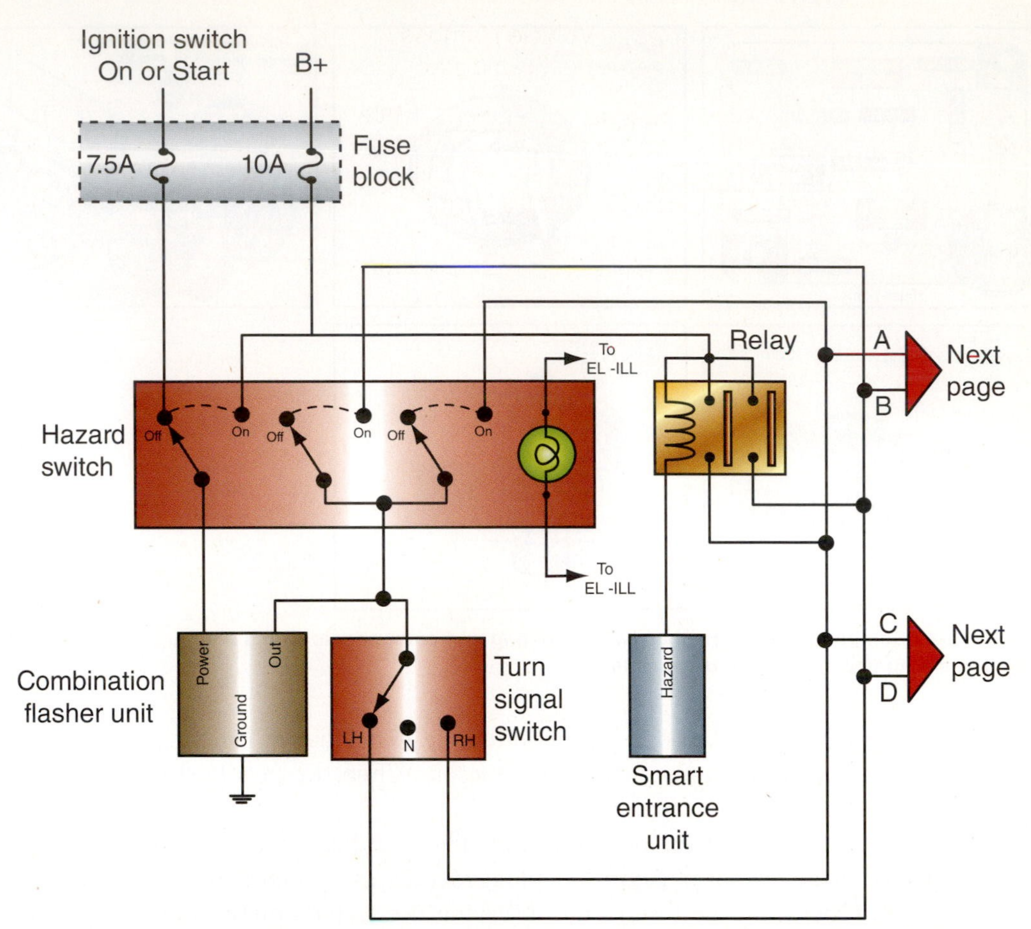

FIGURE 13-20 **The operation of the turn signal switch sometimes results in excessive wear on one set of contacts, causing failure of that circuit.**

FIGURE 13-21 **A trailer's electrical connection can be easily damaged if it is left to hang loose, thereby causing shorts and grounds within the circuits.**

lamps are added to the vehicle's lighting system, a heavy-duty flasher is required. Also, the connection between the trailer harness and the vehicle wiring could be faulty, thereby adding resistance to the circuits. A proper T-connection is shown in Figure 13-22.

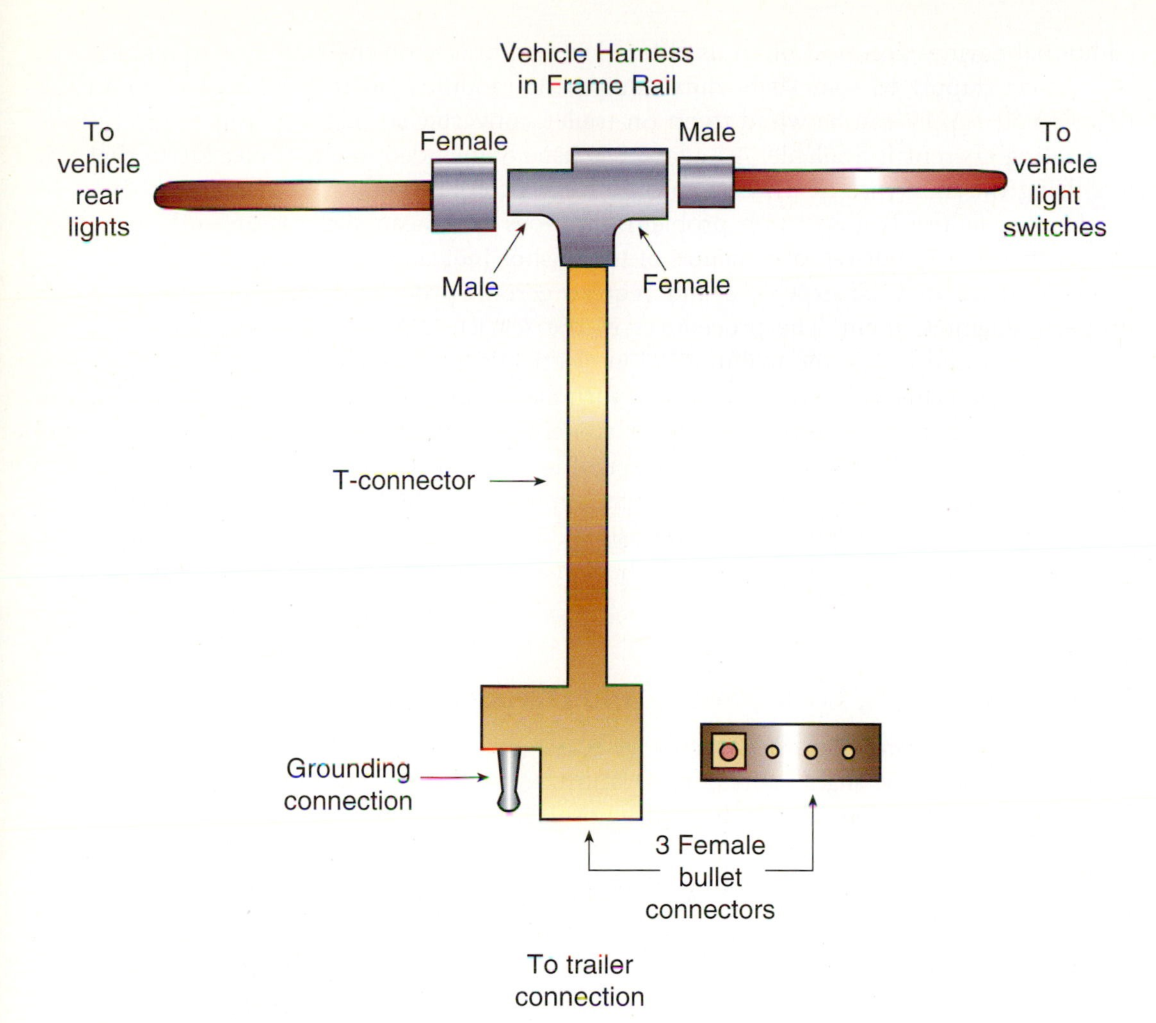

FIGURE 13-22 **Vehicle or trailer light connections should use a T-type harness, which will provide good contacts and weatherproofing. It is usually connected into the vehicle's harness just forward of one of the rear lights.**

Newer trucks and some passenger vehicles may have a trailer kit installed by the dealer or other vendor. This kit has the usual mechanical trailer hitch, but the electrical connection may be different. The vehicle side of the trailer connection can be round and have more cavities than the old four-wire connector commonly used (Figure 13-23). The

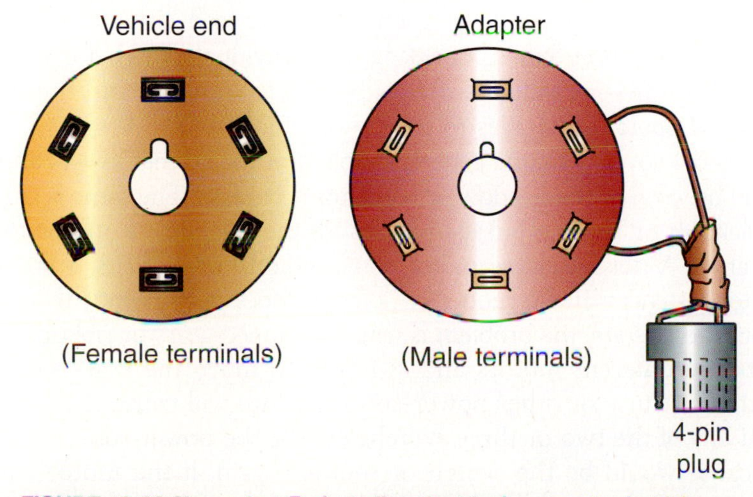

FIGURE 13-23 **Shown is a 7-pin trailer electrical connector (vehicle side) with an adapter that steps it down to the old flat 4-pin connector still being used on most light trailers without electric trailer brakes.**

additional cavities are most often used for electrical brakes on the trailer or to supply a 12-volt power supply to some light-duty equipment mounted on the trailer. In some cases, this 12-volt supply can be wired to an on-trailer converter so that low-amperage, 120-volt alternating current is available. An adapter is usually included in the trailer kit so that trailers using the four-wire electrical connection can be connected.

Flasher failure is a common problem and does not mean there is any other problem within the circuit, but repeated failure of the flasher indicates a circuit problem. New electronic flashers may be expensive and require certain procedures and test equipment to properly diagnose them. The procedures in the service manual must be followed to successfully test and isolate the malfunction in electronic circuits.

Classroom Manual
pages 393, 394, 395, 396, 397

Like all lighting systems, failure of all turn signal lamps at one time points to a breakdown of the protection device, switch, or some other common component of the system. Most turn signals will not work unless the ignition switch is in the RUN position. Four-way or hazard circuits can be checked in the same manner as the turn signals. The hazard light circuit will work in any ignition switch position. A different switch and a heavy-duty flasher are used for hazard warning lights, but the same conductors and lamps are used after the switch.

Power Windows, Locks, Seats, and Defogger Systems

Service of the power windows, door locks, seats, and defogger consists of checking operation, fuses, and some slight lubrication. Almost any repair to these systems requires disassembly of the inner door panels or seat removal. In this section, preliminary checks will be discussed.

SERVICE TIP: Most defoggers have a timer that will shut down the system after a specific time. If a customer complains of poor defogger action, operate the system to determine if the system operates for the time limit specified by the service manual. Educate the customer concerning the system's operation if the system does operate properly.

The first step in troubleshooting is to check all operations and determine answers to the following questions:

1. Are all the devices inoperative?
2. Do some of the devices work with one switch but not with the others?
3. Do some or all of the devices work slowly or create noise during operation?
4. Do the noisy devices attempt to move the window, lock, or seat?

Remember from the Classroom Manual that switches for the power windows and door locks are wired parallel to the master switch with a common protection device. If the answer to question 1 is yes, the obvious problem is in the power feed to the master switch. Should the answer to question 1 be no, then the problem is not the power supply.

An example of the situation in question 2 is the operation of the power windows. If the master switch lowers all of the windows, then the power feed, the motors, and the conductors are good. The only thing left to check is the individual switch for each window. The situation is reversed if the individual switches work the windows and the master does not. The same theory of operation applies to the power door locks.

Questions 3 and 4 cover the same general problem. A motor that lowers and raises the window slowly could be worn or the window or regulator is binding. The same is true of power door locks. Worn solenoids or binding linkage may cause the solenoid or motor to hum, hang, or refuse to work at all. A noisy window motor could indicate worn gears, a broken ribbon strip, or loose regulator fasteners. If the motor or solenoid does not attempt to move the glass or lock but continues to operate, the problem is usually stripped gears or ribbon.

Power seats use different motors and switches to move the seat, but a common fuse protects everything. A fuse or other power-source failure will cause the failure of all operational modes. If any of the two or three switches work, the power source and fuse are good. The probable cause would be the switch or motor. Again, if the motor can be heard but the seat component does not move, the problem will probably be found in the cables or screwnut.

Climate Control Systems

Normally, an entry-level technician will not perform full repairs on climate control systems, but he or she could perform basic, troubleshooting procedures and possibly correct a minor problem. Some shops will not allow noncertified technicians to do any air-conditioning (A/C) work. Since this system relies on the transfer of heat and the movement of air, many times the malfunction may be traced to a switch or component. The repair may not require the opening of the A/C system, which must be done by an Environmental Protection Agency (EPA)-certified technician. Some common complaints of manual heating and A/C systems that do not require the opening of the A/C system and a full systems performance test will be discussed.

Classroom Manual
pages 398, 399, 400, 401, 402, 403

SPECIAL TOOLS

Jumper wire

Multimeter or voltmeter

No Air from Vents at the Halt

The most common problem here is a failed blower motor or motor circuit. If the customer states that air comes from the vents while driving but not at the halt, the blower motor is inoperative for some reason. Airflow during driving results from air entering the intake just below and forward of the windshield. Any time the vehicle stops or slows, there is no intake. Perform the following test on the blower motor circuit.

WARNING: Do not use a test light on automatic climate control or electronic systems. The electronic circuits can be damaged using test lights. Consult the service manual for proper test procedures and equipment.

WARNING: Do not use a test light on blower motors that are on automatic climate control systems unless directed to do so by the service manual. The motor may not be considered electronic, but its controls will be, and most do not have a resistor block for motor speed control. Damage to the circuit can result if test procedures are violated.

As usual, check the fuse first. If it is good, locate the feed wire to the blower motor and check for voltage with a multimeter or voltmeter. With the meter connected between the feed and a good ground, turn the ignition switch to ON and place the blower control switch in each position (Figure 13-24). If there is voltage in one or more speeds but not on all, check the service manual to locate any resistors. Most vehicles have a **resistor block** between the switch and the motor. The resistor block controls the amount of voltage reaching the motor, thus adjusting its speed (Figure 13-25). At the resistor block connection, use a wiring diagram to determine which of the three or four wires are coming from the switch. Test for battery voltage in each switch position. Voltage on each or some of the conductors indicates that the resistor block is open. Voltage on none of them indicates there is a problem within the switch. The switch and block are not repairable and must be replaced individually.

A **resistor block** is a series resistor with multiple inputs and a single output. Usually used to control speed on an electric motor.

No voltage at the resistor block output or at the motor indicates an open in the circuit. Some manufacturers use a **thermal limiter** in the resistor block. The limiter acts like a circuit breaker and opens because of excessive current or heat. The thermal limiter is usually part of the resistor block and the block has to be replaced.

A **thermal limiter** is a special circuit protection device that opens the circuit when overheated.

If there is voltage at the motor in all speeds, check the motor's ground before replacing it. Connect the meter between the feed wire and the ground wire on the motor. Move the blower switch to high and ignition to ON. A battery voltage reading indicates that the power and ground are good and the motor must be replaced. No voltage at this point indicates the motor's ground is bad. Repair the ground wire or connection.

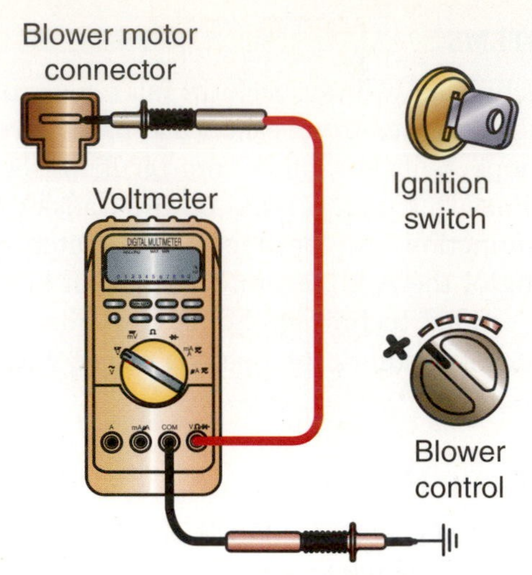

FIGURE 13-24 **This motor has positive power even though the ignition key is on and the blower switch is off. The circuit uses a grounding switch that is typical of those found in electrical circuits.**

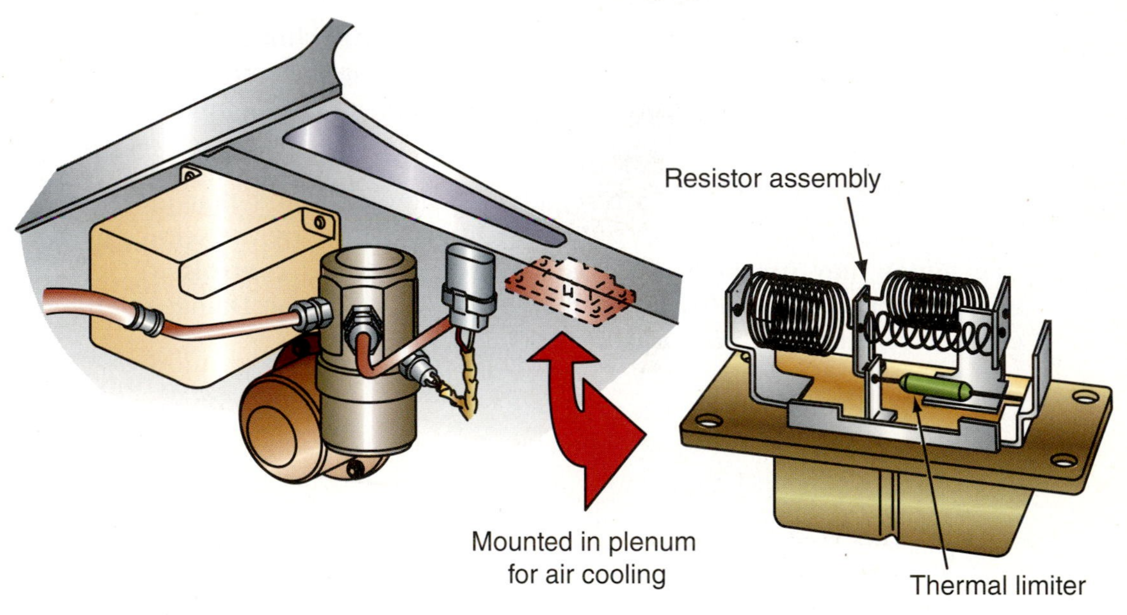

FIGURE 13-25 **Typical location of a resistor for manual heating and A/C systems.**

Automatic climate control systems use small, electric motors to set any or all of the various **blend doors** to almost any position needed to raise or lower the passenger compartment temperature.

Air Comes from the Vent Only

The air is directed to selected vents by moving a lever to different modes: floor, panel, or defrost. The lever is attached to a vacuum switch or connected to the **blend doors** via cable. Before removing the climate control panel, consult the service manual to determine the type of controls. If vacuum operated, start your diagnosis by checking under the hood along the firewall. There should be one or two small vacuum hoses entering the passenger compartment. Locate the nearest vacuum connection in each hose found and check for vacuum with the engine running. It is best to use a vacuum gauge, but placing a finger over the hose end on the engine side will be a quick check. Vacuum will be felt if it is present. If there is no vacuum, follow the hose to its connection with the engine's intake and then repair the problem. Observe the vacuum hose for any check valves that may be used (Figure 13-26). Many times, a loss of heater or air controls within the passenger compartment is a result of rotten or disconnected vacuum hoses or bad check valves.

SPECIAL TOOL
Vacuum gauge

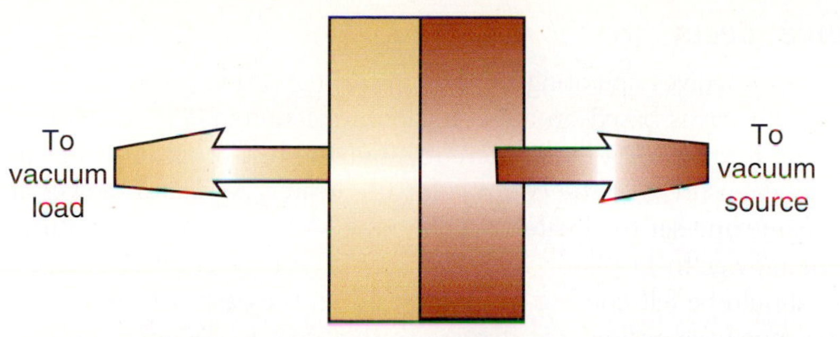

FIGURE 13-26 **Vacuum check valves are used to maintain system vacuum when the engine is under load or shut down.**

CAUTION: Do not remove the radiator cap from a warm or hot engine. Extreme burns could be caused by pressurized, hot coolant. Always allow the engine to cool before opening the cooling system.

Assuming vacuum is present on the hoses or the system is cable operated, it may be necessary to remove a portion of the dash or dash-mounted components to correct the problem. It is best that an experienced technician either performs the repairs or provides supervision.

Heater Does Not Work, A/C Is Good

In this instance, always check the coolant level and engine temperature. The coolant level on most vehicles can be checked through the translucent overflow reservoir. An engine that is still cool or cold after a mile or two of driving indicates a stuck-open or missing thermostat. Low coolant level will cause poor flow through the heater core, whereas an open thermostat will not allow the engine to warm. Both can cause heater problems.

Some older vehicles have a water control valve or heater valve located in the engine compartment. The valve is placed in the hose that delivers hot coolant to the heater core. The valve will be mechanically operated by a cable or vacuum motor. To check a cable control, have coworker move the TEMPERATURE lever from hot to cold and back to hot. The cable and valve lever should move accordingly. If not, the cable or the control lever must be replaced or repaired.

CAUTION: Perform the following test with a cold or warm engine. Some valves are difficult to see easily and may be difficult to reach. A hot engine presents a probability of being burned. It will not make any difference if the engine is not hot because the valve is mechanically operated, not heat controlled.

A vacuum-controlled valve is checked with the engine operating. Again have the coworker move the TEMPERATURE control from hot to cold and back to hot. The valve lever and the arm of the vacuum motor should move accordingly. If they do not, the problem may be in the vacuum supply, temperature control, or the vacuum motor on the valve.

Disconnect the vacuum hose from the vacuum motor and place a finger over it or connect a vacuum gauge. With the engine operating, there should be vacuum in one direction and none in the other (e.g., 18 inches of vacuum in hot, zero in cold). If no vacuum is present in either position, use the procedures to check the vacuum supply outlined in the previous section on vent diagnosis. The entire heater control unit or head is usually replaced as a unit.

Heater Works, but No A/C

WARNING: Do not open the A/C system before recovering the refrigerant using EPA equipment. Releasing refrigerant to the atmosphere could result in heavy fines or injury or damage to the vehicle. Only EPA-certified technicians are allowed to recover refrigerant.

Most repairs on the A/C system are best left to EPA-certified and experienced technicians. However, there are a few checks that can be made quickly and safely by a new technician under the supervision of a certified person. Performing the following performance tests will give the technician a better idea of what is happening within the system.

SERVICE TIP:
Constant or full-flow heater cores do not have the water control valve. The cores on these heaters receive coolant at all times the engine is operating. If there is no heat available, the problem is probably in the blend doors, their control circuits, or the engine coolant system.

CAUTION:
Use eye and face protection while recovering or recharging A/C refrigerant. Refrigerant can cause eye and flesh injuries by freezing.

SPECIAL TOOLS
Service manual
Belt tension gauge

Performance Tests

A performance test involves operating the system through every mode. The following tests are set up using certain terms based on a specified control system. The system being tested may have controls that are labeled a little differently, but the test procedures are the same. Before starting the engine, ensure that the controls for the heater, A/C, and blower are set to OFF.

Start the engine and set the heater–A/C mode to VENT, the temperature to COOL, and then switch the blower to LOW or 1 (Figure 13-27). Air of about the same temperature as the outside air should be felt coming from the vents in the dash. Check each vent for airflow. If one vent does not have airflow, the ducting to that vent is disconnected or missing. Switch the blower to the other speeds, one at a time, and feel for an increase in airflow at each higher speed. Failure to obtain airflow in any speed normally indicates there is a problem with the blower or blower switch. Loss of flow at one or more speeds, but not all, indicates a problem with the blower switch. Leave the blower on HIGH for the next tests.

Move the mode lever (knob) to FLOOR and feel for airflow at the floor (Figure 13-28). Move the lever (knob or button) to DEFROST and feel for airflow at the bottom of the

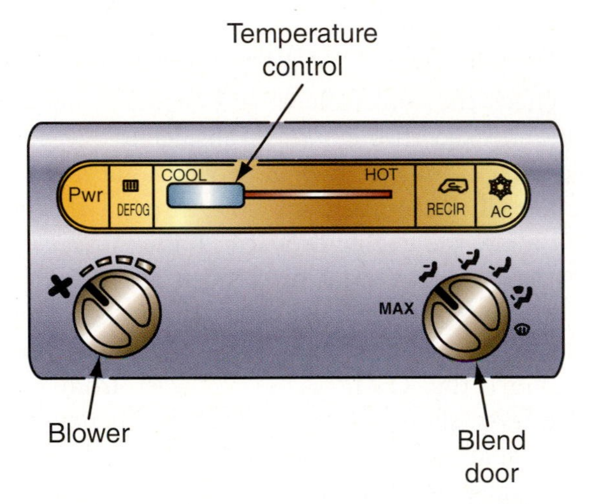

FIGURE 13-27 **Set the temperature to cool (cold), blower to LOWEST speed, and the mode control to dash or instrument panel. Note international pictorial symbols. COOL will usually be colored blue.**

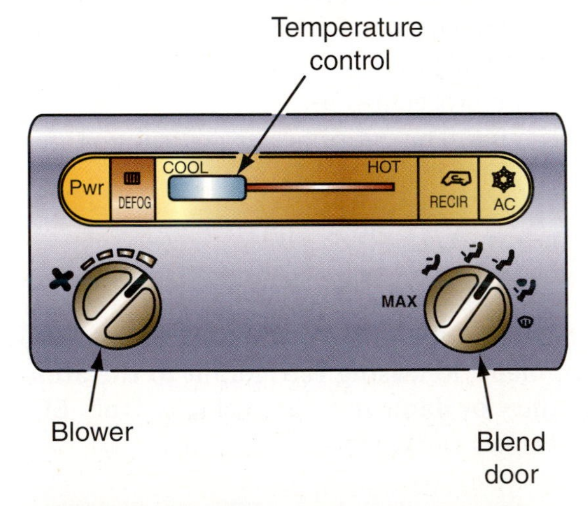

FIGURE 13-28 **Set the temperature to mid-range, mode to HI/LO, and blower to mid-range or high speed for this test. Note that the AC and defrost buttons are shown just to highlight the possible differences between vehicles.**

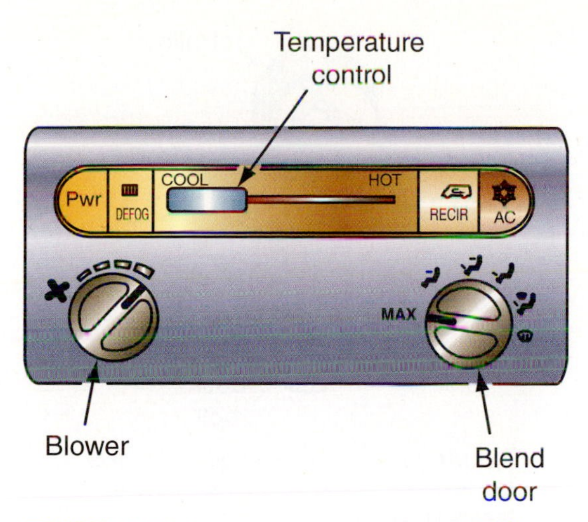

FIGURE 13-29 Set the temperature to cool (mid-range may be selected for checking proper hot/cold air mix), switch on the AC by using the AC button. Note the MAX AC setting on the blend door. This actually causes the incoming air to be drawn from within the passenger compartment instead of outside air.

windshield. If there is a mode setting for HI/LO or mixed, move the lever to that position. The airflow should come from the floor and the vent or from the floor and the defrost. Failure to receive airflow at one of the modes indicates that the controlling cables or lever (knob) is broken or the blend door is broken. Move the mode to VENT for the following test. Keep the blower on HIGH.

Set the temperature lever (knob) to HOT. Within a few seconds, there should be warm or hot air (depending on engine temperature) from the vents. If not, then the heater control valve or control mechanism is damaged. Move the temperature lever (knob) back to COOL and leave the blower on HIGH.

Select the MAX position with the A/C controls (Figure 13-29). The sound of the blower should increase immediately and, within a few seconds, the A/C compressor should switch on, sometimes noticeable by a loud under-the-hood click. A few seconds later, cool air should come from the vents. Switch the mode to NORMAL. The only noticeable difference will be less noise from the blower. Switch to DEFROST mode. There should be cold air at the bottom of the windshield. Failure to obtain cold air after a few minutes indicates that the A/C system is not functioning properly.

The following are some common faults found with climate control systems. In most cases, the suggested cause can occur on manual or automatic temperature control (ATC) systems. However, the typical sensors, actuator, and program commands in an ATC control module can cause what appears to be a mechanical problem. Experienced technicians should perform the final diagnosis and test of ATC systems.

Mechanical Diagnosing

Loose or worn engine drive belts tend to slip and produce a squeal when placed under load. The A/C compressor is one of the heaviest belt-driven loads. The A/C drive belt may drive other components such as the alternator, fan, and water pump. A seized compressor will cause the belt to slip or shed apart.

Check the tension of the belt by pressing down on it about halfway across its longest span or by using a belt tensioner tool. The belt should not deflect more than 1/2 inch on most systems. An oily or worn belt may slip even if the tension is correct. The belt must be replaced if it is oily or worn. The belt is replaced following similar procedures

SERVICE TIP:
Most of the newest vehicles do not use cables or vacuum to operate the blend doors. Small electric stepper motors are used to move the doors and are controlled by a computer. Units of this type require an experienced technician with good electronic expertise to perform the diagnosis and repair.

SPECIAL TOOLS
Belt tension gauge
Service manual

SERVICE TIP:
Although most vehicles now use a serpentine belt to drive most accessory pulleys, some still require a separate belt to drive the A/C compressor. Most serpentine belts do not become loose unless there is a failure with the automatic tensioner; however, they may break from old age, excessive stress, or abuse.

SPECIAL TOOLS

Small pry bar

1/2-inch breaker bar

Belt tension gauge

Classroom Manual
page 394

The **panic** mode button is used to set off the vehicle's alarm in case of personal danger.

SPECIAL TOOL

Service manual

SERVICE TIP: When the panic button is activated or the alarm is set off, most systems require the door to be unlocked with the key and the engine started before the alarm will shut down. Some aftermarket systems have a separate disarming procedure that can be difficult to determine if the operator manual for the system is not available.

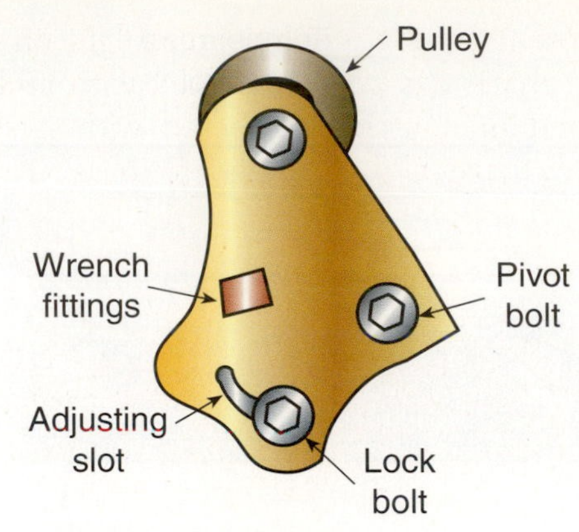

FIGURE 13-30 **A 1/2-inch-square breaker bar is used to make belt tension adjustments in this unit.**

used on other drive belts. The most common problem with replacing A/C compressor drive belts is locating and accessing the fasteners. There may be up to four or more fasteners.

Loosen the pivot adjustment fasteners. Before forcing the component to move, check for a square, 1/2-inch hole somewhere at the front. The hole is used for inserting the drive end of 1/2-inch-drive breaker bar (Figure 13-30). If there is no hole, then the component must be forced to move with a small pry bar because the belt tension cannot be set using hand and arm power alone. Use extreme caution in the placement of the bar. One end of the bar must be placed on the engine block without catching any wires or hoses. The portion of the bar touching the compressor must be placed against a solid, metal portion. Ensure that the bar is not placed against the wires, hoses, or lines. In most cases when a pry bar must be used, the best place is beside the mounting bracket. This is the strongest part of the component outside the housing.

Force the component toward the engine and remove the belt. Install the new belt, ensuring that it is properly seated in each pulley. Apply outer force against the component until the correct belt tension is achieved. Hold the component in place while tightening the adjustment or the easiest fastener to access. Remove the bar and tighten all other fasteners. Sometimes, and whenever possible, it is necessary to tighten the adjustment and pivot fasteners before removing the bar to prevent the compressor from twisting and loosening the belt. Measure the belt tension after all fasteners have been tightened.

Antitheft and Keyless Entry System Function Testing

Before testing the antitheft system, alert everyone in the shop. If possible, move the vehicle outside and away from the shop doors. A remote control that has lock, unlock, **panic,** and trunk buttons will be discussed.

Close and lock all vehicle doors using the remote-control button. Leave one window down in case the door will not unlock. An audible signal or a light should indicate that the system is armed and the alarms are set. Test the panic mode next. The vehicle's alarm should sound and some of the exterior lights should flash. Disarm the system by pressing the unlock button. Rearm the system and press the button to unlock the trunk. The trunk lock should function, but the alarm should not. Close the trunk and ensure that the system is armed. Next, test the ability of the system to recognize tampering. With the system armed, violently rock the vehicle. The alarms should sound and flash. Disarm the system.

Rearm the system and note if there is a red or green light on the instrument panel. The red light indicates that the vehicle has been tampered with and the green light indicates that the system is armed and no one has tampered with it since the last time it was armed. Different systems may have different lights with different meanings. Consult the service manual.

CASE STUDY

This is a problem that was national in scope, but it only concerned one brand of vehicle and only for a model year or two. The customers complained that the vent air was not cool when the air conditioning was on in MAX or NORMAL. The heater worked fine. This system used stepper motors on the blend doors. Routine electrical tests did not locate the problem. The doors could be heard moving back and forth as the controls were moved. Studying the service manual's system operation section revealed a possible cause. When the ignition key was moved to ON, the A/C computer ran a test on its circuits. Part of this test was the search for a 35-mA load on each of the motors as they move the doors through their entire range of movement. A DMM set to amps and connected between the heat blend door motor and the computer showed 35 mA at full cold and at full hot. Supposedly the system was functioning correctly. Further physical investigation found the problem on top of the air box. The arm from the motor was held in place to the door lever with a small plastic clip. The clip had broken and was hanging down. As the motor moved the door toward the cold position, the clip would hook a ridge of the air box. The motor would try to push the door further, but the 35-mA load was reached and the computer assumed this was the limit of door travel and stopped the motor. The end result was the door never reached its cold position even though the ampere reading was correct. Replacing the clip with a better connection solved the problem.

TERMS TO KNOW

Blend doors
Instrument voltage regulator (IVR)
KOEO
KOER
Lamp check
Memory saver
Panic
Pigtail
Resistor block
Thermal limiter

ASE-STYLE REVIEW QUESTIONS

1. Climate control system diagnosis is being discussed.
   Technician A says an open blower motor fuse could cause a compressor clutch malfunction.
   Technician B says an inoperative blower motor may be caused by a poor ground connection.
   Who is correct?
   A. A only
   B. B only
   C. Both A and B
   D. Neither A nor B

2. Lighting systems are being discussed.
   Technician A says inoperative marking lights may be caused by an open circuit breaker.
   Technician B says testing a lighting system without relays can normally be done with a test light.
   Who is correct?
   A. A only
   B. B only
   C. Both A and B
   D. Neither A nor B

3. A customer complains that the headlights turn on and off, generally when traveling on a rough road.
   Technician A says the circuit breaker may be weak or failing.
   Technician B says a loose ground on the relay may be the fault.
   Who is correct?
   A. A only
   B. B only
   C. Both A and B
   D. Neither A nor B

4. A power window is stuck about halfway up and will not move up or down. There is a "clunk" each time the switch is operated.
   Technician A says the drive gear is stripped.
   Technician B says the window motor switch is shorting to ground.
   Who is correct?
   A. A only
   B. B only
   C. Both A and B
   D. Neither A nor B

5. Instrument gauge diagnosis is being discussed.

   Technician A says a test light can be used to test digital gauge systems.

   Technician B says supplying 12 VDC to a gauge will cause it to read full.

   Who is correct?

   A. A only C. Both A and B
   B. B only D. Neither A nor B

6. Warning light systems are being discussed.

   Technician A says grounding an oil pressure warning light wire at the sending unit will cause the gauge to read low.

   Technician B says grounding the engine temperature sending unit will cause the warning light to come on and the gauge to read hot.

   Who is correct?

   A. A only C. Both A and B
   B. B only D. Neither A nor B

7. A vehicle with low engine oil pressure is being discussed.

   Technician A says using a manual pressure gauge can test the oil pump output.

   Technician B says the low oil pressure reading may be caused by a bad gauge.

   Who is correct?

   A. A only C. Both A and B
   B. B only D. Neither A nor B

8. The alternator warning light is on at idle and off at speeds above idle.

   Technician A says the problem may be a misadjusted throttle stop screw.

   Technician B says the battery connections could cause this problem.

   Who is correct?

   A. A only C. Both A and B
   B. B only D. Neither A nor B

9. Climate control is being discussed.

   Technician A says a noncertified technician can perform preliminary testing.

   Technician B says an EPA-certified technician must perform any repairs requiring removal and replacing of the A/C refrigerant.

   Who is correct?

   A. A only C. Both A and B
   B. B only D. Neither A nor B

10. A vehicle power seat has forward and backward movement but not tilt. There is a low, "whirring" sound when the tilt switch is operated.

    Technician A says the motor drive gear may be stripped.

    Technician B says the forward and upward cable may be broken.

    Who is correct?

    A. A only C. B only
    B. Both A and B D. Neither A nor B

JOB SHEET 42

Name ______________________________ Date ______________

Accessories Performance Tests

CAUTION: Before beginning a performance test, study the service or operator's manual and the system controls for familiarization and expected results.

NATEF Correlation

This job sheet addresses the following NATEF tasks:

Diagnose incorrect operation of motor-driven accessory circuits; determine necessary action.

Diagnose the cause of false, intermittent, or no operation of antitheft systems.

Objective

Upon completion and review of this job sheet, you should be able to conduct a performance test on selected accessories.

Tools and Materials Needed

Service manual

Procedures

Task Completed

Remote-Control Antitheft Systems

1. Confirm that the vehicle has a remote-control antitheft system. ☐
2. Shut all doors and arm the system from the outside with the remote. ☐

   Did the door locks function? ____________

   Did a horn or light function to indicate that the alarm is armed? ____________

   If not, check the remote's battery before proceeding.
3. Press the panic mode button. Did the alarms sound? ____________ ☐
4. Disarm and rearm the system. ☐
5. Look through the driver's window without touching the vehicle. Is there a light in the dash indicating that the alarm is armed? ____________ ☐
6. Ensure that the remote is available to disarm the system. ☐
7. Attempt to open each door. If all are locked, rock the vehicle from side to side. Disarm the system if the alarm sounds. ☐
8. Did the alarm operate correctly? ____________ ☐
9. Explain. ____

   ____

   ____

Task Completed

Power Windows

☐ **10.** Block the wheels and apply the parking brake.

☐ **11.** Open all doors and switch the ignition to RUN.

☐ **12.** Operate each window in turn using the master switch.

Did each window work smoothly up and down? ________________

Did each window stop smoothly at the top and bottom? ________________

☐ **13.** Operate each window using the individual switches on the door.

Did each window work smoothly up and down? ________________

Did each window stop smoothly at the top and bottom? ________________

☐ **14.** Set the window lock. Attempt to operate each window, first with the master switch and then with the individual switches.

Did each window work with the master switch? ________________

Did the right front or either of the rear windows work using their individual switches? ________________

Comments

____

____

____

Power Seats

☐ **15.** Sit in the driver's seat with the driver's door open. Switch the ignition to RUN.

☐ **16.** Operate the switch to move the entire seat fully forward and fully backward.
Did the seat move smoothly and quietly? ________________

☐ **17.** With the seat fully back, operate the switch for the backrest. Do not raise or lower the backrest completely.
Did the backrest perform as expected? ________________

☐ **18.** Set the backrest to a comfortable position and operate the tilt switch.
Did the seat tilt up and down at the rear? ________________
Comments

____

____

____

☐ **19.** If the other front seat is equipped with power, repeat steps 10 through 18 for that seat.
Did the seat perform as expected? ________________

Instructor's Response ____

____

____

____

JOB SHEET 43

Name ______________________________ Date ______________

Diagnosing an Oil Pressure Warning Light Circuit

NATEF Correlation

This job sheet addresses the following NATEF tasks:

Inspect the test gauges and gauge sending units for cause of intermittent, high, low, or no gauge readings; determine necessary action.

Diagnose the cause of incorrect operation of warning devices and other driver information systems; determine necessary action.

Objectives

Upon completion and review of this job sheet, you should be able to diagnose a nonelectronic oil pressure warning light circuit.

Tools and Materials Needed

Hand tools

Manual oil pressure gauge

Jumper wire

Service manual

CAUTION: Do not use the following test procedures or equipment on electronic circuits. Damage to the circuit could result.

Procedures

Task Completed

1. Locate the service data. ☐

   Make and model of vehicle ______________

   Engine ______________

   Type of instruments ______________

   Warning lights only or lights and gauges ______________

   Minimum and maximum oil pressures ______________

2. Is the oil level correct? ______________ ☐
3. Are there signs of oil leakage? ______________ ☐
4. Proceed with this step only after correcting the oil level. Turn the key to KOEO. ☐
5. Did all the warning lights come on? ______________ If not, replace defective lamps and retest. ☐
6. Start the engine and idle. ☐
7. Is the oil light still on? ______________ If yes, slightly speed up the engine. ☐
8. Did the light go out? ______________ If no, proceed with the circuit test. If yes, proceed with the manual oil pressure test starting with step 13. ☐
9. Stop the engine, locate the oil pressure sending unit under the hood, and disconnect the wire from the sending unit. ☐
10. Use jumper wire to ground the sending unit wire. ☐

Task Completed

☐ **11.** Set the ignition to RUN (KOEO). Is the light on? ________________ If yes, go to the next step. If no, follow the circuit test routine in the service manual.

☐ **12.** Remove the wire from the ground. Did the light go out? ________________ If yes, proceed with the next step. If no, follow the circuit test in the service manual.

☐ **13.** Remove the oil pressure sending unit.

☐ **14.** Install the manual oil pressure gauge.

☐ **15.** Start the engine and idle. Record the oil pressure. ________________

☐ **16.** Does the oil pressure meet minimum specifications? ________________ If yes, proceed with the next step. If no, stop the engine and install the sending unit. Consult with the customer on all test results and recommended actions.

☐ **17.** Remove the pressure gauge. Install a new oil pressure sending unit and connect its harness. Check the light operation. Return the vehicle to the customer.

Dual-Purpose Light-Oil and Temperature Combined

☐ **18.** Locate the temperature sending unit under the hood.

☐ **19.** Disconnect the wire. Set the ignition to RUN (KOEO).

☐ **20.** Did the light go out? ________________ If yes, check the condition of the engine cooling system and sending unit. If no, check the oil pressure warning circuit following steps 8 through 16.

Instructor's Response ____

____

____

____

GLOSSARY
GLOSARIO

Note: **Terms are highlighted in color**, followed by **Spanish translation in bold**.

Accountant The person or persons who receive payments, bill for payment, and manage the financial statement for the department or business.

Contador La persona o las personas que reciben los pagos, mandan las cuentas, y administran los resumenes financieras del departamento o el negocio.

Angle indicator Experience is the best method to learn how to set up and use an **angle indicator**. The manual one is awkward to use in tight places.

Indicador de ángulo La experiencia es el mejor método para aprender a preparar y utilizar un **indicador de ángulos**. El manual es complicado de usar en espacios estrechos.

Antiseize A chemical compound applied between two mating fasteners to prevent seizing.

Anti-adherente Compuesto químico aplicado entre dos sujetadores para evitar que se adhieran.

Axle stub Outer splined end of a FWD drive axle that extends through the hub delivering transaxle output to the hub and wheel.

Muñón del eje Extremo externo dentado del eje propulsor de tracción delantera que se extiende a través del cubo y que transmite la potencia del transeje al cubo y a la rueda.

Baking soda A household base chemical used to neutralize acids. Commonly used to remove corrosion on a battery.

Bicarbonato de sodio Químico de uso familiar que sirve para neutralizar ácidos. Comúnmente se utiliza para sacar la corrosión de la batería.

Blend door A door or doors used to direct incoming air inside the interior of heating and air conditioning duct work.

Mariposa Puerta o puertas utilizadas para dirigir el aire que ingresa hacia el interior del conducto del aire acondicionado.

Blind hole A hole that does not go all the way through a part.

Agujero ciego Agujero que no atraviesa completamente una pieza.

Bore A hole drilled into or through a part, but commonly used to describe the cylinder bore in an engine.

Cavidad interior Orificio tallado en una pieza, pero comúnmente utilizado para referirse al diámetro del cilindro en un motor.

Brinelling A load stress defect causing a bearing or bearing race to flake and crack.

Endurecimiento Defecto a raíz de excesiva carga que provoca que los cojinetes se escamen o se resquebrajen.

Burned area A damaged area of a bearing resulting from excessive heat; usually appears to have a bluish tint.

Área quemada Área dañada de un cojinete a causa de calor excesivo; por lo general presenta un tinte azulado.

Burnishing Is an operation used to heat and seat the brake pads to the rotor.

Bruñido El bruñido es una operación empleada para calentar y asentar las pastillas de freno del rotor.

Camber wear Abnormal tire wear caused by excessive negative or positive camber.

Desgaste del ángulo de caída Desgaste anormal del neumático como consecuencia de un ángulo de caída excesivo, positivo o negativo.

Capacitor A storage device for electricity; discharges stored electricity on command or a programmed condition.

Capacitor Un dispositivo de almacenamiento para la electricidad, que descarga la electricidad almacenada ante una señal de control o una condición programada.

Carburetor A prefuel injection air/fuel mixing component. Still found on many gasoline-fueled vehicle and small engine equipment.

Carburador Componente que contiene una mezcla de inyección de aire/combustible. Se encuentra aun en varios vehículos a gasolina y en pequeños equipos a motor.

Castellated nut A nut with castle-like turrets on top for insertion of a cotter key as a locking device.

Tuerca de corona Una tuerca con las torrecitas parecidas a un castillo en la parte superior para insertar un pasador de chaveta como un dispositivo de cerrojo.

C-clamp A small portable clamping tool so named because of its shape.

Presilla Una pequeña herramienta de apriete que se llama así en inglés por su forma de C.

Center bearing A bearing used to support the front section of a two-piece RWD driveshaft.

Cojinete central Cojinete que se usa para soportar la sección frontal de un eje de transmisión de tracción trasera.

Commission The percentage of a sale or service that is paid to a technician for his or her labor.

Comisión Porcentaje sobre la venta o el servicio que se abona al técnico como contraprestación por su trabajo.

Compressor A pump that compresses air or vapor. Commonly refers to an air compressor or an air conditioning compressor.

Compresor Bomba que comprime aire o vapor. Comúnmente hace referencia al compresor de aire o al del aire acondicionado.

Condition seizing The unintentional seizing of two adjacent components caused by temperature, corrosion, or other environmental conditions.

GLOSSARY
GLOSARIO

Adherencia acondicionada Adherencia no intencional de dos componentes adyacentes causada por la temperatura, la corrosión u otras condiciones medioambientales.

Control arm A part of the suspension that attaches the wheel assembly to the vehicle while allowing radial and vertical movement of the wheel assembly.

Brazo de control Parte de la suspensión que une las ruedas al vehículo y permite su movimiento radial y vertical.

Cooling fan The fan usually located at the front of the engine compartment for drawing air over the radiator fins.

Ventilador de enfriamiento Ventilador generalmente situado en la parte frontal del compartimiento del motor para llevar aire a las laminillas del radiador.

Corrosion Deterioration caused by acid or other acidic material on metal. Commonly associated with electrical conductor, but includes rusting of metal.

Corrosión La deterioración causado por el ácido u otra materia o materiales acídicas. Comunmente asociado con los conductores eléctricos, pero incluye la oxidación del metal.

Cotter key A nonthreaded fastener normally used to prevent a fastener from turning or to connect two or more low-stress components. Also called a cotter pin.

Pasador de chaveta Sujetador sin rosca que normalmente se utiliza para evitar que el sujetador gire o para conectar dos o más componentes de menor tensión.

Coupler A connection between two components; typically flexible and coupler two shafts.

Acoplador una conexión entre dos componentes que, por lo general, es flexible y que acopla dos árboles.

Cupping A tire wear condition typically caused by out-of-balance tires or lack of tire rotation.

Acopamiento Una condición de desgaste del neumático que por lo general es consecuencia de neumáticos desbalanceados o de falta de rotación de los neumáticos.

Current draw test The current draw test measures the amount of current the starter draws when actuated. It helps to determines the electrical and mechanical condition of the starting system.

Prueba de consumo La prueba de consumo de corriente mide la cantidad de corriente que consume el motor de arranque cuando se acciona. Ayuda a determinar el estado eléctrico y mecánico del sistema de arranque.

Database A computerized information file.

Datos Un archivo de información computerizado.

Diagnostic Trouble Codes (DTCs) Are digital alphanumeric fault codes that represent a circuit failure in a monitored system.

Códigos de error de diagnóstico (DTCs, en sus siglas en inglés) Los códigos de error de diagnóstico son códigos digitales y alfanuméricos de avería que representan un fallo del circuito en un sistema monitorizado.

Diesel exhaust fluid (DEF) The engine computer uses a strategy to spray a fine mist of diesel exhaust fluid in the exhaust during certain periods to help clean the exhaust emissions and burn the soot in the diesel particulate filter.

Fluido de escape de diésel (DEF, en sus siglas en inglés) La computadora de un motor ejecuta la estrategia de difuminar fluido de escape de diésel en el propio escape durante determinados periodos para ayudar a limpiar las emisiones del escape y para quemar el hollín del filtro de partículas de diésel.

diesel particulate filter A diesel particulate filter is used to capture soot and incinerate it during regeneration cycles using DEF.

Filtro de partículas de diésel Se usa un filtro de partículas de diésel para retener el hollín e incinerarlo durante los ciclos de regeneración en los que interviene el DEF.

Dipstick In automotive applications a dipstick is a thin flexible rod used to check fluid levels in certain components.

Varilla de nivel En las aplicaciones de automoción una varilla de nivel es una vara flexible y delgada usada para comprobar los niveles de fluido de determinados componentes.

Diverter valve A valve used with Secondary Air Injection systems. *See* "Secondary Air Injection" in the Classroom Manual.

Válvula desviador Una válvula que se usa con los sistemas de inyección de aire secundaria. *Ver tambien* la sección de inyección de aire secundaria ("Secondary Air Injection") en La Manual de Clase.

Drifting A slight continuous pull to one side; typically caused by low tire pressure, dragging brakes, or wheel misalignment.

Desplazamiento lateral Fuerza constante y leve hacia un lado; suele estar causado por una presión baja en los neumáticos, por un arrastre de los frenos o por un mal equilibrado de las ruedas.

Drive axles Transaxle the engine's power from the transaxle to the drive wheels.

Ejes de tracción Los ejes de tracción transmiten la energía desde el conjunto motor a las ruedas de tracción.

Entry-level technician A technician at his or her first job in a technical field.

Técnico de nivel elemental Técnico que realiza su primer trabajo en su especilidad.

Fading A condition where the brakes initially work but lose their stopping power because of heat or other conditions affecting the friction material.

Fading El fading es un efecto que se produce cuando los frenos funcionan inicialmente pero pierden su poder de frenada a causa del calor o de otras condiciones que afectan al material de fricción.

Female coupling The receiving end of a connection. Usually refers to hydraulic or air connection between hoses or between a hose and tool.

Acoplamiento hembra Parte receptora de una conexión. Comúnmente hace referencia a la conexión hidráulica o de aire entre las mangueras o entre una manguera y una herramienta.

Filter wrench A special tool for removing and installing oil filters. May also be required on some fuel and water filters.

Llave de filtro Herramienta especial para quitar e instalar filtros de aceite. También puede ser necesaria en algunos filtros de combustible y de agua.

Flat rate An automotive pay scale when the technician is paid by "book" or labor rate time as opposed to actual clock hours.

Tasa plana Escala de pago automotriz utilizada para abonar al técnico según "el registro" o según tasa de tiempo de trabajo, en oposición a los relojes de jornada.

Flux A protective coating on solder or welding rods to assist in cleaning areas as they are joined together either by soldering or welding.

Fundente Revestimiento protector en soldaduras o varillas para ayudar a limpiar áreas mientras se encuentran adheridas debido a que fueron soldadas.

Free travel The amount of clutch pedal movement until the clutch release bearing contacts the release levers of the clutch's pressure plate.

Juego libre La cantidad de movimiento del pedal de embrague hasta cuando el cojinete de desconexión del embrague toca las palancas de reposición de la placa de presión del embrague.

Front pipe/Muffler/Tailpipe The front pipe, muffler, and tailpipe are the three main components used to channel exhaust gases from the engine to the rear bumper and reduce exhaust noises.

Tubo de escape frontal/silenciador/tubo de escape El tubo de escape frontal, el silenciador, y la salida del tubo de escape son los tres componentes empleados para canalizar los gases de escape del motor hacia el parachoques trasero y para reducir el ruido de la emisión de gases.

Fuel rail A common fuel routing device for multiport fuel injections.

Guía de combustible Dispositivo común que dirige el combustible para inyección de combustible de puerto múltiple.

Gas-charged shocks Have some nitrogen gas in them to increase the shock's action.

Amortiguadores de gas Los amortiguadores de gas contienen nitrógeno en su interior para aumentar la acción amortiguadora.

Glazing A condition when the friction material and rotor or drum develops a slick nearly nonfriction surface. Usually caused by heat.

Glaseado El **glaseado** se produce cuando el material de fricción y el rotor o el tambor acaba desarrollando una superficie bruñida y sin apenas fricción. Esto suele estar causado por el calor.

Glycol-based A type of commonly used brake fluid.

A base de glicol Un tipo de fluido de freno muy común.

Grabbing brakes Are usually caused by oil or brake fluid on the friction material.

Frenos patinan Generalmente los frenos patinan debido a la presencia de aceite o líquido de frenso en el material de fricción

Grease A thick near-solid lubricant used for wheel bearings and vehicle chassis components.

Grasa Un lubricante espeso, casi sólido que se utiliza para los cojinetes de las ruedas y para los componentes del chasis de los vehículos.

Grease fitting The grease fitting on some aftermarket U-joints are placed in the bearing cup instead of the joint cross.

Engrasador El engrasador en algunas juntas universales de postventa están en el tazón del cojinete en vez de en la junta en cruz.

Ground reference voltage testing Used to determine the voltage within a circuit.

Prueba de referencia de tierra se La prueba de referencia de tierra se usa para determinar el voltaje en un circuito.

Hard pedal A hard pedal condition is when high driver input braking force does not sufficiently slow the vehicle.

Pedal duro Se dice que un freno tiene pedal duro cuando una fuerza intensa aplicada por el conductor no basta para frenar el vehículo.

Hot spots Hot spots are formed by excessive heat on the weakest area of the disc or drum.

Puntos calientes Los puntos calientes se forman por un exceso de calor en la parte más débil del disco o tambor.

Hourly rate Technician pay based on actual clock hours worked; typical pay rate for non-automotive repair employees.

Tarifa por hora Salario que se paga a los técnicos sobre la base de las horas reloj reales trabajadas Tarifa de pago habitual para empleados de reparaciones no vehiculares.

Ignition timing Refers to the timing of spark plug firing in relation to position of the piston.

sincronización del encendido La sincronización del encendido se refiere a la coordinación de la chispa de la bujía con la posición del pistón.

Infinite Infinite means forever or too large. In automobile circuit testing, it means the value being measured is too great for the test device being used.

Infinito Infinito quiere decir para siempre o demasiado grande. En las pruebas de circuitos de automóviles, se usa este término para referirse a un valor demasiado grande para el dispositivo de medición que se esté utilizando.

Instrument voltage regulator (IVR) Limits the voltage supplied to the gauges to 12 VDC so the gauges are more accurate.

Instrumento regulador de tensión (IVR, en sus siglas en inglés) Un instrumento regulador de tensión limita el voltaje que entra a los medidores a 12 VDC de manera que las mediciones sean más precisas.

GLOSSARY
GLOSARIO

Isolators Isolators are made of rubber or rubber-like material to reduce noise.

Aislantes Los aislantes están hechos de goma o materiales similares a la goma con el fin de reducir el ruido.

Joint stud The threaded stud of a ball joint or steering linkage that fits tightly into a mated bore on adjacent components.

Espiga de la junta El poste enroscado de una junta esférica o una biela de dirección que se ajusta apretadamente en un agujero emparejado en los componentes adyacentes.

Journal The machined and treated area on a shaft where the bearings are placed.

Muñón Zona mecanizada y tratada de un eje donde se colocan los cojinetes.

KOEO KOEO means ignition key on, engine off, a test condition required to perform certain tests.

KOEO Las siglas inglesas KOEO quieren decir llave puesta (*key on*) motor apagado (*engine off*), una condición requerida para realizar ciertas pruebas.

KOER KOER means ignition key on, engine running, another test condition required for certain tests.

KOER Las siglas inglesas KOER quieren decir llave puesta (*key on*) motor encendido (*engine running*), otra condición requerida para realizar ciertas pruebas.

Labor guide A manual listing the time required to perform a specific mechanical service such as replacing a starter motor. Flat rate pay is calculated based on the labor guide.

Guía de trabajo Manual que indica la cantidad de tiempo necesaria para efectuar un servicio mecánico, por ejemplo el reemplazo del motor de arranque. La tasa plana se calcula en base a esta guía de trabajo.

Labor time The amount of labor used to complete a repair. May be actual time or the time shown in a labor guide.

Tiempo de trabajo La cantidad de trabajo para completar una reparación. Puede ser tiempo actual o el tiempo indicado en una guía de trabajo.

Limited slip A limited slip differential is designed to reduce wheel spin using some type of clutch mechanism. Using incorrect differential fluid will damage the clutches.

Deslizamiento limitado Un diferencial de deslizamiento limitado está diseñado para reducir el patinaje de la rueda mediante algún tipo de mecanismo de embrague. El uso de un fluido de diferencial incorrecto daña los discos del embrague.

Locking brakes Locking brakes may be caused by misadjusted shoes, blockage in the brake line, or slick tires or road surface.

bloqueo de los frenos El bloqueo de los frenos puede estar causado por un desajuste de las zapatas de freno, por un bloqueo en el tubo del sistema de frenado, por el desgaste de los neumáticos o por una superficie de carretera demasiado lisa.

Logo A symbol or phase promoting a business.

Logotipo Un símbolo o una frase que promociona un negocio.

Male coupling The male part of a quick disconnect connection commonly found in air and hydraulic systems.

Acoplamiento macho Pieza macho de un sistema de desacoplamiento rápido que se encuentra generalmente en los sistemas aéreos e hidráulicos.

Mass airflow (MAF) sensor The mass airflow measures the amount of air entering the engine's intake system.

Flujo de la masa de aire (MAF, en sus siglas en inglés) El flujo de la masa de aire mide la cantidad de aire que penetra en el sistema de toma de aire del motor.

Negative camber An alignment angle where the top of the wheel is tilted towards the engine.

Ángulo de caída negativo Un ángulo de alineación, cuya rueda tiene una parte superior que se inclina hacia el motor.

Neutral flame A gas-fired adjustable flame emitting its highest heat with minimum soot.

Llama neutra Llama ajustable encendida a gas que emite su máximo calor con menor hollín.

Nut splitter A tool used to remove rounded or seized nuts.

Bifurcador de tuerca Herramienta utilizada para extraer tuercas redondas o adheridas.

On-vehicle brake lathe An on-vehicle brake lathe is designed to machine a brake rotor on the vehicle.

Torno de frenos "on car" está diseñado para trabajar en un rotor de frenos en el propio vehículo.

Out-of-round A hole or shaft that is not perfectly round. Usually referring to the cylinder bore.

Deformado Orificio o eje que no es perfectamente redondo. Comúnmente hace referencia al diámetro del cilindro.

Panic The physical and psychological reaction of a person to danger. In the context of this text, the extremely hard braking effort by the driver to avoid an accident or an attempt to regain control of the vehicle.

Pánico Reacción física y fisiológica de una persona cuando se encuentra en peligro. En el contexto de este texto, el esfuerzo de freno extremadamente arduo que hace el conductor para evitar un accidente o el intento para retomar el control del vehículo.

Pedal pulsation A brake failure where the pedal tries to move up and down while the driver is applying force to the pedal. Usually caused when the brake rotors' friction areas are not parallel or when a rotor is warped.

Pulsación del pedal Un fallo del freno en el cual el pedal se mueva verticalmente mientras que el conductor aplica la fuerza al pedal. Suele ser causado cuando las áreas de fricción de un rotor no son paralelas o cuando un rotor es alabeado.

Penetrating oil A light weight oil used to loosen rust or corrosion on a fastener.

Aceite de penetración Un aceite liviano que se utiliza para aflojar la herrumbre o la corrosión en un sujetador.

Pigtail A pigtail is a replacement connector with a short length of the wiring for connection to the vehicle wiring.

Empalme pigtail Un empalme pigtail es un cable corto y flexible de conexión que se conecta al cableado del vehículo.

Positive camber An alignment angle where the top of the wheel is tilted away from the engine.

Ángulo de caída positivo Un ángulo de alineación en el que la parte superior de la rueda se inclina hacia el lado contrario del motor.

Positive crankcase ventilation (PVC) A system used to ventilate and reduce pressure in the engine's crankcase. Part of the emission control systems.

Ventilación positiva del cárter (PVC) Sistema utilizado para ventilar o reducir la presión en el cárter. Parte de los sistemas de control de emisión.

Pressure bleeder A tool used to pressure bleed a brake system by forcing brake fluid through the master cylinder into the brake lines.

Purgador de presión Una herramienta que se usa para purgar un sistema de frenos bajo presión por medio de forzar el fluido de freno através del cilindro maestro a las lineas de freno.

Pulling Steering symptom usually caused by defect found in tires or steering system causing the vehicle to pull hard left or right and requiring driver to steer in opposite direction.

Arrastre lateral Síntoma del sistema de dirección causado por defectos en los neumáticos o en el sistema de dirección en sí que se caracteriza por la tendencia del vehículo de tirar fuertemente hacia la derecha o hacia la izquierda y que para contrarrestarla el conductor debe halar el volante hacia la dirección contraria.

Quick disconnect coupling A coupling employing a male and a female connection in hydraulic and air systems.

Desacoplamiento rápido Acoplamiento que emplea conexiones hembra y macho en sistemas hidráulicos y aéreos.

Race A machine area or component on which bearing rollers or balls move on.

Anillo de rodadura El área o el componente de una máquina sobre los que se mueven los rodillos con cojinetes o bolas.

Radiator Engine cooling component; located behind front grill.

Radiador Componente de refrigeración del motor del radiador, que está ubicado detrás de la parrilla delantera.

Ratio The proportional amount of one item to another. Commonly used to express gear relationship in size or number of teeth.

Relación La cantidad proporcional de una cosa a otra. Suele usarse para expresar la relación de tamaño o número de dientes de los engranajes.

Recall A manufacturer's term used when certain vehicles must be returned to a service center, usually a dealership, for correction of some sort. It is not a warranty and usually involves all of certain brand, model, and year of vehicle. It is paid by the manufacturer and may be ordered by the federal government based on safety data.

Retiro Un término del fabricante que se usa cuando un cierto vehículo debe regresarse al centro de servicio, tipicamente al distribuidor, para algún reparación. No esta bajo garantía, y normalmente involucra todos los vehículos de cierta marca, modelo y año. Las refacciones se pagan por el fabricante y pueden ser por orden del gobierno federal basado en los datos de seguridad.

Repair order A form that is filled out to identify the car, customer, and complaint on a car to be serviced.

Solicitud de reparación Formulario que se llena para identificar el vehículo y el cliente, donde se detalla el problema de un vehículo que necesita servicio.

Resistor block A resistor block is a series resistor with multiple inputs and a single output. Commonly used to control speed on an electric motor.

Bloque de resistencia Un bloque de resistencia es una resistencia en serie con múltiples entradas y una sola salida. Se suele utilizar para controlar la velocidad de los motores eléctricos.

Ring travel area The area of the cylinder bore covering the area of top dead and bottom dead limits of piston travel.

Área de recorrido del anillo Área las paredes del cilindro que cubren los límites muertos superiores e inferiores del traslado del pistón.

Runout The measurable side-to-side variation or wobble in a part as it rotates.

Desviación Cantidad de desalineación o bamboleo de un lado al otro de una pieza mientras ésta gira.

Sealing tool A pliable or solid closed-pipe shaped tool used to block fluid leakage from a RWD transmission when the driveshaft is removed.

Herramienta de sellado Herramienta en forma de tubo cerrado sólido flexible o rígido para bloquear fugas de líquido de transmisión, en vehículos de tracción trasera, cuando se desmonta el eje de transmisión.

Seepage A slow leak that does not produce a liquid drop or a run. Most seals require seepage to function properly. Should not be considered a leak at seals and gaskets.

Infiltración Un goteo despacio que no produce gotas ni chorro líquido. La mayoría de los sellos requieren una infiltración para funcionar correctamente. No se debe considerar como un fuga en los sellos y empaques.

Self-powered A self-powered test light is connected across a portion of the circuit. If the lamp lights then there is continuity between the test points.

Autogenerador En una sección del circuito hay conectado un piloto de prueba autogenerador. Si la luz se enciende, quiere decir que hay continuidad entre los puntos de la prueba.

GLOSSARY
GLOSARIO

Serpentine belt A single drive belt used to drive all engine-mounted accessories.

Correa serpentina Una correa de un sólo mando que sirve para mandar todos los accesorios montados en el motor.

Shafts A machined rod upon which wheels, bearings, and gears are mounted. Most commonly made of steel, but may be made of other materials.

Árboles Un vástago maquinado sobre el que se montan las ruedas, los cojinetes y los engranajes. Por lo general, están fabricados en metal, pero pueden ser de otros materiales.

Slime Chemicals in hydraulic that congeal forming a corrosive agent. Usually present in hydraulic or water systems that have not been maintained properly.

Limo Químicos en la hidráulica que congelan formando un agente corrosivo. En general están presentes en los sistemas hidráulicos y de agua que no contaron con apropiado mantenimiento.

Service manager The ranking administrator in an automotive service shop.

Gerente de servicio Administrador de un establecimiento de servicio automotriz.

Service writer The administrator who initiates a repair order and usually the customer's contact for vehicle repairs.

Generador de servicio Administrador que inicia la orden de reparación y generalmente establece contacto con el cliente para la reparación del vehículo.

Solder A soft pliable metal used for joining of low tension connections. Most commonly used to join two pieces of wire or a terminal to a wire.

Soldador Metal flexible suave utilizado para unir conexiones de baja tensión. Comúnmente usadas para unir dos piezas de cable o una terminal a un cable.

Soldering gun A tool with an electric heating element used to melt solder.

Pistola de soldadura Herramienta con un elemento eléctrico de calentamiento utilizada para fundir la soldadura.

Soldering iron A tool with an electric heating element used to melt solder.

Soldador Herramienta con un elemento eléctrico de calentamiento utilizada para fundir la soldadura.

Spalling A bearing defect resulting from heat and lack of lubrication.

Resquebrajamiento Defecto del cojinete que se presenta a raíz del calor o falta de lubricación.

Spongy pedal A spongy pedal is the result of air in the hydraulic system. Feels like stepping on an inflated balloon.

Pedal esponjoso El **pedal esponjoso** se produce por la presencia de aire en el sistema hidráulico. La sensación es la de pisar un globo inflado.

Spout A means to transfer liquid without splashing or dripping if used properly. May refer to Ford's ignition system also.

Boquilla de llenado Método para transferir líquido sin salpicarlo o derramarlo si se usa de forma apropiada. También puede hacer referencia al sistema de encendido Ford.

Spring-loaded switch A switch that will return automatically to the OFF position when released.

Interruptor accionado por resorte Un interruptor que regresa automáticamente a la posición APAGADO al soltarlo.

Steering wheel free play Amount of movement at the steering wheel without corresponding movement of the steering linkage.

Juego del volante Cantidad de movimiento en el volante in movimiento correspondiente del varillaje de la dirección.

Striker A tool used to ignite the flame on an oxyacetylene torch.

Cebador Una herramienta que se utiliza para encender la llama de un soplete de oxiacetileno.

Stripped threads Threads that have been damaged by cross threading.

Filetes roidos Filetes que han sido averiados debido al cruzamiento de roscas.

Stripping The procedure used to remove insulation from an electrical wire.

Desaislación Procedimiento a través del cual se remueve el aislamiento de un alambre eléctrico.

Stud remover A tool that grips a stud and can be driven by a wrench to unscrew a stud.

Removedor de espárragos Herramienta que prende un espárrago y que puede ser accionado por una llave para destornillar el mismo.

Subframe A small frame used to support the power pack in a unibody-constructed vehicle. May be used on some vehicles to support the suspension or drive components.

Bastidor auxiliar Una pequeña armazón que se usa para soportar la batería común en un vehículo de construcción monocasco. Puede usarse en algunos vehículos para soportar los componentes de suspensión o tracción trasera.

Suction pump Any pump use to suck liquid or other material to move it to a different location. Normally not a pressure pump.

Aspirador Cualquier bomba que sirve para aspira al líquido u otra material moviendolo a otro lugar. Normalmente no es bomba de presión.

Supplemental Restraint System (SRS) Technically correct term for the air bag systems on vehicles.

Sistema de contención suplementario (SRS, en inglés) Término técnicamente correcto para los sistemas de bolsas de aire en los vehículos.

Tachometer A gauge that translates electronic pulses from an engine sensor into data that is read as revolution per minutes (rpm). Usually scaled in 100 rpm.

Taquímetro Un medidor que traduce los impulsos electrónicos de un sensor de motor a los datos que se interpretan como revoluciones por minuto (rpm). Suele tener la escala de 100 rpm.

Tag Used to highlight a condition or provide technical information.

Etiqueta Se utiliza para resaltar una condición o brindar información técnica.

Technical Service Bulletin (TSB) Repair instructions from a vehicle manufacturer detailing how certain repairs or corrections are to be made.

Boletín de Servicios Técnico (TSB) Las instrucciones de reparaciones de un fabricante de vehículos detallando como se deben efectuar ciertas reparaciones o correcciones.

Test light A test light is connected between a positive conductor and ground. It will only show voltage presence not amount.

Piloto de prueba Hay un piloto de prueba conectado entre el conductor positivo y la tierra. Sólo muestra la presencia de voltaje, no la cantidad del mismo.

Thermal limiter A thermal limiter is a special circuit protection device that opens the circuit when overheated.

Limitador térmico El limitador térmico es un dispositivo especial de protección del circuito que abre dicho circuito cuando éste se sobrecalienta.

Threadlocking compound A material used on threads to prevent them from unscrewing from vibration.

Compuesto de sujeción para fieltes Material utilizado en los filetes para prevenir que se aflojen debido a la vibración.

Thread sealing compound A material used on threads to prevent liquid-like coolant from damaging the threads.

Compuesto sellador para filetes Material utilizado en los filetes para prevenir que un líquido, como por ejemplo un refrigerante, averie los filetes.

Throttle body The part of the intake system where the amount of air entering the intake is controlled through the throttle linkage. Usually houses the throttle valve, idle speed control, and may have other electronic or mechanical component.

Cuerpo de acceleración La parte del sistema de entrada en donde la cantidad del aire entrando a la entrada se controla por medio de la biela de acceleración. Suele incluir la válvula de acceleración, el control de marcha mínima, y puede incluir otro componente electrónico o mecánico.

Through hole A hole that goes all the way through a part.

Agujero pasante Agujero que atraviesa completamente una pieza.

Thrust wear Wear created by two components rubbing together as a result of some force being applied. Normally associated with cylinder wear when the piston is thrust against the cylinder wall during the power stroke.

Desgaste del empuje El desgaste creado por dos componentes que se frotan como resultado de una fuerza aplicada. Normalmente se asocia con el desgaste del cilindro cuando el piston se oprima contra el muro del piston durante la carrera de potencia.

Toe wear Tire wear created by improperly set toe. Usually causes a sawtooth pattern across the tread, and is worse on one side of the tread.

Desgaste de divergencia o convergencia El desgaste del neumático debido a una divergencia o convergencia indebido. Suele causar un daño de aparencia de un diseño parecido a dientes de sierra en la banda del rodadura del neumático, y es peor de un lado.

Toothed wheel or ring (tone ring) A device that repeatedly causes an electrical impulse to be generated each time a tooth passes through a magnetic field.

Anillo dentado (aro reluctor) Dispositivo que origina en forma intermitente un impulso eléctrico que debe generarse cada vez que un diente pasa a través de un campo magnético.

Top off The process of adding liquid to bring the level to specifications.

Llenar El proceso de añadir el líquido para que el nivel queda a las especificaciones.

Transaxle The transmission for a front-wheel drive vehicle. May also be applied to certain rear-wheel drives. It combines the transmission and final drive or differential into one housing.

Transmisión transversal Transmisión de los vehículos con tracción delantera. También puede aplicarse a ciertos vehículos con tracción trasera. Combina la transmisión y tracción final o diferencial en una caja.

Transverse The typical installation of an engine in a front-wheel drive vehicle.

Transversal Instalación típica de un motor en vehículos de tracción delantera.

Upwind Being upwind of an object; facing an object with wind coming from behind the body is an upwind condition.

Contra el viento Estar contra el viento de un objeto, es decir, de cara a un objeto con el viento soplando desde detrás del cuerpo es estar contra el viento.

Vacuum pump A vacuum pump is a belt or gear-driven pump to remove air from a system.

Bomba de vacío Una bomba de vacío es una bomba accionada por correas o por engranajes que se usa para extraer aire de un sistema.

Vehicle identification number (VIN) A set of numbers and letters on a plate used to identify information about a car.

Número de identificación del vehículo Juego de números y letras en una placa utilizado para identificar un vehículo.

Voltage drop Voltage drop is a measurement of the voltage across a portion of the circuit.

GLOSSARY
GLOSARIO

Caída de tensión Una caída de tensión es una medida del voltaje a lo largo de una sección determinada de un circuito.

Water pump A belt-driven pump used to move coolant from the bottom of the radiator back to the engine block.

Bomba de agua Bomba con correa utilizada para mover refrigerante desde la parte inferior del radiador la caja del motor.

Wave A tire wear condition typically caused by out-of-balance tires or lack of tire rotation.

Onda Una condición de desgaste del neumático que, habitualmente, es consecuencia de que haya neumáticos desbalanceados o de falta de rotación de los neumáticos.

Welding tip The point at which the gases in an oxyacetylene torch are mixed and ignited.

Punta de soldeo El punto en el que los gases de un soplete de oxiacetileno se mezclan y se encienden.

Wheel bearing A bearing installed between the stationary spindle (shaft) and the rotating wheel. See roller bearing and ball bearing.

Cojinete de rueda Un cojinete instalado entre el husillo fijo (árbol) y la rueda de rotación. *Véase, cojinete de rodillo y cojinete de bola.*

Yoke Part of a driveshaft used to support the U-joint bearings and provide a connection between two sections of a driveshaft.

Brida Parte del eje de transmisión que sostiene los cojinetes de la junta del cardan y serve de conexión entre dos secciones de un eje de transmisión.

A

B

INDEX

E

F

INDEX

T

INDEX